D0875802

SWITCHING POWER CONVERTERS

SWITCHING POWER CONVERTERS

Peter Wood
Advisory Engineer
Power Electronics Laboratory
Westinghouse Electric Corporation
Research and Development Center
Pittsburgh, Pennsylvania

VNR VAN NOSTRAND REINHOLD COMPANY
NEW YORK CINCINNATI ATLANTA DALLAS SAN FRANCISCO
LONDON TORONTO MELBOURNE

Van Nostrand Reinhold Company Regional Offices:
New York Cincinnati Atlanta Dallas San Francisco

Van Nostrand Reinhold Company International Offices:
London Toronto Melbourne

Library of Congress Catalog Card Number: 80-19296

ISBN: 0-442-24333-2

Manufactured in the United States of America

Published by Van Nostrand Reinhold Company
135 West 50th Street, New York, N.Y. 10020

Published simultaneously in Canada by Van Nostrand Reinhold Ltd.

15 14 13 12 11 10 9 8 7 6 5 4 3 2 1

Library of Congress Cataloging in Publication Data
Wood, Peter, 1932–
Switching power converters.

Includes index.
1. Electric current converters. 2. Semiconductor switches. I. Title.
TK7872.C8W66 621.3815'322 80-19296
ISBN 0-442-24333-2

To all those who, like Edmund Burke, believe that "good order is the foundation of all good things."

Preface

The field of switching power converters has assumed growing importance since silicon-based semiconductor switching devices were first developed. Utility, industrial, military, and commercial applications have expanded rapidly, and a large group of professionals now practice the art. A number of academic institutions have added the subject to their graduate electrical engineering curricula, and more are expected to follow.

Like most of the established practitioners of power electronics, I came to the field of switching power converters more by chance than intent. In 20 years of research-and-development work for a major corporation, I have had the good fortune to be exposed to virtually all aspects of and applications for switching power converters, and to be part of a group of first-class workers who were assembled to further knowledge of the art. The understanding of converters that I developed in that environment is presented in *Switching Power Converters*.

Most traditional views of the field have seemed somewhat disjointed; converters were largely regarded as related only because they all use semiconductor switches and have certain topological similarities. This did not seem logical to me or to my coworkers. Gradually, from the cumulative experience of our work, the view expounded herein was developed: that is, switching power converters are related by function and behavior; their basic characteristics do not in any way depend on the types of switches used, nor on the applications to which they are put, nor on the topologies in which they are realized.

This view has proven valuable in the working environment, creating a rational basis for the conceptual process in converter development and for the comparison of diverse converters proposed for a given application. In addition, the coherent, ordered, and totally unified converter theory presented has been well proven in the halls of academe. I have taught a graduate course at Carnegie-Mellon University in Pittsburgh for four years, and I spent two quarters as visiting McKay Lecturer at the University of California at Berkeley. I found that students (many at Carnegie-Mellon were themselves working professionals in the field) and faculty alike appreciate the simplicity and completeness of the approach. Moreover, those meeting the subject for the first time were able to easily grasp the essentials. Those with much prior experience and well-ingrained traditional views were quick to perceive the value and logic of applying a unified analytic approach to all types of converters.

Consequently, *Switching Power Converters* is intended for both the graduate student first meeting switching power converters and the practicing engineer well-versed in design subtleties. It is not a design manual; no one should expect to be able to design a converter merely through reading, or completing a course based on, the material presented here. The intent is to provide an understanding of the basic principles, relationships, and characteristics of switching power converters, in order to provide a firm basis from which design competence may later be acquired. The book addresses the whole field; it is the first to do so. From it, the novice can gain an understanding of all switching converters, their capabilities, and their limitations. The experienced practitioner should gain a much clearer appreciation of the relationships of converters in the field and to electrical engineering as a whole.

Finally, I would like to acknowledge those who have contributed to the book's gestation and production. First, my thanks go to Deborah Hutnyan, who typed the complete manuscript in an incredibly short time and with an accuracy that defies explanation in view of my appalling handwriting. To my friend and erstwhile colleague Brian Pelly, my thanks for the many fruitful discussions over the years and for his having "lit the lamp" by his original use of existence functions in the analysis of the naturally commutated cycloconverter. To my friend and colleague Dr. Laszlo Gyugyi, who first extended the analytic technique to the family of ac-to-ac converters, my appreciation for the many helpful discussions of the past 15 years and for his many perceptive comments and suggestions for improving the manuscript, which he read in its entirety. My wife Rita, daughter Kim, and son Grant exhibited patience while I was writing this book; without their understanding, I could not have succeeded.

PETER WOOD
Pittsburgh, Pennsylvania

TERMS, SYMBOLS, AND ACRONYMS

Defined quantity. A current or voltage assumed to exist at an ideal switching power converter's terminals.

Dependent quantity. A current or voltage produced by the converter's switches acting on a defined quantity.

V_S, I_S. Defined (source) dc voltage and current.

v_S, i_S. Defined oscillatory voltage and current.

v_D, i_D. Dependent voltage and current, which always contain oscillatory components.

Wanted component. Constituent of a dependent quantity that the user desires the converter to produce.

Unwanted components. Those oscillatory constituents of dependent quantities that are the inevitable consequence of switching converter operation but that are not involved in the power transfer sought.

V_{DD}, I_{DD}. The dc wanted (desired) components of voltage and current.

v_{DD}, i_{DD}. Oscillatory wanted components.

v_{DU}, i_{DU}. Unwanted components.

Existence function. A mathematical function representing the behavior of a switch in a converter.

H. Existence function.

e, i. Instantaneous values of voltage and current in circuits during transient phenomena.

E, I. Initial values of voltage and current in circuits prior to switch closing or opening.

$\bar{x}$. Laplace transform of the time-dependent quantity x.

L. Inductance.

C. Capacitance.

R. Resistance.

s. The complex variable of the Laplace transformation.

S. A switch.

Q. An active (controllable) semiconductor switch.

D. A diode.

f. Frequency.

ω. Angular frequency.

t. Time.

ϵ. The base for Naperian (natural) logarithms.

Line voltage. Voltage measured between lines of a polyphase system.

Phase voltage. Voltage measured between line and neutral in a polyphase system.

Line current. Current flowing in a line in a polyphase system.

Phase current. Current flowing in a load connected from line to line in a polyphase system.

M, N, m, n, p, q, l, k, K. Integers used in Fourier expansions.

x. Fractional variable effecting control of converter operation; modulation index.

α. Phase delay or advance controlling converter operation.

ψ. Phase angle of converter load.

ϕ, ρ. Phase angles for certain methods of converter operation and control.

Modulating function. A mathematical function introduced to existence functions to represent cyclic variations in the operation of the switches.

$M(t)$, $M(t,x)$, $M(t,x,\phi)$. Modulating functions.

HVDC. High-voltage dc transmission.

SVG. Static VAR generator.

PWM. Pulse-width modulation.

CAM. Conduction-angle modulation-variation of the conduction periods of switches in a converter in a nonperiodic manner.

Contents

SWITCHING POWER CONVERTERS

1

Conversion Functions and The Switching Matrix

1.1 Rationale for Switching Power Converters

Power electronics is the branch of electrical and electronic engineering concerned with the analysis, design, manufacture, and application of switching power converters. This book is concerned mainly with analysis and design. Manufacture and application, the ends to which all engineering is ultimately directed, necessarily depend on the ability to execute predictable designs. In turn, whereas empiricism is often the foundation for design early in the growth of an industry and is always the true foundation of theory, maturity comes only with the development of an adequate body of theory and analysis.

That power converters exist at all is due to the fact that very few people use electricity. Electricity has proven to be a very convenient way of distributing energy to a wide variety of geographically dispersed consumers. However, at the point of consumption, almost all electrical energy is converted into other forms—principally, to light, heat, mechanical work, and information. In those instances where electrical energy is used in its original form, mainly in the electrochemical and metals industries, the usual low-frequency ac distribution is generally not suitable for direct application.

The energy-converting devices used are vast in number, and, for the most part, they existed well before switching power converters became at all prominent. To convert electrical energy to light, incandescent bulbs, fluorescent tubes, and a variety of electrically excited plasma devices are used. In generating heat, resistive heating elements, induction heating coils, and dielectric heating devices are employed. The conversion of electrical energy to mechanical work is provided by ac and dc machines, while it becomes information in computers, data-processing systems, and communications networks of many types.

These energy-converting artifacts almost all share at least one of two characteristics. Many are unable to work efficiently, and some cannot work at all, if fed with the fixed-voltage 50 or 60 Hz ac that has become the standard means of distributing electrical energy. Many are also inherently incapable of controlling the rate of energy conversion. For instance, incandescent bulbs produce a fixed illumination and resistive heating elements produce a fixed number of calories per second (cal/sec) when connected directly to the appropriate level of the common supply; ac motors run at an approximately constant speed.

Many energy consumers, with many applications, require or desire that they be able to control the rate of energy conversion, the brightness of the bulb, the temperature of the heater, or the speed of the motor. Those wishing to use forms of electrical energy other than low-frequency ac in order to make their particular conversion devices work must find a means for changing the common supply to the form they need. Generally, they will also need to control the rate of conversion. From these needs arose the power converter industry. Power converters perform two functions:

1. They are almost always capable of controlling the rate of power flow from the common supply to the load(s) they serve.
2. In many instances, they are also capable of modifying the supply's form to one more suitable for their loads.

Since power converters are devices interposed between sources of electrical energy and loads that usually convert that energy to some other form (or, at the very least, consume it to make some product), efficiency has always been a cardinal consideration in their design and application. Energy represents a contribution to the cost of all products and services, and, when a power converter is employed, any energy that it consumes while performing its function is, in principle, an unnecessary additional cost. Thus, to satisfy the total application requirements, power converters must be made as efficient as possible. In the electrical and electronic world, linear active devices are notoriously inefficient. The maximum theoretical efficiency of a Class A amplifier, for example, is 50%; that of a Class B amplifier is only 78.5%; and, although Class C amplifiers are capable of higher efficiencies, sometimes exceeding 90%, they are of distinctly limited usefulness as power converters. Linear dc regulators, which are discussed in Chapter 3, are also extremely inefficient.

In contrast, a perfect switch is lossless; it carries current at zero voltage while it is closed and supports voltage at zero current while it is open. Practical switches are not so pristine, of course, but they generally make for a much better approximation to the goal of no losses than can any linear device. Thus, most practical power converters use switches as the active devices that make conversion possible, and this book is concerned almost entirely with switching power converters.

1.2 History of Switching Power Converters

The term *power electronics* was not coined until 1970 or thereabouts. Switching power converters were investigated and applied, however, long before the discovery and development of semiconductors, and had used a variety of switching elements.

Switching power converters were first developed in the early 1920s. However, until the 1960s, power conversion was accomplished mainly by rotary converters consisting of two or more standard electrical machines. Typical examples of these are (1) the motor-generator sets, 50-or-60-Hz induction motor direct-driving a multipole high-frequency alternator, used as supplies for induction heating, and (2) the Ward-Leonard sets, ac induction motor–dc generator–dc motor, used to provide variable-speed four-quadrant dc-motor drives.

The first switching power converters seriously considered for widespread application used mercury-arc rectifiers and grid-controlled mercury-arc tubes. Initially, only the ac-to-dc conversion function was attempted, using what later came to be termed "current-sourced, line-commutated" circuit arrangements. A wide variety of topologies were explored, and it was quickly realized that such converters could also provide the dc-to-ac conversion function. Following this, attempts were made to develop ac-ac converters based on the same technology. Although fairly successful from a technical viewpoint, the mercury-arc converters that emerged found only limited commercialization due to a number of inherent problems. These included performance limitations directly attributable to the switching devices being used, efficiencies that were not as high as desired because of tube conducting voltage drops, high costs, questionable reliability, and excessive maintenance requirements.

At the same time, work was proceeding on power converters using electromechanical switching devices, mainly rotary "commutators." While cheaper, more efficient, and perhaps more reliable than the mercury-arc tubes, the high maintenance requirements and poor performance of these devices inhibited their widespread application.

Not until the 1930s was the possibility of alternate circuit approaches thoroughly explored. Attempts to implement the "voltage-sourced, force-commutated" dc-to-ac conversion function, as it came to be known, suffered from a lack of understanding of its relationship to the technology previously established. More important, it suffered from the very severe limitations of mercury-arc tubes and hydrogen thyratrons, which were the switching devices then available. This interesting part of the power converter field later became very important. At the time, however, it never got beyond the stage of a laboratory curiosity.

Although the lack of commercial acceptance caused interest in mercury-arc converters to wane in the late 1930s, the years of activity with them produced some important results. The basic operating principles of the current-sourced converters were clearly established and well understood, at least for the ac-to-dc conversion function. It was evident that variable-speed two- and four-quadrant dc-motor drives could be implemented using such converters, and that given better performance, lower cost, and lower maintenance needs, the Ward-

Leonard set might have an effective competitive technology. The possibilities of ac-to-ac and voltage-sourced dc-to-ac technologies had emerged, and the former had in fact produced a few installations even though neither technology was fully understood.

Except for a few specialized applications, such as very high power ac-to-dc converters for the electrochemical and metals industries, the mercury-arc converter business then sank to a low ebb for some 15 years. It was revived by a power conversion application of enormous importance to the field at large: high-voltage direct-current transmission (HVDC). In the 1950s, the development of reliable tubes at sufficiently high voltage and current ratings led to their use in converters for transmitting power, first under water and shortly thereafter over land. This was accomplished by converting ac to high-voltage dc at the generating or sending end of a link and then reinverting into the ac system at the receiving end. Originally operating at a few tens of kilovolts and handling a few tens of megawatts, HVDC links were later developed that operated at several hundred kilovolts and handled powers of several hundred megawatts. The ability of the links, using switching power converters as "terminals" at both ends, to provide reversible and controllable power interchange between two asynchronous ac systems excited the imaginations of electric utilities and power-converter specialists alike. The mercury-arc converter's reign was short-lived, however; in the late 1960s, semiconductor switches began to replace the tubes as switches, and ultimately completely supplanted them.

During the period of abeyance in mercury-arc converter development (ca. 1940 to 1955), another type of switch became prominent in power conversion technology. During World War II, the magnetic amplifier, or controllable saturating reactor, was intensively developed and fairly well exploited for a variety of power conversion needs. Since its most natural implementation was as an ac switch, the device found most usage in ac regulators. These were used both to control power flow to ac loads and as precursors to and controllers for ac-to-dc converters using mercury-arc rectifier tubes or those primitive semiconductor rectifiers, selenium and copper oxide.

In the decade following the war, converters using magnetic amplifiers were somewhat commercialized. However, high cost, lower-than-desired efficiency, and performance limitations again restricted market penetration, and by the late 1950s such converters made up a stable but not overly significant industry.

The expansion and diversification of the power converter industry, which began in the late 1950s and continued apace thereafter, originated in the event that most influenced electrical and electronic technology in the twentieth century—the discovery of the transistor by John Bardeen, Walter H. Brattain, and William B. Shockley in 1948. Although transistors themselves were not to become significant to power converters for 25 years, their discovery spawned the semiconductor industry, which produced other switching devices that had great impact very quickly.

First came germanium, and shortly thereafter silicon, PN-junction high-current rectifiers. These rectifiers proved to be much more efficient than mercury-arc, selenium, or copper oxide. In little more than a decade, the current and voltage ratings increased and their costs decreased to such an extent that they conquered the bulk of consumer, commercial, and industrial low-to-medium-voltage applications. Many ac-to-dc conversion applications, which were only marginally economic before, become clearly so with these devices. In 1957 the single most important switching device of the succeeding 20 years was developed; the thyristor or silicon-controlled rectifier (SCR). This solid-state ''equivalent'' of the grid-controlled mercury-arc tube expanded rapidly in capability and underwent marked cost reduction. Its superior performance characteristics—lower conducting drop, faster switching, greater reliability, and lower maintenance needs—permitted it to penetrate markets that had been impossible for mercury arc. Within 15 years, it had essentially supplanted the Ward-Leonard set in multiquadrant dc-motor drive applications, had swept aside the mercury-arc tube in virtually all the traditional controlled ac-to-dc converter applications, and had largely driven out magnetic amplifiers. Moreover, it was used to develop other switching converter technologies in this two-decade period, to the point of possible or actual commerical exploitation. AC-motor drives, computer and transmitter power supplies, induction heating, lighting, and heating controls in general—all either fell before its onslaught or were brought to the point of readiness to do so. As previously mentioned, the thyristor also displaced mercury arc in HVDC transmission, and by 1975 it had further penetrated the electric utility arena with the development of the static VAR generator (SVG).

Further developments in and market potential for switching power converters look promising. It is possible that a host of new energy technologies will emerge, both for generation and storage—for example, batteries, fuel cells, solar photovoltaic, magnetohydro-dynamic, superconducting magnetic, wind power, tidal power, and so on. Many are of inherent and most of possible dc character, giving rise to the vision of a future need for vast amounts of power conversion equipment for their interface with the existing ac transmission and distribution systems.

The thyristor, although a technical and commercial success, was not free from limitations. Its latching nature and various switching problems caused power converter designers to seek alternatives for a variety of applications, mainly higher frequency converters and force-commutated converters of all types. During 1967–1977, several close relatives of the thyristor emerged in an attempt to overcome these problems, including the gate-controlled switch (GCS), gate-assisted turnoff device (GATT), and reverse-conducting thyristor (RCT). None were entirely successful; as thyristor limitations were overcome, other equally serious problems arose. Toward the end of the 1970s, power transistors were sufficiently improved in capability to challenge the thyristor in low-to-medium-power (up to a few kilowatts) converters. They replaced thyristors in the lower power applications (up to a few hundred watts) and began to be seriously considered for higher power high-frequency converters and for the more sophisticated converter technologies. In turn, starting in 1978, transistors were challenged by the metal-oxide semiconductor field-effect transistor (MOSFET), which had by then become the staple technology of digital and analog logic and signal-processing circuitry. Of course, it remains to be seen which devices will ultimately fill what applications. Improvements are constantly being made in all: diode, thyristor, GCS, GATT, RCT, transistor, and MOSFET.

1.3 Conversion Functions

In the strict sense, there exists only one type of electrical distribution—polyphase ac—and hence only one possible type of power conversion—polyphase ac, of some given frequency and voltage, to polyphase ac of some other frequency and voltage. All specific distributions, converters, and conversion functions are special—and usually restricted—examples of this all-embracing case. However, because of the special and distinct properties possessed by certain power distributions, electrical engineers have traditionally recognized sharp divisions into specific types. Thus, they generally recognize three types of power distribution, namely:

- polyphase ac (usually three-phase)
- single-phase ac
- dc

Although these distinctions are valid, there are only four basic conversion functions that can be implemented:

- ac to ac
- ac to dc
- dc to ac
- dc to dc

Within each conversion function, of course, lies the possibility of a great variety of converter topologies and implementations. Nonetheless, the functions listed are the only ones that can possibly exist, and they are related to each other.

The most general case, obviously, is polyphase ac to polyphase ac. In both function and converter topology, single-phase ac arises when the number of phases at either or both sides of a converter is restricted to one. Such a restriction will produce polyphase ac to single-phase ac converters; single-phase ac to polyphase ac converters; and single-phase ac to single-phase ac converters. The dc conversion functions then arise when the frequency at a single-phase converter's terminals is reduced to zero. Doing this will give rise to polyphase ac-to-dc converters; dc to polyphase ac converters; single phase ac-to-dc converters; and dc to single-phase ac converters. Now, if frequency at the single-phase ac terminals of these converters is reduced to zero, then dc-to-dc conversion results.

Thus, there exists a converter and a conversion function hierarchy, which arises from the successive application of restrictions to the most general conversion function. Converters and conversion functions belong to a clearly recognizable family, and within each of the functions listed, there are clearly recognizable and related subdivisions leading to those specific switching converter topologies that can implement the conversion functions.

The functional hierarchy developed is shown in Table 1-1. The left-hand side of the table has been arbitrarily considered to be input and the right-hand side to be output. This implies that power flow in the converters implementing these functions would be from left to right, but in fact many switching converters exhibit reversible power flow capability. Any ac-to-ac converter must do so, of course, to function at all, and most fully controllable ac-to-dc converters are also dc-to-ac converters and vice versa. In fact, as will be seen, the single-quadrant converter (unidirectional power flow out of or into dc terminals) is a special restricted case of the two-quadrant (bidirectional power flow) converter,

Table 1-1. Switching converters functional hierarchy.

Polyphase AC to Polyphase AC (Usually three-phase)	
Restrict input to single phase	Restrict output to single phase
Single-Phase AC to Polyphase AC (Usually three-phase)	Polyphase AC to Single-Phase AC (Usually three-phase)
Restrict input to zero frequency	Restrict output to zero frequency
DC to Polyphase AC	Polyphase AC to DC
Restrict output to single phase	Restrict input to single phase
DC to Single-Phase AC	Single-Phase AC to DC
Restrict output to zero frequency	Restrict input to zero frequency
DC to DC	

which in turn is a special restricted case of the four-quadrant converter, which is identical to an ac-to-ac converter.

From this hierarchy of conversion functions and the corresponding converters, one can deduce that if a converter topology is postulated to implement the most general function, polyphase ac to polyphase ac, then converter topologies capable of implementing all other functions can be derived therefrom. This deduction is correct; the topology postulated (and used) to fulfill the most general function is called the *general switching matrix;* all other converter topologies can be derived from it by applying appropriate functional restrictions leading to topological simplifications.

1.4 Switching Matrix

The general switching matrix is depicted in Figure 1-1. It represents the simplest conceivable switching converter topology that can perform the polyphase-ac to polyphase-ac conversion function, comprising a single switch placed, for example, at each intersection of the M input lines and the N output lines. Consider how such a converter will operate and what the switches will be required to do.

It can be predicated that one side of the converter—for example, the input—will be connected to a polyphase set of ac voltages. A switch can only be open, in which case it will carry no current but support voltage, or closed, in which case it will carry current but support no voltage. Now consider the output line (labeled A). It is connected to M switches, each of which connects to one of the input lines. Clearly, only one of these can be closed at any given time. If two or more were closed simultaneously, they would short-circuit the input voltage sources to which they are connected. In practical terms, the switches or protective devices such as fuses would then be destroyed; analytically, this condition is said to violate Kirchoff's law for voltage (KLV).

For a voltage other than simply an input voltage to appear on line A, more than one of its M switches must be used. It can be expected that all will come into play, and that because only one can be closed at any time, they will be closed sequentially in some pattern, each opening as the next in sequence is closed. The voltage on line A will then consist of a sequence of segments taken from the M input sinusoidal voltages. This output voltage clearly cannot be a pure sinusoid (or pure dc), but it can be postulated that a pattern of closings and openings exists for the M switches, which makes the largest amplitude component of the Fourier expansion for this voltage a sinusoid at some wanted output frequency. If this is so, then the same is true for all the other $N-1$ output lines. Furthermore, the switching patterns for these lines can be time-displaced from the pattern for the reference (line A) so that the major wanted components of these output voltages form a complete polyphase set.

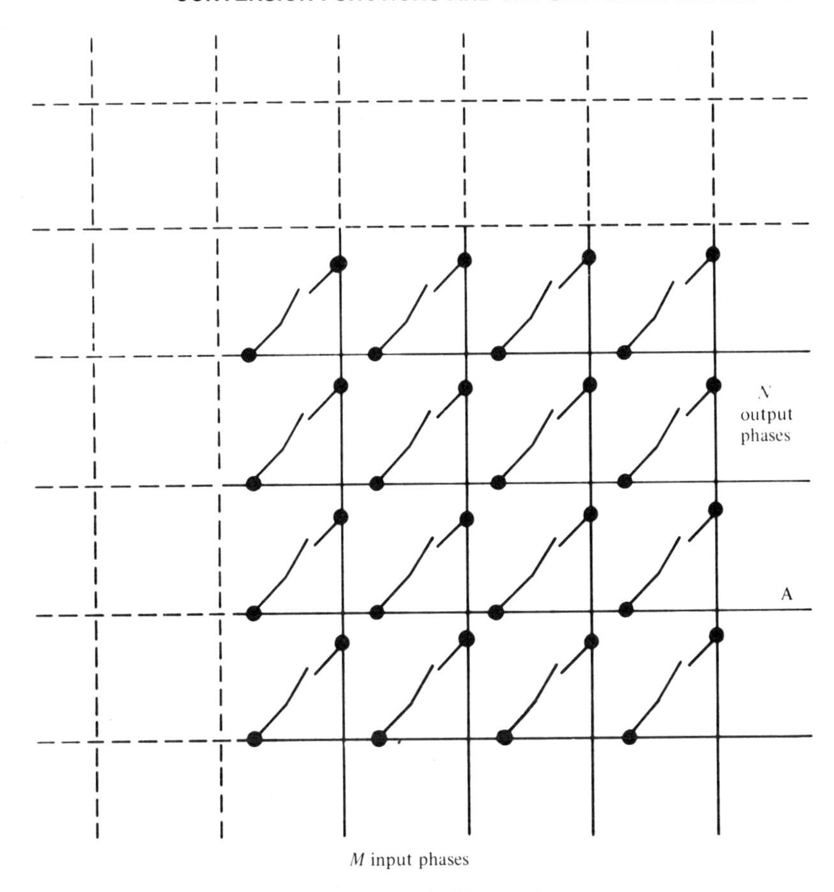

Figure 1-1. General switching matrix.

The polyphase input voltages are called *defined* or *assigned* quantities, since their existence has been arbitrarily assumed. The output line voltages, including their wanted components, are *dependent* quantities. They depend on two factors: the input voltages and the switching patterns. Currents flowing in the output lines will then be determined by these voltages and the load network connected to those lines. If the wanted components of the output voltages are sinusoids at some frequency, it is reasonable to assume that the major component of current flowing should be a sinusoid at the same frequency. In fact, if active (real) power at the wanted frequency is to be transferred to or from the output lines, such a current component is necessary. Hence, it can be postulated that the network connected to the output terminals of this switching matrix would ideally permit only wanted frequency components of current to flow. Thus, the network may be replaced by a polyphase set of ac current sinks (or sources).

If this is done, consider now the current that will flow in any input line. Each input line is connected to N switches; each switch connects to an output line in which a sinusoidal current has been arbitrarily defined except that it must have the same frequency as the wanted component of the output voltage. Thus, the input-line current will consist of a number of segments of the output currents. Note that it is possible for two or more of the N switches to be simultaneously closed, since this does not violate KLV. However, at least one switch must be closed at all times, or Kirchoff's law for current (KLC) will be violated. The pattern of switch operations for the N switches can be derived from the patterns that enable each group of M to produce the output voltages. Hence, the input line currents are dependent quantities, dependent on both the defined output currents and the switching pattern.

In operating this switching matrix as an idealized ac-to-ac converter, there are three basic requirements:

1. a defined set of ac input voltages
2. a defined set of ac output currents
3. a switch-operating pattern that will make the major components of the output voltages a polyphase set of the same frequency and progressive phase displacement as the defined currents.

Given these prerequisites, the output voltages and input currents are dependent quantities—they depend on the defined quantities, voltage and current, and the switching pattern. If a formal means of describing that pattern is available, then the dependent quantities can be formally derived and described. In this manner, the external terminal performance of the converter can be precisely defined, as can the internal behavior.

In the preceding discussion, there is an implicit assumption that power flow is from input to output. Nothing in the operational description that has evolved requires it to be so. Power could equally well flow from output to input, with the defined currents acting as energy sources and the defined voltages acting as energy sinks. Thus, for this switching matrix, the terms *input* and *output* do not define the direction of power flow—which they normally do in electrical engineering. Here, they are merely labels for convenient identification of the terminal sets of the matrix. Nothing in the operational description has limited the relationship of the input and output frequencies either. Since the wanted output frequency depends solely on the switching pattern, it is independent of and may be lower than, equal to, or higher than the input frequency. Furthermore, there is no inherent limitation placed on the displacement factor (power factor at the fundamental or wanted frequency) at either set of terminals. The displacement factor at one set, customarily the output, can be arbitrarily defined. The displacement factor at the other then becomes a dependent function, dependent on that defined and the switching pattern.

There are two complementary operating restrictions:

1. The instantaneous value of an output voltage clearly can never exceed the peak value of one of the input voltages.
2. The instantaneous value of an input current can never exceed the peak value of one of the output currents unless two or more switches connected to an input line are simultaneously closed.

Thus, the matrix cannot produce wanted components of output voltage that are of higher amplitude than the input voltages. Presumably, with appropriate variation of the switching pattern, it can produce any wanted output level from zero to the crest value of the input. These various properties viewed together allow this switching matrix to be regarded as a completely generalized transformer. It can transform frequency, phase angle, and, within the stipulated limits on maximum output voltage and input current, impedance. These properties are explored quantitatively and in depth in Chapter 5.

The basic performance required of the switches can also be deduced from the operational description. Since both voltages and currents at both sets of terminals are ac, the switches must be able to carry current in either direction when closed and to support voltage of either polarity when open. They may be characterized as needing bidirectional current-carrying, bidirectional voltage-blocking capabilities. In addition, they must be capable of being closed and opened in a repetitive pattern that produces the desired output voltages or at least the wanted components thereof. So far, it is not possible to say whether any limitations of switching behavior are acceptable, since the pattern has been presumed to exist and its details have not been established. It can be presumed that, in all probability, the switches will be required to close on command regardless of the instantaneous polarity of voltage withstood immediately before closing and regardless of the direction of current flow immediately afterward. Concomitantly, they will also probably be required to open on command regardless of the direction of current flow before opening or the polarity of voltage applied after it.

The complete reversibility of power flow exhibited by the matrix, together with the arbitrary assignment of phase angle at one set of terminals, suggests yet another property. The description so far assigns voltages at the input and currents at the output. These definitions can be reversed, with an appropriate transposition of the operating patterns of the individual switches. Of course, the input voltages and output currents then become dependent quantities. More important, the converter is now at least a two-port reciprocal of that described previously. Obviously, it cannot be a complete dual since the topology of the matrix is not planar. However, it exhibits more than two-port reciprocity, since voltage and current transpositions occur not only at the input and output terminals but also within the converter itself. There is also the transposition of the switching pattern, which indicates more than simple reciprocity.

This property of the matrix is rather trivial as far as the ac-to-ac conversion function is concerned. The total transposition can also be regarded as simply changing to a similar matrix with N input and M output lines, which obviously must operate in essentially the same manner, possess the same basic properties, and have the same basic requirements. Reciprocity, and ultimately duality when topology becomes planar, are much more important for the more restricted conversion functions, as will be seen.

Obviously, if the number of input or output terminals, or both, is reduced to two, then the single-phase ac converters result. Reducing output to two or input to two will produce polyphase to single phase or single phase to polyphase, respectively. Both configuration and reciprocity then indicate that the only difference between these two converters lies in the patterns assigned individual switches, and that the overall pattern needed for one will be a transposition of that applicable to the other. It is also obvious that the matrix retains all its basic properties and still makes the same demands on the switches. This restriction of the conversion function, from polyphase to single-phase ac at one set of terminals, can be viewed as a degeneration of the general switching matrix. Clearly, if the matrix is degenerated further to single phase at the other set, a single-phase-to-single-phase converter results. In addition, all the properties are still maintained; reciprocity now becomes truly trivial, for a single-phase-to-single-phase converter and its reciprocal are, obviously, identical. The two degenerations are depicted in Figures 1-2 and 1-3.

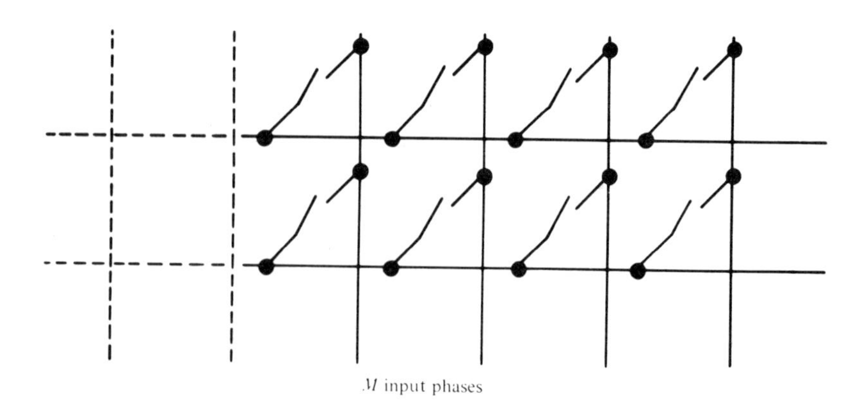

Figure 1-2. Polyphase-to-single-phase or dc switching matrix.

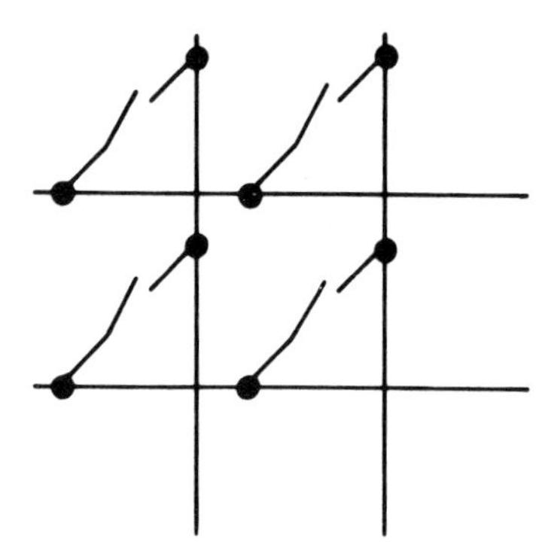

Figure 1-3. Single-phase to single-phase or dc switching matrix.

Suppose now that at the output terminals of the matrix of Figure 1-2, for the polyphase-ac to single-phase-ac function, the frequency of the wanted component is reduced to zero. For the converter to continue to fulfill a useful function (i.e., to transfer power), the frequency of the defined quantity there must also be reduced to zero. Now the matrix will act as a polyphase ac-to-dc converter. The property of reversibility means it will also act as a dc-to-ac converter. Moreover, the property of reciprocity means that the defined quantity restricted to zero frequency may be either current or voltage. Thus, it can be concluded that there are two types of ac-to-dc/dc-to-ac converter: referring to the defined dc terminal quantities, they are known as current-sourced and voltage-sourced. Since power flow is reversible, they are both inherently capable of two-quadrant operation at their dc terminals; the current-sourced is a unidirectional-current bidirectional-voltage converter, and the voltage-sourced is a unidirectional-voltage bidirectional-current converter. Their mutual reciprocity is quite evident from this fact alone.

These converters have the same topology (their switching matrices are identical), and they share it with the polyphase-ac-to-single-phase-ac converter. How do they differ from each other and from the ac-to-ac converter? It can be deduced that the differences will lie in three areas: switching patterns, switch requirements, and displacement factor. First, consider switching patterns. With the ac-to-ac converter, as the frequency of the single-phase quantity is reduced, the switching pattern will change in some way; when the frequency reaches zero to give an ac-to-dc/dc-to-ac converter, the pattern will differ from that which produces an ac output (or input). Moreover, the precise nature of the change will depend on whether the converter is voltage- or current-sourced. If current-sourced, then the pattern must not violate KLV when all the switch connections viewed from the ac terminals are considered and must not violate KLC when the view from the dc terminals is considered. If voltage-sourced, the converses apply, and reciprocity is again evident. One pattern cannot possibly satisfy both sets of conditions simultaneously, and so two different patterns are needed.

If the converter only has to produce unidirectional current at the dc terminals, then the switches only have to carry current in one direction when closed. However, they will still have to withstand voltage of either polarity when open. If unidirectional dc voltage is defined, the switches will need bidirectional current-carrying capability but will only have to withstand voltage of one polarity. Hence, the required switch characteristics have the same restriction that is applied to the defined quantity at the dc terminal, and once more reciprocity emerges.

Displacement factor is not applicable to dc quantities. For these converters and conversion functions, the displacement factor at the ac terminals is a dependent function, depending solely on the switching pattern. Thus, through appropriate maniuplation of the pattern, the displacement factor can be made anything desired.

In both the polyphase-ac to single-phase-ac converter and the current-sourced ac-to-dc converter, the switching matrix can be simplified even further. The full matrix depicted in Figure 1-2 is known as the *bridge configuration;* it has no direct connection between input and output lines—all connections are made via the switches. If the ac voltage source possesses a neutral, and the ac or dc current source's current can be allowed to flow therein, then the "midpoint" topology depicted in Figure 1-4 results. Switch requirements and basic converter properties remain unaffected, but the external terminal performance of this further degenerated switching matrix may be degraded as compared to the performance of the full matrix. Quantitative assessments of its behavior are found in Chapters 4 and 5.

It is not possible, however, to use this configuration with the ac-to-dc voltage-sourced converter. If the dc terminals are connected to a voltage source, then

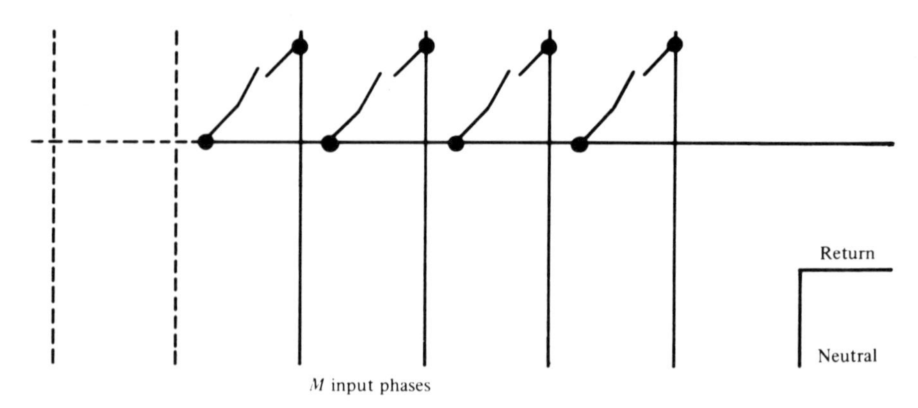

Figure 1-4. Midpoint switching matrix.

the ac terminals must be connected to current sources. In this event, all the switches of the midpoint topology of Figure 1-4 would have to remain permanently closed in order to avoid violation of KLC. The resulting voltages impressed on the ac current sources would all be continuous dc, and no power transfer could take place. This restriction can be lifted if coupled ("OR'ed") current sources are postulated. In practice, this is so difficult to achieve, and the converter performance is so poor, that the configuration is not used except for the two-phase realization.

As depicted in Figure 1-3, reducing the number of ac input terminals to two in the matrix of Figure 1-2 clearly produces the single-phase-ac to single-phase-ac converter-switching matrix. This is identical to the single-phase ac-to-dc/dc-to-ac converters, and in all cases the same switch requirements and basic converter properties are obtained as for the polyphase-ac input versions. Note that the midpoint configuration is not feasible for these cases unless discontinuous current is allowed. As discussed in Chapters 3 through 5, some converters do operate in such a manner, but their properties are usually so inferior to those that do not operate that way, that they are only met in low-power or highly specialized and economically constrained applications.

The ac-to-dc/dc-to-ac converters discussed so far all have two-quadrant (reversible power flow) capability. At times, only single-quadrant (unidirectional power flow) operation is needed; in this case, some modification of the switching matrix might be expected to accompany that operating restriction. Clearly, neither of the defined quantities is subject to change. The only way in which the current-sourced converters can be restricted to one quadrant is by restricting them to produce only one polarity of dc terminal voltage. The only way to constrain the voltage-sourced converters is to have them produce only unipolar dc terminal current. Changes in switching matrix topology cannot effect these restrictions, so the only mechanisms available to achieve single-quadrant capability alone are changes in switching patterns and in switch capabilities. These two factors are not totally independent, since any restriction of switch performance to less than fully bilateral would clearly obviate the implementation of switching patterns to produce ac-to-ac converters. Moreover, further restriction of switch capabilities below those demanded by the two-quadrant converters can be deduced to interfere with the ability to implement the switching patterns needed for those conversion functions.

The converse does not apply. It would be possible to limit the switching pattern while retaining the switch capabilities. However, such a course would produce a totally artificial restriction on converter operation, for the matrix employed would still be capable of two-quadrant operation if the switches were controlled differently. Hence, to make an inherently single-quadrant converter, the proper action is to restrict switch capability in order to force the appropriate limitation of switching patterns and dc terminal behavior. The only other basic restriction possible on the switches in both current- and voltage-sourced converters is to render them simply unilateral—i.e., to have them unidirectional current-carrying and unidirectional voltage-blocking in either case. Obviously, doing so will restrict permissible switching patterns. However, deduction of the properties of the converters that thus arise is not possible without considering the actual switching patterns and internal converter operation. These converters are discussed in detail in Chapter 4; at this point, it is sufficient to state that single-quadrant converters do result, some of which have the capability of controlling power flow.

Returning to the process of matrix degeneration for progressively simpler conversion functions, consider the matrix of Figure 1-3 acting as a dc to single-phase-ac converter. In performing this function, the matrix will operate at some time in every one of the four quadrants at the ac terminals, be it current- or voltage-sourced at the dc terminals. It is therefore quite capable of performing the function of four-quadrant dc-to-dc conversion, either current- or voltage-sourced. This conversion function, although rarely needed in practice, is introduced here as the beginning of matrix degradation to those dc-to-dc converters that are widely used.

As for the ac-to-dc converters, restriction of the number of operating quadrants is possible only by restricting switch capabilities. If two of the switches, diagonally placed in this 2×2 matrix, are made unilateral, then two-quadrant dc-to-dc converter matrices result, as depicted in Figures 1-5 and 1-6. Note the reciprocity of these two switching matrices; in the current-sourced case, the

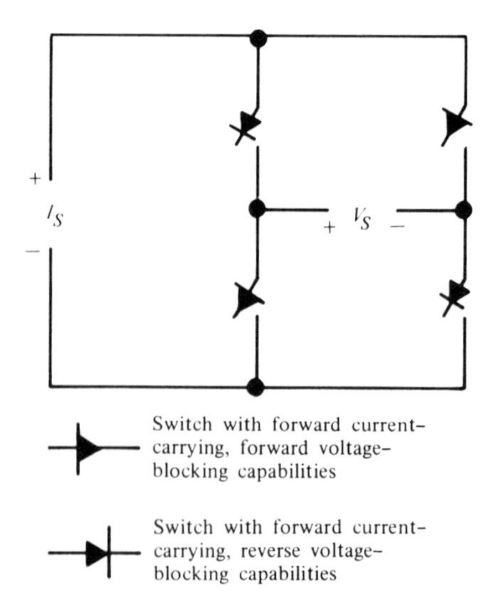

Figure 1-5. Two-quadrant current-sourced dc-to-dc converter.

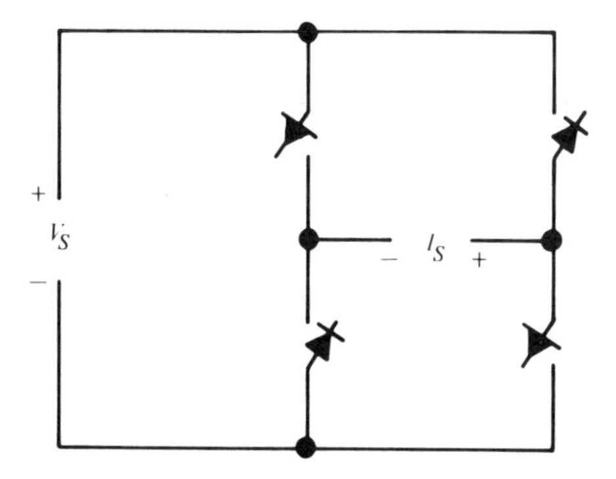

Figure 1-6. Two-quadrant voltage-sourced dc-to-dc converter.

switches restricted to unilateral capability lie on the conjugate diagonal of the matrix, while in the voltage-sourced case they lie on its principal diagonal, and, moreover, the polarity of voltage to be withstood is reversed.

Figures 1-7 and 1-8 show the final degenerations, to single-quadrant dc-to-dc converter-switching matrices. This is accomplished by eliminating one unilateral and one bilateral switch from each of the two-quadrant matrices. Once again, there is reciprocity, and the end products are obviously full duals. For the current-sourced case, both switches in the upper row are eliminated, the unilateral one being replaced by a solid connection. For the voltage-sourced, both switches in the right column are eliminated, and the bilateral one is replaced by a solid connection.

The hierarchy of conversion functions shown in Table 1-1 can be repeated as a hierarchy of converter-switching matrices with some expansion to accommodate the consistent reciprocity that occurs. The tabulation of ac-to-ac converters is shown in Table 1-2.

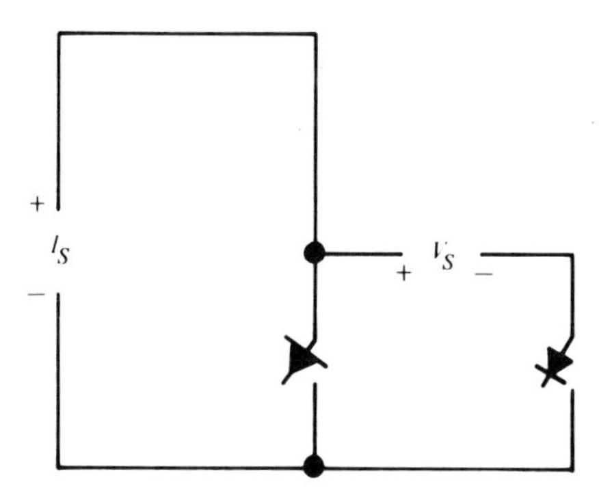

Figure 1-7. Single-quadrant current-sourced dc-to-dc converter.

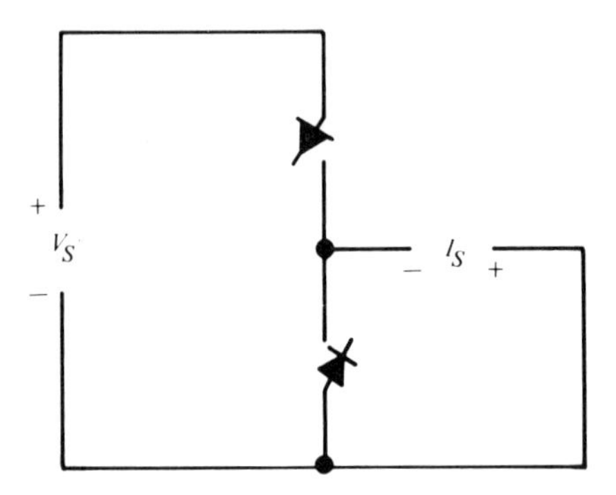

Figure 1-8. Single-quadrant voltage-sourced dc-to-dc converter.

The derivation of ac-to-dc, dc-to-ac, and dc-to-dc converters is shown in Table 1-3.

The single-quadrant (nonreversible) ac-to-dc and dc-to-ac converters are, as was discussed, derived from the two-quadrant versions by restricting switch capabilities and thereby forcing changes in the switching patterns. Current and voltage in Tables 1-2 and 1-3 refer to the defined quantities at the converters' terminals.

In developing the qualitative operating description of the general matrix, it was observed that the dependent quantities consisted of sequential segments of the defined quantities as determined by the switching pattern of the converter matrix. This remains true, of course, for all the converters and conversion functions. Thus, in no case can a dependent function be simply the wanted

Table 1-2. AC-to-AC converter hierarchy.

Polyphase ac voltage to (reversible) Polyphase ac current	Is the reciprocal of	Polyphase ac current to (reversible) Polyphase ac voltage
	Reduce number of input phases	
Single-phase ac voltage to (reversible) Polyphase ac current	Is the reciprocal of	Single-phase ac current to (reversible) Polyphase ac voltage
	Reduce number of output phases	
Single-phase ac voltage to (reversible) Single-phase ac current	Is the reciprocal of	Single-phase ac current to (reversible) Single-phase ac voltage

Table 1-3. Derivation and hierarchy of ac-to-dc and dc-to-dc converters.

Single-phase ac voltage to (reversible) Polyphase ac current	Is the reciprocal of	Single-phase ac current to (reversible) Polyphase ac voltage
	Reduce input frequency to zero	
dc voltage to (reversible) Polyphase ac current	Is the reciprocal of	dc current to (reversible) Polyphase ac voltage
	Reduce output phases	
dc voltage to (reversible) Single-phase ac current	Is the reciprocal of	dc current to (reversible) Single-phase ac voltage
	Change switching pattern	
dc voltage to (reversible) Four-quadrant dc current	Is the reciprocal of	dc current to (reversible) Four-quadrant dc voltage
	Restrict capabilities of half of switches	
dc voltage to (reversible) two-quadrant dc current	Is the reciprocal of	dc current to (reversible) two-quadrant dc voltage
	Remove half of switches	
dc voltage to dc current		dc current to dc voltage

component thereof, and an inevitable and undesirable property of switching-power converters is deduced. That is, the wanted component of a dependent quantity will always be accompanied by unwanted components. Dependent ac voltages and currents thus consist of wanted components (generally sinusoids and then also termed *fundamental*) accompanied by a spectrum of unwanted components, which may or may not be harmonics of the fundamental. Similarly, dependent dc voltages and currents consist of ideally smooth wanted components accompanied by a spectrum of unwanted ac components collectively referred to as *ripple*.

The existence of unwanted components in the dependent quantities is an inherent and inevitable consequence of the use of switching, as opposed to linear, power converters. The unwanted components generally do not participate in the power transfer process, which is usually a converter's primary function. They are usually a major factor in limiting switching converter performance and application, and they are always the determining factor in the design of interfaces between a practical converter switching matrix and its real sources and loads.

Problems

1. Sketch switching matrices for the following converters:

a. Three-phase ac to three-phase ac
b. Three-phase ac-to-dc defined voltage, bridge
c. Three-phase ac-to-dc defined current, midpoint
d. Single-phase ac-to-dc defined voltage, bridge
e. Single-phase ac-to-dc defined current, bridge

2. Indicate on V-I plane sketches in which quadrants the four ac-to-dc converters of Problem 1 will operate.

3. Formulate the switching pattern restrictions necessary to avoid violation of KLC and KLV for all the converters of Problem 1.

4. Explain the properties of reciprocity and reversibility as they apply to converter switching matrices.

5. Can a compound converter, consisting of an ac-to-dc converter feeding a dc-to-ac converter, provide all the properties of a direct ac-to-ac converter. Explain the reasons for your answer.

6. Explain why the dependent quantities of switching converters always contain unwanted components.

2
Existence Functions

2.1 Nature and Use

In approaching the design of a switching power converter for any application, two related sets of parameters are of paramount interest. The external terminal performance, defined by the dependent quantities, will determine how well the converter meets application needs and what will be needed to interface it successfully with the actual sources and loads involved. The internal currents and voltages of the converter's switching loops will determine the selection of the active switching devices to be used and also any auxiliary passive components needed to enable the devices to function properly.

Both sets of parameters can be obtained analytically from the dependent quantities, which in turn are determined by the defined quantities and the switching pattern of the converter matrix, as discussed in Chapter 1. To obtain a precise, quantitative definition of the dependent quantities, some means of formally and quantitatively describing the switching pattern is needed. The mathematical expressions used for that purpose are herein called *existence functions*. In addition to permitting accurate mathematical formulations for the dependent quantities, they also allow dependent quantity and internal converter waveforms to be constructed graphically.

The existence function for a single switch assumes unit value whenever the switch is closed and is zero whenever the switch is open. In a converter, each switch is closed and opened according to some repetitive pattern; hence, its existence function will take the form shown in Figure 2-1—a train of pulses of unit amplitude. Neither the pulses nor the intervening zero-value periods will necessarily all have the same time duration; however, the requirement that a repetitive switching pattern exist means that the function must at least consist of repeated groups of pulses. The simplest, or *unmodulated,* existence functions (see Figure 2-1) have pulses all of the same time duration and zero intervals with the same property. The more complex variety, which has differing pulse durations and various interspersed zero times, is called a *modulated* existence function.

The use of existence functions to derive dependent quantities and internal converter stresses is very simple. Consider Figure 2-2, which shows an isolated

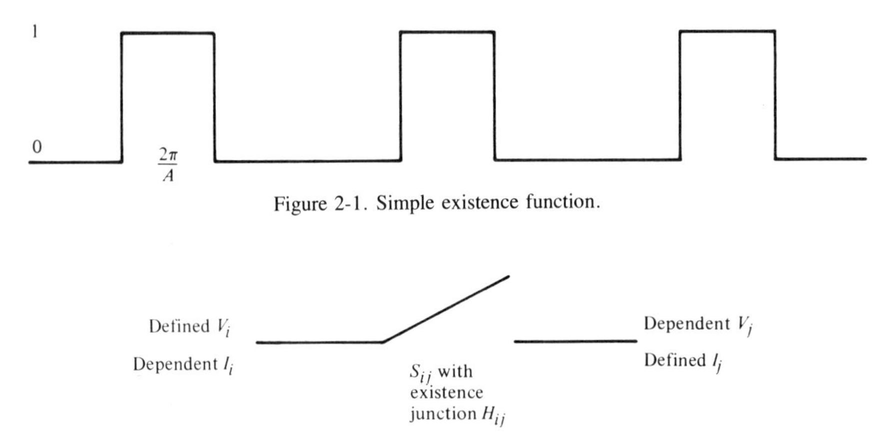

Figure 2-1. Simple existence function.

Figure 2-2. Isolated switch of converter matrix.

switch in a converter matrix. This switch is connected to V_i, the ith of a set of M-defined voltages, and to I_j, the jth of a set of N-defined currents. Its existence function is H_{ij}, a train of unit-value pulses separated by zero-value intervals as previously described. Now, whenever the switch is closed, V_j, the voltage impressed on I_j, must be V_i because the switch supports no voltage and no other switch connected to I_j can be closed without violating KLV. Whenever the switch is open, voltage V_i is removed from I_j with the switch supporting the difference between V_i and V_j existing at that time, which is determined by some other defined voltage and the corresponding switch. Hence, the equation for V_j can take the form:

$$V_j \text{ (dependent)} = H_{ij} V_i + \text{Contributions of remainder of sources and switches} \quad (2.1)$$

This is obviously valid since H_{ij} has unit value when the switch is closed and zero value when it is open. A similar argument can be used to establish the current in the switch as:

$$I_S = H_{ij} I_j \quad (2.2)$$

The current flowing in V_i is clearly the current in this switch plus the currents in all other switches connected to this same defined voltage. Hence it can be written as:

$$I_i = H_{ij} I_j + \text{Contributions of remainder of sources and switches} \quad (2.3)$$

The simple equations, 2.1 and 2.3, give the contributions of this individual switch to the dependent quantities. This switch is not special in any way, however; its position in the matrix has not been restricted in any way, and, except for the noncoicidence of closure with other switches in the same row, neither has its pattern. Thus, the same equations apply no matter what specific position the switch occupies, and hence they apply for all of the switches. The complete equation for the voltage impressed on I_j can be written as:

$$V_j = \sum_{i=1}^{i=M} H_{ij} V_i \tag{2.4}$$

The equation for the total current flowing in V_i can be written as:

$$I_i = \sum_{j=1}^{j=N} H_{ij} I_j \tag{2.5}$$

The voltage impressed on a switch, V_S, is the difference between its voltage source, V_i, and the voltage impressed on its current source while it is open. Thus, the equation for V_S becomes:

$$V_S = V_i - V_j \tag{2.6}$$

Thus, the dependent quantities and the switch stresses can be completely defined using the existence functions. For precise mathematical expressions, appropriate mathematical expressions are needed to define the existence functions. For waveform construction, the equations show that if all the source waves and existence-function pulse trains are plotted together with a common time scale, dependent quantity and switch-stress waveforms can be drawn from those segments of the source waves coinciding with unit-value periods of appropriate existence functions. This is illustrated by the simple example of Figure 2-3.

Now, observe that just as there was nothing special about the particular switch chosen to derive Eqs. 2.1 through 2.6, neither was there anything special about the particular defined voltage and current sources, V_i and I_j. For instance, V_i could be any one of the M voltage sources defined for the converter-switching matrix, and I_j any one of the N current sources. Thus, Eqs. 2.4 and 2.5 will hold for all dependent quantities; Eq. 2.4 is valid regardless of the particular j, from 1 to N; and Eq. 2.5 is valid regardless of the particular i, from 1 to M. If the defined voltage sources, V_1 to V_M, are expressed as the M-element column vector $[V_{in}]$ (called the *defined voltage vector*), and all the existence functions

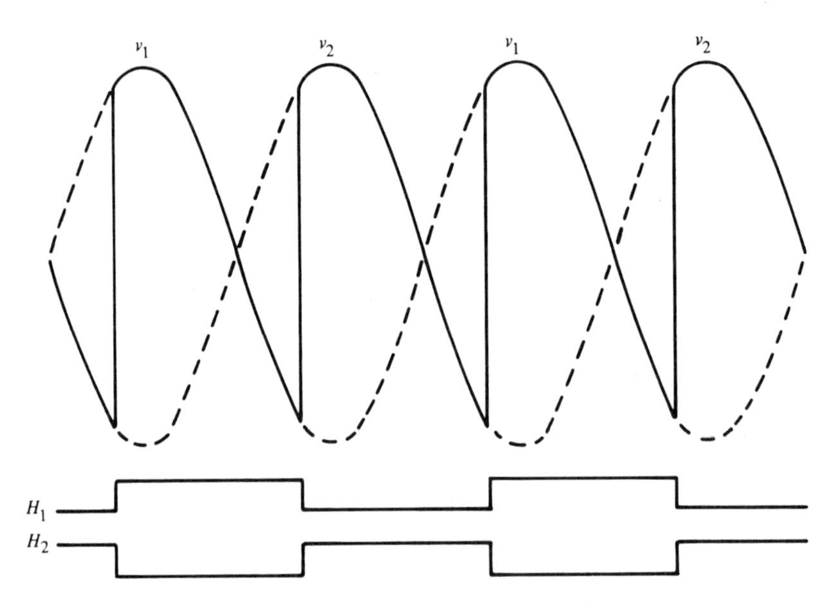

Figure 2-3. Example of waveform construction.

are expressed as the $M \times N$ matrix $[H]$, then the dependent output voltage vector, consisting of all the voltages impressed on the defined current sources, can be expressed:

$$[V_{\text{out}}] = [H]\,[V_{\text{in}}] \tag{2.7}$$

Similarly, if $[I_{\text{out}}]$ is the N-element defined current vector, then the set of M-input currents can be defined as the M element vector:

$$[I_{\text{in}}] = [H]^T\,[I_{\text{out}}] \tag{2.8}$$

where $[H]^T$ is the transpose of $[H]$. The sets of switch currents and voltages may be expressed by summations of similar matrix products.

Equations 2.7 and 2.8 represent the most elegant formulations of the dependent quantities arising from the application of this basic analytic technique to switching converters. In practice, converters are rarely complex enough in topology to call for full use of the matrix expressions. Even when they are, they can usually be broken into a number of subconverters, each of which replicates the behavior of the others. Thus, in this book, as in the field in general, more use is made of Eqs. 2.2 through 2.6, giving the dependent quantities for individual lines in a converter and the stresses for individual switches. It is usual to find that the remaining dependent quantities and stresses are merely phase-displaced (time-displaced) replicas of those individual ones that are so derived.

2.2 Mathematical Representation of Existence Functions

Before using these techniques to attempt analyses of converters and conversion functions, two questions must be answered. First, what types of switching pattern pertain to the various converters and functions? That is, what are the unit-value and zero-value periods of the existence functions for a converter of a given topology performing a given function? Second, what mathematical representation is to be used for the existence functions to yield readily interpretable expressions for the dependent quantities and switch stresses?

The first question is addressed progressively in Chapters 3 through 5. As will be seen, for the simpler converters and functions, simple existence functions can be postulated and tested. From such results, the more complex switching patterns and existence functions needed for the more complicated converter topologies and functions can be deduced. The second question will be addressed here.

As has been seen, existence functions are trains of unit-value pulses interspersed by periods of zero value. Since switching patterns are invariably repetitive, existence functions are periodic. They can be mathematically represented in various ways; Laplace transforms can be used, as can Z-transforms. However, the expressions that result from these transforms do not give readily interpretable results for the magnitudes and temporal parameters of the dependent quantities and switch stresses. However, if Fourier expansions are chosen as the vehicle for expressing existence functions, these problems disappear. The expressions that then arise for dependent quantities and switch stresses are readily interpretable; the only problem is that, like all such closed-form solutions to complex problems, they are somewhat cumbersome. At the very least, infinite-series summations reflecting the Fourier expansions result; when modulated existence functions are involved, double and triple infinite-series summations result.

Fourier-series expansion of a simple existence function such as that depicted in Figure 2-1 is, for present purposes, best accomplished by setting the zero time reference at the midpoint of one of the unit-value periods.

Let the repetition frequency of the pulses be f with a time period $T = 1/f$; define the angular frequency $\omega = 2\pi f$, then $\omega T = 2\pi$ rad. If the angular duration of the unit-value period is $2\pi/A$ rad, where $A \geqslant 1$, then the boundaries of the unit-value period with respect to the time zero reference are $-\pi/A$ and π/A rad. Now express this function as:

$$H(\omega t) = \sum_{n=0}^{n=\infty} [C_n \cos(n\omega t) + S_n \sin(n\omega t)] \tag{2.9}$$

From the standard determination of coefficients for a Fourier expansion:

$$\begin{aligned} S_n &= (2/T)\int_{-T/2}^{T/2} H\sin(n\omega t)\,dt \\ &= (1/\pi)\int_{-\pi}^{\pi} H\sin(n\omega t)\,d\omega t \\ &= (1/\pi)\int_{-\pi/A}^{\pi/A} \sin(n\omega t)\,d\omega t \\ &= 0 \end{aligned} \tag{2.10}$$

$$\begin{aligned} C_0 &= (1/T)\int_{-T/2}^{T/2} H\,dt \\ &= (1/2\pi)\int_{-\pi}^{\pi} H\,d\omega t \\ &= (1/2\pi)\int_{-\pi/A}^{\pi/A} d\omega t \\ &= 1/A \end{aligned} \tag{2.11}$$

$$\begin{aligned} C_n\ (n\neq 0) &= (2/T)\int_{-T/2}^{T/2} H\cos(n\omega t)\,dt \\ &= (1/\pi)\int_{-\pi}^{\pi} H\cos(n\omega t)\,d\omega t \\ &= (1/\pi)\int_{-\pi/A}^{\pi/A} \cos(n\omega t)\,d\omega t \\ &= (2/\pi)\sin(n\pi/A)/n \end{aligned} \tag{2.12}$$

Thus

$$H(\omega t) = 1/A + (2/\pi)\sum_{n=1}^{n=\infty} [\sin(n\pi/A)/n]\cos(n\omega t) \tag{2.13}$$

The average value term, $1/A$, can be brought within the summation to give an alternate form of this expansion:

$$H(\omega t) = (1/\pi)\sum_{n=-\infty}^{n=\infty} [\sin(n\pi/A)/n]\cos(n\omega t) \tag{2.14}$$

This form is sometimes more useful for particular analyses than that of 2.13.

In both of these expansions for the simple unmodulated existence function of Figure 2-1, the only restriction on A is that already stipulated; $A \geqslant 1$. A may be an integer or a rational or irrational number. Where A is in fact an integer, two consequences of interest arise. If n is an integer multiple of A, then $\sin(n\pi/A) = 0$, and so all frequency components for which this condition is satisfied disappear from the expansion. Also, with A an integer, there exists the possibility of a complete set of existence functions as depicted in Figure 2-4, shown for the case $A = 6$. The progressive phase displacement of the existence functions of such a set is $2\pi/A$ rad, the same as that for a complete set of A equal-amplitude equi-angularly displaced phasors. Thus, the kth member of such a set numbered from 0 to $A-1$ can be represented as:

$$H_{Ak} = 1/A + (2/\pi) \sum_{n=1}^{n=\infty} [\sin(n\pi/A)/n]\cos(n(\omega t - 2k\pi/A)) \qquad (2.15)$$

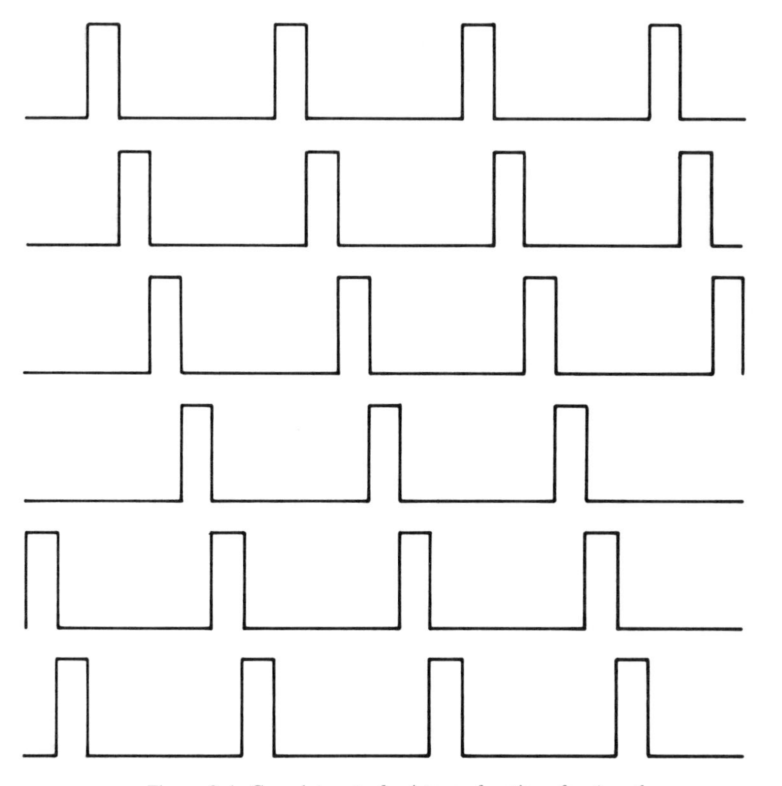

Figure 2-4. Complete set of existence functions for $A = 6$.

This representation assumes the overall time-zero reference is applied to the existence function designated 0 and that the functions designated 1 to A−1 are phase (time) delayed therefrom. Now, it is obvious from Figure 2.4 that the sum of such a complete set is unity, i.e.,

$$\sum_{k=0}^{k=A-1} H_{Ak} = 1 \tag{2.16}$$

2.3 Properties of Complete Phasor Sets

The foregoing brings to light a property of complete phasor sets that is important to the performance of switching power converters. From Eqs. 2.15 and 2.16 it might be deduced that:

$$\sum_{k=0}^{k=A-1} \cos(n(\omega t-2k\pi/A)) = 0 \tag{2.17}$$

for all n that are not integer multiples of A, A an integer, because

$$\sum_{k=0}^{k=A-1} (1/A) = 1 \tag{2.18}$$

and the coefficient $\sin(n\pi/A)$ vanishes in all H_{Ak} for all n that are integer multiples of A.

If Eq. 2.17 does hold, then it implies that the sum of an equiangularly displaced set of A phasors is zero and remains so when all the angular displacements of the phasors are multiplied by any integer that is not an integer multiple of A. This is in fact so; it can be proven, and the value the sum assumes when n is an integer multiple of A can be found by the following manipulation. The kth phasor of a complete set of A will be given by:

$$V_k = V\cos(\omega t-2k\pi/A) \tag{2.19}$$

When its displacement is multiplied by some integer n, it becomes:

$$V_{k,n} = V\cos(\omega t-2nk\pi/A) \tag{2.20}$$

Now introduce the vector operator defined by:

$$b = \epsilon^{-(j\ 2\pi/A)} \qquad (2.21)$$
$$= \cos(2\pi/A) - j \sin(2\pi/A)$$

where $j = (-1)^{1/2}$. Then, Equation 2.19 can be rewritten as:

$$V_k = b^k\ V \cos(\omega t) \qquad (2.22)$$

since

$$b^k = \epsilon^{-(j\ 2k\pi/A)} \qquad (2.23)$$
$$= \cos(2k\pi/A) - j \sin(2k\pi/A)$$

Now the phasor sum is given by:

$$S_A = \sum_{k=0}^{k=A-1} V_k \qquad (2.24)$$

From Equation 2.22, this becomes:

$$S_A = \left(1+b+b^2+. . . b^{A-1}\right) V \cos(\omega t) \qquad (2.25)$$

The multiplier in Eq. 2.25 is a finite geometric progression having A terms and the sum:

$$S_b = (1-b^A)/(1-b) \qquad (2.26)$$

From Eq. 2.23:

$$b^A = e^{-(j\ 2\pi)}$$
$$= \cos(2\pi) - j \sin(2\pi) \qquad (2.27)$$
$$= 1$$

From Eq. 2.27, $S_b = 0$, and hence $S_A = 0$. Now consider multiplication of the displacements by n. Equation 2.20 can be rewritten:

$$V_{k,n} = b^{nk}\ V \cos(\omega t) \qquad (2.28)$$

since

$$\begin{aligned} b^{nk} &= \epsilon^{-(j\ 2nk\pi/A)} \\ &= \cos(2nk\pi/A) - j\sin(2nk\pi/A) \end{aligned} \tag{2.29}$$

Thus, the new phasor sum $S_{A,n}$ is given by:

$$S_{A,n} = (1+b^n+b^{2n}+.\ .\ .\ b^{(A-1)n})\ V\cos(\omega t) \tag{2.30}$$

Again, the multiplier is a finite geometric progression with A terms, and its value is:

$$S_{b,n} = (1-b^{An})/(1-b^n) \tag{2.31}$$

This sum is again zero, since b^A is 1, unless b^n is also 1. In that case, Equation 2.31 is indeterminate, but:

$$\begin{aligned} S_{b,n} &= (1+1+1.\ .\ .\ 1) \\ &= A \end{aligned} \tag{2.32}$$

Now $b^n = 1$ if, and only if, n is an integer multiple of A. Thus, the sum of the phasor set is still zero if n is not an integer multiple of A but is $AV\cos(\omega t)$ if n is such a multiple. This is a basic property of complete phasor sets, and an important one in regard to the dependent quantities of switching power converters.

2.4 Introduction of Control Variables to Existence Functions

So far, only fixed existence functions and sets thereof have been considered. This means that only converters operating with fixed switching patterns have been considered. Now, a converter with such a restriction will perform a conversion function, but will do so in an uncontrolled manner. In Chapter 1, it was pointed out that control of power flow is a major function of converters. The means for implementing such control will now be explored.

Clearly, control of power flow in switching power converters is not to be accomplished by varying the defined quantities. They are usually subject to some variation, and in some instances may be controlled in magnitude, frequency, or phase by some means external to the converter. However, the converter—the switching matrix—is generally expected to control power flow, for that is its main purpose in most applications. In so doing, it will inevitably exert control on some of its terminal quantities. Consider, for example, a typical ac-to-ac converter operting with defined voltage and current sets at its terminals. In the real, as opposed to the idealized analytic, world, the voltage set usually will be truly defined. The current set is, however, the result of the action of the converter's dependent voltage set on the impedances connected to it, because, as was seen in Chapter 1, defining the currents is an anlytic convenience. Thus, when controlling power flow, the converter will control the magnitudes and perhaps the frequencies and phases of both the defined and dependent functions at one set of terminals and the magnitudes of the dependent quantities at the other.

Since the converter consists solely of switches, which have only two possible states, such control can be effected only by changes in switching pattern. Of course, these changes must not affect the conversion function being performed, but merely alter the rate of power flow through the converter-switching matrix. To find out how this can be done, the nature of the existence functions used to describe the switching pattern must be examined. We have seen that these existence functions are of fixed magnitude—they can assume only the values unity and zero. Hence, the only variables available to effect control are temporal ones. Therefore, by altering the temporal relationships of the switching patterns, power flow is controlled.

Consider what temporal variables are available to implement such control. By examining the simple existence function of Figure 2-1, it can be seen that three such variables exist:

1. the duty cycle of the pulse train—the duration of the period of unity value relative to the total period. This corresponds to varying the parameter A in the mathematical formulation of Equation 2.13.
2. the repetition frequency of the pulse train can be varied, corresponding to a variation of ω in Equation 2.13.
3. the phase, or time displacement, of the existence function relative to some fixed external reference can be varied. A very important technique, this corresponds to the introduction of a phase-modulating function into the arguments of all the oscillatory terms of Equation 2.13.

These variables can be used singly or in any combination needed to effect the desired control. The most important techniques in practical use, and their application areas, will now be discussed in more detail.

In dc-to-dc converters, the defined quantities are perfectly smooth dc. The dependent quantities are obtained by multiplying the defined quantities by the existence functions of the switches. The wanted components of the dependent quantities must also be perfectly smooth dc, for such are needed if real power transfer is to exist. The only term of the expression of Equation 2.13, which, when multiplying a perfectly smooth dc, will yield a perfectly smooth dc, is $1/A$. All other terms are oscillatory and will yield unwanted oscillatory components. Thus, the only way control can be effected in dc-to-dc converters is by exerting control on A. However, even with this restriction, there are three ways this may be accomplished. With the repetition frequency and total period constant, A can be varied by varying the ratio of unit-value period to zero-value period, or duty cycle, of the pulse train. If the repetition frequency is allowed to change, then A may be varied either by keeping a constant unit-value period and changing the zero-value period or by keeping a constant zero-value period and changing the unit-value period. Thus, while control is effected only by variation of A, it can be accompanied by a variation of ω, which is a consequence of the technique used to vary A.

It might be argued that an alternate control technique exists, because a modulating function $M(t)$ could be introduced into the arguments of the oscillatory terms of the existence functions, and the cross product of one of them with the dc-defined quantities would yield dc-wanted components. If this were done, the term in question would have to be rendered nonoscillatory by $M(t)$. As such, it would become a part of the average term, $1/A$ of the modulated existence function, which would then replicate a simple existence function with control of A. Thus, there is no point in pursuing such a method of control.

In dc-to-ac and ac-to-dc converters, two basic techniques for controlling power flow are available. First, consider that if one defined quantity is dc and the other ac, then, when these are multiplied by the existence functions, the wanted component of the dependent quantity at the ac terminals must have the same frequency as the defined quantity if power transfer is to occur. Such a wanted component will exist if one of the oscillatory terms in Equation 2.13 has the same frequency as the defined ac quantity. Since the term for $n = 1$ has the largest amplitude, it is logical to choose it. When the defined-ac quantity is multiplied by the existence functions, perfectly smooth dc wanted components must result in the resultant dependent quantity. The term $1/A$ will produce oscillatory components, and the sum of some number of such components can only be an oscillatory component or zero (if the summed components form a complete phasor set). All other terms will form products of the type $\cos(n\omega t)$ $\cos(\omega_s t)$, where ω is from the existence functions and ω_s is the angular frequency of the defined ac quantity.

Decomposing such a product yields terms containing $\cos[(n\omega-\omega_s)t]$ and $\cos[(n\omega+\omega_s)t]$. The latter are clearly oscillatory whatever the value of n and regardless of the relationship between ω and ω_s; the former will produce a smooth dc component if $n\omega = \omega_s$, since $\cos(0) = 1$. Thus, if we have chosen $\omega = \omega_s$, the wanted component of the dc-terminal dependent quantity is produced by the first term of the summation in Equation 2.13, just as was that of the ac dependent quantity.

It can be said that ω cannot be varied, or else the wanted components will no longer be generated. Varying A would exert some control over the amplitude of those that are generated with $\omega = \omega_s$, because of the amplitude multiplier,

$\sin(\pi/A)$. However, the existence functions for ac-to-dc and dc-to-ac converters must be complete sets in order to avoid violation of KLC and KLV, and so A becomes an invariant integer. Therefore, the inescapable conclusion is that control, if it is possible at all, must be effected by introducing some sort of modulating function into the arguments of the oscillatory terms of the existence function. In doing so, a modulated existence function will be created.

Suppose the modulating function is arbitrary and termed M. Then the oscillatory terms contain $\cos[n(\omega t+M)]$, and their cross products with the ac defined quantity contain $\cos[n(\omega t+M)]\cos\omega_s t$. These decompose into terms $\cos(n\omega+nM+\omega_s t)$ and $\cos(n\omega+nM-\omega_s t)$; as before, if $n = 1$ and $\omega = \omega_s$, then a smooth dc component is produced by the latter if M is not a time-varying function. Moreover, the amplitude of the wanted component contains the multiplier $\cos(M)$ and thus can be varied by varying M. It is clear that the required modulating function, M, is simply a phase shift (delay or advance, since $\cos(M) = \cos(-M)$) of the existence functions with respect to an external time reference. Usually, the ac defined quantity is taken as the reference, and the advance or delay angle employed for control is designated α.

It is also clear that the wanted component of the dependent ac quantity does not suffer amplitude modification when this control technique is employed. The cross product of the dc defined quantity and the oscillatory terms of the existence function has elements containing $\cos[n(\omega t+M)]$. This, for $n = 1$ when $\omega = \omega_s$, becomes $\cos(\omega_s t+M)$; the wanted component of the ac-dependent quantity is phase-shifted by M (i.e., α), but its amplitude is unaffected.

The second technique for implementing control in ac-to-dc and dc-to-ac converters is perhaps more obvious but leads to rather complex expressions for the existence functions and dependent quantities. It consists of making A time-dependent so that the term $1/A$ in Equation 2.13 assumes the form $(1/A_0)\ [1\pm k\cos\ (\omega_s t+\phi)]$ where k is a modulation index, and ϕ an arbitrary phase angle. KLC and KLV will not be violated if modulation of the complete set is effected so that $\Sigma(1/A)=1$, which condition is satisfied if the modulating functions are a complete phasor set. This method, since it involves cyclical variation of the unit value period, is termed *pulse-width modulation* (PWM). The modulation index, k, which controls the amplitude of the wanted components, is varied by varying the degree to which the unit value period is varied.

Obviously, both ac and dc wanted components are now produced by the products of this term of the existence function and the defined dc and ac quantities, respectively, and both their magnitudes are controlled by changing the modulation index, k. However, there are a number of unfortunate consequences attached to the use of PWM. When $k = 0$, A is constant and no wanted components should be produced. Thus, ω cannot equal ω_s; in fact, information theory shows that in order for $1/A$ to contain $k\cos(\omega_s t)$, the relationship $\omega > 2\omega_s$ must be satisfied. From an anlytical viewpoint, this is not important.

However, in practice, the need for $\omega > 2\omega_s$ (actually, $>> 2\omega_s$) means that the converter's switches will operate at a much higher frequency than that of the defined ac quantity. Since practical switches have losses associated with their openings and closings, the losses with PWM control will be higher than those for phase-shift control, given the same switches.

The second major consequence of PWM is the complex spectra of unwanted components that are generated at both sets of terminals. In modulating to make $1/A$ contain $k \cos\omega_s t$, the oscillatory terms of the existence functions are transformed; a modulating function that is an oscillatory function of time appears in their arguments. In general, they then contain $\cos(n\omega \pm x \cos m\omega_s t)$, where m is an integer independent of n and x is related to k. The spectrum produced by any one of these terms contains an infinite series of sidebands, of all angular frequencies $n\omega \pm m\omega_s$. If ω is an integer multiple of ω_s, these sidebands are harmonics of ω_s; if it is not, they are not. In the latter case, the spectrum will contain some components of angular frequency less than ω_s, and such components are usually extremely deleterious to a converter's application. Since each oscillatory component of the existence function gives rise to an infinite series of sidebands, and there are an infinite series of oscillatory components, the total resulting spectrum consists of an infinite summation of infinite series of sidebands.

The generation of such a double Fourier expansion by PWM was first discovered by Bennett[1] in the early 1930s; its presence can be a serious problem in switching power converters, particularly if ω_s is variable. Note, however, that if ω is an integer multiple of ω_s, the spectrum will condense into a single Fourier expansion, consisting of the infinite series of harmonics of ω_s.

Since PWM involves no essential phase shift with respect to the defined ac quantity, a PWM converter may be operated at any phase delay or advance desired. If it is operated at other than zero advance or delay, then phase-angle control will occur in addition to PWM control—the first instance of a double control function occurring.

The two control means described are the only ones that to date have been implemented in dc-to-ac and ac-to-dc switching power converters, either singly or in combination. It is mathematically conceivable that a modulating function of time $M(t)$ can be found and introduced into the arguments of the oscillatory terms of the existence functions, as $\cos\{n[\omega t \pm M(t)]\}$, so that the result of the cross product of a term with $\cos(\omega_s t)$ is dc, and the result of the cross product of the same term with dc contains $\cos(\omega_s t)$. However, the particular term chosen must be such that $n[\omega t \pm M(t)]$ must be equivalent to $\omega_s t \pm M$, and the modulated existence function becomes a phase-shifted simple existence function. Hence, there is no point in attempting to implement such a control.

When ac-to-ac converters are considered, the possibilities are found to be similar to, but somewhat expanded in scope over, those for ac-to-dc converters.

Consider the dc-terminal dependent quantity of a phase-controlled converter; it contains, as has been seen, the multiplier cos α, where α is the phase shift and is also the modulating function introduced into the arguments of the oscillatory terms of the existence functions. Now suppose α is not a constant, but is a function of time $M(t)$. Then the dependent quantity at what were the dc terminals contains $\cos[M(t)]$; if $\cos[M(t)]$ is oscillatory, then this term becomes an oscillatory wanted component whereof the angular frequency (for example, ω_o), is completely independent of the switching rate of the existence functions defined by ω. For this to be so, $\cos[M(t)]$ must have the form $\cos(\omega_0 t)$, and it is seen immediately that a trivial implementation is had by putting $M(t) = \pm \omega_0 t$. If this is done, the oscillatory terms of the existence functions will contain $\cos[n(\omega_s \pm \omega_o)t)]$, ω_s being the angular frequency of one defined ac quantity and ω_o that of the other. This is rather surprising—simply switching at a higher or lower rate than the supply frequency will make the converter produce a wanted component in a dependent quantity with a frequency equal to the difference between the switching rate and the supply frequency. However, magnitude control has been lost—the component so produced is clearly of fixed amplitude, since its amplitude multipler, $(2/\pi) \sin(\pi/A)$, is not subject to variation.

Before exploring the way in which magnitude control could be instituted for the ac-to-ac converter with $M(t) = \pm \omega_0 t$—that is, a converter with a linear modulating function—consider another function $M(t)$, which will yield $\cos[M(t)] = \cos(\omega_0 t)$. This is when $M(t)$ takes the form $\arccos[\cos(\omega_o t)]$. Those familiar with the peculiarities of trigonometric functions will see that although $M(t) = \pm \omega_0 t$ is the trivial solution, there is a nontrivial solution taking the form of a periodic triangular wave. Depicted in Figure 2-5, this class of modulating function has an important property; if it is possible to realize $M(t) = \arccos[\cos(\omega_o t)]$ as a periodic function, then it is also possible to realize $M(t) = \arccos[x \cos(\omega_o t)]$. In this case, x is a modulating index, $0 \leqslant x \leqslant 1$, and

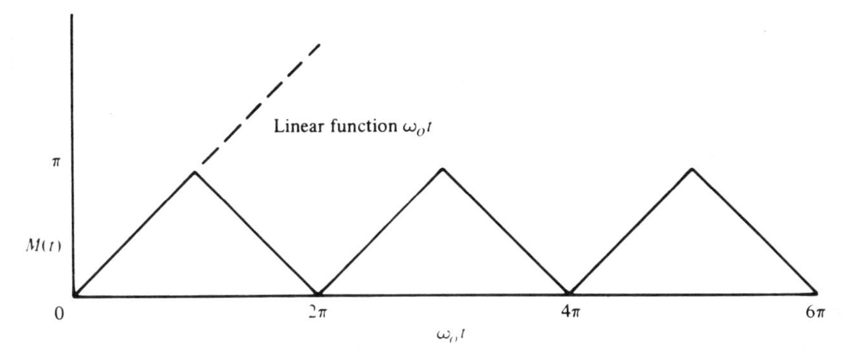

Figure 2-5. Triangular modulating function, $M(t) = \arccos[\cos(\omega_0 t)]$.

the control scheme that has evolved for ac-to-ac converters also has magnitude-control capability since the wanted component of a dependent quantity contains $x\cos(\omega_o t)$.

This particular modulating function, known now as the triangular or nonlinear class, has historic and practical importance too. It is, perhaps surprisingly, very easy to realize in practice, and when applied to the creation and control of ac-to-ac converters in one specific fashion, it yields a converter with very special properties. As a result, it was the first ac-to-ac converter control technique ever tried, and it led to some technically (if not commercially) successful developments in the late 1920s and early 1930s. It is still the only control technique, and the aforementioned ac-to-ac converter the only one, of any great practical importance.

Returning now to the linear modulating function, there remains the question of magnitude control. It is not possible to effect such control by phase shift, for the cross product of $\cos(\omega_s t)$ and $\cos[(\omega_s \pm \omega_o)t \pm \alpha]$ contains wanted components with $\cos(\omega_o t \pm \alpha)$—phase-shifted, but that are not affected as regards amplitude. To produce a variable amplitude multipler, A must be made variable so that $\sin(\pi/A)$ becomes variable. It has already been seen that A can be varied; here, to avoid further unwanted components in the dependent quantities and problems with violation of KLC, the variation of A must be made nonperiodic and implemented in such a fashion that KLC is respected. The technique used to do this might be termed *synchronous pulse-duration modulation*. It consists of modifying the existence functions so that the converter operates in two modes during each switching cycle—as an ac-to-ac converter with a linear modulating function for part of the time and as an ac-to-dc converter producing zero wanted component for the rest of the time. Varying the respective times will vary the magnitude of the wanted ac component; the technique is discussed in detail in Chapter 5.

It might be expected that if PWM can be used to implement control for an ac-to-dc converter, it could also be used for an ac-to-ac converter. It can, and in principle can do so quite simply; if PWM makes $1/A = (1/A_o)[1 \pm k\cos(\omega_s t)]$, then, as was seen previously, its cross product with a defined ac quantity containing $\cos(\omega_s t)$ yields dc. Obviously, if the modulating function makes $1/A = (1/A_o)(1 \pm k\cos[(\omega_s \pm \omega_o t]$, the dependent quantity's wanted component contains $k\cos(\omega_o t)$. It should not be surprising that the unwanted component spectra of ac-to-ac converters with this type of control, PWM, are more closely related to those of ac-to-dc/dc-to-ac converters than to those of ac-to-ac converters with linear or triangular modulating functions. As a result, they have been called *inverter type* ac-to-ac converters, after the name in common usage for dc-to-ac converters. Gyugyi and Pelly,[2] the prime authorities of the ac-to-ac converter field, have suggested that such PWM converters do not really belong to the same family as those with linear and nonlinear modulating functions. Admitting the differences in control principle, and in spectral character, this is still a moot point—the difference is akin to that between phase-controlled and PWM ac-to-dc/dc-to-ac converters. One final note on this subject; just as the relationship $\omega > 2\omega_s$ must hold for PWM to be implementable with ac-to-dc/dc-to-ac converters, so the relationship $\omega > 2(\omega_s \pm \omega_o)$ must hold for it to be effected for an ac-to-ac converter.

Finally, we must briefly consider a very important class of ac-to-ac converters that operate with discontinuous defined current. They are important because of practical rather than analytic considerations and are introduced here to exemplify the limitations and peculiarities of this type of conversion function and converter. If the general ac-to-ac converter matrix is degraded, as depicted in Figure 2-6, so that each input is connected to one and only one output by one switch, it is impossible to have continuous current flow at the output and still have the converter control power transfer—the switches must be closed all the time. This type of converter is very rarely used to attempt ac-to-ac conversion with an output frequency differing from the input frequency, because its performance is then so poor, with such a preponderance of unwanted components in the dependent quantities, that it is unusable for most purposes. However, it is widely used as an ac regulator controlling power flow at the input frequency.

To do so, the output current must be discontinuous. Hence, one mode of control is for a switch to be closed for only a portion of each half cycle of the input voltage to which it is connected. The corresponding output voltage will, then, consist of segements of the input-voltage sinusoid. The resulting output current cannot be defined, for it is the response of the load network connected at the output to this excitation; if a certain very common switch restriction is applied (i.e., the switch automatically opens when the current goes to zero), the very existence functions themselves become load-dependent. It will be seen in Chapter 5 that this control mode, with this particular switch restriction, is closely analogous to phase-shift control of ac-to-dc/dc-to-ac converters.

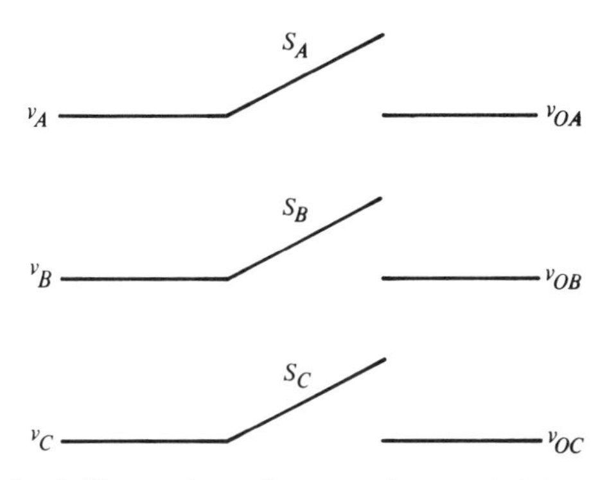

Figure 2-6. Degraded switching matrix used as ac regulator (typical three-phase arrangement).

Another control mode for this type of converter, which has gained great but to some extent unjustified popularity, is to have a switch remain closed for a number of complete cycles of the ac input and then remain open for a number of complete cycles, varying the closed and open periods to effect control but always maintaining a number of complete cycles in each period. This technique is termed *integral cycle control.*

The spectral characteristics of this type of ac-to-ac converter with either type of control are hardly attractive. It is the simplicity of the converter coupled with a large demand for its basic function that has led to its widespread application.

References

1. Bennett, W.R. New results in the calculation of modulation products. *The Bell System Technical Journal* **12:** 228–243 (1933).
2. Gyugyi, L., and B.R. Pelly. *Static Power Frequency Changers.* New York: John Wiley & Sons, 1976.

Problems

1. Show that the existence functions of the complete set for $A = 2$ contain only odd-order oscillatory components and that the two functions can be expressed as:

$$H_o = X + Y$$

$$H_\pi = X - Y$$

2. Sketch the complete set of existence functions for $A = 3$. Write an expression for the reference function of the set. Calculate and tabulate the amplitudes of the average component and the first six oscillatory components of that function.

3. Control of dc-to-dc converters is obtained by varying A in the existence function of the active switch. Calculate the values of A which give the maximum and minimum amplitudes of the first three oscillatory components in the existence function. (Note: There are multiple maxima and minima for $n > 1$.)

4. Explain how the introduction of a controlled phase shift into the arguments of the oscillatory terms of the switch existence functions in an ac-to-dc/dc-to-ac converter will produce amplitude control of a wanted component in a converter dependent quantity.

5. Show that a complete set of M existence functions can satisfy the requirements, not violating KLC and KLV, for the switches in an M phase ac-to-dc current-sourced midpoint converter. Deduce that two related complete sets can satisfy the requirements for a current-sourced bridge, but cannot for the reciprocal voltage-sourced bridge.

6. Deduce the restrictions, if any, on ω_o when an ac-to-ac converter is created by introducing the modulating functions $\omega_o t$ or $-\omega_o t$ into the arguments of the oscillatory terms of the existence functions of the switches.

3
DC-to-DC Converters

3.1 Linear DC Regulators

It is now time to begin quantitative assessments of switching converter performance and characteristics. The study will begin with the simplest converters, the dc-to-dc converters. As the switching patterns, or existence function matrices, for these converters are derived, the patterns for the more complex converters and functions will also become clear.

Before exploring switching dc-to-dc converters, however, consider the disadvantages of linear dc regulators. Figure 3-1 shows an elementary schematic of a typical series pass regulator of this type. The linear device, Q, typically supports any voltage within its ratings, regardless of the current flowing through it, and allows a current flow directly proportional (hence, *linear*) to the voltage applied or current injected at its control electrode, C. In other words, the characteristics of a pentode vacuum tube or bipolar transistor are generally found in the regulating device, Q. Now, since this device can only absorb voltage, not generate it, one disadvantage becomes apparent immediately: this power converter, or regulator, will always have an output-terminal dc voltage lower than that of its source. Second, the power dissipated in Q as a result of the load current flowing through it and the difference between source and load voltages, which it supports, is power lost as a result of the conversion (regulatory) process.

Explaining this latter point further, assume that Q is an ideal device with a minimum voltage drop, at any current, of zero. Now assume a regulated output voltage V_O and an uncontrolled source with nominal voltage V_S and a per-unit tolerance of $\pm x$ $(0 \leqslant x \leqslant 1)$. Then the minimum source voltage is $(1-x)V_S$, and

$$V_O = (1-x)V_S \tag{3.1}$$

for the limiting case with a perfect device. Now if the load current at some time is I_L and the source voltage, within the range $(1 \pm x)V_S$, is V_T, then the loss in Q is given by

$$W_A = (V_T - V_O)I_L \tag{3.2}$$

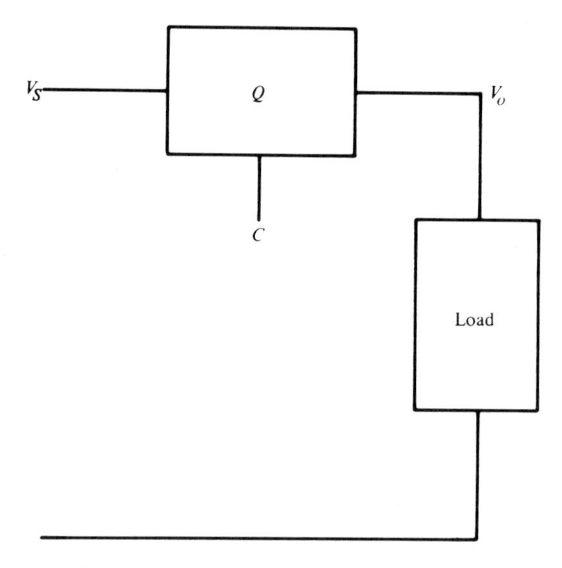

Figure 3-1. Series pass linear dc regulator.

while the power delivered to the load is

$$W_L = V_O I_L \tag{3.3}$$

Thus, the fractional efficiency at any operating point is

$$\eta_T = W_L/(W_L + W_A) = V_O/V_T \tag{3.4}$$

which, for this idealized case, ranges from 1 to $(1-x)/(1+x)$. If V_T has a normal distribution over some long period of time, the average fractional efficiency over that period will be $1-x$. In any event, severe power loss can ensue, and usually does, from the use of this type of regulator.

Similar arguments apply to the common shunt regulator depicted in Figure 3-2. Again, the regulated or controlled output voltage must always be lower than the source voltage; moreover, there are now two loss generating elements—the series resistor and the linear device. In this case, at least one is always dissipating power even when no load is present at the output terminals.

The only advantage possessed by linear regulators, as compared to switching dc-to-dc converters, is that no unwanted oscillatory components are created in either the output voltage or input current. Thus, linear regulators can be connected directly to any type of dc source and load without concern for interfacing; they quite often play an active role in removing oscillatory components present

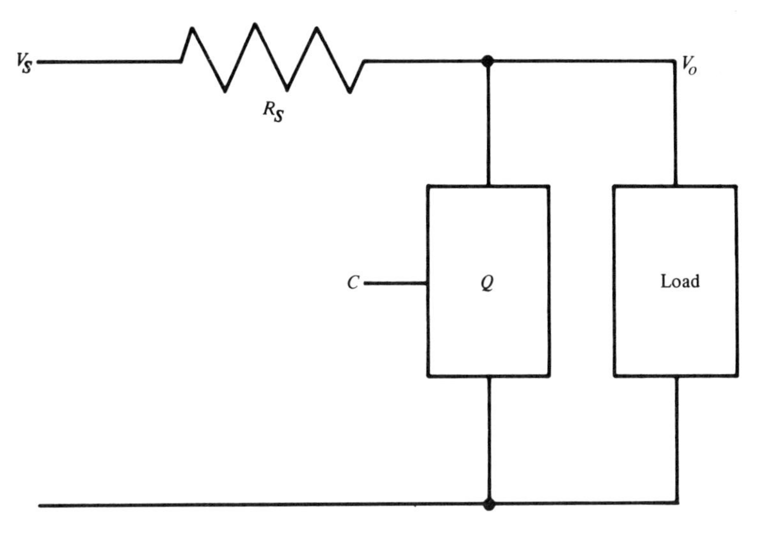

Figure 3-2. Shunt dc regulator.

in the dc source voltage or, more rarely, load current. Such "power active filters" have found quite extensive use—for example, where the dc source is a rectified alternator voltage.

3.2 Voltage-Sourced Single-Quadrant DC-to-DC Converter

Consider the single quadrant dc-to-dc switching power converter of Figure 3-3. Since neither S_1 nor S_2 can absorb power (when idealized), this converter is ideally lossless. Suppose then the existence function H_1 defined by

$$H_1 = 1/A + (2/\pi) \sum_{n=1}^{n=\infty} [\sin(n\pi/A)/n]\cos(n\omega t) \tag{3.5}$$

is arbitrarily assigned to S_1. Then, to avoid violation of KLC at the output terminals and KLV at the input terminals, S_2 must have the existence function.

$$H_2 = 1 - H_1 = \overline{H}_1 \tag{3.6}$$

It is clear that the output terminal voltage is given by

$$v_D = H_1 V_S + H_2 0 \tag{3.7}$$

and the source current by

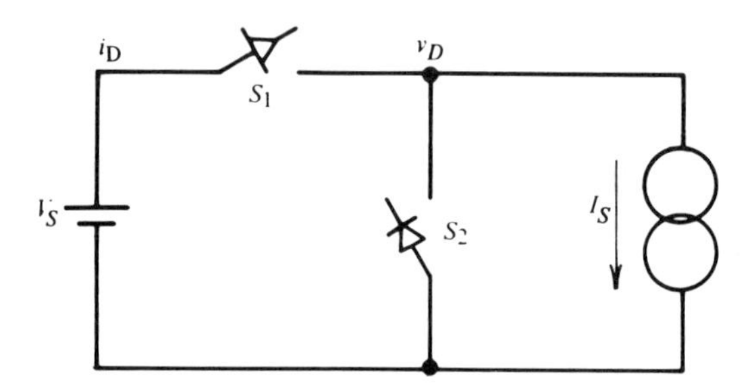

Figure 3-3. Voltage-sourced (buck) dc-to-dc converter.

$$i_D = H_1\, I_S + H_2\, 0 \tag{3.8}$$

where V_S and I_S are the defined source voltage and output current, respectively; v_D and i_D are, of course, this converter's dependent quantities.

From Equation 3.7, it can be seen that the wanted component of v_D is given by

$$V_{DD} = (1/A)V_S \tag{3.9}$$

and that of i_D by

$$I_{DD} = (1/A)I_S \tag{3.10}$$

Since the oscillatory unwanted components of v_D transfer no power to I_S, the output power is,

$$W_O = V_{DD}\, I_S = (1/A)V_S\, I_S \tag{3.11}$$

The input power is

$$W_S = V_S\, I_{DD} = V_S\, (1/A)I_S \tag{3.12}$$

Equations 3.11 and 3.12 illustrate an important, though obvious, property of idealized switching power converters—that the real powers at the input and output terminals must be equal.

Now, V_{DD} for the converter of Figure 3-3 can be controlled by varying A. Since A is subject to the constraints $1 \leqslant A \leqslant \infty$, $V_{DD} \leqslant V_S$, this converter suffers from one of the disadvantages of linear regulators; that is, the wanted component of the converter's output terminal voltage must always be lower than its defined source voltage. For this reason, it is commonly called the *buck* dc-to-dc converter.

The unwanted components of v_D and i_D, collectively termed *ripple,* are given by

$$v_{DU} = (2V_S/\pi) \sum_{n=1}^{n=\infty} [\sin (n\pi/A)/n] \cos(n\omega t) \tag{3.13}$$

and

$$i_{DU} = (2I_S/\pi) \sum_{n=1}^{n=\infty} [\sin (n\pi/A)/n] \cos(n\omega t) \tag{3.14}$$

The waveforms of v_D and i_D are simply replicas of the existence function H_1's pulse train with amplitudes V_S and I_S, respectively. The largest amplitude unwanted component will be that for $n = 1$, since $-1 \leqslant \sin (n\pi/A) \leqslant 1$ for all A; this component's amplitude will maximize when $A = 2$, for then $\sin(\pi/A) = 1$. At this operating point,

$$(v_{DU1})_{\max} = 2V_S/\pi \tag{3.15}$$

$$(i_{DU1})_{\max} = 2I_S/\pi \tag{3.16}$$

while

$$(V_{DD})_{A=2} = V_S/2 \tag{3.17}$$

and

$$(I_{DD})_{A=2} = I_S/2 \tag{3.18}$$

In both instances, the largest unwanted component at this worst-case operating point has an amplitude $4/\pi$ times that of the wanted component. This fact highlights the potential severity of the problem of unwanted components produced by switching power converters. The interfacing techniques necessary to deal with such problems are discussed in Chapter 6.

The stresses on the switches in this converter can readily be deduced. The maximum voltage to be withstood by S_1, when open, is obviously V_S, and the same is true for S_2. The peak current in both switches is I_S; the average current in S_1 is I_{DD}, and in S_2 it is $I_s - I_{DD}$. The root mean square (rms) currents in S_1 and S_2, the conductors used to connect them, and V_S are of interest because of joule heating in a practical converter and source. The rms current in S_1 and V_S can be seen from Figure 3-4 to be

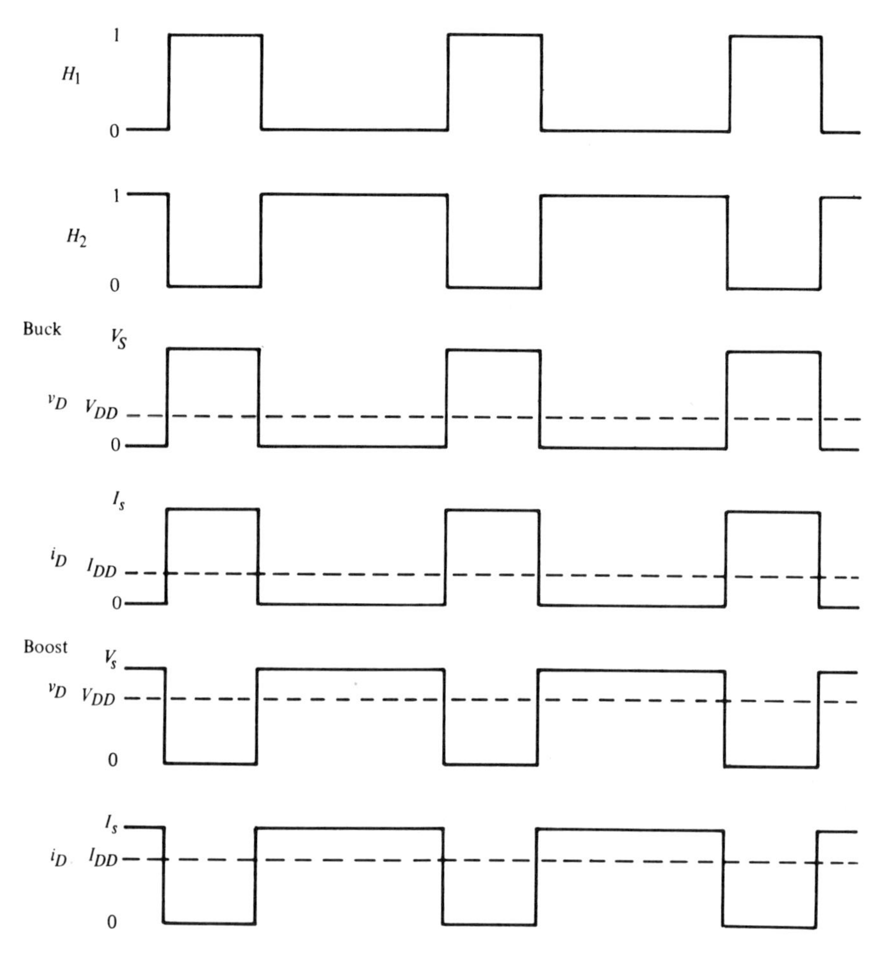

Figure 3-4. Dependent quantities of buck and boost converters.

$$I_1 = I_S/\sqrt{A} \tag{3.19}$$

while that in S_2, is by examination of the waveform in Figure 3-4,

$$I_2 = I_S \sqrt{(A-1)/A} \tag{3.20}$$

3.3 Current-Sourced Single-Quadrant DC-to-DC Converter

In Chapter 1, it was shown that the converter of Figure 3-3 has a dual, shown in Figure 3-5. Since the buck converter always produces a wanted component

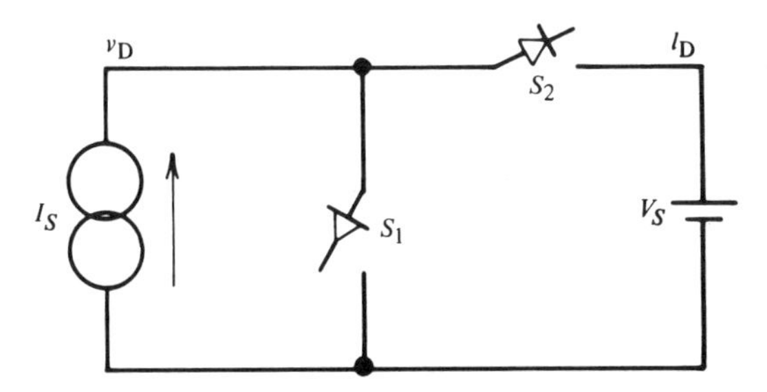

Figure 3-5. Current-sourced (boost) dc-to-dc converter.

of output voltage lower than the defined input voltage, it might well be expected that reciprocal behavior would be exhibited by the converter of Figure 3-5. The output voltage and input current are now defined; if H_1, defined by Equation 3.5, is assigned to S_1 in this converter, then H_2 (Equation 3.6) must be assigned to S_2 to avoid violation of KLC at the input terminals and KLV at the output terminals. The reciprocity of this statement concerning that obtaining for the buck converter is quite evident.

The equations:

$$v_D = H_1\,0 + H_2\,V_S \tag{3.21}$$

and

$$i_D = H_1\,0 + H_2\,I_S \tag{3.22}$$

for the dependent quantities, now output current and input voltage, also illustrate the complete reciprocity (duality) of the converter of Figure 3-5 and the buck converter of Figure 3-3. The wanted components are given by

$$V_{DD} = (1-1/A)V_S \tag{3.23}$$

$$I_{DD} = (1-1/A)I_S \tag{3.24}$$

Now, transpose Equation 3.23; the result is

$$V_S = [A/(A-1)]V_D \tag{3.25}$$

Since $1 \leqslant A \leqslant \infty$, it is clear that $V_S \geqslant V_{DD}$; i.e., the output voltage is always greater than the wanted component of input voltage. This property has given rise to the common use of the name *boost* dc-to-dc converter for the circuit of Figure 3-5.

That equality of powers, input and output, still holds is evident from Equations 3.23 and 3.24. The result obtained when Equation 3.23 is multiplied by I_S is also produced when Equation 3.24 is multiplied by V_S. Also, the expressions for the ripple produced are identical to Equations 3.13 and 3.14 save for the exchange of V for I and I for V and the multiplication of all terms in the summation by -1. Thus, for the boost converter:

$$v_{DU} = -(2V_S/\pi) \sum_{n=1}^{n=\infty} [\sin\,(n\pi/A)/n]\,\cos(n\omega t) \tag{3.26}$$

and

$$i_{DU} = -(2I_S/\pi) \sum_{n=1}^{n=\infty} [\sin\,(n\pi/A)/n]\,\cos(n\omega t) \tag{3.27}$$

The waveforms of v_D and i_D will be the complements, at the same value of A, of those for v_D and i_D in the buck converter. All are depicted together, for $A \sim 3$, in Figure 3-4.

Since the unwanted component expressions match except for reciprocity and multiplication by -1, their largest amplitude terms are again those for $n=1$, with absolute maxima occurring again at $A=2$. Once more, the largest unwanted components have amplitude $4/\pi$ times the wanted components; again, the potential problems created by the oscillatory components are indicated.

The boost converter allows removal of the restriction shared by the linear regulators and buck converter, for it allows generation of a dc output voltage higher than the available dc input voltage (V_{DD}). In effect, the boost converter acts as a step-up dc-to-dc auto transformer.

As for the buck converter, switch stresses and rms currents are easily deduced in the boost converter. The maximum voltage stress on S_1 and S_2 is V_S, the defined output voltage. The peak current is S_1 and S_2 is I_S; the average current is S_2 is I_{DD}, and that in S_1 is $I_S - I_{DD}$. The rms current in S_2 and V_S is $I_S\sqrt{(A-1)/A}$; that in S_1 is $I_S/\sqrt{A}$. Note that for both buck and boost converters, the average and rms currents in any branches have maximum values equal to the converters' defined currents, maximizing either as $A \to 1$ or as $A \to \infty$.

3.4 Buck-Boost DC-to-DC Converter

Since both buck and boost dc-to-dc converters exist, it is proper to enquire whether ther is a topology capable of giving a dc output voltage lower than, equal to, or higher than the available source voltage—i.e., a buck-boost converter. Some applications may call for such a capability, and in any event the properties so far uncovered suggest that it should be possible.

Obviously, no modification of the topologies of Figures 3-3 and 3-5 is going to suffice for this purpose. However, suppose that a cascade connection of buck and boost converters is ventured, with the defined current load, or sink, of the buck acting as the defined current source of the boost as depicted in Figure 3-6a. Of course, the dc polarity of the boost must be inverted to accomplish

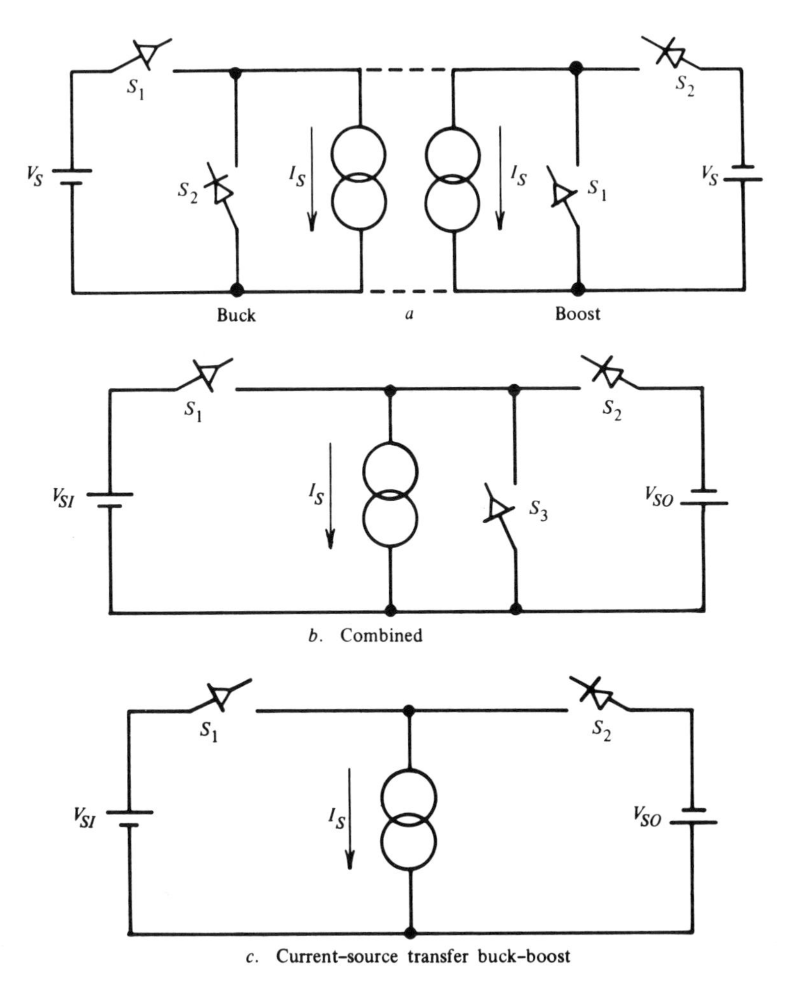

Figure 3-6. Derivation of buck-boost converter from cascade connection of buck and boost converters.

this. Now, there is a superfluity of switches in Figures 3-6a. S_2 of the buck and S_1 of the boost both act across the current sink/source, and either one of them will suffice for the purpose. The configuration of Figure 3-6b is then produced; however, observe that only one of the three switches therein could be closed at any given time. If S_1 and S_2 were both closed, KLV would be violated around the loop V_{SI} $-S_1$ $-S_2$ $-V_{SO}$; if S_1 and S_3 were both closed, then KLV would be violated with regard to V_{SI}; and if S_2 and S_3 were both closed, then KLV would be violated with regard to V_{SO}. Now observe that in any period when S_3 only is closed, the converter is doing nothing but circulating the defined current—no energy is being absorbed from V_{SI} or delivered to V_{SO}, since both S_1 and S_2 must be open. This strongly suggests that S_3 is superfluous; remove it, and Figure 3-6c is the result. Provided either S_1 or S_2 is closed, but never both, neither KLC or KLV is violated, and all periods are active—energy is either being extracted from V_{SI} (S_1 closed) or delivered to V_{SO} (S_2 closed). If H_1 (Equation 3.5) is assigned to S_1, then H_2 (Equation 3.6) must be assigned S_2. Note that there are now three dependent quantities. The voltage across the defined current is given by

$$v_D = H_1 V_{SI} + H_2 V_{SO} \tag{3.27a}$$

The output current is given by

$$i_{DO} = H_1 0 - H_2 I_S \tag{3.28}$$

The source current is given by

$$i_{DS} = H_1 I_S + H_2 0 \tag{3.29}$$

It is the avowed purpose of this converter to transfer energy from V_{SI} to V_{SO}. The current source/sink, I_S, is to be used merely as an energy-transfer device and should neither absorb nor deliver energy over the long term. Thus,

$$0 = V_{DD} = (1/A)V_{SI} + (1-1/A)V_{SO} \tag{3.30}$$

can be written to define the average direct voltage across I_S. Transposing,

$$V_{SO} = -V_{SI}/(A-1) \tag{3.31}$$

Note two important aspects of Equation 3.31. V_{SO} must be opposite in polarity to V_{SI} and can have magnitude less than ($A > 2$), equal to ($A = 2$), or greater than ($A < 2$) that of V_{SI}. The topology of Figure 3-6c is indeed a buck-boost dc-to-dc converter.

The wanted components of i_{DO} and i_{DI} are given by, from Equations 3.28 and 3.29,

$$I_{DDO} = (1/A - 1)I_S \tag{3.32}$$

$$I_{DDI} = (1/A)I_S \tag{3.33}$$

Equations 3.32 and 3.33, when multiplied by V_{SO} and V_{SI}, respectively, confirm the necessary equality of input and output powers. The unwanted components of i_{DO} and i_{DI} are, from Equations 3.28 and 3.29,

$$i_{DOU} = i_{DSU} = (2I_D/\pi) \sum_{n=1}^{n=\infty} [\sin(n\pi/A)/n]\cos(n\omega t) \tag{3.34}$$

That i_{DOU} and i_{DSU} must be equal is readily apparent; I_S is defined as a smooth dc-current source/sink, and hence all oscillatory current components must flow in the loop V_{SI} $-S_1$ $-S_2$ $-V_{SO}$. The largest amplitude constituent, as for the unwanted components of the buck and boost converters, is that for $n=1$ and maximizes when $A=2$.

The unwanted, oscillatory components of v_D (its "wanted" component, V_{DD}, is zero) can be derived by first substituting Equation 3.31 in 3.27a to give

$$\begin{aligned} v_D &= [(AH_1 - 1)/(A-1)]V_{SI} \\ &= (1 - AH_1)V_{SO} \end{aligned} \tag{3.35}$$

Making appropriate substitution from Equation 3.5 in 3.35 yields

$$\begin{aligned} v_{DU} &= \{2AV_{SI}/[(A-1)\pi]\} \sum_{n=1}^{n=\infty} [\sin(n\pi/A)/n]\cos(n\omega t) \\ &= -(2AV_{SO}/\pi) \sum_{n=1}^{n=\infty} [\sin(n\pi/A)/n]\cos(n\omega t) \end{aligned} \tag{3.36}$$

Maximum unwanted component amplitude derivation is not possible directly from Equation 3.36. Considering the second realization, the coefficient of all terms can be rearranged to $-(2\ V_{SO})\ [\sin(n\pi/A)/(n\pi/A)]$, which has a limiting value of $-2V_{SO}$ as $A \rightarrow \infty$; however, as $A \rightarrow \infty$, $V_{SO} \rightarrow 0$ from Equation 3.31. A similar result is obtained from the first realization of Equation 3.36. The maximum amplitudes of the oscillatory components in v_{DU} can be expressed in terms of $V_{SI} - V_{SO}$, the sum of the magnitudes of V_{SI} and V_{SO}. From Equation 3.31,

$$V_{SI} - V_{SO} = V_{SI}\ [1 + 1/(A-1)] = V_{SI}\ [A/(A-1)] \tag{3.37}$$

Substituting in Equation 3.36,

$$v_{DU} = [2(V_{SI} - V_{SO})/\pi] \sum_{n=1}^{n=\infty} [\sin(n\pi/A)/n]\cos(n\omega t) \tag{3.38}$$

Thus, it is clear that the maximum amplitude component of v_{DU} occurs for $n=1$ and $A=2$. Since, at $A=2$, $V_{SO} = -V_{SI} = (V_{SO} - V_{SI})/2$, this maximum amplitude unwanted component in v_{DU} has a magnitude $4/\pi$ times V_{SI} (or $-V_{SO}$).

Again, switch voltage stresses—peak, average, and rms currents—are easily deduced. The peak voltage seen by S_1 and S_2 is $V_{SI} - V_{SO}$, which is the sum of the magnitudes of V_{SI} and V_{SO}. The peak current in S_1 and S_2 is I_S, while the average currents are I_S/A (I_{DDS}) and $I_S - I_{DDS}$, respectively. The rms current in V_{SI} and S_1 is $I_S/\sqrt{A}$, that in V_{SO} and S_2 is $I_S\ \sqrt{(A-1)/A}$.

The buck and boost converters are mutual duals and the buck-boost converter just analyzed was derived by cascade-connecting them. Therefore, it might be anticipated that this converter has a dual, which might be derived by reversing the order of cascade connection. This is so, and the resulting topology is depicted in Figure 3-7. Known variously as the *Cuk* (after a supposed inventor) and *optimum topology* converter (for reasons that are far from clear), its performance may be analyzed as follows: if H_1 is assigned to S_1, then H_2 must be assigned to S_2 to avoid violation of KLV and KLC. There are three dependent quantities again, defined by

$$v_{DS} = H_1\ 0 + H_2\ V_S \tag{3.39}$$

$$v_{DO} = -H_1\ V_S + H_2\ 0 \tag{3.40}$$

$$i_D = H_1\ I_{SO} + H_2\ I_{SI} \tag{3.41}$$

The duality of this set and the defining set (Equations 3.27 through 3.29), for the buck-boost is evident. It follows that all succeeding equations relating to the buck-boost may be applied to its dual by transposing voltage and currents, and source and sink quantities. This, of course, is one of the major benefits of the observation of duality or reciprocity between two networks. Thus, for the converter of Figure 3-7,

$$I_{DD} = 0 = (1/A)I_{SO} + (1-1/A)I_{SI} \tag{3.42}$$

giving

$$I_{SO} = -(A-1)I_{SI} \tag{3.43}$$

$$V_{DDO} = -(1/A)V_S \tag{3.44}$$

$$V_{DDI} = (1-1/A)V_S \tag{3.45}$$

from Equation 3.45,

$$V_{DDO} = -V_{DDI}/(A-1) \tag{3.46}$$

Equation 3.46 shows the buck-boost capability of this topology, since $1 \leqslant A \leqslant \infty$. Also,

$$v_{DOU} = v_{DIU} = -(2V_S/\pi) \sum_{n=1}^{n=\infty} [\sin(n\pi/A)/n]\cos(n\omega t) \tag{3.47}$$

Finally, i_D can be expressed as

$$i_D = [(AH_1-1)/(A-1)]I_{SO} = (1-AH_1)I_{SI} \tag{3.48}$$

giving

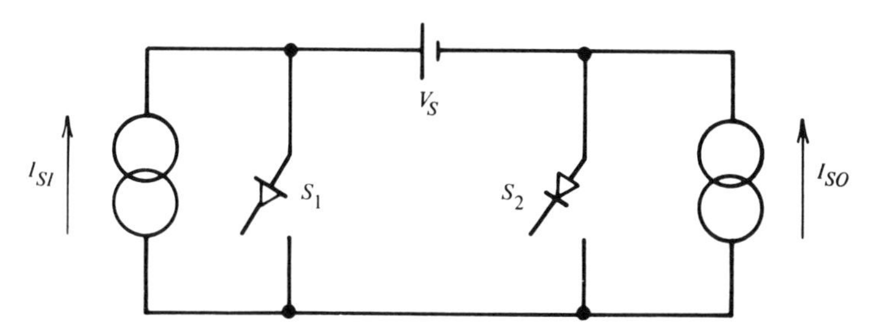

Figure 3-7. Voltage source transfer buck-boost converter.

$$i_{DU} = \{2AI_{SO}/[(A-1)\pi]\} \sum_{n=1}^{n=\infty} [\sin(n\pi/A)/n]\cos(n\omega t)$$

$$= -(2AI_{SI}/\pi) \sum_{n=1}^{n=\infty} [\sin(n\pi/A)/n]\cos(n\omega t) \qquad (3.49)$$

As with Equation 3.36, Equation 3.49 will not give the maximum amplitude component of i_{DU} directly, but if i_{DU} is expressed in terms of $I_{SI}-I_{SO}$, then

$$i_{DU} = -[2(I_{SI}-I_{SO})/\pi] \sum_{n=1}^{n=\infty} [\sin(n\pi/A)/n]\cos(n\omega t) \qquad (3.50)$$

Power equality is thus maintained, and the current in V_S is purely oscillatory. The peak voltage seen by S_1 and S_2 is obviously V_S; peak current is $I_{SI}-I_{SO}$. Average currents are $(I_{SI}-I_{SO})/A$ in S_1 and $(I_{SI}-I_{SO})(1-1/A)$ in S_2 with rms currents of $(I_{SI}-I_{SO})/\sqrt{A}$ *and* $(I_{SI}-I_{SO})\sqrt{(A-1)/A}$, respectively. The rms current in V_S is, $I_{SI}-I_{SO}$ since the peak and rms values of any rectangular wave are identical.

A new and rationally based terminology will now be used to describe these two buck-boost dc-to-dc converters. The converter shown in Figure 3-6c will be termed the *current-source transfer buck-boost,* while that in Figure 3-7 will be termed the *voltage-source-transfer buck-boost*.

3.5 Two-Quadrant DC-to-DC Converters

So far, all the converters considered have been of the single-quadrant variety; energy transfer has been unidirectional from a defined source to a defined sink. All the switches employed need only unidirectional voltage-blocking, unidirectional current-carrying capacity. This was deduced in the discussion in Chapter 1. However, two distinct varieties of unipolar switch are required. Suppose all switches labeled S_1 in the buck, boost, and both buck-boost converters are termed *active* and those labeled S_2 are termed *complementary*. The voltage applied to an active switch when open has the same polarity as the current flow through that switch when it is closed. For the complementary switch, the voltage applied when open has the opposite polarity to the current flowing when the switch is closed. This distinction has an important practical consequence, for the simplest of all switching power devices, the diode, possesses precisely the qualities required of the complementary switch. Not surprisingly, it is used universally as that switch, and thus automatically assumes the complementary existence function H_2 without needing control input (for which it has, in any event, no provision). This "auto-complementary" property is also useful in single-quadrant ac-to-dc converters, as will be discussed in Chapter 4.

There are a number of applications for dc-to-dc converters in which operation in more than one quadrant of the $V-I$ plane is desired. To accomplish this, the wanted components of the dependent quantities must be reversible. In addition, the energy source must be capable of acting as a sink, and the energy sink must be capable of acting as a source. Analytically, since both are defined perfect sources (one voltage, one current), this requirement presents no problem. In practice, the application would not call for two-quadrant operation unless the entity normally acting as an energy sink is capable and is needed to act as an energy source at least for some periods of time. However, the designer must be sure that which is normally an energy source is in fact capable of absorbing the energy fed back to it during such intervals.

As shown in Chapter 1, the basic configuration is the same for both voltage-sourced (buck) and current-sourced (boost) two-quadrant dc-to-dc converters. The differences lie in the positioning of the active switches and the direction of voltage blocking for the complementary switches. Figure 3-8 shows the topology for a two-quadrant buck converter. On examination, it is found that there are two possible modes of operation for this converter, which will be designated "mode 1" and "mode 2."

In mode 1 operation, the active switches S_1 and S_{1A} are both operated with existence function H_1; the complementary switches S_2 and S_{2A} will then assume the complementary existence function H_2 ($1-H_1$ or $\overline{H}_1$). The dependent quantity expressions are:

$$v_D = H_1 V_S - H_2 V_S = (2H_1 - 1)V_S \tag{3.51}$$

and

$$i_D = H_1 I_S - H_2 I_S = (2H_1 - 1)I_S \tag{3.52}$$

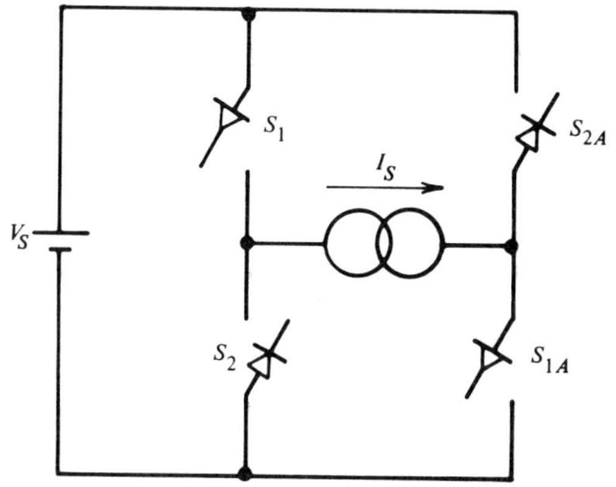

Figure 3-8. Two-quadrant buck converter.

The wanted components are

$$V_{DD} = (2/A-1)V_S \tag{3.53}$$

$$I_{DD} = (2/A-1)I_S \tag{3.54}$$

Equality of powers is maintained. Moreover, observe that for $A=2$, both V_{DD} and I_{DD} are equal to zero; no power is transferred. For $1 \leqslant A < 2$, both V_{DD} and I_{DD} are positive, and the converter operates in the first quadrant of the $V-I$ plane; both output voltage and output current are positive, and energy is transferred from V_S to I_S. For $A > 2$, both V_{DD} and I_{DD} are negative, and the converter operates in the second quadrant of the $V-I$ plane with energy being transferred from I_S to V_S. In either event, the magnitudes of V_{DD} and I_{DD} are less than those of V_S and I_S—the converter is of the buck variety.

As with all dc-to-dc converters, the switch stresses are easy to deduce. The peak voltage applied to any open switch is V_S, the peak current through any switch I_S. Average currents are I_S/A for S_1 and S_{1A}, and $I_S(1-1/A)$ *for* S_2 and S_{2A}. Corresponding rms currents are $I_S/\sqrt{A}$ and $I_S\sqrt{(A-1)/A}$, respectively. The absolute maximum value of average and rms current in any loop is, of course, I_S.

The unwanted oscillatory components of v_D and i_D are given by

$$v_{DU} = (4V_S/\pi) \sum_{n=1}^{n=\infty} [\sin(n\pi/A)/n]\cos(n\omega t) \tag{3.55}$$

$$i_{DU} = (4I_S/\pi) \sum_{n=1}^{n=\infty} [\sin(n\pi/A)/n]\cos(n\omega t) \tag{3.56}$$

As in the single-quadrant buck and boost converters, the largest amplitude components occur for $n=1$ and $A=2$. The magnitudes in question must still be related to V_S and I_S, since for $A=2$ both wanted components of the dependent quantities are zero. Thus, the maximum amplitudes of the first-order unwanted component of the dependent quantities of this converter operated in mode 1 are $4/\pi$ times the defined quantities.

A significant feature of mode 1 operation is the smooth transition from operation in either quadrant to operation in the other. This is primarily due to the fact that the converter, except in limiting conditions ($A=1$ or ∞), never operates solely in one quadrant. It spends a portion of each operating cycle, while S_1 and S_{1A} are closed, in the first quadrant, and the remainder, while S_2 and S_{2A} are closed, in the second quadrant. When the times spent in each are equal ($A=2$), net operation is in neither quadrant. When the time spent in the first quadrant is greater than that spent in the second, net operation is in the first quadrant and vice versa.

Mode 2 operation does not involve such averaging between periods of first- and second-quadrant operation in each operating cycle. In mode 2, first-quadrant operation is achieved by keeping either S_1 or S_{1A} closed all the time and assigning H_1 to the other. S_{2A} or S_2 automatically assumes $1-H_1$; the other remains open. Second-quadrant operation is achieved by keeping S_1 or S_{1A} open all the time and assigning H_1 to the other. S_{2A} or S_2 then automatically assumes $1-H_1$; the other remains closed. The dependent quantity expressions for this mode then become:

In the first quadrant:

$$v_D = H_1 V_S \tag{3.57}$$

$$i_D = H_1 I_S \tag{3.58}$$

In the second quadrant,

$$v_D = -(1-H_1)V_S \tag{3.59}$$

$$i_D = -(1-H_1)I_S \tag{3.60}$$

Equations 3.57 and 3.58 show that the dependent quantities for first-quadant operation are identical in all respects to those of the simple buck converter. The only point of difference is that S_1 or S_{1A}, whichever is kept closed, carries I_S continuously. Equations 3.59 and 3.60 show that the dependent quantities for second-quadrant operation are the negatives of those for a buck converter operating with H_1 and $\overline{H}_1$ interchanged; however, S_1 or S_{1A}, whichever is kept open, carries no current but is under continuous voltage stress. As in mode 1, the magnitudes of V_{DD} and I_{DD} are always less than those of V_S and I_S—this is a buck converter.

It is apparent that the maximum unwanted component amplitudes generated in mode 2 operation are exactly one-half of those generated in mode 1 operation—a considerable benefit. On the other hand, the equilibrium condition with net operation in neither quadrant does not exist—a potential drawback. Preference for one of the two modes of operation depends on particular application requirements.

The topology for the two-quadrant current-sourced boost converter is shown in Figure 3-9. Two modes of operation are possible, which ought to have been expected in view of the reciprocal relationship with the two-quadrant buck converter. In mode 1, S_1 and S_{1A} are both assigned H_1, with S_2 and S_{2A} automatically assuming $1-H_1$, as their existence functions. The dependent quantity expressions are

$$i_D = (1-H_1)I_S - H_1 I_S = (1-2H_1)I_S \tag{3.61}$$

$$v_D = (1-H_1)V_S - H_1 V_S = (1-2H_1)V_S \tag{3.62}$$

When compared with 3.51 and 3.52, it is clear that equations 3.61 and 3.62 indicate reciprocity. Thus, the wanted components are given by

$$I_{DD} = (1-2/A)I_S \tag{3.63}$$

$$V_{DD} = (1-2/A)V_S \tag{3.64}$$

Again, for $A=2$, an equilibrium condition exists with both I_{DD} and V_{DD} equal to zero. For $A > 2$, operation is in the first quadrant, while for $1 \leqslant A < 2$, operation is in the second quadrant. In either event, the magnitudes of V_S and I_S are greater than those of V_{DD} and I_{DD}—this is a boost converter. As for the two-quadrant buck converter, the maximum unwanted components of i_D and v_D are those for $n=1$ and occur at $A=2$ with magnitudes of $4/\pi$ times those of the defined quantities.

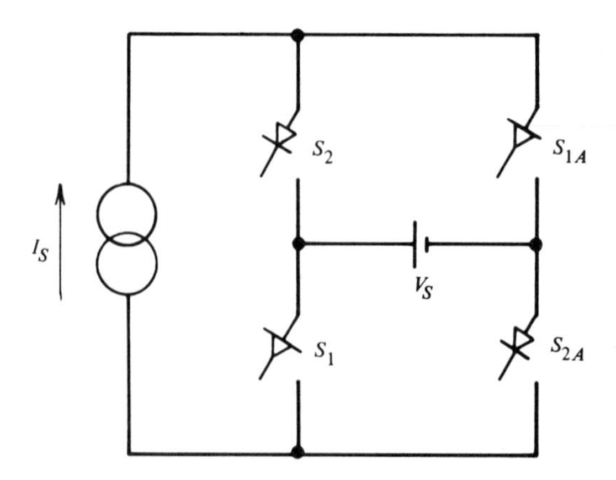

Figure 3-9. Two-quadrant boost converter.

Mode 2 operation again involves different situations for the two quadrants, without an equilibrium condition of zero power transfer. For operation in the first quadrant, either S_1 or S_{1A} is left open and the other is assigned H_1. Then, S_2 or S_{2A} will assume $1-H_1$; the other remains closed. For operation in the second quadrant, S_1 or S_{1A} is kept closed while the other is assigned H_1; S_2 or S_{2A} assumes $1-H_1$, and the other remains open. By now, it may be belaboring the point to emphasize the reciprocity of these statements and those used to describe mode 2 operation for the two-quadrant buck converter.

The dependent quantity expressions are, in the first quadrant,

$$i_D = (1-H_1)I_S \tag{3.65}$$

$$v_D = (1-H_1)V_S \tag{3.66}$$

and in the second quadrant,

$$i_D = -H_1 I_S \tag{3.67}$$

$$v_D = -H_1 V_S \tag{3.68}$$

The relationship of this converter in mode 2 to the simple single-quadrant boost converter is quite clear if Equation 3.65 and 3.66 are compared with Equations 3.21 and 3.22, and if Equations 3.67 and 3.68 are similarly compared after interchanging H_1 and H_2. Thus, as for the two-quadrant buck, the worst-case unwanted components in mode 2 operation are only one-half the amplitudes of those generated in mode 1, but the converter operated in mode 2 has no equilibrium condition.

Since it is possible to create two-quadrant buck and boost converters, it should also be possible to create two-quadrant buck-boost converters. Before attempting to do so, consider again the operating principles of both current-source-transfer and voltage-source-transfer, single-quadrant buck-boost converters. In both cases, energy is extracted from a dc source, temporarily stored in a transfer medium, and then released to a dc sink in each operating cycle. The wanted component of the transfer medium's dependent quantity is zero in each case.

For the energy transfer to be reversed (i.e., for the sink to become the source and vice versa), the defined quantity of the transfer medium must be reversed. This is the only avenue open to achieve reversal of the wanted components of the source-and-sink dependent quantities, and such a reversal is necessary if second-quadrant operation is desired. Given the possibility of a reversal in defined quantities for the transfer media, the reversals of source-and-sink dependent-quantity wanted components can be accomplished if switch characteristics are expanded. For the current-source transfer buck-boost to be capable of two-quadrant operation, both the active and complementary unilateral switches must be replaced by bidirectional current-carrying, unidirectional voltage-blocking switches. In the voltage-source-transfer buck-boost, both must be replaced by unidirectional current-carrying, bidirectional voltage-blocking switches. In both cases, of course, both switches become active.

The resulting topologies are depicted in Figures 3-10 and 3-11. There are few, if any, applications for these converters, and they will not be pursued here. It should be clear, however, that the analytic technique used for all converters so far can also be exploited to deal with these if analyses are desired. However, it should be noted that these buck-boost two-quadrant converters operate only in mode 2; they have no capability for mode 1 operation, with an equilibrium condition, since that would necessitate zero-value defined quantities for the energy-transfer media.

It is perhaps ironic that the only "two-quadrant" dc-to-dc converter that has found reasonably common application is not, in reality, a two-quadrant converter at all. In fact, it is a scheme, depicted in Figure 3-12, which relies on the characteristics of its load, a dc machine armature, to produce second-quadrant operation purely as a result of the induced voltage of the armature and which enters the third quadrant by reversing the defined quantity, the armature current. The present discussion will be confined to a brief, qualitative, operating description.

The machine is motored, in the first quadrant, by operating S_1 and S_2 of Figure 3-12 as a buck converter. S_{1A} and S_{2A} remain inactive. To achieve operation in the second quadrant, S_1 is then kept open while the induced voltage of the armature acts to decrease the defined quantity, the armature current, to zero; S_2 conducts until the zero current condition is reached. Reversal of armature current (and development of braking torque) is then achieved using S_{1A}

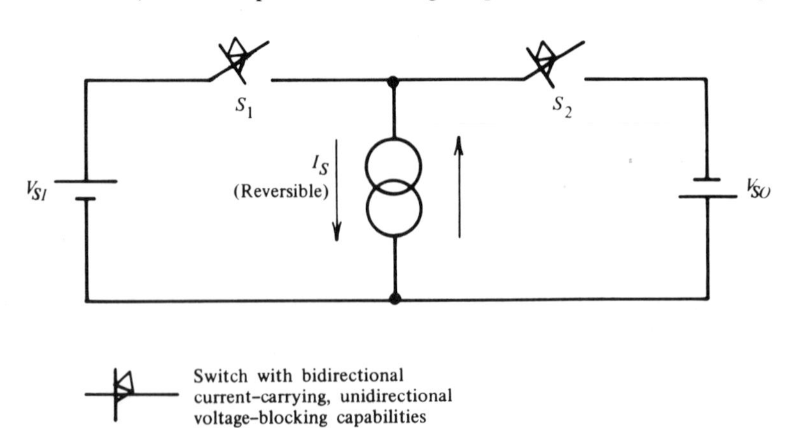

Figure 3-10. Two-quadrant current-source-transfer buck-boost converter.

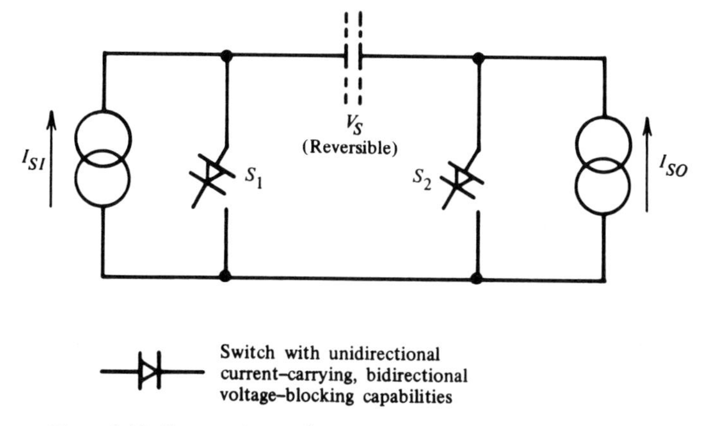

Figure 3-11. Two-quadrant voltage-source-transfer buck-boost converter.

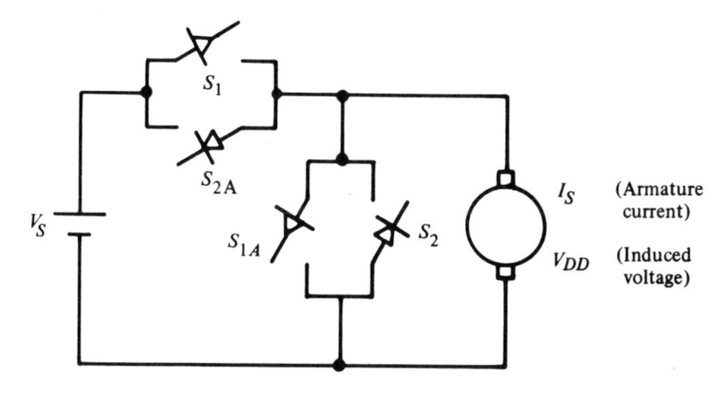

Figure 3-12. Two-quadrant machine-drive converter.

and S_{2A} as a boost converter, with S_1 and S_2 remaining inactive. Once the induced armature voltage falls to zero, of course, operation will cease—the converter cannot enter the fourth quadrant and motor the machine in the reverse direction.

3.6 Four-Quadrant DC-to-DC Converters

Although there are no known applications of four-quadrant dc-to-dc converters, they are introduced here both to complete the dc-to-dc converter family and to set the stage for the ac-to-dc/dc-to-ac converters that will be discussed in Chapter 4. As mentioned in Chapter 1, four-quadrant dc-to-dc converters have the same topologies as their two-quadrant degenerated relatives, but the switch capabilities are expanded.

Thus, in the four-quadrant voltage-sourced (buck) converter of Figure 3-13, all four switches have bilateral-current, unilateral-voltage capability. In the current-sourced (boost) converter of Figure 3-14, all four switches have bilateral-voltage unilateral-current capability. However, since two-quadrant operation was achieved by reversing the wanted components of the dependent quantities, entry to the third and fourth quadrants can only be secured by reversing the "load's" defined quantity. Hence, to achieve four-quadrant operation, the defined current source must be reversible in the buck version, and in the boost converter, the defined voltage source must be reversible. Therefore, since reversals of the energy-transfer media's defined quantities were needed for two-quadrant buck-boost operation, and since the amplitudes of the wanted com-

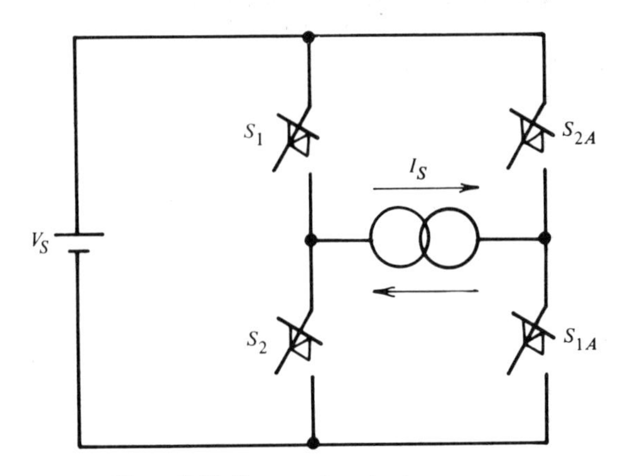

Figure 3-13. Four-quadrant buck converter.

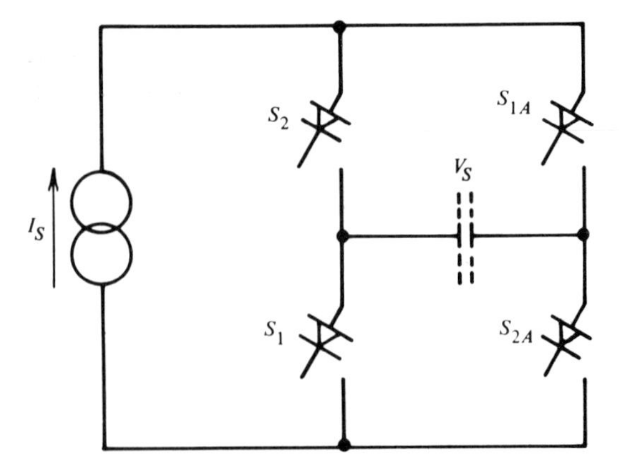

Figure 3-14. Four-quadrant boost converter.

ponents of the dependent quantities are zero and thus not subject to reversal, there is no possibility of a four-quadrant buck-boost converter.

Like the two-quadrant converters, these four-quadrant converters can be operated in mode 1 or mode 2. In mode 1, at equilibrium, the dependent output quantity consists solely of oscillatory components, with zero dc-wanted component. Now, since the defined output quantity has been made reversible, to achieve four-quadrant operation, it can also be made oscillatory. If it oscillates at the appropriate frequency, then power transfer will take place—a single-phase ac-to-dc/dc-to-single-phase converter results, just as predicted in Chapter 1, in reverse order of derivation.

For these four-quadrant converters, analysis of mode 2 operation exactly parallels that of the two-quadrant converters. In the buck converter of Figure 3-13, mode 2 operation in the first and second quadrants is achieved by having S_1, S_{1A}, S_2, and S_{2A} play exactly the same roles as in the two-quadrant converter. Operation in the third and fourth quadrants, with the defined current reversed, is obtained by having S_1 and S_2 exchange roles, as do S_{1A} and S_{2A}. Similarly, in the boost four-quadrant converter of Figure 3-14, first- and second-quadrant mode 2 operation occurs when the switches are operated exactly as they would be in the two-quadrant versions. The third and fourth operating quadrants are entered by reversing the defined voltage and exchanging the roles of S_1 and S_2, with S_{1A} and S_{2A}.

Mode 1 operation is also essentially similar to, and anlaysis for it parallels, that in the two-quadrant cases. The same source reversals and switch-role exchanges occur for Mode 1 and Mode 2. In both cases, it should be noted that active and complementary switches are no longer involved. These four-quadrant converters contain only active switches of apposite capabilities, and hence complementary existence functions are not automatically assumed but in general must be definitely assigned.

3.7 Harmonic Neutralization in DC-to-DC Converters

In all members of the dc-to-dc converter family described and analyzed so far, the amplitudes of the unwanted oscillatory components in the dependent quantities can become distressingly large. Not only do these unwanted components not participate in the converters' power-transfer function, they can wreak havoc with its ability to successfully meet an application. The joule heating effects of the unwanted components have already been considered. In addition, the unwanted components can create a number of other undesirable effects in real sources and loads. Oscillatory currents in dc-machine windings can produce oscillatory torques; ripple on supplies for linear electronic devices can produce distortion, offset, hysteresis, and spurious responses; on supplies for digital electronics, they can cause a variety of malfunctions, including illegitimate data.

The interfacing techniques discussed in Chapter 6 eliminate, or at least greatly attenuate, the influence of converter dependent-quantity unwanted components on overall converter system performance. It should be obvious, however, that the smaller in amplitude and higher in frequency the unwanted components are, the easier the interfacing requirements will be.

The amplitudes of the wanted components in a given simple dc-to-dc converter, single-, two-, or four-quadrant, operating over a given range of A are not subject to modification. Their frequencies can be increased by raising the switching angular frequency, ω, but this is not desirable because of the losses that are associated with the openings and closings of practical switches. There is a technique available to accomplish both ends—amplitude reduction and frequency increase of the unwanted components—by using multiple-phase staggered converters. Called *harmonic neutralization,* or more precisely *polyphasic harmonic neutralization,* it is applicable to all types of switching power converters and is based on the fundamental property of complete phasor sets that was discussed in Chapter 2.

As stated, in dc-to-dc converters, harmonic neutralization is achieved by using multiple-phase staggered converters. The unwanted components of a dependent quantity in any dc-to-dc converter, operating in any quadrant or mode, can be given the general form:

$$R_U = C_U \sum_{n=1}^{n=\infty} [\sin(n\pi/A)/n]\cos(n\omega t) \tag{3.69}$$

In Equation 3.69, C_U is a coefficient containing a defined quantity (voltage or current) and one of the multipliers $2/\pi$ or $4/\pi$ depending on converter type and mode. Now, suppose there are M identical converters labeled 0 to $M-1$. If that labeled 0 is the time reference and the remainder are progressively phase-displaced by $2\pi/M$ radians, then the kth member of this group of M converters has unwanted components in the form

$$R_{Uk} = C_U \sum_{n=1}^{n=\infty} [\sin(n\pi/A)/n]\cos(n(\omega t \pm 2k\pi/M)) \tag{3.70}$$

If the dependent quantities of these M converters are combined by summing, then the resulting summed unwanted components are obviously given by

$$R_{UM} = C_U \sum_{k=0}^{k=M-1} \sum_{n=1}^{n=\infty} [\sin(n\pi/A)/n]\cos(n(\omega t \pm 2k\pi/M)) \tag{3.71}$$

Since n and k are independent variables, the order in which the double summation is taken is not important. Thus, each component of order n consists of the summation over k, i.e.,

$$R_{UM,n} = C_U \sum_{k=0}^{k=M-1} [\sin(n\pi/A)/n]\cos[n(\omega t \pm 2k\pi/M)] \qquad (3.72)$$

This is the sum of a complete set of M phasors with all displacements multiplied by n. It was seen in Chapter 2 that such a sum is zero except when n is an integer multiple of M. Thus, Equation 3.73 can be rewritten, putting $n=pM$ for the nonzero components:

$$R_{UM,pM} = MC_U [\sin(pM\pi/A)/pM]\cos(pM\omega t) \qquad (3.73)$$

The spectrum of unwanted components in the dependent quantities of the M summed converters is seen to contain only harmonics of order pM. Thus, the lowest frequency present is M times the switching frequency, and the maximum amplitude of that component is $1/M$ times the maximum amplitude of the switching frequency component in an individual converter.

An interesting case arises for any operating condition where A is an integer and M is an integer multiple of A. In this event, all the components $R_{UM,pM}$ have zero amplitude since $\sin(pM\pi/A)=0$ for this condition. The combination of M switching converters will then operate with no unwanted components at all in the dependent quantities. Obviously, the condition can only exist for a limited number of discrete operating points, and so it is of little practical concern.

Now, how can M-phase staggered converters be combined in this fashion to give the very marked improvement in unwanted component spectra that is evident from Equation 3.73? Consider first the defined voltage terminals of the converters; they will all be operating with the same defined voltage magnitude to produce the same wanted component of dependent voltage when operated at the same A, and the dependent quantity at these terminals is current. Thus, the converters act as current sources or sinks at their defined voltage terminals, and current sources may be parallel connected at will. Hence, all that is necessary to combine the converters at their defined voltage terminals is to directly parallel-connect them all to the same defined voltage source or sink. The dependent quantities will sum, resulting in the spectral benefits that have been calculated and in a combined wanted component equal to M times the wanted component of each individual converter. This can also be regarded as a situation in which each of the M converters produces $1/M$th of the total wanted component.

At the defined current terminals, the situation is not quite so simple. All M converters will have the same magnitude and polarity of defined current, of course, and they will produce identical wanted components of voltage in the dependent quantities at these terminals. However, the unwanted components will not be identical—they form complete phasor sets—and since the converters are acting as voltage sources or sinks at these terminals, they cannot be directly parallel-connected without violating KLV. They could be directly series-connected, of course, except that they have been parallel-connected at the defined voltage terminals and so they may not be series-connected elsewhere unless full (transformer) isolation exists. In dc-to-dc converters, there is no such isolation, and so series connection at the defined current terminals cannot be effected.

Some means of effecting the parallel connection must be sought, therefore, despite the differences in unwanted voltage components. Note that the wanted components of the dependent voltages are all equal, so no violation of KLV is involved by parallel connection as far as they are concerned. The resultant wanted component will be identical to the wanted component of each individual converter; therefore, effecting parallel connection at the defined current terminals, where the converters act as voltage sources, is tantamount to summing and averaging their dependent quantities. The technique for accomplishing such a summation and averaging depends on the basic properties of multilimb iron-cored reactors and transformers.

Consider the M limb core depicted in Figure 3-15, with a single winding on each limb and with all windings identical (a "wye" winding for an M phase

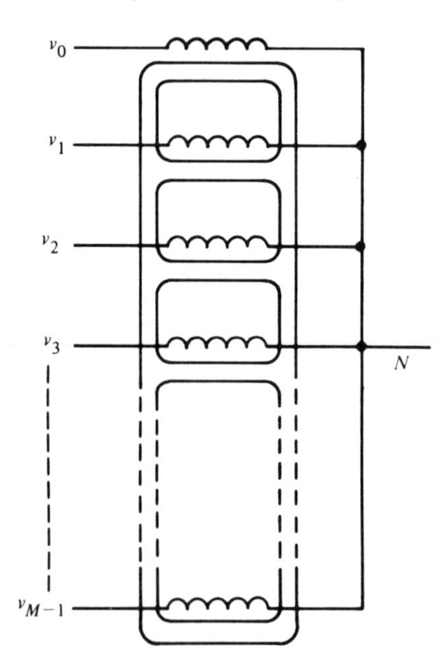

Figure 3-15. Interphase reactor on M-limb core.

transformer, if you will). Now, suppose equal dc currents are caused to flow through all M windings and out of the common, neutral, point N. There will be no magnetizing force exerted on the core, for the algebraic sum of the ampere turns on each limb is zero. If this be true for dc, then it must also be true for any set of in-phase ac currents caused to flow in the windings. Thus, if the windings are energized by a set of in-phase ac voltages, the reactor is transparent to that; it does not offer the impedance of its magnetizing inductance, since no magnetizing force can be exerted by in-phase ac currents. In such a case, the voltage at point N is the same as that of any of the applied voltages, which is, of course, the same as their average.

The same situation does not exist if a complete set of M equiangularly displaced equal-amplitude phasors are applied. When such a set of currents flows in the windings, magnetizing force is exerted on the core, for each limb is then subjected to the ampere turns created by the winding thereon. The result is that the reactor does offer its magnetizing inductance as impedance to complete voltage phasor sets. Moreover, the sum of all M magnetizing currents at point N is zero, and the voltage there is zero, which is both the sum and average of the applied voltages.

These properties are well known as regards the wye windings of three-phase, three-limb transformers. However, they do apply no matter how many limbs and windings a core has, and cores of up to nine limbs are used, although not commonly.

Now, consider what happens when such a reactor is connected between the dependent voltage terminals of M appositely phase-staggered dc-to-dc converters and a common defined dc-current source or sink. The dc currents flow unimpeded by the reactor without causing magnetization of the reactor core. The unwanted components of the dependent voltages that are not of order pM (i.e., that sum to zero) are complete sets and hence are in effect summed by the action of the reactor; no voltage at these frequencies will be impressed on the defined dc current. Those unwanted components of the dependent voltages that do not sum to zero (order pM) are in-phase, or zero-sequence, sets. Therefore, those voltages find the reactor transparent, and voltage at those frequencies is impressed on the defined dc current source or sink, as is the wanted smooth dc voltage.

Used in this manner, these magnetic components are termed *interphase reactors* (sometimes *interphase transformers*). Since cores with more than three limbs are not very common, interphase reactors are often cascade-connected when more than three converters are combined. For example, six converters would first be connected in two groups of three to two three-limb reactors, and then the neutral points of those connected to a two-limb reactor. Alternatively, the six converters might be connected in pairs to three two-limb reactors, the neutral points of which would be connected to a three-limb reactor. When the number of converters can be expressed as 2^N, such as 4 or 8, a binary cascade of two-limb reactors results. This arrangement is, rather whimsically, referred to as a *whiffletree connection,* after the yoke used in harnessing teams of horses or oxen.

The general technique of combining phase-staggered converters to obtain spectral improvements in the dependent quantities, known as *harmonic neutralization* or *cancellation,* is clearly not restricted in application to dc-to-dc converters. The unwanted components in the dependent quantities of all types of converter are oscillatory; those from phase-staggered converters of any kind can be made to form complete phasor sets and thus give rise to the phenomenon of neutralization when the converters are combined. Analytically, the process can be extended to any desired degree. Practically, a variety of imperfections in converters, sources, and interphase reactors limits the extent to which harmonic neutralization can be used. It is rare to find more than six converters so combined, and it is not at all common for more than four to be used. Usually, only two or three are combined, for the increasing practical difficulties and diminishing benefits of larger numbers tend to inhibit further expansions.

Problems

1. Sketch the single-quadrant dc-to-dc converter that would be used to supply a constant 5-V output if the available dc source had a range of 7 to 10.8 V. What range of A would be used for control? Calculate the maximum and minimum amplitudes, relative to the 5-V output, of the first three oscillatory components in the dependent voltage.

2. For the converter of Problem 1, calculate the maximum and minimum average and rms currents in both switches if the output current is 10 A dc.

3. Repeat Problem 1 using a constant dc output voltage of 15 V.

4. Repeat Problem 2 for the converter of Problem 3.

5. Repeat Problem 1 for a dc output voltage of 15 V and a source-voltage range from 10.5 to 16.2 V.

6. Repeat Problem 2 for the converter of Problem 5.

7. Show with sketches how the voltage-source-transfer buck-boost converter can be derived from a combination of the boost and buck converters.

8. Sketch a three-phase harmonic neutralized boost converter. Compare the frequencies and maximum amplitudes of the first three oscillatory components in the dependent current and voltage of this converter with those of a simple boost converter operating at the same switching frequency. (You may assume any frequency and any values of the wanted components of current and voltage to facilitate the comparisons, if you wish.)

4
AC-to-DC and DC-to-AC Converters

4.1 Phase-Shift Controlled Current-Sourced Two-Quadrant Midpoint Converters

The midpoint converter configuration was introduced in Figure 1-4. It was deduced there that the switches for this current-sourced configuration require unidirectional current-carrying, bidirectional voltage-blocking capability. All switches connect to one dc terminal with a defined current, I_S, flowing therein. They also individually connect each to one of the defined voltages. These voltages form a complete phasor set, of which the kth of M is defined as $V\cos(\omega_s t - 2k\pi/M)$, where ω_s is their common angular frequency. The return for I_S is connected to the neutral of the defined set of source voltages. The complete topology is depicted in Figure 4-1.

Observe that a single-phase version of the midpoint converter is not useful, for if but a single switch is employed, the defined dc current will force it to remain permanently closed. Thus, the defined ac voltage will be applied to the defined dc current that flows in the ac voltage source; no power transfer takes place. This can be remedied if a discontinuous dc terminal current is permitted. It is highly unusual to find a two-quadrant single-phase converter operated in this manner; however, single-quadrant single-phase uncontrolled converters with this mode of operation are quite common in low-power applications, and controlled versions are also sometimes used. Their behavior and performance are discussed in Section 4.3.

In Chapter 2, it was shown that the existence functions of switches S_0 to S_{M-1} have to contain an oscillatory term with angular frequency ω_s if a wanted dc component is to be produced in the dependent voltage applied to the defined dc current. There is nothing special about any of the individual defined ac voltages, v_0 to v_{M-1}, or any of the corresponding switches. It is therefore reasonable to assume that the existence functions, H_0 to H_{M-1}, should all be identical except for phase displacement. To avoid violation of KLC with regard to the defined current, at least one switch must be closed at all times, whereas to avoid violation

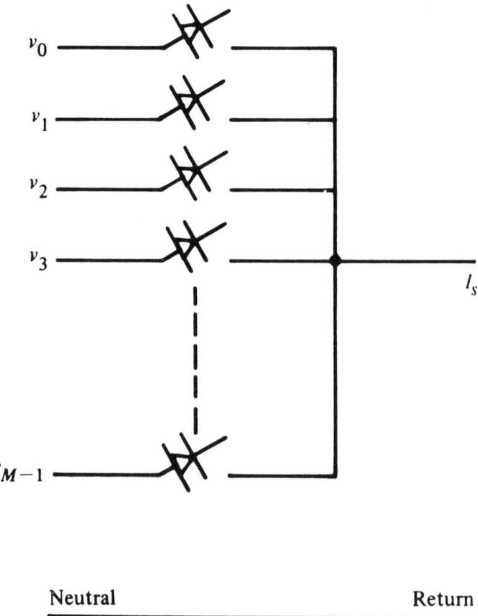

Figure 4-1. Midpoint converter (positive group).

of KLV with regard to the defined voltages, only one of the switches may be closed at any time. The foregoing strongly suggests a complete set of M existence functions would be appropriate, with the kth defined as

$$H_k = 1/M + (2/\pi) \sum_{n=1}^{n=\infty} [\sin(n\pi/M)/n]\cos[n(\omega_s t - 2k\pi/M)] \tag{4.1}$$

Now the dc terminal contribution of the kth switch is $H_k\ v_k$, and the complete dependent voltage is given by:

$$v_D = \sum_{k=0}^{k=M-1} H_k\ v_k \tag{4.2}$$

The structure of Equation 4.2 has two distinct parts. The first is easily disposed of, for

$$v_D(1/M) = \sum_{k=0}^{k=M-1} (1/M)\ V \cos(\omega_s t - 2k\pi/M) = 0 \tag{4.3}$$

Thus, v_D consists in fact only of the second part of Equation 4.2, giving

$$v_D = (2V/\pi) \sum_{k=0}^{k=M-1} \sum_{n=1}^{n=\infty} [\sin(n\pi/M)/n]\cos[n(\omega_s t - 2k\pi/M)] \cos(\omega_s t - 2k\pi/M) \quad (4.4)$$

Using the trigonometric expansion,

$$\cos A \cos B = [\cos(A+B) + \cos(A-B)]/2 \quad (4.5)$$

the cross products of the double summation in Equation 4.4 become

$$P_n = \{\cos[(n+1)(\omega_s t - 2k\pi/M)] + \cos[(n-1)(\omega_s t - 2k\pi/M)]\}/2 \quad (4.6)$$

Now the wanted dc component in v_D can only come from the cross product P_1—to be exact, from its second part. After summing M such terms, the result is

$$V_{DD} = (MV/\pi) \sin(\pi/M) \quad (4.7)$$

The oscillatory contributions of the cross products P_n, when summed over 0 to $M-1$, are the sums of complete phasor sets with displacements (and frequencies) multiplied by $n+1$ and $n-1$. Hence, they result in zero contributions to v_{DU}—the unwanted components of the dependent voltage—except when $n+1 = pM$ or $n-1 = pM$ and p = any integer from 0 to ∞. Thus, v_{DU} contains oscillatory components only of angular frequencies $pM\omega_s$; each of the components has two contributions:

$$v_{DU}(n+1) = (MV/\pi)\sin[(pM-1)\pi/M]\cos(pM\omega_s t)/(pM-1) \quad (4.8)$$

and

$$v_{DU}(n-1) = (MV/\pi)\sin[(pM+1)\pi/M]\cos(pM\omega_s t)/(pM+1) \quad (4.9)$$

Now observe that

$$\sin[(pM\pm 1)\pi/M] = \sin(p\pi \pm \pi/M) = \pm\cos(p\pi)\sin(\pi/M) \quad (4.10)$$

In consequence, v_{DU} is given by

$$v_{DU} = V_{DD} \sum_{P=1}^{P=\infty} \cos(p\pi)[\cos(pM\omega_s t)/(pM+1) - \cos(pM\omega_s t)/(pM-1)] \quad (4.11)$$

So far, control of the wanted component has not been considered. As shown in Chapter 2, this will be effected by introducing α, a phase delay or advance, into the arguments of the existence functions. Adopting the normal convention in which a positive α is an advance and a negative α a delay, the kth existence function becomes

$$H_k = 1/M + (2/\pi) \sum_{n=1}^{n=\infty} [\sin(n\pi/M)/n]\cos[n(\omega_s t - 2k\pi/M + \alpha)] \quad (4.12)$$

Observe that in the sum of the products $H_k\ v_k$, the presence of α has absolutely no effect on the zero value of the first part. In the second part, α appears in the arguments of the existence functions only, not in those of the defined voltages. Thus, the cross products of Equation 4.6 are modified to

$$P_{n,\alpha} = \{\cos[(n+1)(\omega_s t - 2k\pi/M) + n\alpha] + \cos[(n-1)(\omega_s t - 2k\pi/M) + n\alpha]\}/2 \quad (4.13)$$

As a result, the wanted component of V_D, V_{DD} becomes

$$V_{DD,\alpha} = (MV/\pi)\sin(\pi/M)\cos\alpha \quad (4.14)$$

as was forecast in Chapter 2. The unwanted components acquire phase advances or delays; from $n+1 = pM$, $n = pM-1$, while from $n-1 = pM$, $n = pM+1$ and v_{DU} is modified to

$$v_{DU,\alpha} = V_{DD} \sum_{p=1}^{p=\infty} \cos(p\pi)\{\cos[pM\omega_s t + (pM+1)\alpha]/(pM+1) - \cos[pM\omega_s t + (pM-1)\alpha]/(pM-1)\} \quad (4.15)$$

To determine the amplitudes of these unwanted oscillatory components in the dependent voltage, it is necessary to expand to cosine and sine terms and then to take the square root of the sum of the squares of their coefficients. The coefficients in question are

$$C_C = V_{DD}\cos(p\pi)\{\cos[(pM+1)\alpha]/(pM+1) - \cos[(pM-1)\alpha]/(pM-1)\} \quad (4.16)$$

and

$$C_S = -V_{DD}\cos(p\pi)\{\sin[(pM+1)\alpha]/(pM+1) - \sin[(pM-1)\alpha]/(pM-1)\} \quad (4.17)$$

Observing that

$$\cos[(pM+1)\alpha]\cos[(pM-1)\alpha] + \sin[(pM+1)\alpha]\sin[(pM-1)\alpha] = \cos(2\alpha) \quad (4.18)$$

Squaring and summing C_c and C_s yields:

$$v_{DU,\alpha,pM} = V_{DD}\,\{1/(pM+1)^2+1/(pM-1)^2 - 2\cos(2\alpha)/[(pM+1)(pM-1)]\}^{1/2} \quad (4.19)$$

Before exploring the implications of Equations 4.15 and 4.19, consider the range of α and its effect on $V_{DD,\alpha}$. At $\alpha = 0$, V_{DD} is given by Equation 4.7; as α is increased in magnitude from 0 to $\pi/2$ rad, $\cos\alpha$ declines from 1 to 0, and hence $V_{DD,\alpha}$, given by Equation 4.14 declines from V_{DD} to 0. Over this range of α, $V_{DD,\alpha}$ has the same polarity as the defined current, and the converter will transfer power from the defined voltage sources to the defined current sink. As α is increased in magnitude from $\pi/2$ to π rad, $\cos\alpha$ declines further from 0 to -1 and $V_{DD,\alpha}$ declines from 0 to $-V_{DD}$. Over this range, $V_{DD,\alpha}$ has the opposite polarity to the defined current, and the converter will transfer power from the defined current source to the defined voltage sinks. The useful range of $|\alpha|$ is clearly 0 to π rad since $\pi+\theta \equiv -(\pi-\theta)$.

The region $|\alpha| = 0$ to $\pi/2$ rad has the converter operating as an ac-to-dc converter and is termed the *rectification quadrant* (or *region*). The region $|\alpha| = \pi/2$ to π rad had the converter operating as a dc-to-ac converter and is termed the *inversion quadrant*. Thus, a phase-controlled midpoint current-sourced converter is a two-quadrant unidirectional-current bidirectional-voltage converter at its dc terminals. It is capable of rectifying (ac-to-dc) or inverting (dc-to-ac) operation, and the rate of power transfer in either quadrant is controlled by $|\alpha|$. These observations clearly apply whether α is positive or negative—i.e., whether control is accomplished by phase-advancing or -delaying the existence functions with respect to the defined ac voltages.

The operating point $\alpha = 0$, for this and all other phase-controlled current-sourced ac-to-dc converters, is termed the *rectification end stop* because $\alpha=0$ represents the limiting condition for rectifying operation, with the maximum ac-to-dc power transfer the converter can produce. Similarly, the operating point

$\alpha = \pm\pi$ is termed the *inversion end stop,* since it represents the limiting condition for inverting operation with maximum dc-to-ac power transfer.

Now consider the effect of α on the unwanted oscillatory components. Putting α=0 or π in Equation 4.19 yields

$$|v_{DU,0,pM}| = |v_{DU,\pi,pM}| = V_{DD}[1/(pM+1)-1/(pM-1)] \qquad (4.20)$$

while putting $|\alpha| = \pi/2$ yields

$$|v_{DU,\pm\pi/2,pM}| = V_{DD}[1/(pM+1)+1/(pM-1)] \qquad (4.21)$$

Equation 4.21 gives the maximum amplitudes ever assumed by these unwanted oscillatory components, collectively termed *ripple,* while Equation 4.20 gives the smallest amplitudes they assume. Hence, at maximum power transfer, $|\alpha| = 0$ or π, the converter generates least ripple, while at zero power transfer, $\alpha = \pm\pi/2$, the ripple is at its maximum.

There are two values of M of great practical importance, $M = 2$ and $M = 3$. The first value yields the two-phase midpoint converter, also known as the center-tapped single-phase circuit; the latter yields the three-phase midpoint converter, which, in addition to its inherent importance, is the basic building block of the great majority of current-sourced ac-to-dc/dc-to-ac converters. It is of some interest to evaluate the maximum wanted and unwanted components of the dc terminal voltages for these cases:

- For $M = 2$, $V_{DD,0,2} = (2/\pi)V$, $|v_{DU,0,2}| = (2/3)V_{DD,0,2}$ and $|v_{DU,\pi/2,2}| = (4/3)V_{DD,0,2}$
- For $M = 3$, $V_{DD,0,3} = 3\sqrt{3}V/(2\pi)$, $|v_{DU,0,3}| = (1/4)V_{DD,0,3}$, and $|v_{DU,\pi/2,3}| = (3/4)V_{DD,0,3}$.

These show that as M is increased, the maximum wanted component magnitude more closely approaches the peak value of the defined voltages, as can be deduced from Equation 4.7 since $\sin\theta/\theta \to 1$ as $\theta \to 0$. Also, they show that the ripple can be a severe problem with ac-to-dc/dc-to-ac converters, but that the severity of the problem will decrease progressively as M is increased. The improvement is twofold; not only do the maximum and minimum amplitudes decrease, but the basal frequency of the ripple increases. However, the ratio of maximum-to-minimum amplitudes increases with M—in fact, from Equations 4.20 and 4.21, it can be seen that this ratio is equal to M.

The dependent voltage waveform for any given M can be constructed by the technique described in Chapter 2. Figures 4-2 and 4-3 show waveforms for a three-phase midpoint converter for α's of 0, $\pi/6$, $\pi/3$, $\pi/2$, $2\pi/3$, $5\pi/6$, and π rad. Figure 4-2 shows the waveforms for positive α, phase-advance control,

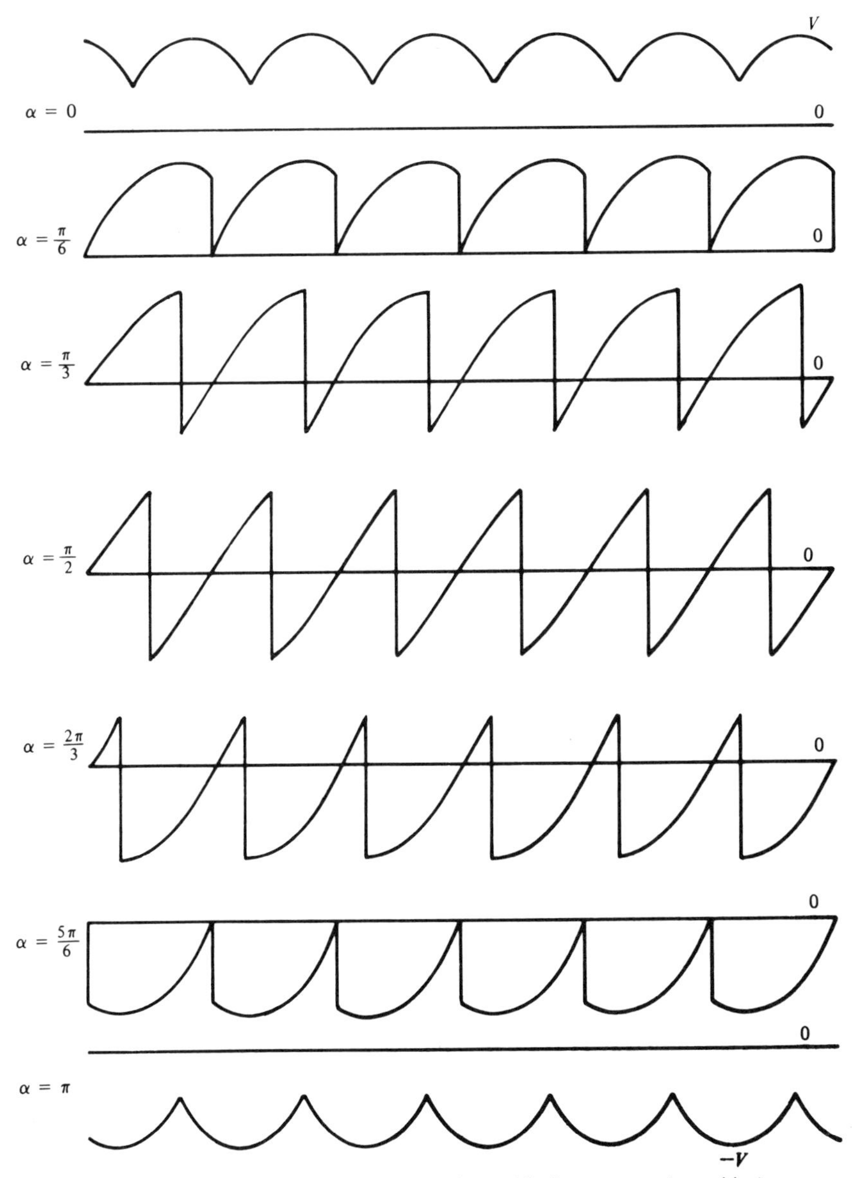

Figure 4-2. DC terminal voltage of three-phase midpoint converter (α positive).

while Figure 4-3 shows waveforms for negative α. The mirror image, or reversed time, relationship of the two cases is quite evident. Also, it should be observed that for the case of positive α, all switching transitions are to a less positive

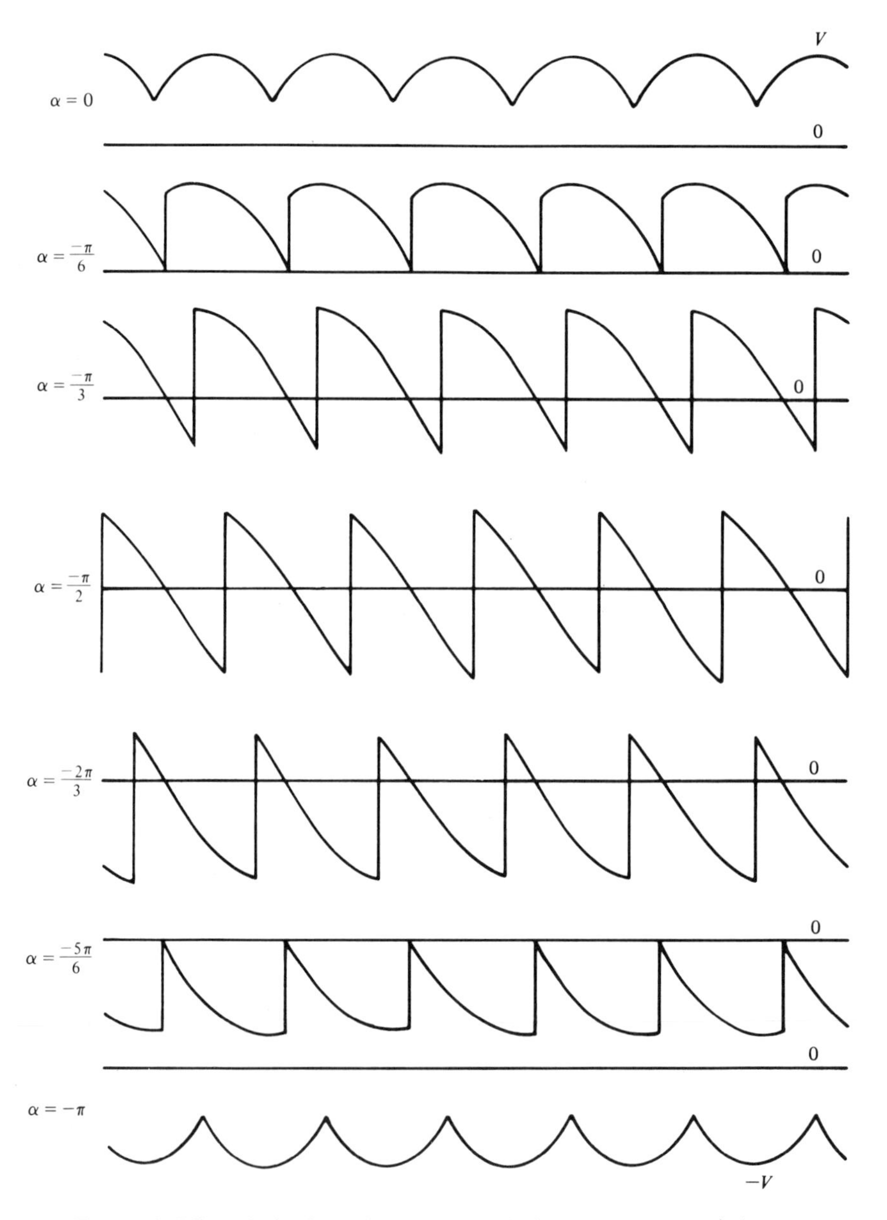

Figure 4-3. DC terminal voltage of three-phase midpoint converter (α negative).

defined voltage. The converse is true for negative α, in which case all switching transitions are to a more positive defined voltage. This remains the case, obviously, for any value of M; the two types of waveform produced have been named *negative type* (all transitions to a more negative voltage, α positive) and *positive type* (all transitions to a more positive voltage, α negative). These wave types are associated with certain characteristics of converter behavior, as will be seen shortly.

Consider another property of the waveforms of Figures 4-2 and 4-3. From $|\alpha| = 0$ to $\pi/6$ rad, V_D, the dependent voltage, is always positive and is of the same polarity as the defined current; thus, for this range of $|\alpha|$, the converter operates solely in the rectifying quadrant. Similarly, from $5\pi/6$ to π rad delay or advance, the converter operates solely in the inverting quadrant. However, between $\pi/6$ and $5\pi/6$ rad, the converter shows periods of operation in both quadrants with net operation depending on the relative time spent in each; when the times are equal, net operation is the equilibrium condition with zero power transfer. This operating regime produces conditions analagous to those for mode 1 operation of the two- and four-quadrant dc-to-dc converters. Indeed, except for the one-quadrant dc-to-dc and ac-to-dc/dc-to-ac converters, it is a universal phenomenon in switching power converters. For any midpoint current-sourced ac-to-dc/dc-to-ac converter, the region for operation in both quadrants in each switch period can be shown to be from $|\alpha| = (M-2)\pi/2M$ to $|\alpha| = (M+2)\pi/2M$. There is one important aspect to this behavior: in order for the converter to operate, the defined current must be capable of acting as a source, and the defined voltages must be capable of acting as sinks for the short times involved, even if net operation is always in one quadrant only.

Attention now turns to the dependent currents produced by this type of converter. Clearly, each is simply a waveform replica of the corresponding switch existence function with amplitude I_S, the defined current; that flowing in S_k and v_k may be written as

$$i_{Dk} = I_S/M + (2I_S/\pi) \sum_{n=1}^{n=\infty} [\sin(n\pi/M)/n]\cos[n(\omega_s t - 2k\pi/M)] \qquad (4.22)$$

for the case where $\alpha = 0$. Since α is introduced in the arguments of the oscillatory terms of the existence functions, it also appears in those of the dependent currents. As deduced in Chapter 2, it is the phase of the oscillatory components of i_{Dk} relative to the defined voltages that is affected by α, not their magnitudes. The equation

$$i_{Dk,\alpha} = I_S/M + 2I_S/\pi \sum_{n=1}^{n=\infty} [\sin(n\pi/M)/n]\cos[n(\omega_s t - 2k\pi/M + \alpha)] \qquad (4.23)$$

gives the dependent current for any α and k.

Since all the currents are identical and have identical phase displacements relative to the defined voltage sources in which they flow, it is necessary to study one only to appreciate the impact of the converter on its ac supply except for one term. Note that $i_{Dk,\alpha}$ contains an unwanted dc component, I_S/M. This is, in fact, a pronounced disadvantage for the midpoint scheme. If M is small, as it usually is in practice, this unwanted dc component is large in amplitude and contributes substantially to joule heating in the ac network while not participating in real power development therein. Also, practical imperfections in a real system make such a dc component a potential cause of saturation in transformer cores and machine yokes. Ideally, M identical dc currents, flowing in M identical windings on the M identical limbs of an M phase transformer or in an M phase generator or motor, will cause no net dc magnetization—the alegebraic sum of the dc ampere turns is zero everywhere in the magnetic structure. The real world tends not to be so forgiving—the currents, the windings, and the magnetic structure are all likely to evidence imperfections leading to the application of dc magnetizing force and the possibility of saturation as a result.

The wanted component of the dependent current, for the 0th line, which is the one chosen for convenient study, is clearly that for $n=1$, namely

$$i_{DD,\alpha} = (2I_S/\pi)\sin(\pi/M)\cos(\omega_s t+\alpha) \tag{4.24}$$

Since this flows in a defined voltage, $V\cos(\omega_s t)$, real power is produced only by its cosine component; the total average real power at the ac lines is the sum of M identical contributions giving

$$W_{ac} = (MI_S/\pi)\sin(\pi/M)\ V\cos\alpha = V_{DD,\alpha}I_S = W_{dc} \tag{4.25}$$

so that the necessary equality of powers is satisfied. However, the real power at the ac terminals is accompanied by reactive power generated by the sine component of $i_{DD,\alpha}$ in addition to that generated by the unwanted oscillatory components and the dc component. From Equation 4.24, the quadrature component of fundamental frequency current is

$$(i_{DD,\alpha})_Q = -(2I_S/\pi)\sin(\pi/M)\sin\alpha\ \sin(\omega_s t) \tag{4.26}$$

This is a leading current if α is positive (phase-advance control) and a lagging current if α is negative (phase-delay control). Thus, the use of α as a control variable maintains the magnitude of the current constant by introducing an increasing quadrature current component as the real component is reduced. At $\alpha = \pm\pi/2$, when zero power transfer is produced by the converter, the converter burdens its defined ac sources with a purely reactive load at the fundamental frequency equal in magnitude to the purely real one it generates if $\alpha = 0$ or $\pm\pi$.

From Equation 4.23, it is seen that unwanted oscillatory components exist in the ac line currents at all frequencies except those that are integer multiples of M—since $\sin(n\pi/M) = 0$ for that condition. Hence, the frequencies present in the dc-terminal ripple voltage are not present in the ac line currents, but all others are for these midpoint converters. The amplitude of the nth-order unwanted oscillatory component relative to that of the fundamental is, obviously: $(1/n) \sin(n\pi/M)/\sin(\pi/M)$; for $M = 2$, this reduces to $1/n$, as it does for $M = 3$. However, for $M>3$, this ratio can assume values other than $1/n$ for some harmonics—the simple relationship is not immutable. The harmonic currents also separate into in-phase and quadrature components when α is not 0 or $\pm\pi$, while maintaining the same amplitude. Since no nonfundamental current can ever produce real power, their contribution is wholly a reactive power consumption regardless of their phase relationship.

The rms value of $i_{Dk,\alpha}$ is obviously $I_S|\sqrt{M}$. The amplitude of its fundamental component is given by Equation 4.24, and thus the distortion index for the ac line current, the fractional total harmonic distortion, is given by

$$\mu_I = [\pi^2/2M\sin^2(\pi/M) - 1]^{1/2} \tag{4.27}$$

Note that as M increases, μ_I increases; as the dc-terminal performance of the converter improves, the ac-terminal performance is degraded. Therefore, other converter topologies should be sought that would not exhibit such an unfortunate trait; they are encountered before long.

There is an interesting sidelight to Equation 4.27. The square of the rms value of $i_{Dk,\alpha}$ is $I_S{}^2/M$; the square of the dc component is $I_S{}^2/M^2$. From these facts, it can be concluded that

$$\sum_{n=1}^{n=\infty} \sin^2(n\pi/M)/n^2 = (\pi^2/4)(1/M - 1/M^2) \tag{4.28}$$

and various functions of the Bernoulli and Euler numbers may be evaluated in closed form using this relationship.

Finally, for the midpoint converter of Figure 4-1, consider the statements that develop when the dc-terminal wave types are associated with the displacement factors (power factors at fundamental frequency) of the converter. Positive α produces a negative-type wave and a leading displacement factor; negative α produces a positive-type wave and a lagging displacement factor. The defined dc current is positive, and so it might be hypothesized that *if the product of current polarity and wave type is positive, the displacement factor is lagging; if that product is negative, the displacement factor is leading*. This hypothesis will now be verified by examining the behavior of the complementary version of this converter.

4.2 Phase-Shift-Controlled Current-Sourced Two-Quadrant Bridge Converters

It was seen in Chapter 1 that the midpoint converter is a degeneration of the full switching matrix, that is, the bridge configuration. A corollary viewpoint is that the bridge, in fact, consists of two midpoint converters fed by the same defined ac voltages but with defined dc currents of opposite polarity and with an effective series connection of their dc-terminal dependent voltages. Such a view involves the use of a second midpoint converter, which is the complement of that shown in Figure 4-1. Depicted in Figure 4-4a, this converter has the polarity of its defined current reversed and should always produce a wanted component of dependent voltage opposite in polarity to that produced by the original converter if the two are to sum their power-transfer contributions.

It is clear that if the converter of Figure 4-4a is assigned existence functions as defined by Equation 4.1 and is operated over the range $|\alpha| = 0$ to π rad, it will produce exactly the same dc-terminal voltage as does the converter of Figure 4-1. To cause it to produce a complementary dependent voltage so that the combination of the two converters (the bridge configuration of Figure 4-4b) will produce power transfer, its existence functions must be given an initial displacement of $\pm\pi$ rad. Thus,

$$
\begin{aligned}
H'_k &= 1/M + (2/\pi) \sum_{n=1}^{n=\infty} [\sin(n\pi/M)/n]\cos[n(\omega_s t - 2k\pi/M \pm \pi)] \\
&= 1/M + (2/\pi) \sum_{n=1}^{n=\infty} [\sin(n\pi/M)/n]\cos(n\pi)\cos[n(\omega_s t - 2k\pi/M)]
\end{aligned}
\tag{4.29}
$$

Observe that each oscillatory term, of order n, acquires the multiplier $\cos(n\pi)$, and hence so do all the cross products with the defined voltages and defined current. Now, the wanted component of the dependent voltage arose from a portion of the cross product with $n = 1$; it therefore has the multiplier $\cos\pi$, -1, and so

$$V'_{DD,\alpha} = -(M/\pi)\sin(\pi/M)\cos\alpha \tag{4.30}$$

The cross-product terms giving rise to the unwanted components also carry the multiplier $\cos(n\pi)$. Remembering that nonzero unwanted components were developed only for $n+1 = pM$ and $n-1 = pM$, the multiplier becomes $\cos[(pM \pm 1)\pi] = -\cos(pM\pi)$.

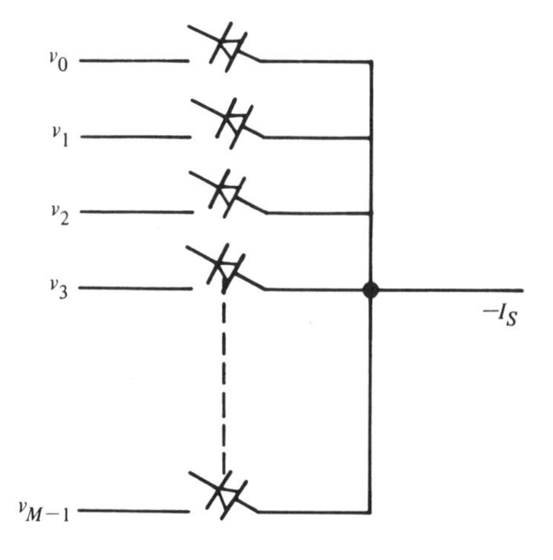

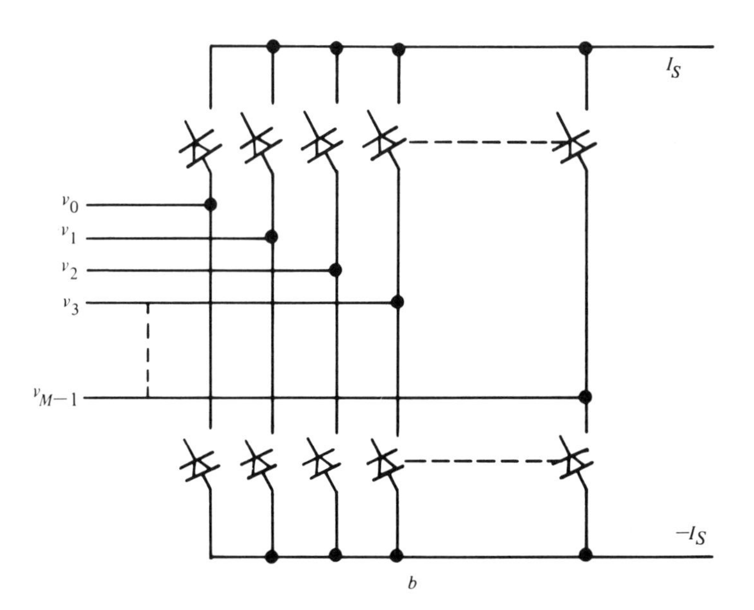

Figure 4-4. *a*, Complementary midpoint converter; *b*, bridge converter.

Thus, the unwanted components of this complementary midpoint converter's dependent voltage are given by

$$v'_{DU} = -\cos(pM\pi)v_{DU} \tag{4.31}$$

where the multiplier $\cos(pM\pi)$ appears, of course, within the summation in the fully expanded form of Equation 4.31 since p is the variable of summation. The waveforms of the dependent voltage for this converter can be constructed using the standard and by now familiar technique. They are depicted, for α positive and negative, in Figures 4-5 and 4-6, and it is observed that not only do the voltage polarities reverse as compared to the waveforms of Figures 4-2 and 4-3—so do the wave types. Positive α, producing a leading displacement factor, is now accompanied by a positive-type wave; negative α, producing a lagging displacement factor, is now accompanied by a negative-type wave. The displacement factors are, of course, still leading and lagging for positive and negative α, respectively, because the introduction of $\cos(n\pi)$ as a multiplier in the line currents creates a -1 multiplier for the fundamental, which, when multiplying the defined current $-I_S$, restores the fundamental current component for this converter to that for the original. Now, the observation as to wave type–current–displacement factor for this converter confirm the hypothesis advanced in Section 4.1—if the cross product of wave type and current is positive, there is a lagging displacement factor, whereas if that cross product is negative, there is a leading displacement factor. As two swallows do not a summer make, so two such closely related and complementary cases perhaps do not a hypothesis verify. The position is strengthened, however, and its universality will become apparent as the text progresses.

The wanted component in the bridge-dependent voltage is given by

$$V_{DDBR,\alpha} = V_{DD,\alpha} - V'_{DD,\alpha} = (2M/\pi)\sin(\pi/M)V\cos\alpha \tag{4.32}$$

The unwanted oscillatory components, the ripple, are given by

$$v_{DUBR,\alpha} = v_{DU,\alpha} - v'_{DU,\alpha} = [1+\cos(pM\pi)]v_{DU,\alpha} \tag{4.33}$$

The multiplier, $[1+\cos(pM\pi)]$, appears within the summation in the fully expanded form of Equation 4.33.

Observe that $\cos(pM\pi)$ is 1 if pM is even and -1 if pM is odd; a curious phenomenon emerges. If M is even, then all pM are even and $1 + \cos(pM\pi) = 2$ for all p. The spectrum of the bridge's dependent voltage is identical to that for the midpoint converter, save that all amplitudes, wanted and unwanted

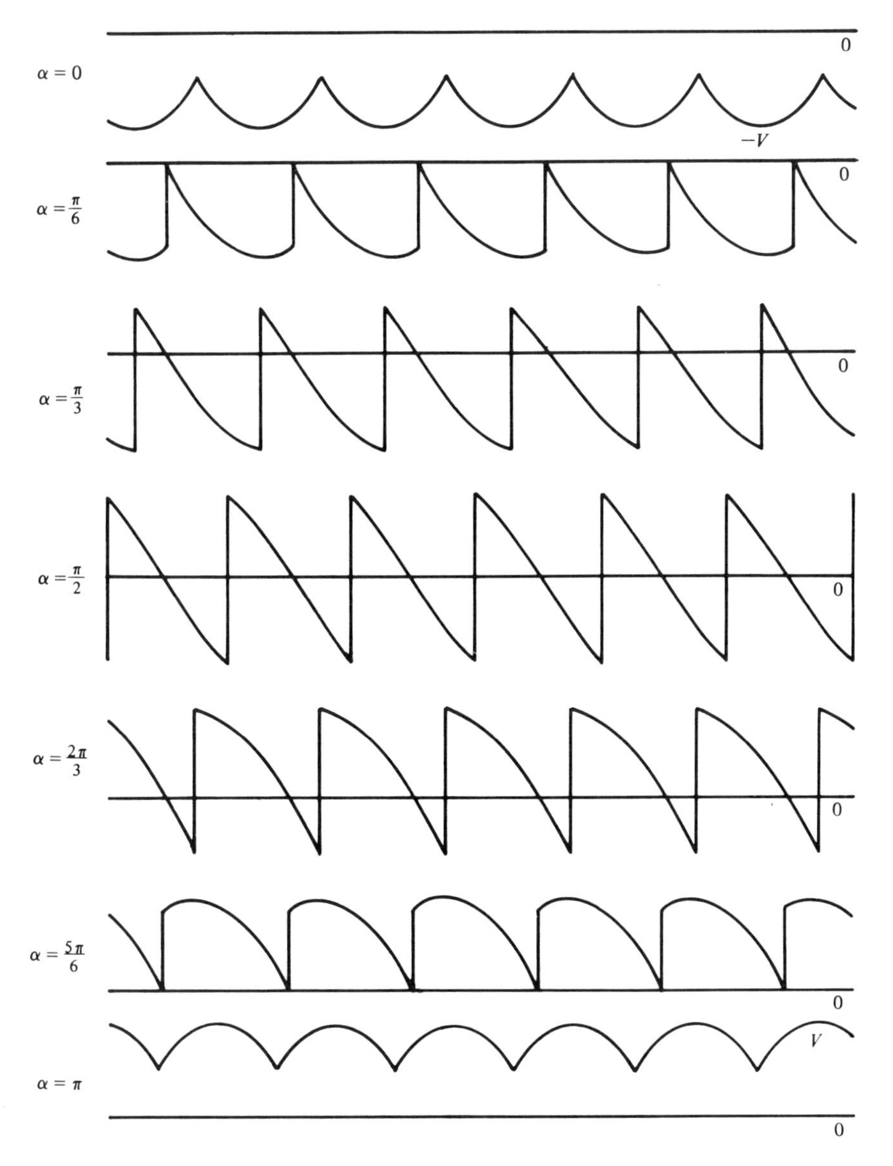

Figure 4-5. DC terminal voltage of three-phase complementary midpoint converter (α positive).

components alike, are doubled. If M is odd, however, then only those pM with p even are even—those with p odd are odd. Hence, all unwanted components for p odd disappear, only those for p even remain with amplitudes double, and the effect is as if M had been doubled:

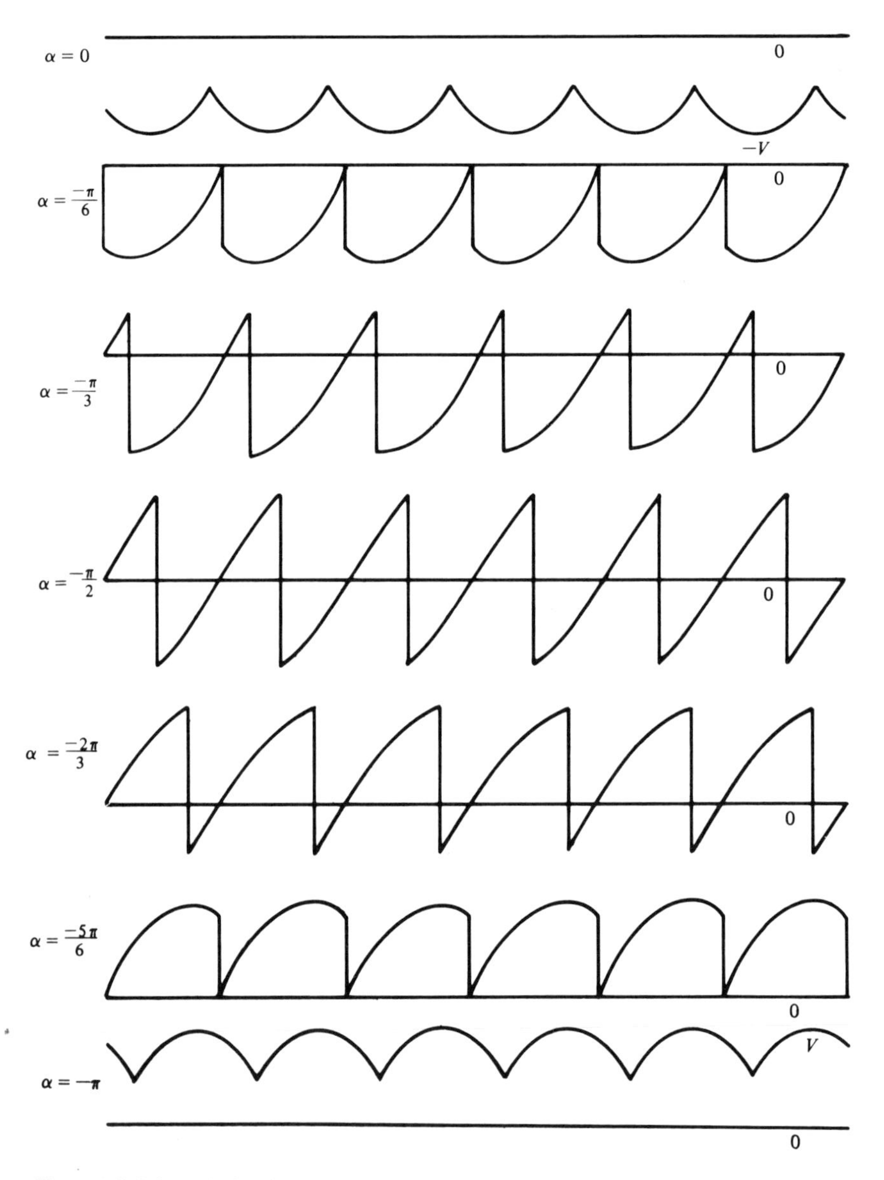

Figure 4-6. DC terminal voltage of three-phase complementary midpoint converter (α negative).

$$v_{DUBR,\alpha} = 2V_{DD} \sum_{p=1}^{p=\infty} \{\cos[2pM\omega_s t + (2pM+1)\alpha]/(2pM+1) - \cos[2pM\omega_s t + (2pM-1)\alpha]/(2pM-1)\} \tag{4.33}$$

The $\cos(p\pi)$ multiplier of Equation 4.15 becomes, of course, simply unity and positive when p can only be even in that equation; hence, it is removed from Equation 4.33. Thus, the spectrum of the dependent voltage for an M phase bridge with M odd is the same as that for a $2M$-phase midpoint converter—a rather dramatic improvement. Perhaps surprisingly, the elimination of the multiplier $\cos(p\pi)$, which alternates in polarity, does not affect the type of ripple wave produced. This is because a $\cos(p\pi)$ multiplier's only significance is in the position of the wave with regard to the time zero reference. The value of $\cos(p\pi)$ is -1 for all odd p; hence, as far as phase relationships within the wave are concerned, it makes no difference whatsoever to all components with odd p—it merely reverses all their polarities. All even-order components have an apparent phase relationship that depends on the time-zero reference. They are all in phase or antiphase—i.e., multiplied by $+1$ or -1, depending on the reference, changing sign over the angle of π rad at the basal frequency of the wave. Hence, whether that wave contains the multiplier $\cos(p\pi)$ has no significance as regards to shape, polarity, or wave type. Figure 4-7 shows the bridge waveforms, constructed by the standard technique, for the important case, $M = 3$.

A rather curious situation occurs when a two-phase midpoint converter ($M=2$) is converted to a bridge by adding a complementary two-phase midpoint converter. The elimination of the neutral connection from a two-phase set, two phasors at π rad displacement, reduces the defined voltage to a single phase. The resulting converter, then, is a single-phase bridge; a two-phase bridge cannot exist.

The phenomenon that a bridge with an odd number of phases exhibits in its dependent voltage spectrum introduces another term in the lexicon of switching power converters: the *pulse number* of a converter, P. In the 1920s, workers in the switching converter field using mercury-arc rectifiers first recognized that a number of different circuit topologies gave the same dc terminal ripple. They coined the term *pulse number* since they characterized the dc terminal performance in terms of the number of pulses, or fluctuations, in the dc terminal voltage in one cycle of the ac supply. As will be seen, the term as now used has even deeper implications than identity of dependent dc voltage spectra.

Turning now to the dependent currents of the complementary and bridge converters, observe that since Equation 4.29 defines H'_k, the complementary converter's line current for $k = 0$ is

$$i'_D = -I_S/M - (2I_S/\pi) \sum_{n=1}^{n=\infty} [\sin(n\pi/M)/n]\cos(n\pi)\cos(n\omega_s t)] \qquad (4.34)$$

(at $\alpha = 0$). The bridge line current is, obviously, the sum of the two midpoint converter currents. First note that the dc components cancel, eliminating one of

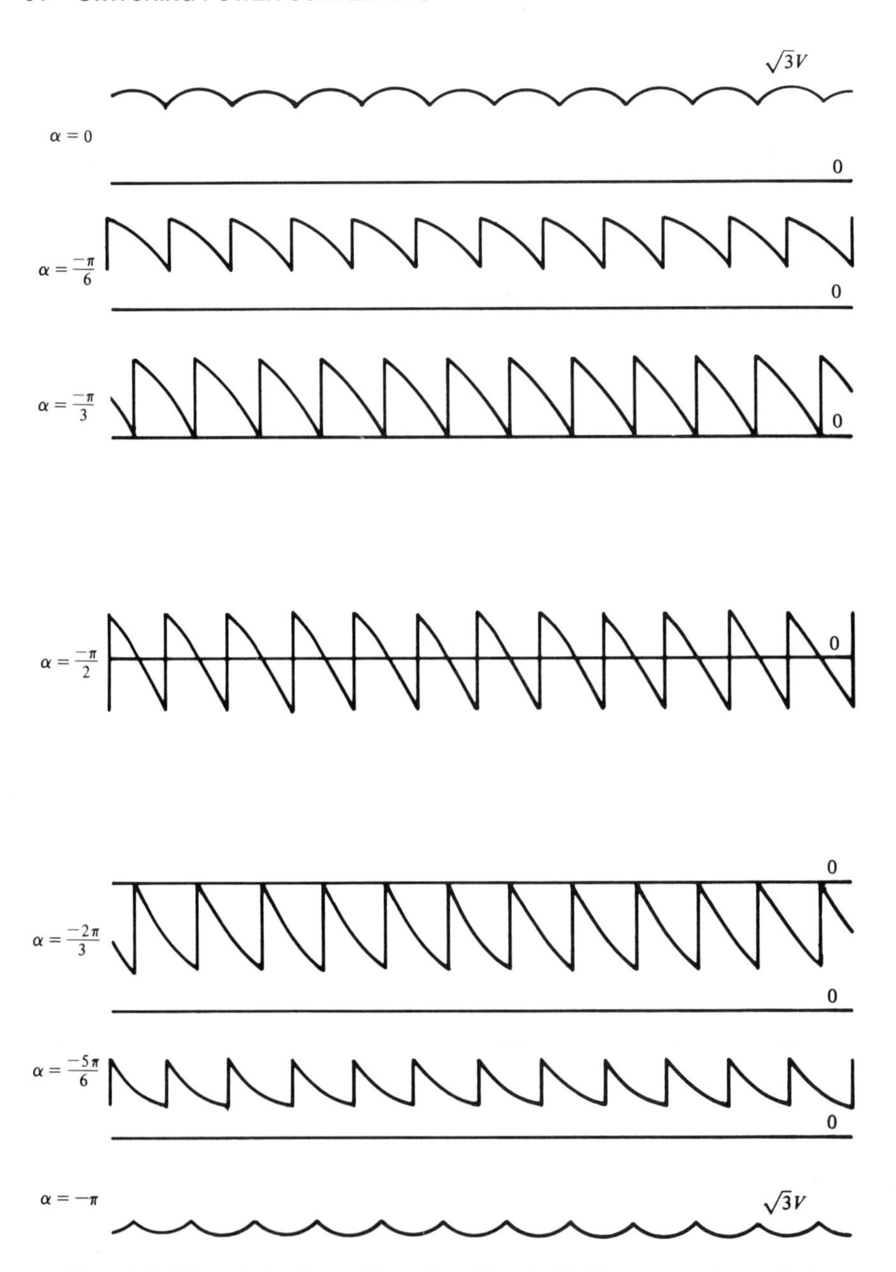

Figure 4-7. DC terminal voltage of three-phase (six-pulse) bridge converter (α negative).

the problems associated with the midpoint converter.

The total bridge current is given by

$$i_{DBR} = (2I_S/\pi) \sum_{n=1}^{n=\infty} [\sin(n\pi/M)/n](1-\cos n\pi[n(\omega_s t+\alpha)] \qquad (4.35)$$

Now, $1-\cos(n\pi)$ is 0 for n even and 2 for n odd. Hence, no even-order harmonics are present in the bridge-dependent currents; only odd orders that are not integer multiples of M are present, regardless of whether M is even or odd. If M is even, all odd-order harmonics will be present, since an odd integer cannot be an integer multiple of an even one; only if M is odd will certain harmonics be missing from the line current spectrum. Rewriting Equation 4.35 to reflect this,

$$i_{DBR} = (4I_S/\pi) \sum_{p=1}^{p=\infty} \{\sin[(2p-1)\pi/M]/(2p-1)\}\cos[(2p-1)(\omega_s t+\alpha)] \qquad (4.36)$$

For the important case where $M = 3$, the odd numbers that are not integer multiples of 3 can be expressed as the series $6p+1$ and $6p-1$; this is because all numbers (except 1) that are not integer multiples of 3 can be found in the series $3m+1$ and $3m-1$. Such an integer is clearly even if m is odd and odd if m is even; putting $m = 2p$ yields $6p\pm1$, the common shorthand form for the series. Note the multiplier 6 appearing in this harmonic series; just as the dc terminal voltage of the bridge is said to have a six-pulse character because it contains only harmonics of order $6p$, so the ac line currents are said to have a six-pulse character because they contain only harmonics of order $6p\pm1$ with relative amplitudes $1/(6p\pm1)$. This latter statement is valid since $\sin[(6p\pm1)\pi/3] = \sin(2p\pi\pm\pi/3) = \pm\sin(\pi/3)$. Extending this principle, a P pulse converter (current-sourced, ac-to-dc/dc-to-ac) is defined as one for which the dependent voltage contains only harmonics of orders mP, m any integer, and where the dependent currents contain only harmonics of orders $mP\pm1$.

The waveform of the classic six-pulse current is depicted in Figure 4-8a. It is not, however, the only wave to exhibit this particular spectrum. The waveform of Figure 4-8b—obtained by taking the difference of two of the waves of Figure 4-8a, one at $-4\pi/3$ rad displacement relative to the other—is also a six-pulse wave. This wave is produced in the lines of a three-phase supply system when a six-pulse (three-phase) bridge is fed via a delta-wye transformer. The change in wave shape is the result of the different phase shifts that the various harmonics undergo as they pass through the transformer. It is generally true that there is no unique P-pulse ac-current wave shape produced by current-sourced converters, although the P-pulse dc dependent-voltage wave is unique. It is also gen-

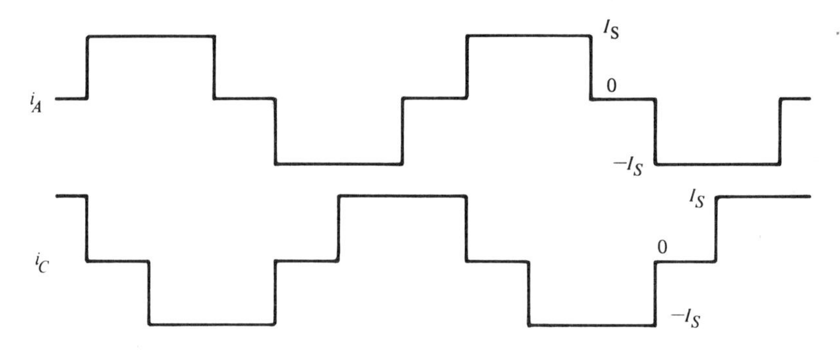

a. Line currents of three-phase bridge.

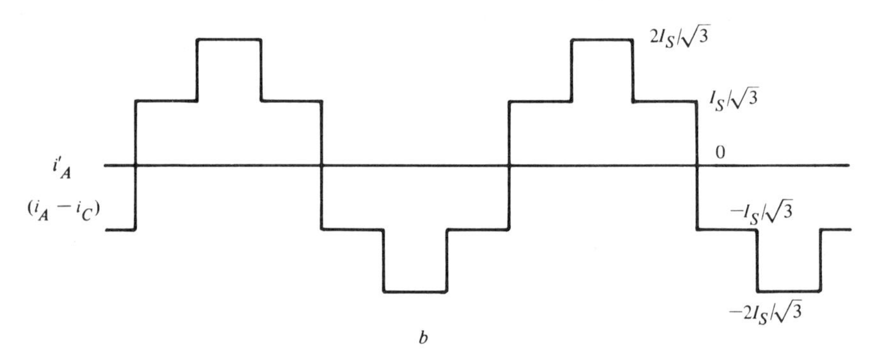

b

Figure 4-8. Line current in three-phase supply for three-phase bridge fed by delta-wye transformer.

erally true that while the pulse number of the dependent dc voltage wave can be identified by a cursory examination of the wave, the pulse number of the current wave cannot be identified from such an examination. In fact, apart from simple waves that become familiar, identification of pulse number by visual inspection is not possible.

The classic six-pulse wave of Figure 4-8a is often termed a *quasi-square wave* in the literature, particularly when it is the voltage wave produced by a voltage-sourced converter (see Section 4.6). Such imprecise nomenclature is to be eschewed.

The dependent-current spectra of all bridges are seen to be improved over those of the corresponding midpoint converters by the elimination of the dc component and all even-order harmonics. The resulting distortion index, defined as it was in deriving Equation 4.27, is

$$\mu_I = [\pi^2/4M\sin^2(\pi/M) - 1]^{1/2} \tag{4.37}$$

As for the midpoint case, this distortion index increases when M becomes large. Thus, in both cases, the dependent-current spectrum is degraded as the dependent-voltage spectrum improves, a highly undesirable situation. However, in both instances, the minimum μ_I occurs when $\tan(\pi/M) = 2\pi/M$. The value of M needed to satisfy this equation is 2.695; the nearest integer (M must be an integer) is 3, and hence the most important practical case is also the one leading to minimum μ_I for both midpoint and bridge converters.

4.3 Phase-Shift-Controlled Current-Sourced Single-Quadrant Bridge and Midpoint Converters

In the preceding discussion, one important internal converter quantity has been ignored. The switch-voltage stresses have been left, deliberately, for it is by examining them that the means for implementing single-quadrant converters, controlled and uncontrolled, may be deduced.

Figure 4-9 shows the voltage waveforms on one switch of the midpoint converter of Figure 4-1 at phase-advance angles corresponding to those of Figures 4-2 and 4-3. Figure 4-10 shows a similar set of waveforms for phase delay. In both cases, the switches are required to block voltage of either polarity when open, just as predicted in Chapter 1, except at $\alpha = 0$ or $\pm\pi$. At $\alpha = 0$, only negative- or inverse-polarity voltage stress is observed; switch-voltage polarity is specified with regard to the direction of current flow in the switch. At $\alpha = \pm\pi$, only forward voltage stress occurs. The same is true for the complementary converter of Figure 4-4a—only inverse voltage stress on the switches at $\alpha = 0$, only forward voltage stress at $\alpha = \pm\pi$.

Thus, if a midpoint converter is implemented with unipolar switches having only inverse voltage-blocking capability, the only possible operating point is $\alpha = 0$. If it is implemented with switches having only forward blocking capability, $\alpha = \pm\pi$ are the only operating points. The latter case is of no practical importance, but the former is; the switches in question are the complementary switches of the dc-to-dc converters in Chapter 3.

We will now examine the implications of the two different unilateral switch characteristics called for at α=0 and α=$\pm\pi$. Even if the diode were not yet known, a switching device with forward current-carrying and reverse voltage-blocking capability would be quite plausible. There is no inconsistency in these capabilities—the device either carries current or blocks voltage according to the circuit conditions of the moment. A unilateral forward voltage-blocking, forward current-carrying device is most implausible. If it does not block reverse voltage, why does it not carry reverse current? Furthermore, what makes it decide whether to block forward voltage or carry forward current, since external circuit conditions will surely favor the flow of forward current when forward voltage is applied? The latter question can be answered by postulating a control electrode and an external decisionmaking agency. The former is unanswerable—if a de-

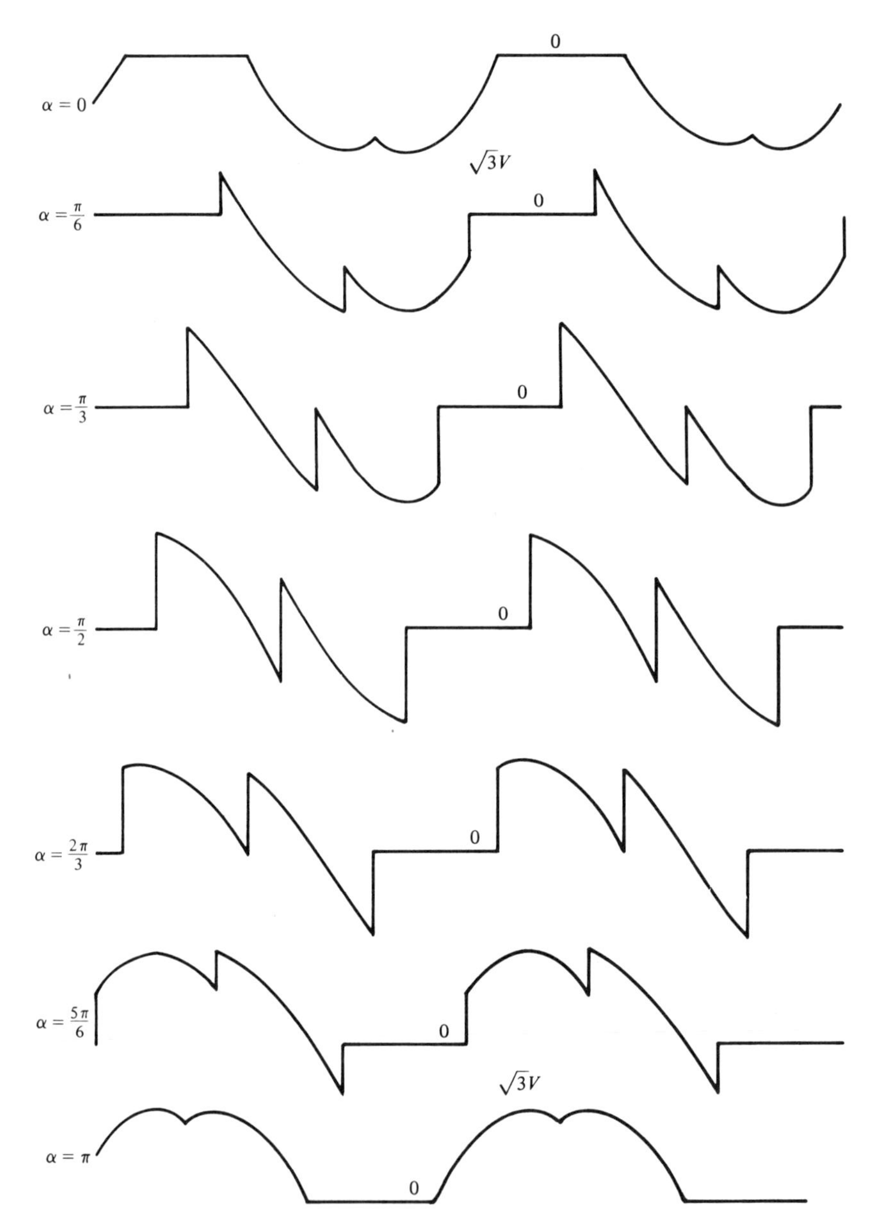

Figure 4-9. Switch voltages, three-phase midpoint converters with α positive.

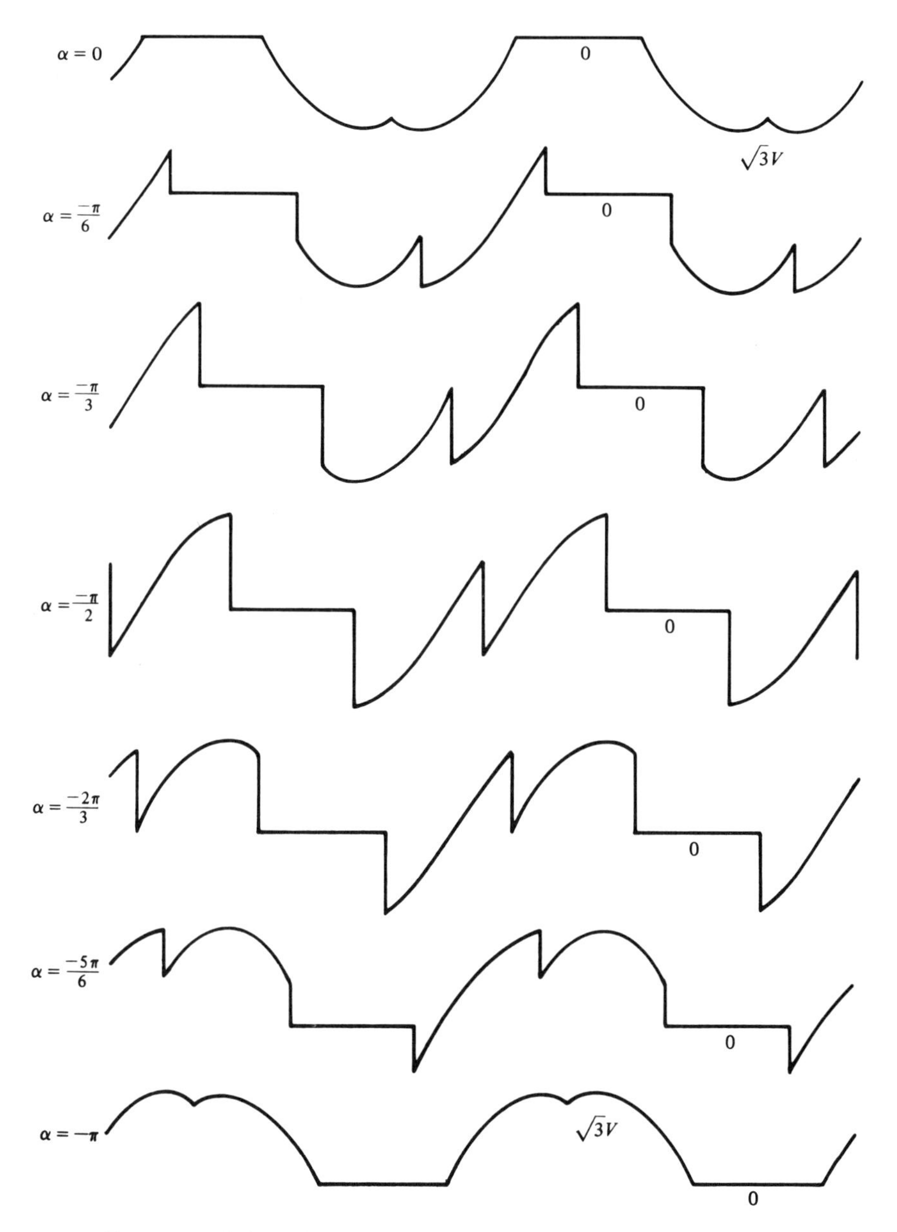

Figure 4-10. Switch voltages, three-phase midpoint converters with α negative.

vice will not support reverse voltage, it must carry reverse current, for an external circuit capable of producing reverse voltage across a switch must surely be capable of driving reverse current through it. Hence, the statement that the case would demand a forward blocking, forward current-carrying switch is of no practical importance; such a device does not and cannot exist. If a device possesses both these qualities, then it must also be capable either of blocking reverse voltage or of carrying reverse current. The only plausible fully unilateral device is that which will carry forward current and support reverse voltage—the common diode was inevitable.

A midpoint converter implemented with such switches will operate automatically at $\alpha = 0$; no control of power flow is possible. Historically, of course, such converters—simple rectifier sets—were the first ever to be implemented. This led to the view that controlled converters, especially the two-quadrant variety, were obtained by extending the capabilities of rectifiers through the introduction of controllable bidirectional voltage-blocking switches. This reasoning might be likened to the mathematical process of deducing the general from the specific. Although the former view was not quite as reprehensible as that process, the converse view presented here—that the rectifier is a special, restricted case of the controlled converter—is more akin to reducing from the general to the specific and, in consequence, may be regarded as more rational.

Obviously, this special, restricted case will have the same dependent-quantity spectra as the controlled converter shows at $\alpha = 0$. In addition, if a complementary converter of the same type is added, an uncontrolled bridge will be created. This will exhibit the same dependent quantities as the corresponding controlled bridge operating at its rectification end stop.

These uncontrolled converters are single-quadrant power processors, since they operate only at the rectification end stop. They have many applications, despite their inability to control power flow or operate in the second quadrant. However, a number of applications call for single-quadrant ac-to-dc converters with the ability to control power flow in their one quadrant of operation. Consider now the performance that is achieved when an uncontrolled midpoint converter is combined with a complementary controlled midpoint converter. The midpoint uncontrolled converter will produce a wanted dc voltage V_{DD}, as defined by Equation 4.7. The controlled converter will produce a wanted component $-V_{DD,\alpha}$, defined by Equation 4.14. The bridge's wanted dependent-voltage component will then be

$$V_{DDBR} = (1+\cos\alpha)V_{DD} \tag{4.38}$$

At $\alpha = 0$, the bridge is at rectification end stop. As $|\alpha|$ is increased, V_{DDBR} declines; for $|\alpha| = \pi/2$ rad, $V_{DDBR} = V_{DD}$ and at $|\alpha| = \pi$ rad $V_{DDBR} = 0$. Thus, this bridge is a controlled single-quadrant converter. Its performance, as regards unwanted components, is considerably degraded from that of the two-quadrant bridge with the same M, as will now be shown. The unwanted components of the uncontrolled half are given by Equation 4.11; those of the controlled half are given, from Equation 4.15 and 4.31, as

$$v_{DUCC} = -V_{DD} \sum_{p=1}^{p=\infty} \cos(p\pi)\cos(pM\pi)\{\cos[Pm\omega_s t+(PM+1)\alpha]/(PM+1) - \cos[PM\omega_s t+(PM-1)\alpha]/PM-1)\} \quad (4.39)$$

The total unwanted components in the dependent voltage of this single-quadrant converter are, of course, given by the difference between the expressions of Equations 4.11 and 4.39. Except at $\alpha = 0$, there will be no elimination of components for p odd when M is odd (i.e., no doubling of the pulse number) because of the phase shift between the two sets of unwanted components comprising the total. At $\alpha = \pm\pi$, since $\cos[(pM\pm 1)\pi] = -\cos(pM\pi)$, the two sets are identical, and the converter produces no unwanted components at the same time as the wanted component is zero. For any other values of α, this single-quadrant converter has unwanted dependent-voltage components of all orders pM regardless of whether M is even or odd. The amplitudes of these components are generally rather less than those of the fully controlled bridge if M is even, but considerably greater if M is odd. The very cumbersome amplitude expression that develops is

$$\begin{aligned} A_{DUBR} = \Big(&2\{1+\cos(pM\pi)\cos[(pM+1)\alpha]\}/(pM+1)^2 \\ &+2\{1+\cos(pM\pi)\cos[(pM-1)\alpha]\}/(pM-1)^2 \\ &-2(1+\cos 2\alpha)/[(pM+1)(pM+1)] \\ &-2\cos(pM\pi)\{\cos[(pM+1)\alpha]+\cos[(pM-1)\alpha]\}/[(pM+1)(pM-1)]\Big)^{1/2} \end{aligned} \quad (4.40)$$

Waveforms for negative α and $M=3$ are shown in Figure 4-11; those for positive α, as in all other cases, are simply mirror images with regard to the zero-time axis, or time-reversed versions, of the waveforms shown.

Turning to the dependent currents, the contribution of the uncontrolled converter is given by Equation 4.22 with $k = 0$. That of the complementary controlled converter is given by Equation 4.34 with α introduced into the arguments of the oscillatory terms. The dc terms still vanish when these contributions are summed, but except at $\alpha = 0$ and $|\alpha| = \pi$, oscillatory terms in general do not. The sum of the two contributions is

$$i_{DBR} = (2I_S/\pi) \sum_{n=1}^{n=\infty} \{\sin(n\pi/M)/n\}\{\cos(n\omega t)-\cos(n\pi)\cos[n(\omega t+\alpha)]\} \quad (4.41)$$

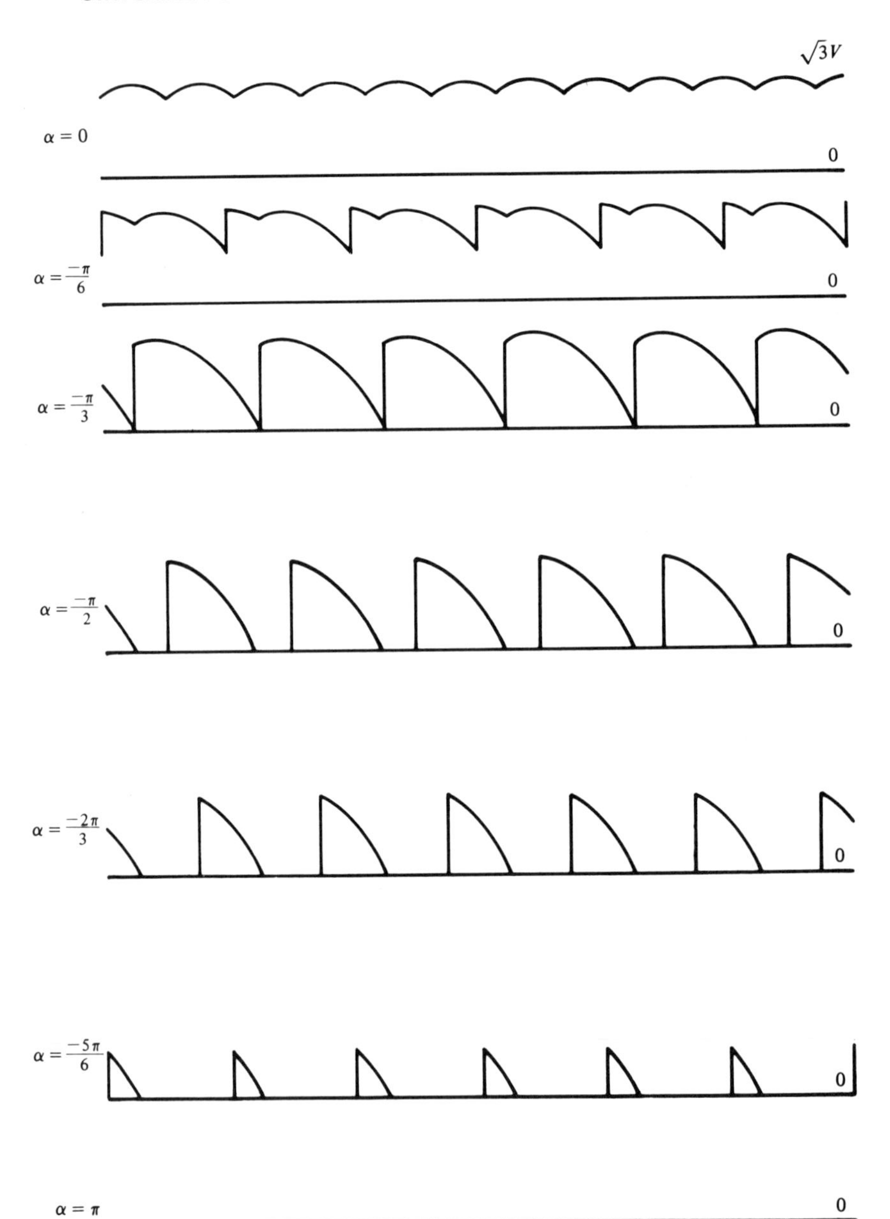

Figure 4-11. DC terminal voltage of three-phase half-controlled bridge (α negative).

For $\alpha = 0$, this is the classic six-pulse current of Figure 4-8a. As $|\alpha|$ increases, however, the wave departs increasingly from this form, as shown in Figure 4-12, with α negative. The periods during which current flows grow progressively shorter and during which, at $|\alpha| = \pi$, no current flows at all. There are no real, quadrature, or harmonic components of current flowing in the defined ac voltage source when this single-quadrant converter is producing no output.

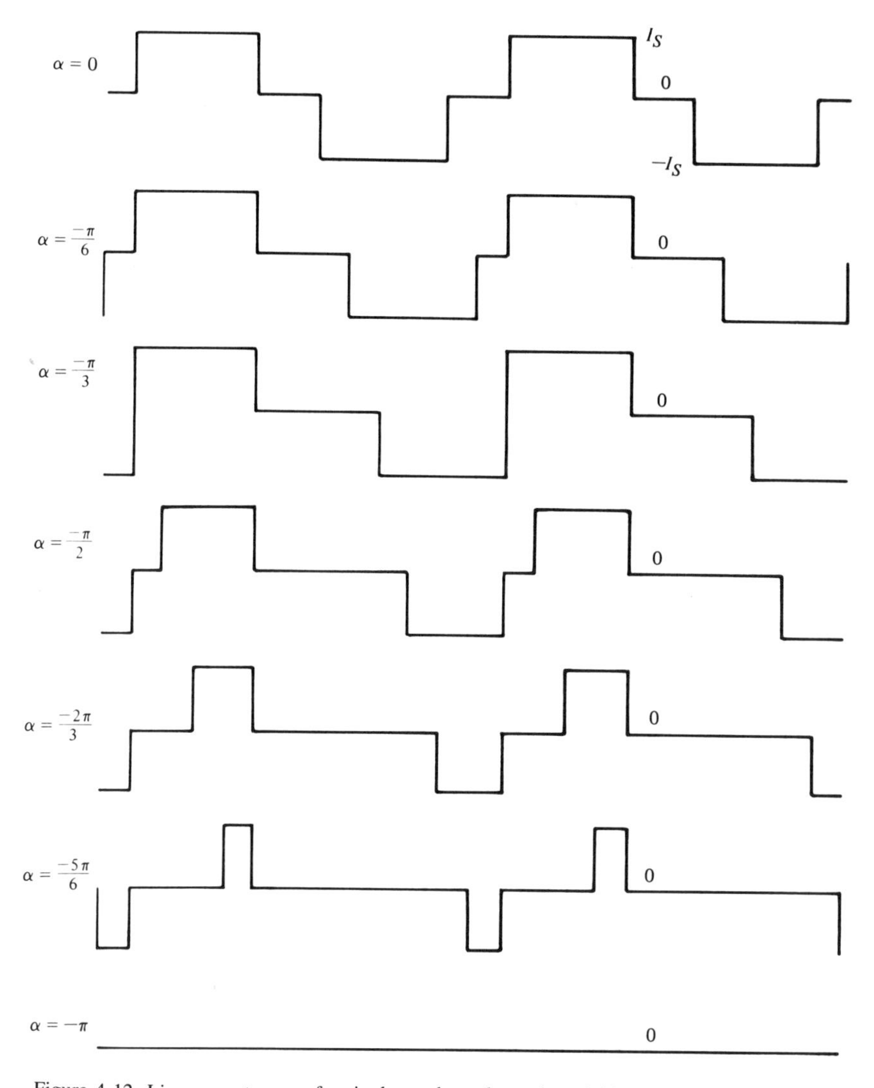

Figure 4-12. Line current waves for single-quadrant three-phase bridge converter (α negative).

Of course, this is in marked contrast to the two-quadrant converters that produce maximum quadrature current and their normal full complement of harmonics when effecting zero-power transfer. However, the single-quadrant converter exhibits dependent-quantity spectra akin to those of the midpoint converter for all operating conditions other than $\alpha = 0$ or $\pm\pi$. Hence, its favorable behavior as zero-power transfer is approached is largely negated, at least as compared to a bridge with M odd, by the spectral degradation that occurs over most of the operating region.

The distortion index for the single-quadrant bridge converter is clearly, from inspection of Figure 4-12, a function of α. For $\alpha = 0$ to $\pm\ (\pi - 2\pi/M)$, there is no shortening of conduction periods in the dependent current, and μ_I is given by

$$\mu_I = [\pi^2/4M\sin^2(\pi/M)\cos^2(\alpha/2) - 1]^{1/2} \tag{4.42}$$

For $|\alpha| > \pi - 2\pi/M$, the conduction periods are shortened, and μ_I is given by

$$\mu_I = [\pi(\pi - \alpha)/8\sin^2(\pi/M)\cos^2(\alpha/2) - 1]^{1/2} \tag{4.43}$$

Equation 4.42 shows that the distortion index, μ_I, steadily increases over that for the corresponding two-quadrant converter, becoming identical to the index for the corresponding midpoint converter at $|\alpha| = \pi/2$. This condition can only be reached, however, for $M \geqslant 4$. Equation 4.43, upon examination, shows that μ_I continues to increase for $|\alpha| > \pi - 2\pi/M$, and in fact has the limiting value ∞ at $|\alpha| = \pi$. Thus, over some portion of its operating range, any single-quadrant bridge converter will exhibit a dependent-current distortion-index exceeding that for the corresponding midpoint converter. These single-quadrant bridge converters, obtained by restricting the capabilities of one-half of the switches in the bridge to reverse voltage-blocking capability only, are often termed *half-controlled bridges* or *semiconverters* in the literature.

It is obviously possible to create an uncontrolled midpoint single-quadrant converter by restricting the voltage-blocking capability of its switches to reverse voltage only. Moreover, its performance will be identical to that of the corresponding two-quadrant converter operating at $\alpha = 0$. It is not possible to create a controlled single-quadrant midpoint converter by switch restrictions; Figures 4-9 and 4-10 show that the switches need bidirectional voltage-blocking capability anywhere in the region $|\alpha| > 0$. However, if the switches do have this capability, then full two-quadrant operation is possible.

It is one of the greatest ironies of the switching power converter field that in order to restrict a controlled midpoint converter to single-quadrant operation, a switch must be added, as shown in Figure 4-13. Less for more—degraded performance as the result of adding a switch—is hardly an attractive technical

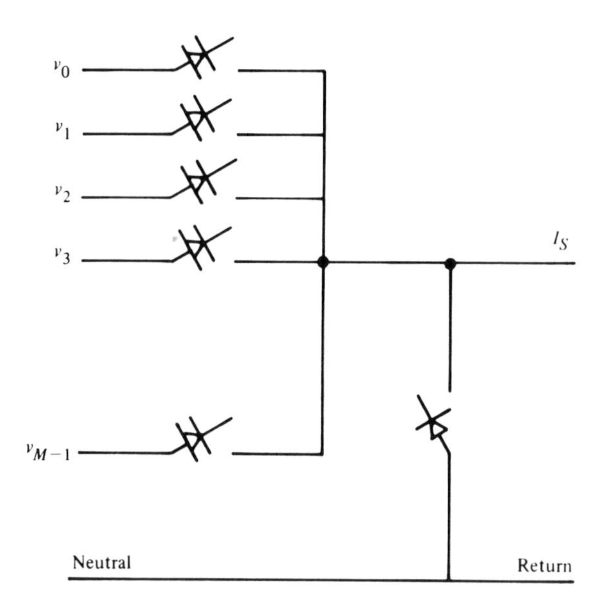

Figure 4-13. Single-quadrant midpoint converter.

or economic proposition. Nonetheless, this version of the "semiconverter" is quite often used because of a single significant fact. If the two-quadrant midpoint converter is restricted to single-quadrant operation through control action, it still operates in the second quadrant for some portion of each operating cycle when $|\alpha| > (M-2)\pi/2M$; the single-quadrant version never enters the second quadrant, since the additional switch prevents it from doing so.

There is a close analogy between the operation of such a converter and that of the buck dc-to-dc converter. The switch added is of the autocomplementary variety, forward current-carrying and reverse voltage-blocking. Control is accomplished by varying the unit-value durations of the existence functions for the active switches. However, in this converter they are automatically shortened to $\pi/2-|\alpha|$ rad by phase control, since the autocomplementary switch will not permit the instantaneous value of the dependent voltage to be less than zero. Thus, for a range of $|\alpha|$ from 0 to $\pi/2-\pi/M$ radians, this single-quadrant midpoint converter behaves exactly like the corresponding two-quadrant converter. However, as $|\alpha|$ increases beyond $\pi/2-\pi/M$, operation changes as depicted in Figure 4-14, showing the dc terminal voltage of such a converter with $M=3$ and α negative. The total range of $|\alpha|$ is 0 to $\pi/2+\pi/M$ rad, for at this latter value the dc terminal voltage and dependent ac current both become zero. In this respect, this converter closely parallels the behavior of the single-quadrant bridge.

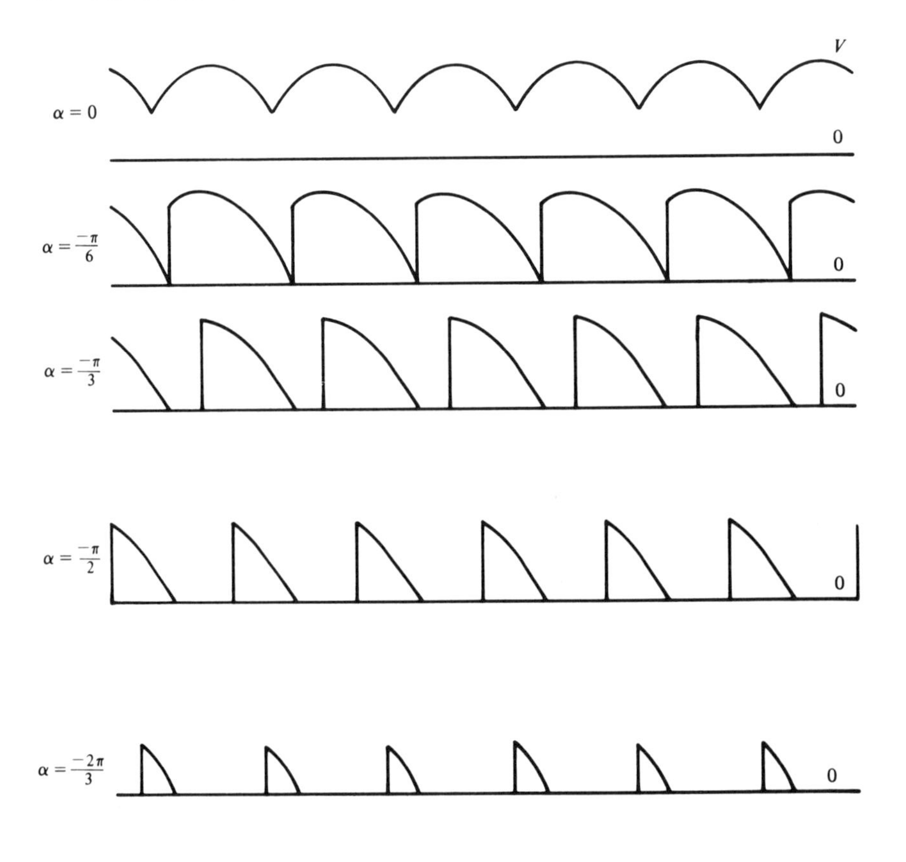

Figure 4-14. DC terminal voltage of three-phase single-quadrant midpoint converter (α negative).

If control is exerted as it is for the two-quadrant version, i.e., by advancing the terminations or delaying the initations of the unit-value periods of the active switch existence functions by $|\alpha|$, according to whether α is positive or negative, then the unit-value period becomes $\pi/2-\alpha+\pi/M$ rad and the effective advance or delay becomes $|\alpha|/2+\pi/4-\pi/2M$ rad. Thus, the existence function for the kth switch can be written as

$$H_k = (\pi/2+\pi/M-|\alpha|)/2\pi+(2/\pi)\sum_{n=1}^{n=\infty}\{\sin[n(\pi/4+\pi/2M-|\alpha|/2)]/n\} \cdot \cos\{n[\omega_s t\pm(\pi/4-\pi/2M+|\alpha|/2)-2k\pi/M]\} \tag{4.44}$$

The dependent dc voltage and unwanted components can be developed as they were for the two-quadrant case. Clearly,

$$\begin{aligned} V_{DD,}\alpha &= (V/2\pi)\sin(\pi/4+\pi/2M-|\alpha|/2)\cos(\pi/4-\pi/2M+|\alpha|/2) \\ &= (V/\pi)[1-\sin(|\alpha|-\pi/M)] \end{aligned} \tag{4.45}$$

The unwanted component expression is, of course, obtained by substituting, in Equations 4.8 and 4.9 $(\pi/4-\pi/2M+|\alpha|/2)$ for $|\alpha|$ and $(pM\pm1)$ $(\pi/4+\pi/2M-|\alpha|/2)$ for $\sin[(pM\pm1)\pi/M]$. The resulting expression, corresponding to Equation 4.11, is extremely cumbersome and will not be given here.

The dependent currents flowing in the defined voltages are, as always for current-sourced midpoint converters, given by simply multiplying the defined current by the existence function. However, there is another dependent current for this converter, namely, that flowing in the autocomplementary switch. This is simply the defined current minus the sum of all the currents flowing in the active switches (and defined voltages).

Thus the dc component of current flowing in the kth defined voltage is

$$I_{DDk} = (\pi/2+\pi/M-|\alpha|)\, I_S/2\pi \tag{4.46}$$

and hence the dc component of current flowing in the autocomplementary switch is

$$I_{DDA} = I_S - M(\pi/2+\pi/M-|\alpha|)\, I_S/2\pi \tag{4.47}$$

Note that $I_{DDk} = 0$ and $I_{DDA} = I_S$ for $|\alpha| = \pi/2+\pi/M$. For $|\alpha| = \pi/2-\pi/M$, the limiting condition at which these equations become valid, $I_{DDA} = 0$ and $I_{DDk} = I_S/M$. The oscillatory components of i_k are given by

$$\begin{aligned} i_{DUk} &= (2I_S/\pi)\sum_{n=1}^{n=\infty} \{\sin[n(\pi/4+\pi/2M-|\alpha|/2)]/n\} \\ &\quad \cdot \cos\{n[\omega_s t\pm(\pi/4-\pi/2M+|\alpha|/2)-2k\pi/M]\} \end{aligned} \tag{4.48}$$

Now the unwanted oscillatory components in the autocomplementary switch must be given by

$$i_{DUA} = \sum_{k=0}^{k=m-1} i_{DUk} \tag{4.49}$$

and it is immediately obvious that the current flowing in that switch contains only components of order pM, p any integer.

Voltage and current waves, for $M=3$ and with α negative, are shown in Figures 4-14 and 4-15 for this converter. Its familial relationships with both the buck dc-to-dc converter and the single-quadrant bridge converter are clearly evident from these synchrograms.

As for the single-quadrant bridge, the distortion index for the dependent ac line current of the single-quadrant midpoint converter is a function of $|\alpha|$. For $0 \leq |\alpha| \leq \pi/2-\pi/M$, of course, it is identical to that for the corresponding two-quadrant converter. With $|\alpha|$ beyond that boundary, Equations 4.46 and 4.48 coupled with the waveforms of Figure 4-15 give μ_I as

$$\mu_I = (\pi\theta/2\sin^2\theta - 1)^{1/2} \tag{4.50}$$

where

$$\theta = \pi/4 + \pi/2M - |\alpha|/2 \tag{4.51}$$

These equations clearly show μ_I increasing as $|\alpha|$ is increased, with a limiting value of ∞ when $|\alpha|$ reaches $\pi/2+\pi/M$.

In contrast to a single-phase version of the two-quadrant converter, the relationship with the buck dc-to-dc converter suggests that a single-phase version of the phase-controlled current-sourced single-quadrant midpoint converter will operate. In fact, it will, and the single-phase version that is sometimes used is governed by the same equations as the polyphase versions, but with M reduced to 1. Obviously, in the single-phase version there is no operating point identical to that for the two-quadrant version—the latter does not exist. In fact, the condition $|\alpha| > \pi/2-\pi/M$ is satisfied for all $|\alpha| > 0$ for $M=2$, in the two-phase version, and for all $|\alpha|$ in the single-phase version of the single-quadrant midpoint converter.

Voltage and current waves are shown in Figure 4-16. An uncontrolled converter, giving only the performance for $\alpha = 0$, is obtained when the active switch is replaced by an autocomplementary switch.

The complete family of current-sourced ac-to-dc and dc-to-ac converters, operating with a smooth, continuous dc terminal current, has now been subjected to analysis. As is discussed in Chapter 6, very few real dc loads or sources are currents of this type, and the converters must be properly interfaced if they are to operate properly. If discontinuous current is allowed at the dc terminals, the dependent quantity spectra will change markedly. These changes come about as a result of two factors:

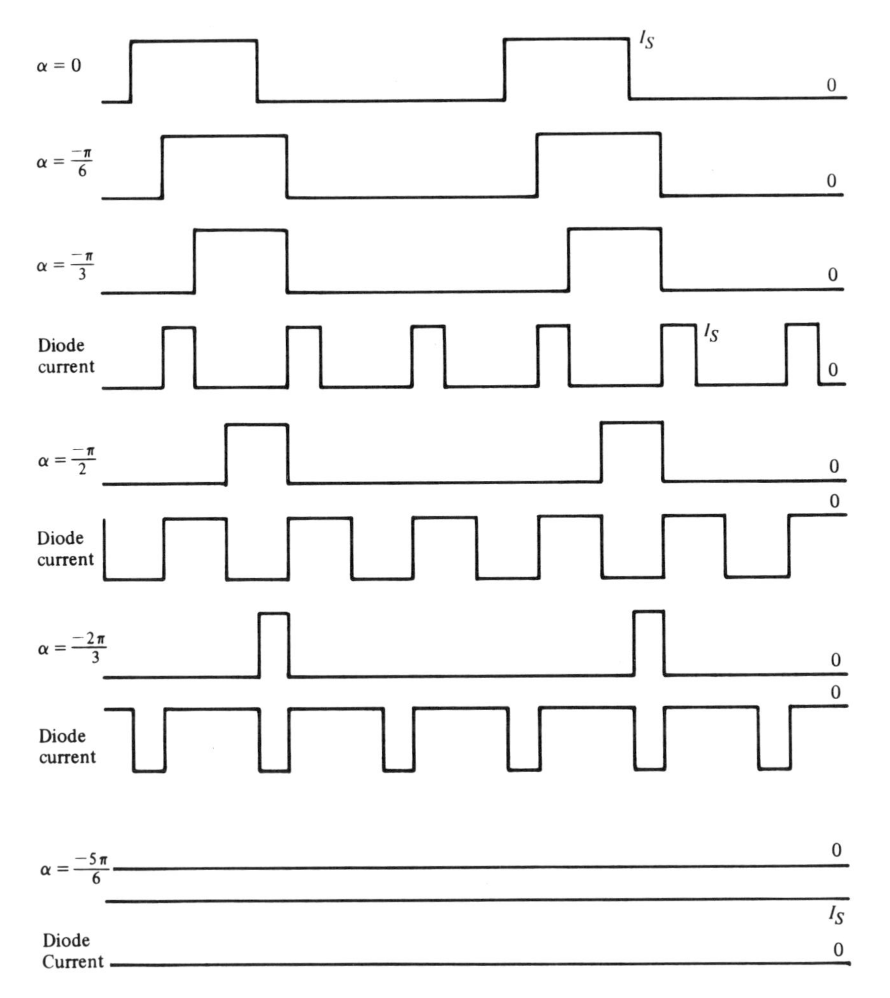

Figure 4-15. Line currents for three-phase single-quadrant midpoint converter (α negative).

1. the introduction of large oscillatory components into the "defined" current, which is now truly the response of the network connected to the converter's dc terminals to the voltage the converter produces,
2. the modification of switch existence functions forced by the character of the defined current

Generally, dependent-quantity spectra are degraded when discontinuous dc terminal current is permitted. Most applications for current-sourced converters call for interfacing with the real dc sources or loads so that the condition does

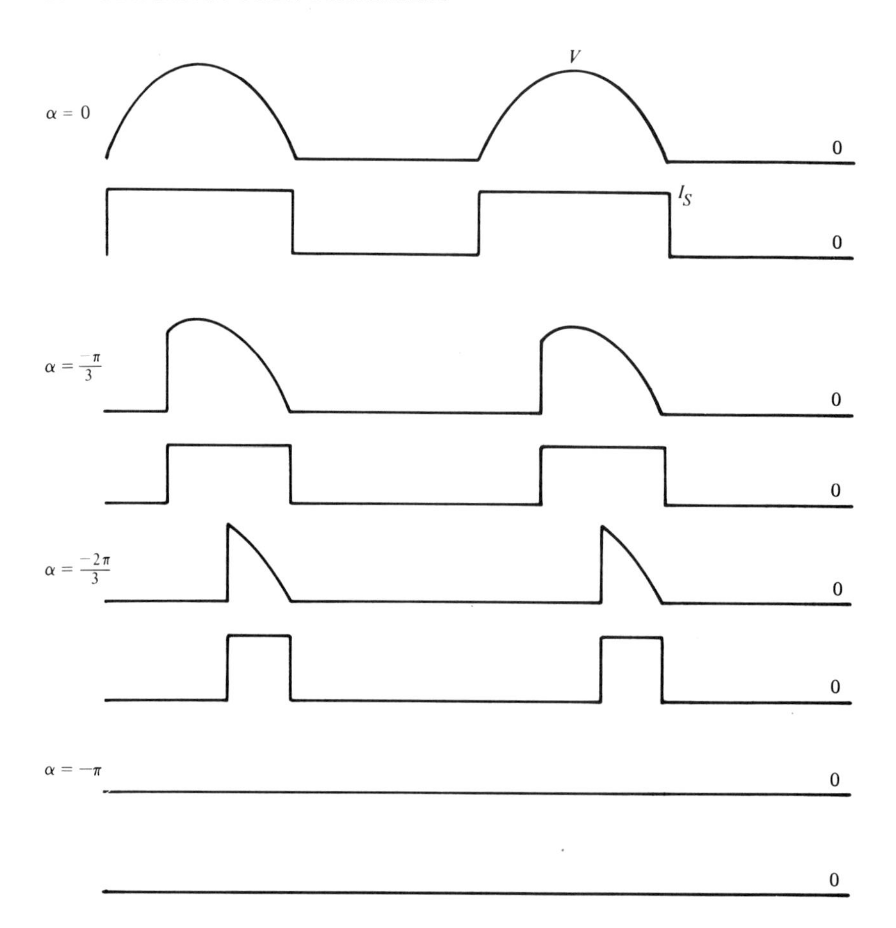

Figure 4-16. DC terminal voltage and line current for single-phase single-quadrant midpoint converter.

not arise over most of the converters' operating ranges. There is, however, one exception. Many uncontrolled single-phase single-quadrant converters of low power rating operate with discontinuous current. This is because poor performance and poor switch utilization can be tolerated, and the interfacing needed for continuous dc terminal current is too expensive. It is rare to find controlled converters operated in the discontinuous current mode; the dc terminal voltage or current transients produced generally lead to unacceptable switch stresses.

All uncontrolled current-sourced ac-to-dc converters are termed *rectifiers* or *rectifier sets*. Midpoint versions are termed *half-wave rectifiers*, and the single-phase version, which must have discontinuous current to operate, is of considerable commercial importance. In most realizations, it takes the form depicted in Figure 4-17a. There, Z_S is the source impedance, a universal characteristic of practical defined ac voltages. It is mandatory, of course, that Z_S exist for this configuration to operate without generating very high peak currents in the ac voltage source and the capacitor. Detailed analysis is postponed until Chapter 6, since it fits more naturally into the framework of a discussion on practical converter interfacing.

4.4 Switch Stresses in Current-Sourced AC-to-DC/DC-to-AC Converters

The nature of the switch stresses arising in all the converters so far analyzed is easily deduced from the synchrograms. Switch current waves universally exhibit the same basic character. Switch peak current is equal to the defined dc current, I_S, switch rms current equals $I_S/\sqrt{M}$, and switch average current is I_S/M for all two-quadrant converters and the single-quadrant bridge converters. The peak current is still I_S, but the rms and average currents become dependent on α for the single-quadrant midpoint converters once $|\alpha| > \pi/2-\pi/M$. They are then given by $I_S(1/4+1/2M-|\alpha|/2\pi)^{1/2}$ and $I_S(1/4+1/2M—|\alpha|/2\pi)$, respectively. One notable feature of all switch current waves is the instantaneous transitions that occur as the switches close and open. These indicate that an infinite $\partial I/\partial t$ capability is required of the switches. In practice, source impedance in the defined voltages mitigates the transitions and $\partial I/\partial t$ requirements somewhat, but converters operating with low-impedance ac sources are likely to have problems in this regard, as discussed in Chapter 8.

The voltage waves of Figures 4-9 and 4-10 apply also to all two-quadrant converters and single-quadrant bridge converters. For the single-quadrant midpoint converters operating with $|\alpha|$ greater than $\pi/2-\pi/M$, switch voltages become simply the defined ac voltages to which the switches are connected for

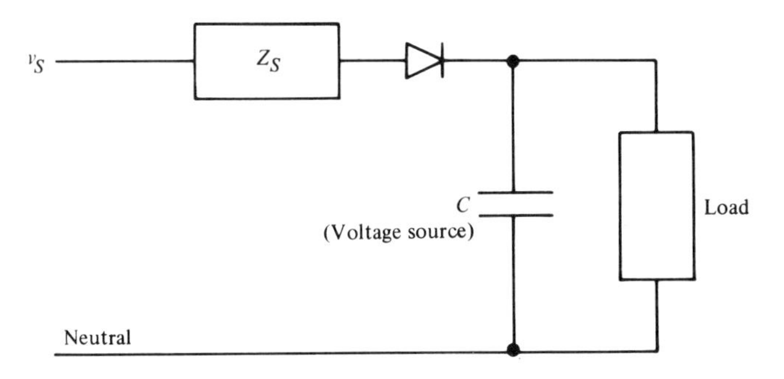

Figure 4-17. Single-phase half-wave rectifier.

those intervals when the dc terminal voltage is zero and remain the same as the appropriate instantaneous values in Figures 4-9 and 4-10 in the intervals when it is not.

Two general observations can be made. First, the peak voltages seen by switches are invariably equal to the peak line-to-line (hereinafter termed *line*) voltages of the defined ac-voltage set. If the peak line-to-neutral (hereinafter termed *phase*) voltage of the set is V, then the peak line voltage is $2V\sin(\pi/M)$. This will be the peak stress, forward and reverse, in two-quadrant converters and single-quadrant bridge converters. In single-quadrant midpoint converters, it is the peak reverse voltage stress; the peak forward voltage stress, and the maximum reverse voltage stress on the autocomplementary switch, are simply V.

Second, with α positive, the rapid transitions in switch voltages as switches open and close are all in the positive direction across a switch. Thus, for any α other than zero and $\pm\pi$, a switch is required to withstand an instantaneous (in the idealized case) forward voltage immediately after opening and is required to close despite having reverse voltage applied. For negative α, the converses are true—immediately after opening, an instantaneous reverse voltage is applied, while immediately prior to closing the switches are blocking forward voltage. These factors are of immense practical importance, as will be seen in Chapter 8.

4.5 Sequential Control of Midpoint Converters in a Bridge

In considering the single-quadrant bridge, it was observed that the dependent currents fell to zero at $\alpha = \pm\pi$. From Equation 4.41, the fundamental, wanted, component of the dependent current of such a converter is given by

$$i_{DDBR1} = (2I_S/\pi)\sin(\pi/M)[\cos(\omega_s t)+\cos(\omega_s t+\alpha_1)] \tag{4.52}$$

and that of a two-quadrant bridge is, from Equation 4.35,

$$i_{DDBR2} = (4I_S/\pi)\sin(\pi/M)\cos(\omega_s t+\alpha_2) \tag{4.53}$$

Examining Equations 4.52 and 4.53, it is seen that the quadrature current demand of the single-quadrant bridge is

$$I_{Q1BR} = -(2I_S/\pi)\sin(\pi/M)\sin\alpha_1 \tag{4.54}$$

while that of the two-quadrant bridge is

$$I_{Q2BR} = -(4I_S/\pi)\sin(\pi/M)\sin\alpha_2 \tag{4.55}$$

Now, in the first quadrant, the wanted component of the dependent voltage for the single-quadrant converter is given, from Equation 4.38, by

$$V_{DDBR1} = (1+\cos\alpha_1)V_{DD} \tag{4.56}$$

while that for the two-quadrant bridge is, from Equation 4.14,

$$V_{DDBR2} = 2V_{DD}\cos\alpha_2 \tag{4.57}$$

From Equations 4.54 and 4.56, it is clear that the maximum quadrature current demand of the single-quadrant bridge is $-(2I_S/\pi)\sin(\pi/M)$ when $|\alpha_1|$ is $\pi/2$ and $V_{DDBR1} = V_{DD}$. From Equations 4.55 and 4.57, the maximum quadrature current demand of the two-quadrant bridge is $(-4I_S/\pi)\sin(\pi/M)$ when $|\alpha_2|$ is $\pi/2$ and $V_{DDBR2} = 0$. In order to reduce V_{DDBR1} to 0, $|\alpha_1|$ must be increased to π, but the quadrature current demand will fall to zero as this is done. Hence, over the first quadrant the maximum quadrature current demand of the single-quadrant bridge is only one-half of that for the two-quadrant bridge.

Now, suppose the single-quadrant bridge did in fact have two-quadrant capability but that it had been operated over the first quadrant with one of its constituent midpoint converters held, by control and intent, at the rectification end stop. If, as the midpoint allowed control reached $\alpha = \pm\pi$ it was held there, at inversion end stop, and that which previously was held was allowed to vary its α, α_1', for example, then the bridge as a whole would be controllable through the inversion quadrant. While traversing that quadrant, it would clearly again give rise to a maximum quadrature current demand of $-(2I_S/\pi)\sin(\pi/M)$ at $\alpha_1' = \pi/2$.

Thus, by sequentially controlling its two constituent midpoint converters, a two-quadrant bridge will operate over its full range while developing a maximum of only one-half of the maximum quadrature current demand it generates when its two midpoint converters are concurrently controlled. The price paid for this benefit is, of course, the degradation of dependent-quantity spectra. Nonetheless, this mode of control is fairly often used, for the quadrature current demands of the phase-controlled converters are one of their biggest disadvantages.

Although not strictly related to sequential control, another option capable of reducing (in fact, eliminating) quadrature current demand is worth mentioning. Suppose two phase-shift-controlled current-sourced ac-to-dc/dc-to-ac converters are connected effectively in parallel at their dc and ac terminals, but one is operated with positive α and the other with negative α while they share the defined dc current equally. The net quadrature current demand of such an arrangement, which is discussed in more detail in Chapter 9, is zero. The dependent quantity spectra, of course, are similar to those of the constituent converters.

4.6 Harmonic Neutralization in Phase-Shift-Controlled Current-Sourced AC-to-DC and DC-to-AC Converters

It was seen in Chapter 3 that if a number of phase-staggered dc-to-dc converters are used, then considerable improvements in dependent-quantity spectra can be obtained when their outputs and inputs are combined appropriately. It is there commented that this basic technique can be applied to all converters, since unwanted components are for the most part oscillatory and thus subject to cancellation (or neutralization) by creating the proper phasor summations.

It has been shown that in their dependent dc voltages, current-sourced ac-to-dc and dc-to-ac converters have unwanted components that are harmonics of the supply frequency of order mP, where m is any integer, 1 to ∞, and P the converter pulse number. P equals the number of phases in the defined voltage set for all converters encountered so far except for the two-quadrant bridge with an odd number of phases, when P equals twice the number of phases, as was seen. The mechanism of harmonic neutralization at the dependent-voltage terminals of these converters, as with dc-to-dc converters, is to operate a number so that their wanted components are identical but their unwanted components form complete phasor sets. They are then combined by summing (series connecting) or summing and averaging (parallel connecting via an interphase reactor) their dependent voltages. Series connection is generally possible with ac-to-dc and dc-to-ac converters because their defined ac voltage sets can be supplied through isolated transformer windings.

Now a complete phasor set of ac quantities is in general formed when ℓ such quantities are set with phase displacements between consecutive members of $2\pi/\ell$ rad. However, in the case under consideration, the phase displacements must be obtained by phase-displacing the defined ac voltage sets feeding the converters and the existence functions controlling their switches. Since all displacements of the unwanted components will be multiplied by mP, as supply frequency is to produce these components, the phase displacements between consecutive sets of defined voltages and existence functions must be $2\pi/\ell P$ rad. If this is so, then the mPth order unwanted component of the kth member of the ℓ converters, $k = 0$ to $\ell - 1$, will have a displacement with respect to reference (the same unwanted component of the 0th member) of $2mk\pi/\ell$. The resulting unwanted component in the summed dependent quantities will take the form

$$v_{USUM} = C_{mP} \sum_{k=0}^{k=\ell-1} \cos\,(mP\omega_s t + \phi_{mP} - 2mk\pi/\ell) \tag{4.58}$$

where C_{mP} is the coefficient appropriate to each of the ℓ converters and ϕ_{mP} is the phase displacement, due to α, appropriate to each of the ℓ converters. The summation of Equation 4.58 is clearly recognizable as that of a complete phasor set with all displacements multiplied by m. As such, it must be zero except when m is an integer multiple of ℓ, and thus the combined dependent quantities will contain only those unwanted components that are harmonics of order $q\ell P$. The effective pulse number is the pulse number of the constituent converters multiplied by ℓ, the number of converters combined, if the phase displacement between the voltages and existence functions of any two consecutive converters is made $2\pi/\ell P$ rad.

Series combination is quite natural provided that all the ac voltage sets feeding the converters are transformer-isolated. Interphase reactors must be used for parallel combination. This can be achieved with nonisolating phase-shifting transformers providing the voltage sets, but a problem arises when the converters in such a case are bridges. If there is no isolation between the ac-voltage sets of two or more bridges to be combined in parallel, then interphase reactors must be provided in both the positive- and negative-defined current connections. Moreover, these reactors will have to support the ripple voltage differences of the bridges' constituent midpoint groups, not just the ripple voltage differences of the bridge voltages. Thus, it is usually disadvantageous not to isolate the ac voltages for bridges that are to be combined.

Now what of the dependent ac currents? It was shown previously that increasing the number of phases, M, improves the dependent-voltage spectra while degrading the dependent-current spectra for all the current-sourced converters. It has just been seen that rather dramatic improvements in dependent-voltage spectra can be achieved by combining a number of appropriately phase-staggered converters even if the number of phases feeding each is low. Hence, by choosing $M = 3$, which gives rise to the best possible individual converter current spectrum, a very high dc-terminal pulse number can be achieved by using the harmonic neutralization technique heretofore described.

This is fortunate, because three is usually exactly the number of supply phases available. However, the spectra of three-pulse (three-phase midpoint converter) and six-pulse (three-phase bridge converter) dependent currents still leave a great deal to be desired. Is it possible that the currents too can be improved by the harmonic neutralization process?

The answer is yes, provided that all the ac-voltage sets are derived from a single three-phase set by phase-shifting transformers, so that the currents from the constituent converters sum, after being phase-shifted by the transformers, in that three-phase supply. The mechanism for neutralization then depends on three properties that phase-shifting transformers possess. The first two are quite obvious:

1. If a voltage applied to the primary of a transformer appears at the secondary phase, shifted by some angle ϕ as a result of transformer action, then any load

current flowing in the primary also acquires a phase shift ϕ when observed in the secondary. This is true because a transformer cannot generate real or quadrature current demand (other than magnetizing current)—it cannot supply or absorb either real or reactive power,

2. If a voltage or load current of any frequency is phase-shifted by ϕ rad in the secondary as compared to the primary of any transformer, then it is phase-shifted by $-\phi$ rad in the primary as compared to the secondary. This is so obvious that it requires no further comment.

The third property is not so obvious, but is nonetheless a universal property of all phase-shifting transformers, whatever the number of exciting phases. Because of the very limited practical importance of transformers and ac systems other than three-phase, only that case will be dealt with here. In general, the property may be stated as:

3. The direction and amount of phase shift for a given phasor set passing through a given transformer depend on the sequence of that phasor set. For a three-phase system, this reduces to the effect that positive and negative sequence components are phase-shifted by equal angles but in opposite directions.

The term *sequence* used in (3) refers to the designation used to describe the symmetrical components of polyphase systems, which are met in Chapter 2. In general, an M phase system possesses M symmetrical components, each a complete phasor set but with phase displacements between consecutive phasors of $2\pi/M$, $4\pi/M$, $6\pi/M$ . . . $2M\pi/M$. Viewed as sets of sinusoids, these phasor sets show differing sequences.

If all sets are labeled A, B, C . . . ,etc., corresponding phasors being the 0th, 1st, 2nd, etc. with phase displacements from reference of 0, $-2p\pi/M$, $-4p\pi/M$, $-6p\pi/M$, etc. and p the integer defining the set, 1, 2, 3 . . . M, then for $p = 1$ the time sequence of positive peaks (or any other specific points on the waves) of the set of sinusoids is A, B, C . . . etc. As p changes, so does this sequence; for $p = M$, all phasors are at zero displacement and there is no sequence.

For a three-phase system, there are only three symmetrical components. They are the positive sequence, ABC, with displacements of $2\pi/3$ rad; the negative sequence, ACB, with displacements $4\pi/3$ rad; and the zero sequence (all three phasors coincide) with displacements $2\pi(\equiv 0)$ rad. The zero sequence component is of no concern here—in fact, it cannot exist in a three-phase three-wire system since it would violate Kirchoff's laws therein.

For the positive sequence of a three-phase system, the phasors A_1, B_1, and C_1 are clearly at displacements of 0, $-2\pi/3$, and $-4\pi/3$ rad, respectively. The negative sequence set A_2, B_2, C_2 are at displacements of 0, $-4\pi/3$, and $-8\pi/$

3 rad, respectively, but since $8\pi/3 = 2\pi + 2\pi/3$, they are at 0, $-4\pi/3$, and $-2\pi/3$ rad. In describing these symmetrical components, it is convenient (and conventional) to define the vector operator a by

$$a = \epsilon^{j2\pi/3} = \cos 2\pi/3 + j\sin 2\pi/3$$

$$= \epsilon^{-j4\pi/3} = \cos 4\pi/3 - j\sin 4\pi/3 \qquad (4.59)$$

Thus,

$$a^2 = \epsilon^{j4\pi/3} = \epsilon^{-j2\pi/3}$$

Now the positive sequence set may be defined as V_1 [$= V_p \cos(\omega_s t)$], a^2V_1 and aV_1; the negative sequence set become V_2 ($=V_n \cos(\omega_s t)$), aV_2 and a^2V_2. These definitions are, of course, quite independent of the frequency of the phasors.

With this background, attention now turns to the manner in which phase-shifting three-phase transformers may be constructed. The most common is the wye-delta (or delta-wye) in which one set of windings is wye-connected—i.e., the three windings, one on each limb, have all finishes (or starts) connected together, and their starts (or finishes) are connected to the three exciting voltages (or loads). The other winding is delta-connected, the three windings are connected in "series"—finish 1 to start 2, finish 2 to start 3, finish 3 to start 1 (or with finishes and starts reversed). They are excited (or loaded) at the three connection points. Thus, the wye voltages of the primary become the delta voltages of the secondary—i.e., A_p becomes AB_s, B_p becomes BC_s, C_p becomes CA_s. The phasor diagram of Figure 4-18a shows that for a positive sequence set, the delta (line) voltages lead the wye (phase) voltages by $\pi/6$ rad. Hence, if the wye is the primary excited by a positive sequence set, the induced secondary delta set is in phase therewith, and thus the corresponding secondary wye set must lag by $\pi/6$ rad. Figure 4-18b shows that for a negative sequence set, the delta voltages lag the wye voltages by $\pi/6$ rad. Hence, if the wye primary of the transformer is excited by a negative sequence set, the corresponding wye secondary voltages will lead by $\pi/6$ rad.

Clearly, from the second transformer property listed, the converses are true if the delta windings are excited. Positive sequence excitation will then produce a secondary wye set leading by $\pi/6$ rad; negative sequence excitation will produce a secondary set lagging by $\pi/6$ rad.

Thus, for this common transformer at least, the third listed property is verified. That it is true of all three-phase phase-shifting transformers may be ascertained by first considering how such transformers generally create the phase-shifted sets of secondary voltages. They do so by adding or subtracting induced voltages from two of the exciting primary phases. The "forked-wye" secondary

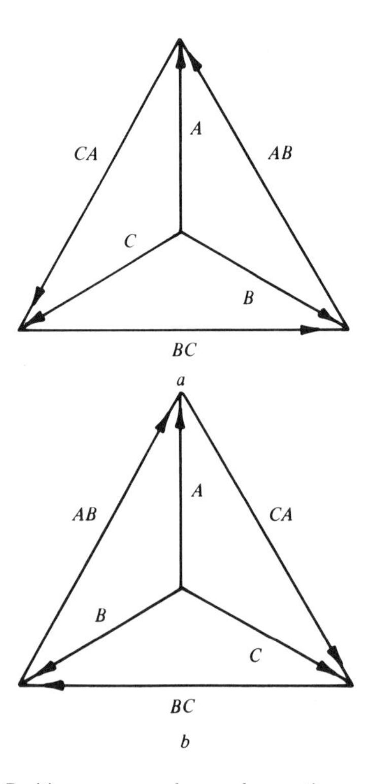

Figure 4-18. *a*, Positive-sequence phasors; *b*, negative-sequence phasors.

of Figure 4-19, for example, produces the leading A', B', and C' phasors by subtracting induced B, C, and A voltages from the induced A, B, and C voltages by the winding connections shown. It produces the lagging A'', B'', and C'' phasors by subtracting induced C, A, and B voltages from the induced A, B, and C voltages. The amount of phase shift produced depends of course, on the magnitudes of the subtrahends relative to that to the minuends—the greater the former, the greater the phase shift.

Similar arguments apply to the ''closed polygon'' and ''extended delta'' secondaries depicted in Figures 4-20 and 4-21. With these, the delta voltages are being manipulated, but each phase-shifted set is again the result of adding or subtracting induced voltages from two of the primary phases. This is also seen to be true of the delta-wye transformer of Figure 4-22 when the secondary delta voltages are considered. This adding and subtracting of induced voltages is the simplest and most economical method, and therefore the universal one, of producing phase-shifted voltage sets.

The phase-shifted voltages can be represented, of course, as phasor sums or differences. Referring again to Figure 4-19, the expressions

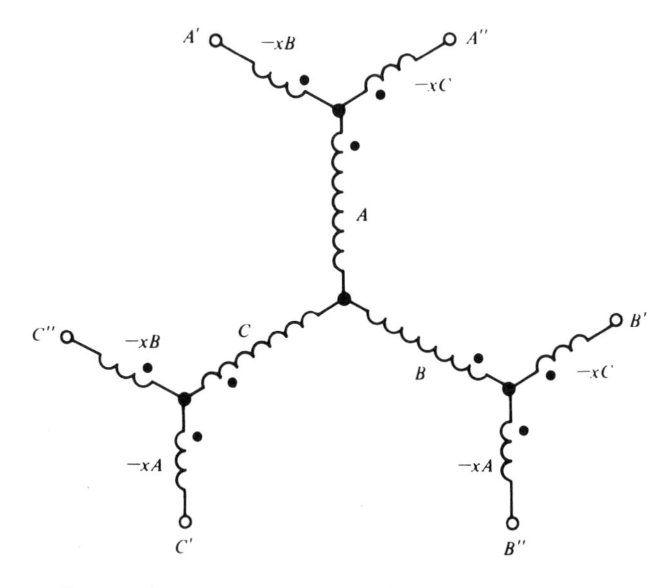

Figure 4-19. "Forked wye" phase-shifting transformer secondary.

$$A' = A - xB$$

$$B' = B - xC$$

$$C' = C - xA$$

$$A'' = A - xC$$

$$B'' = B - xA$$

$$C'' = C - xB$$

can be written, where x is the ratio of the number of turns on the "fork" windings to the number of turns on the wye windings. If the exciting set A_1, B_1, and C_1 are a positive sequence set, V, a^2V, and aV, then the normalized (divided by V) secondary sets may be written as

$$A_1' = 1 - xa^2$$

$$B_1' = a^2 - xa$$

$$C_1' = a - x$$

$$A_1'' = 1 - xa$$

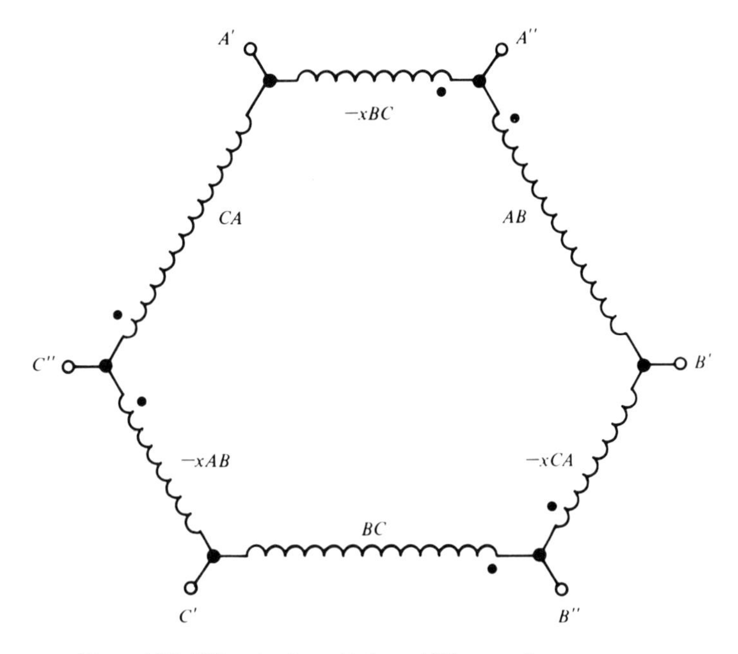

Figure 4-20. "Closed polygon" phase-shifting transformer secondary.

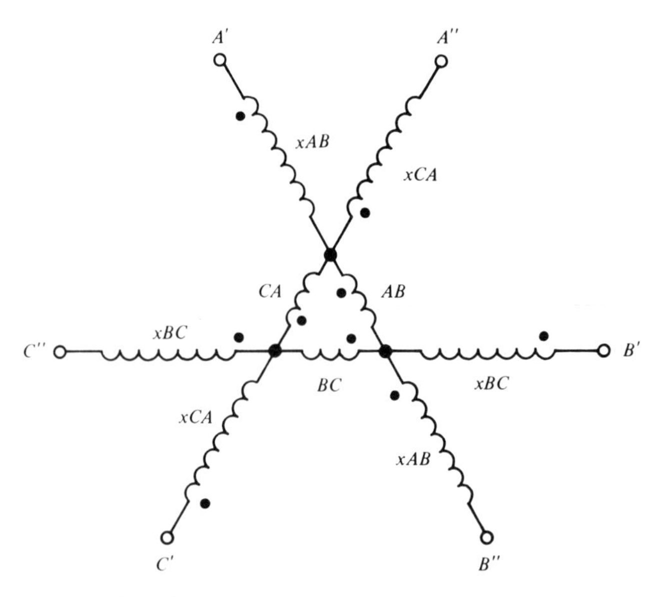

Figure 4-21. "Extended delta" phase-shifting transformer secondary.

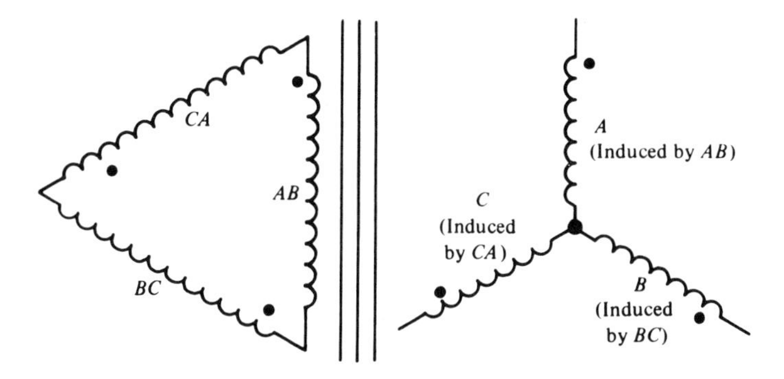

Figure 4-22. Delta-wye transformer.

$$B_1'' = a^2 - x$$

$$C_1'' = a - xa^2$$

Now it is clear from Figure 4-19 that the set A_1', B_1', and C_1' leads the reference set A_1, B_1, and C_1 by some angle that is a function of x—in fact, by the solution of $x = \sin\theta/\sin(\pi/3-\theta)$—and the set A_1'', B_1'' and C_1'' lags by the same angle.

Now suppose the exciting set is made negative sequence, so that A_2, B_2, and C_2 may be represented, when normalized, by 1, a, and a^2. Then the phase-shifted sets can be written as

$$A_2' = 1 - xa$$

$$B_2' = a - xa^2$$

$$C_2' = a^2 - x$$

$$A_2'' = 1 - xa^2$$

$$B_2'' = a - x$$

$$C_2'' = a^2 - xa$$

Observe the relationships of the A phasors of the phase-shifted sets. $A_2'' = A_1'$ and therefore leads A_2; $A_2' = A_1''$ and therefore lags A_2. System symmetry suggests that what is true for the A phasors must also be true for the B and C phasors. This can easily be shown true if multiplication by appropriate vector operators (shifting the phasors to reference position for comparison) is performed. Then

$$a^2B_2' = 1 - xa = aB_1''$$

$$aC_2' = 1 - xa = a^2C_1''$$

$$a^2B_2'' = 1 - xa^2 = aB_1'$$

$$aC_2'' = 1 - xa^2 = a^2C_1'$$

Thus, it is clear that the transformer imparts phase shifts of equal magnitudes but opposite signs to the positive and negative sequence sets. Since all the phase-shifting transformers fabricate their phase-shifted secondary sets in the same manner, this must be true of them all—the third listed property is universal.

Let us apply all of this to converter currents. Suppose again we have ℓ three-pulse (midpoint) converters to feed with ℓ phase-shifted three-phase voltage sets with consecutive displacements of $2\pi/3\ell$ rad. For maximum economy, the secondary system will normally be made symmetrical (except for the special case $\ell = 2$) leading to two cases. If ℓ is even, i.e., $\ell = 2m$, then the phase-shifted sets of secondary voltages will be produced in pairs at $\pm\ \pi/3\ell,\ \pm\ 3\pi/3\ell\ \ldots$ $\pm\ (2m-1)\pi/3\ell$. For ℓ odd, $\ell = 2m+1$, symmetry is achieved by having one set of secondary voltages not shifted and the remainder in pairs at $\pm\ 2\pi/3\ell$, $\pm 4\pi/3\ell\ \ldots\ \pm 2m\pi/3\ell$.

The dc components of the three pulse currents will not be induced in the common three-phase primary (or primaries) and will not exert magnetizing force on the three-limb core (or cores) because they flow in all windings on all limbs and the algebraic sum of the dc ampere turns is zero. The fundamental (wanted) components of the currents all flow at the phase displacement of their voltage sets plus the displacement due to α, the phase-control angle. Hence, by virtue of the second property listed for phase-shifting transformers, all the fundamental current components sum in the primary windings, and this sum is displaced from the primary voltage set by the angle α. This has to be so, in view also of the first property listed for the transformers. What of the harmonics in the three pulse waves? They are all those of order $3k+1$ and $3k-1$, as seen in Section 4.1. Consider now their displacements in any one of three pulse waves, ignoring displacement due to α since it is common to all. For the A phase of any one midpoint, these displacements are

$$(3k+1)\ 0 = 0$$

$$(3k-1)\ 0 = 0$$

For the B phase, they are

$$(3k+1)\,(-2\pi/3) = -2k\pi - 2\pi/3 \equiv -2\pi/3$$

$$(3k-1)\,(-2\pi/3) = -2k\pi+2\pi/3 \equiv 2\pi/3$$

For the C phase, they are

$$(3k+1)\,(-4\pi/3) = -4k\pi-4\pi/3 \equiv -4\pi/3$$

$$(3k-1)\,(-4\pi/3) = -4k\pi+4\pi/3 \equiv 4\pi/3$$

Thus, all $3k+1$ components are positive sequence, and all $3k-1$ components are negative sequence. From second property listed and the third property listed and proved, all $3k+1$ components from a given converter will be phase-shifted by the negative of the phase shift of the converter's voltage set, as they are transferred from the secondary to the primary. All $3k-1$ components will be phase-shifted an equal amount but in opposite direction, i.e., by the phase shift of the converter's voltage set.

For $\ell = 2m$, consider the converters at a phase displacement of $\pm(2n-1)\pi/3\ell$, where $n = 1$ to m. The displacements of the $3k+1$ components of the A phase (which is all that need be considered) are

$$\pm(3k+1)\,(2n-1)\pi/3\ell = \pm k(2n-1)\pi/\ell \pm(2n-1)\pi/3\ell$$

However, on transfer from secondary to primary, they are subjected to a phase shift of $\mp(2n-1)\pi/3\ell$. Thus, their phase displacement in the primary is $\pm k(2n-1)\pi/\ell$. Similarly, the $3k-1$ components have phase displacements at the secondary of

$$\pm(3k-1)\,(2n-1)\pi/3\ell = \pm k(2n-1)\pi/\ell \mp (2n-1)\pi/3\ell$$

However, in transferring from primary to secondary, they are subjected to a phase shift of $\pm(2n-1)\pi/3\ell$, so that their phase displacement in the primary is also $\pm k(2n-1)\pi/\ell$. The resulting primary current is the sum of the ℓ phasors at these displacements, over the range of n. Observe that all have the displacement $\pm\, k\pi/\ell$, which may therefore be ignored—the key is the term $\pm k\; 2n\pi/\ell$, where $n = 1$ to m and $2m=\ell$. A summation of ℓ phasors with these displacements is clearly the sum of a complete phasor set with displacements multiplied by k; therefore, it is zero except for those instances where k is an integer multiple of ℓ, when, of course, it is ℓ times any one of its individual phasors. Thus, the primary current will contain only harmonics of order $3q\ell\pm1$, where q is any integer, or $qP\pm1$ where P is the pulse number previously defined in relation to the combined dc-terminal voltage of the converters, which contains only harmonics of order qP. The same harmonic neutralization technique also works in the primary currents. It is due entirely to the third listed property of the transformers, for that property produces the offsetting phase shift to reduce the phase displacements in the primary to those of complete phasor sets.

For ℓ odd (i.e., $\ell = 2m \pm 1$), consider the converters with voltage sets at $\pm 2n\pi/3\ell$, $n = 1$ to m. The displacements at the secondary of the A phase $3k+1$ components are $\pm\ 2n\pi k/\ell \pm 2\ n\pi/3\ell$. On transferring to the primary, they are phase-shifted by $\mp 2n\pi/3\ell$, and their displacements become $\pm\ 2n\pi k/3\ell$. Similarly, the $3k-1$ components have secondary displacements of $\pm\ 2n\pi k/\ell \mp 2n\pi/3\ell$ and are phase-shifted by $\pm 2n\pi 3\ell$ on transferring to the primary to have displacements there of $\pm\ 2n\pi k/\ell$. The summation of all such components again yields zero except when k is an integer multiple of ℓ. Thus, for this case the primary current also contains harmonics of order $3q\ell \pm 1$—harmonic neutralization has again occurred.

This phenomenon will occur for secondary connected converters of any pulse number that is a multiple of three. Let $P_S = 3K$ and $K =$ any integer. Then the required secondary voltage displacements become $\pm(2n-1)\pi/\ell P_S$, $n = 1$ to m when $\ell = 2m$ and $\pm 2n\pi/\ell P_S$, $n = 1$ to m where $\ell = 2m+1$. The line currents of the converters contain only harmonics of the order $kP_S \pm 1$, from the extended definition of pulse number. All kP_S+1 components will be positive sequence, and all kP_S-1 components will be negative sequence since

$$(kP_S \pm 1)\ (-2\pi/3) = -2k\ K\pi \pm\ 2\pi/3 \equiv \mp\ 2\pi/3$$

$$(kP_S \pm 1)\ (-4\pi/3) = -4k\ K\pi \pm\ 4\pi/3 \equiv \mp\ 4\pi/3$$

The secondary displacements are

$$\pm(kP_S \pm 1)\ (2n-1)\pi/\ell P_S = \pm k \cdot 2n\pi/\ell \pm (\pm(2n-1)\pi/\ell P_S)$$
$$(n = 1 \text{ to } m,\ \ell = 2m)$$

and

$$\pm(kP_S \pm 1) 2n\pi/\ell P_S = \pm 2n\pi k/\ell \pm (\pm 2n\pi/\ell P_S)$$
$$(n = 1 \text{ to } m,\ \ell = 2m+1)$$

with phase shifts from secondary to primary of $\mp(2n-1)\pi/\ell P_S$ or $\mp 2n\pi/\ell P_S$, resulting in primary phase displacements of $\pm 2n\pi k/\ell$. The phasor sums in the primary are thus always zero except for $k = q\ell$, and hence the primary current pulse number is always ℓP_S (i.e., ℓ times the pulse number of the secondary converters).

The residual phase shift $\pm k\pi/\ell$ that occurs in all cases for $\ell = 2n$ represents the displacement that occurs because no secondary voltage set is in phase with the primary. As was stated, it can be ignored, in the phasor summation since it is a common phase displacement to all phasors of the set being summed; it merely indicates the whole set has suffered a shift and does not influence the consecutive displacements.

For the three-pulse case, it was stated that the situation $\ell = 2$ is special insofar as secondary voltage derivation is concerned. This is because the required displacement between the two sets of voltages needed is $2\pi/6 = \pi/3$ rad, which can be accomplished by the very simple transformer secondary shown in Figure 4-23. Here A', B', *and* C' are in phase with the primary wye voltages. The displaced set A'', B'', and C'' lag by $\pi/3$ rad and are such that $A'' \equiv -C$, $B'' \equiv -A$, and $C'' \equiv -B$. This configuration has an interesting practical advantage over and above its simplicity. If the neutral point is disjuncted (i.e., the neutrals of A', B', C' and A'', B'', C'' are separated), then the interphase reactor needed

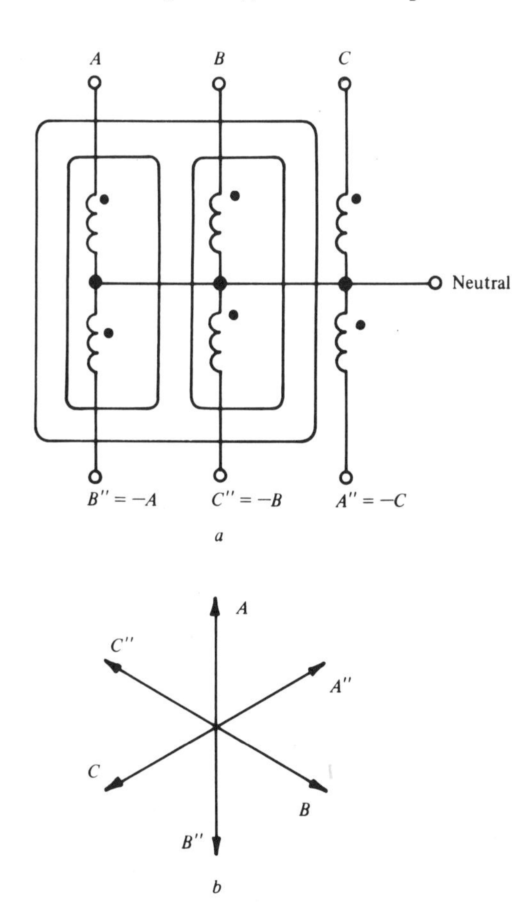

Figure 4-23. *a*, Six-phase secondary winding; *b*, phasor diagram.

to combine the outputs of the two three-pulse converters may be connected between them rather than between the converter terminals. This makes it possible to have the interphase reactor become a part of the same physical assembly as the transformer; in high-power systems, this is a marked advantage.

The case $\ell = 2$ is also special when six-pulse bridges are combined. The phase displacement required between the two voltage sets is then $2\pi/12 = \pi/6$ rad, which can be provided by the wye-delta transformer—wye and delta secondaries on the same transformer with wye or delta primary.

Like the six-phase secondary transformer of Figure 4-23, this transformer is easy to wind and interconnect and has a total secondary volt-ampere rating equal to the load volt-ampere demand. The phase-shifting secondaries of Figures 4-19 through 4-21 have a fourth, not very desirable, property—their secondary volt-ampere rating is greater than the load volt-ampere demand because of the manners in which the phase-shifted voltage sets are produced.

Obviously, the six-phase (for two three-pulse midpoint converters) and wye-delta (for two six-pulse bridges) transformers still effect harmonic neutralization of the primary currents, for they have the same phase-displacement properties as all phase-shifting transformers. In addition, the same phenomenon occurs when an appropriate phase-shift-producing version of any of the other types of winding is used in either case. Therefore, there must be multiple types of current waveshape corresponding to any given pulse number. This point is of no analytic concern. It simply means that a multiplicity of waveshapes can be generated by taking the same harmonics with the same relative amplitudes and modifying their mutual phase relationships. As mentioned in Section 4.1, it is therefore impossible to deduce a wave's pulse number from inspection except for a few simple and familiar cases. There is no essential correlation between the number of level changes in a wave, or their relative amplitudes, and the pulse number of the wave.

Obviously, that which can be done with three-phase systems can also be done with M-phase systems, where M is any integer greater than 2. M-phase transformers can obviously be made to produce phase-shifted sets by mechanisms similar to those used in three-phase transformers. When this is done, the phase shifts for the various symmetrical components of the M-phase set will differ, and proper winding combinations producing the phase-shifted voltages will produce harmonic neutralization in primary currents as well as in dc-terminal voltages. The mathematical proof for a three-phase system has been seen to be a trifle long-winded. It gets more so as the number of phases increases, and because of the very limited practical importance of systems with more than three phases, no other cases, or general proofs will be treated here.

As in the case of dc-to-dc converter harmonic neutralization, it is analytically possible to extend the process in current-sourced ac-to-dc and dc-to-ac converters to any desired degree. Once again, however, practical imperfections limit the degree to which it can be employed usefully. Those imperfections listed in Section 3.8, also must clearly apply to the ac-to-dc/dc-to-ac converters discussed so far in this chapter. In addition, there is another problem for these converters. To produce the phase shifts needed of the secondary voltage sets, irrational turns-ratios are needed. For example, the wye-delta arrangement needs a turns ratio of $\sqrt{3}$ to create equal-amplitude voltage sets at $\pi/6$ rad phase shift. Only rational ratios are possible, and the more voltage sets are required, with more and smaller incremental phase shifts, the more difficult it becomes to arrive at close enough rational approximations to the irrational ratios needed. This problem is compounded as system and transformer ratings increase, for then the turns per volt decrease, finally becoming increasing volts per turn; only the quotients of quite small integers are available to make the approximations. Some are quite reasonable—for example, 7:4 is only just over 1% in error from $\sqrt{3}$—but some irrational ratios are near impossible to approximate accurately with such small integers to use for quotients.

Single- and two-phase ac systems do not afford the opportunity for harmonic neutralization by the mechanism described, since phase-shifted voltage sets cannot be produced by single-phase transformers, except for the complementary phasor of the two-phase set. Thus, strictly speaking, the phenomenon is reserved for polyphase systems. This ought to be expected, since its realization depends on the properties of complete phasor sets and polyphase transformers.

Finally, from this discussion of harmonic neutralization, it should be obvious why the three-pulse midpoint group is regarded as the basic building block for the phase-angle-controlled current-sourced ac-to-dc and dc-to-ac converters. The six-pulse bridge comprises two such groups, and all the three-phase converters with harmonic neutralization are comprised of two or more three-pulse groups or bridges. This fact has been recognized since the 1920s. Its importance cannot be overestimated, for in all such converters each three-pulse group operates without influencing or being influenced by the others over most of the operating range. In the idealized situations used so far for analysis, no three-pulse-group ever knows the others are present. In the real world, the practical effect known as *commutation overlap* causes interactions to develop for some operating conditions.

4.7 Phase-Angle-Influenced Voltage-Sourced AC-to-DC and DC-to-AC Converters

To begin, the use of *influenced* rather than *controlled* should be explained. Analytically, in the world of ideal switches and voltage and current sources, the phase-angle-influenced voltage-sourced ac-to-dc and dc-to-ac converters are still phase-angle controlled, as will be seen shortly. In the practical world, however, ac current sources do not exist; thus, the variables that have to be controlled in the real-world applications of these converters are their dependent ac-voltage sets. These are not controlled by the operating phase delay or advance; the phase angle, in fact, becomes a consequence of the relationship of the dependent voltages of the converters to their ac loads. The phase angle influences power transfer but does not control it, in practical rather than analytic situations; hence the change of terminology.

Consider first what happens with the midpoint converters of Figures 4-1 and-4-4a if a dc voltage and an M-phase current set are defined. The switches clearly-have to be changed to the bidirectional current-carrying, unidirectional voltage-blocking devices first met in the two-quadrant buck-boost dc-to-dc converters of Section 3.5. Even with this change, the converter is not useful. To satisfy KLC with regard to the defined ac currents, the switches must remain permanently closed. No power transfer will take place, much less control thereof. It might be concluded that there is no midpoint version of the voltage-sourced converter, but this conclusion is a little suspect. For the current-sourced converters, the midpoint arises when a neutral is present in the defined ac voltage set. A ''neutral'' in the defined dc current is, of course, not plausible. The preceeding discussion of the apparently nonexistent voltage-sourced midpoint converter involves the tacit assumption that the neutral should belong to the defined ac currents. But why should this be so? does the neutral of the current-sourced midpoint converters belong to the ac or the voltage? Since both are the same in that case, it is impossible to tell.

If the neutral in fact belongs to the ac, then the conclusion that a midpoint voltage-sourced converter does not exist, based on the opening argument, is correct. If, however, the neutral belonged to the voltage, is a neutral in a dc voltage plausible? Indeed it is, but only a two-phase dc-voltage source—positive and negative from a reference—can be created. The midpoint voltage-sourced converter that emerges is depicted in Figure 4–24; it is commonly known as the *half-bridge*, for reasons that will soon become apparent.

In Figure 4-24, it is clear that if H_1 is assigned to S_1 then the existence function of S_2 must be $H_2 = 1-H_1$ in order that neither KLC nor KLV be violated. The resulting dc component of voltage applied to I_S is given by

$$\begin{aligned} V_{DD} &= V_S/2A - V_S(1-1/A)/2 \\ &= V_S\,(1/A-1/2) \end{aligned} \tag{4.60}$$

It is stated in Section 4.1 that dc current components flowing in ac voltage sources are undesirable. This is even more true of dc voltage components applied to ac current sources. Very often, the practical approximations to those sources are realized with the aid of transformers and reactors. For both components, applied dc voltage is disastrous, since dc current flow is limited only by winding

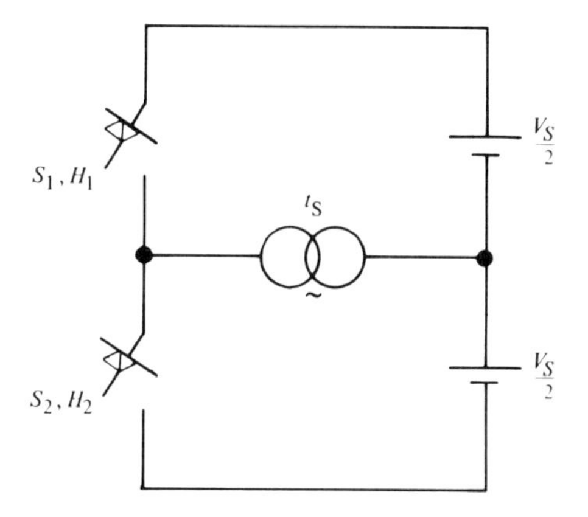

Figure 4-24. Half-bridge voltage-sourced converter.

resistance(s), and saturation is inevitable. Thus, the only viable operating condition is considered to be when $A = 2$ and

$$H_1 = 1/2+(2/\pi) \sum_{n=1}^{n=\infty} [\sin(n\pi/2)/n]\cos[n(\omega_s t+\alpha)] \tag{4.61}$$

$$H_2 = 1/2-(2/\pi) \sum_{n=1}^{n=\infty} [\sin(n\pi/2)/n]\cos[n(\omega_s t+\alpha)] \tag{4.62}$$

The voltage applied to I_S then becomes

$$\begin{aligned} v_D &= H_1\, V_S/2 - H_2\, V_S/2 \\ &= (2V_S/\pi) \sum_{n=1}^{n=\infty} [\sin(n\pi/2)/n]\cos[n(\omega_s t+\alpha)] \end{aligned} \tag{4.63}$$

Equation 4.63 shows this voltage to be a square wave, of amplitude $V_S/2$. Thus, it is identical to the current wave induced in a single-phase source (from which two phases are derived by using a transformer with a center-tapped secondary) by a two-phase two-quadrant midpoint current-sourced converter. Like that current, its amplitude is not affected by α; only its phase, relative to the defined current i_S, is affected.

That defined current has the form $I \cos(\omega_s t)$. There are two currents flowing at the converter dc terminals, i_{D1} and i_{D2}, of Figure 4-24. They are given by

$$i_{D1} = H_1\, i_S \tag{4.64}$$

$$= I\cos(\omega_s t)/2 + (2I/\pi)\sum_{n=1}^{n=\infty}[\sin(n\pi/2)/n]\cos[n(\omega_s t+\alpha)]\cos(\omega_s t)$$

$$i_{D2} = H_2\, i_S \tag{4.65}$$

$$= I\cos(\omega_s t)/2 - (2I/\pi)\sum_{n=1}^{n=\infty}[\sin(n\pi/2)/n]\cos[n(\omega_s t+\alpha)]\cos(\omega_s t)$$

As *KLC* demands, one-half of the defined current flows in each half ("phase") of the defined dc voltage source. The wanted, dc components of i_{D1} and i_{D2}, along with all unwanted oscillatory components other than the fundamental ones just discussed, have equal magnitudes but opposite signs, as *KLC* demands. The wanted component of i_{D1} is

$$I_{DD1} = (I/\pi)\sin(\pi/2)\cos\alpha \tag{4.66}$$

which is identical in form to the wanted component of dc-terminal voltage for the two-phase two-quadrant midpoint current-courced converter, except that a constant multiplier 2 is missing. The unwanted components arise also from the cross products

$$P_n = \cos[n(\omega_s t+\alpha)]\cos\omega_s t$$

$$= \frac{[\cos[(n+1)\omega_s t+n\alpha] + \cos[(n-1)\omega_s t+n\alpha]}{2}$$

and thus, because $\sin(n\pi/2)$ is zero for even n, only even-order components exist. They are given by

$$i_{DU} = (I/\pi)\sum_{m=1}^{m=\infty}\cos(m\pi)\{\cos[2m\omega_s t+(2m+1)\alpha]/(2m+1) - \cos[2m\omega_s t+(2m-1)\alpha]/(2m-1)\} \tag{4.67}$$

which is exactly the same form as the unwanted components of the current-sourced converter, with a multiplier of 2 removed. That multiplier disappears for this voltage-sourced midpoint converter because the dc-dependent quantity, I_{DD}, appears in two separate parts—I_{DD1} and I_{DD2}. Note that the ac-dependent voltage of this converter appears as an inseparable single wave, whereas the dependent current of the two-phase two-quadrant midpoint current-sourced converter occurs in two parts, the separate dependent currents in the two phases.

The reciprocity of the two converters is thus established beyond a shadow of a doubt.

There is also no doubt that this voltage-sourced midpoint converter is a two-quadrant converter. The magnitudes of the wanted component of both i_{D1} and i_{D2} contain $\cos\alpha$. The range of $|\alpha|$ is 0 to π, giving the range of $\cos\alpha$ as 1 to -1, and so the converter acts as an ac-to-dc converter ($\cos\alpha = 1$ to 0) or a dc-to-ac converter ($\cos\alpha = 0$ to -1).

Now suppose a complementary converter is added, with defined current $-i_s$ and existence functions displaced by π rad, as depicted in Figure 4-25a. This defined current is also obtained for this converter if the current of the converter first considered is returned to it, and the single-phase bridge configuration of Figure 4-25b emerges. This is the reciprocal of the single-phase current-sourced bridge in all respects. The reason for the term *half-bridge* applied to the midpoint converter should also now be clear.

More than two dc "phases" cannot be created; hence, polyphase versions of the voltage-sourced converter must operate with polyphase ac-current sources and a single or center-tapped dc-voltage source. However, the center-tapped dc-voltage source returned to the neutral of the set of ac-current sources is a redundant connection, because the defined ac-current sources sum to zero at the neutral point and no current will flow therein. Moreover, the existence of such a connection permits the zero-sequence unwanted components of the converter's dependent voltages to be impressed on the defined ac currents, which is undesirable. Hence, the polyphase bridge voltage-sourced converter takes the form shown in Figure 4-26. In this case, each pair of switches in a *pole* (i.e., one of the M midpoint switch pairs making up the bridge) must have existence functions summing to unity and with $A = 2$ but phased properly with respect to the defined current phasor it serves. Thus, the existence functions of the switches in the kth pole of M, where $k = 0$ to $M - 1$, are

$$H_{k1} = 1/2+(2/\pi)\sum_{n=1}^{n=\infty}[\sin(n\pi/2)/n]\cos[n(\omega_s t+\alpha-2k\pi/M)] \tag{4.68}$$

$$H_{k2} = 1/2-(2/\pi)\sum_{n=1}^{n=\infty}[\sin(n\pi/2)/n]\cos[n(\omega_s t+\alpha-2k\pi/M)] \tag{4.69}$$

The dependent current, i_D, is given by

$$i_D = \sum_{k=0}^{k=m-1} H_{k1}\cdot i_{Sk} \tag{4.70}$$

where $i_{Sk} = I\cos(\omega_s \mathrm{t}-2k\pi/M)$.

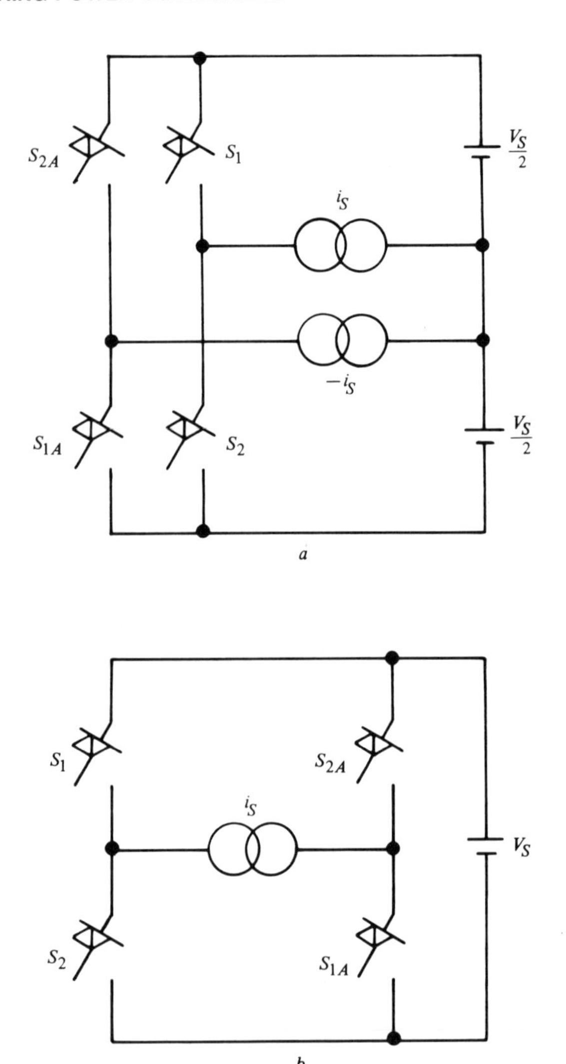

Figure 4-25. *a,* Two complementary half-bridges; *b*, full bridge.

Examination of Equations 4.69 and 4.70, in light of the manipulations of Sections 4.1 and 4.2, leads to the conclusion that the wanted component of i_D is

$$I_{DD} = (M/\pi)\, I \cos\alpha \tag{4.71}$$

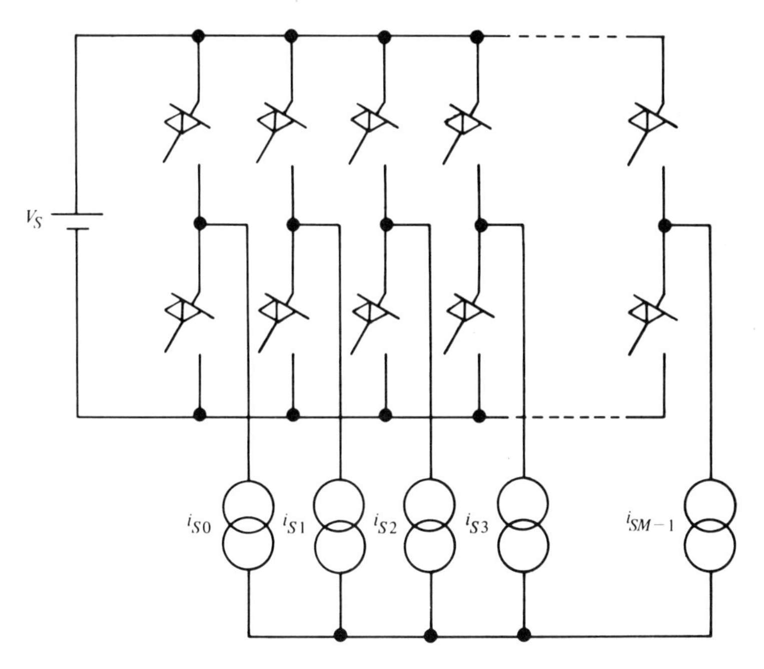

Figure 4-26. M-phase bridge.

and the only unwanted components existing will be those for which $n+1$ and $n-1$ are multiples of M. However, n can only be odd, since $\sin(n\pi/2)$ is zero for n even, and thus the multiples pM can only be even. Hence, as for the case of the unwanted components in the dc terminal voltage of the current-sourced converter, all multiples of M appear if M is even but only even multiples of M appear if M is odd. There are two cases, then:

M even

$$i_{DU} = (MI/\pi) \sum_{p=1}^{p=\infty} \cos(p\pi)\{\cos[pM\omega_s t+(pM+1)\alpha]/(pM+1) - \cos[pM\omega_s t+(pM-1)\alpha]/(pM-1)\} \tag{4.72}$$

M odd.

$$i_{DU} = (MI/\pi) \sum_{p=1}^{p=\infty} \{\cos[2pM\omega_s t+(2pM+1)\alpha]/(2pM+1) - \cos[2pM\omega_s t++(2pM-1)\alpha]/(2pM-1)\} \tag{4.73}$$

The only difference between these expressions and those for the unwanted components in the dc-terminal voltages of current-sourced bridges is the absence of the multiplier $2\sin(\pi/M)$. Everything else is identical, including a pulse number $P = 2M$ when M is odd. The absence of that multiplier should not be surprising, for it is the multiplier for translating phase voltages to line voltages in an M-phase system. In the current-sourced converters, the phase voltages were defined and yielded the dependent dc voltages. For this voltage-sourced converter, it is the line currents that are yielding the expression for dependent dc current. If that current is expressed in terms of the phase currents, then the multiplier reappears. Note the reversal of sense of the terms *line* and *phase* when applied to currents and voltages in an M-phase system. Phase voltage is that from line to neutral; line voltage is that from line to line. Phase current is that flowing from line to line; line current is that flowing from line to neutral. The converter's phase voltages, of course, are given by Equation 4.63 after the phase shift $-2k\pi/M$ is introduced into the arguments of all terms for the kth such dependent voltage of M.

A line voltage is the difference between two consecutive phase voltages in the system. Hence, the line voltage of the M phase system has the form

$$v_{DL} = (2V_S/\pi) \sum_{n=1}^{n=\infty} [\sin(n\pi/2)/n]\{\cos[n(\omega_s t+\alpha)] - \cos[n(\omega_s t+\alpha-2\pi/M)]\} \tag{4.74}$$

with an appropriate phase shift introduced to both oscillatory term arguments to specify a particular member of the set.

Observe that the difference of oscillatory functions is of the form

$$\cos\theta - \cos(\theta-\gamma)$$

which can be expanded to

$$\cos\theta\,(1-\cos\gamma) - \sin\theta\sin\gamma$$

where $\theta = [n(\omega_s t+\alpha)]$ and $\gamma = 2n\pi/M$. Thus, the magnitude of this difference is given by

$$\begin{aligned} A_n^2 &= (1-\cos\gamma)^2 + \sin^2\gamma \\ &= 2(1-2\cos\gamma) \\ &= 4\sin^2(\gamma/2) \end{aligned} \tag{4.75}$$

yielding

$$A_n = 2\sin(n\pi/M)$$

The phase angle that these voltages possess,

$$\begin{aligned}\psi_n &= \arc\tan\left[\sin\gamma/(1-\cos\gamma)\right] \\ &= \arc\tan\left[\cot(n\pi/M)\right]\end{aligned} \tag{4.76}$$

is simply the phase angle by which a line voltage of an M-phase system leads the corresponding phase voltage, multiplied of course by the order of the harmonic involved. Thus, the line voltage becomes

$$v_{DL} = (4V_S/\pi)\sum_{n=1}^{n=\infty}[\sin(n\pi/2)/n]\sin(n\pi/M)\cos[n(\omega_s t+\alpha+\psi)] \tag{4.77}$$

where $\psi = \pi/2-\pi/M$. This expression is in effect identical to that for the dependent ac currents of a current-sourced bridge—odd harmonics only, and only those that are not multiples of M when M is odd.

Since the expressions for dependent quantities are identical, with transposition of current for voltage and adjustment of amplitude in the case of the voltage-sourced converters' dependent dc terminal current, the waveforms will be identical except for transposition of current and voltage. Thus, Figure 4-7 is valid for the dc terminal current of a three-phase voltage-sourced bridge; Figure 4-8a is valid for its line voltages. Reciprocity is demonstrated. It can be deduced from the properties of phase-shifting transformers (discussed in Section 4.6) that if the line voltages of a three-phase voltage-sourced bridge converter are used to excite the delta primary of a delta-wye transformer, the line voltages of the wye secondary will have the waveform of Figure 4-8b.

The switch waveforms in the voltage-sourced converter are not simply transpositions of those in the current-sourced converter, however, except for the two-pulse (single-phase bridge) case. Obviously, all switches in an M-phase voltage-sourced converter will see the same voltage and current waves as those in the single-phase bridge. The current waves are depicted, for α positive, in Figure 4-27. The voltage waves are rectangular, with zero value when current is present and the value V_S when no current is flowing. In other words, they are replicas of the complements of the switches' existence functions with amplitude V_S.

Since there are no midpoint converters apart from the half-bridge, the two-quadrant voltage-sourced converters are dealt with apart from the rather odd version depicted in Figure 4-28a. It is obviously a realization of a two-phase midpoint voltage-sourced converter with the neutral belonging to the defined ac

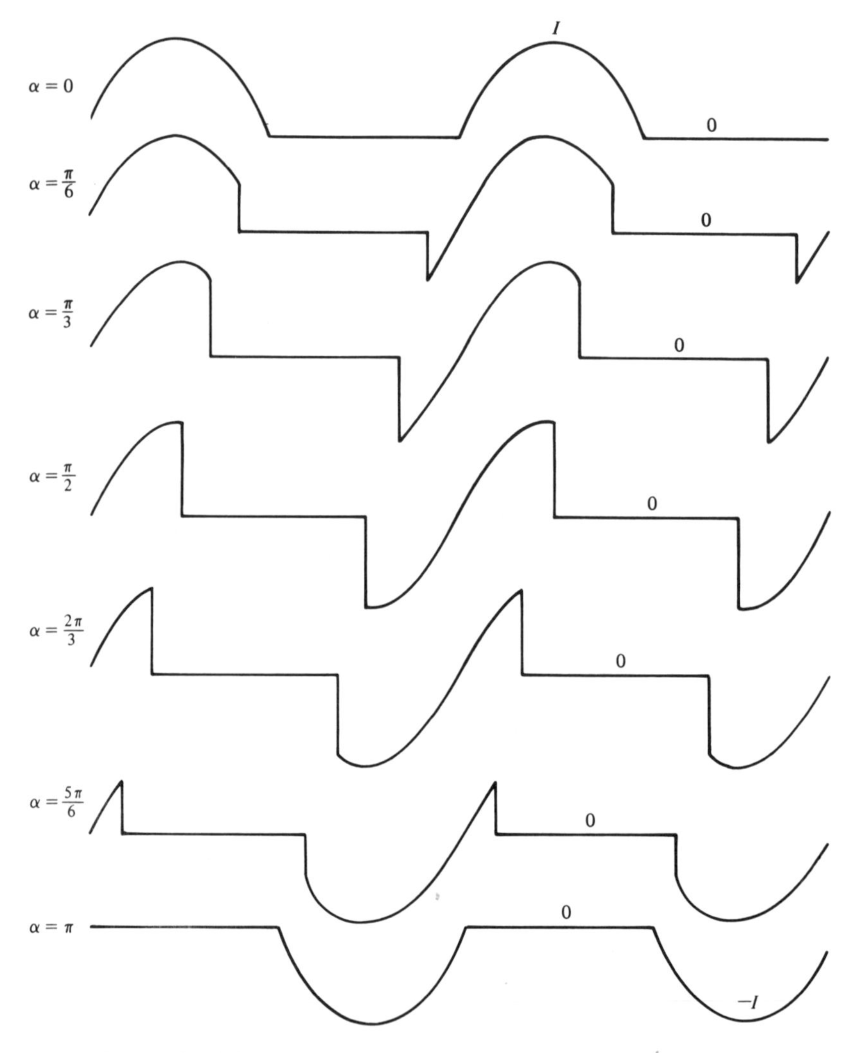

Figure 4-27. Switch current waves in voltage-sourced converters (α positive).

currents, not to the defined dc voltage. It can operate only if these currents are coupled and OR-ed, i.e., if only one of the two current sources is active at any given time. This condition is met when a single-phase current source feeds through a transformer in which the center-tapped secondary forms the coupled sources, as depicted in Figure 4-28b. The analysis is straightforward, and the properties are as for all the two-quadrant voltage-sourced converters. Note that the dependent voltages are prevented from exhibiting dc components by their mutual coupling by the transformer.

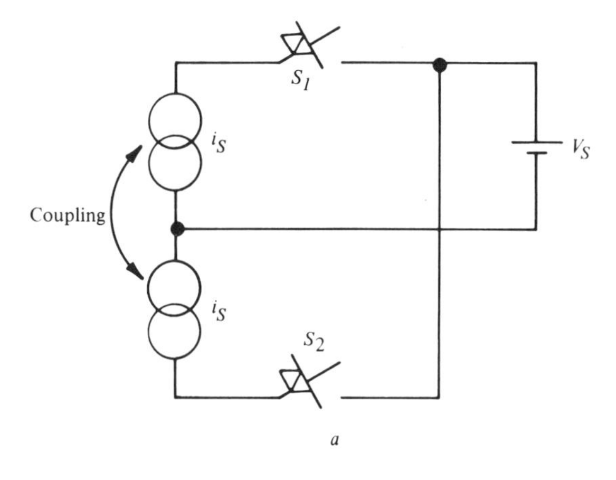

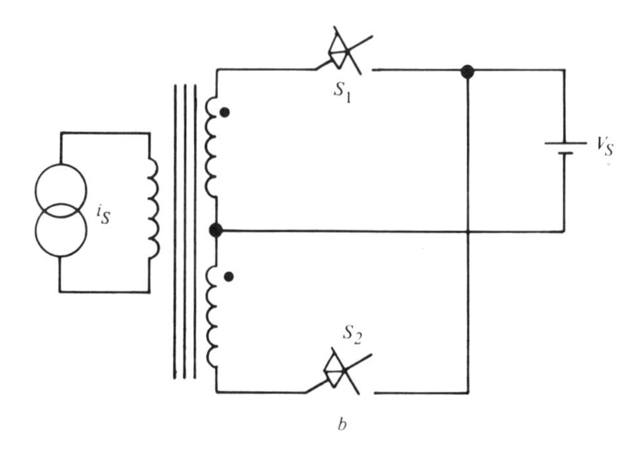

Figure 4-28. *a*, Two-phase midpoint voltage-sourced converter; *b*, realization using center-tapped transformer secondary.

Because this topology is simple, uses only two switches, and produces a higher amplitude-dependent ac-voltage wanted component (across the whole transformer secondary) from a given dc than does either the half bridge or the single-phase bridge, it has found extensive application. In fact, it was the first voltage-sourced converter ever investigated,[1,2] and thus it is of historical importance too. Its existence suggests that a "half-bridge" current-sourced converter might be found if a similar coupled and OR-ed current-source postulate is used. Such a topology is shown in Figure 4-29a, with a possible source realization via coupled reactors depicted in Figure 4-29b. This converter has seen no practical use. Its analysis is quite straightforward. However, note that the coupled dc-current sources are presumed to offer no impediment to the flow of ac current by virtue of their coupling—i.e., to exhibit no common-mode impedance to ac current. It is clear that the coupled reactor fulfills this requirement.

It should not be assumed that the abbreviated treatment given two-quadrant voltage-sourced ac-to-dc/dc-to-ac converters reflects a lack of importance on their part. The existence of the reciprocity relationship with the current-sourced converters makes it unnecessary to expend much effort and space on their description and analysis.

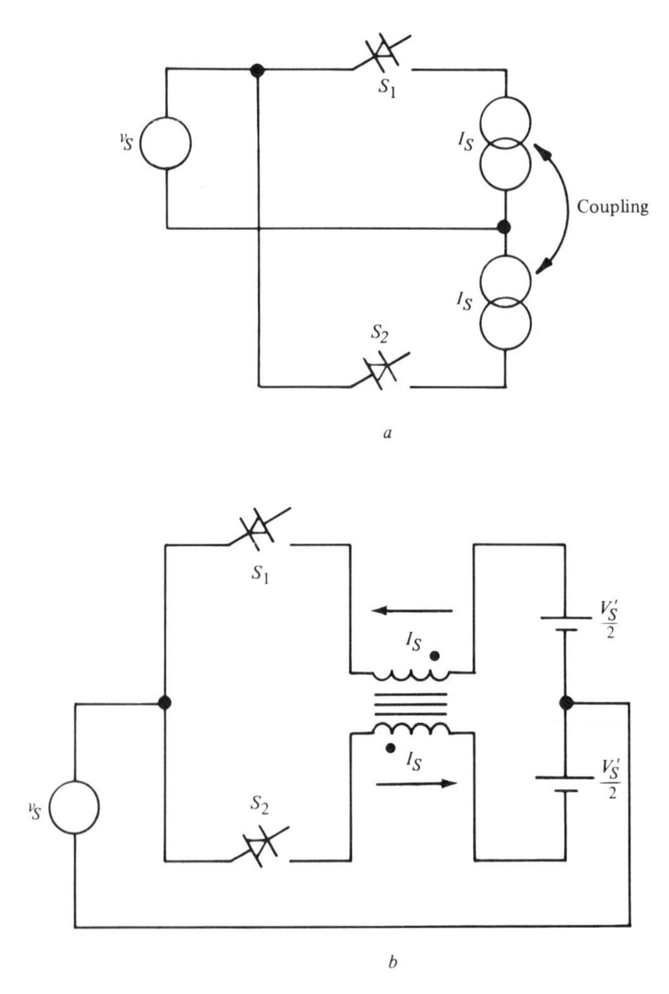

Figure 4-29. *a*, Half-bridge current-sourced converter; *b*, realization using coupled reactor to approximate current sources.

4.8 Single-Quadrant Voltage-Sourced AC-to-DC and DC-to-AC Converters

It can be observed from the waveforms of Figure 4-27 that reciprocity with the current-sourced converters extends to yet another situation. The switches in voltage-sourced converters have only one polarity of voltage impressed upon them. However, they are required to carry current of both polarities (which was seen to be a plausible combination of switch properties) for all circumstances save $\alpha = 0$ or $|\alpha| = \pi$. In the former case, the autocomplementary unilateral switches can be employed to produce an uncontrolled single-quadrant voltage-sourced ac-to-dc converter. This type of converter is sometimes used as part of a compound dc-to-dc converter—dc-to-ac converter followed by ac-to-dc converter—but it is never used when the primary energy source is ac because that source is then invariably a defined voltage, not the defined current required by this type of converter. However, the single-phase converter of Figure 4-17 can be regarded equally well as a dc-voltage-sourced converter operating with a discontinuous defined ac current source. In fact, its analysis is easier if it is approached from this view.

Controlled single-quadrant operation can be achieved in voltage-sourced converters much as it is in current-sourced converters. Consider the single-phase bridge of Figure 4-25b. If both switches in one pole are reduced to unidirectional current capability, that pole will operate at rectification end stop only, while the other can still operate over the full two-quadrant range. The net result is fully controlled operation over the rectification quadrant, just as for the current-sourced case. The analysis is very similar to that for the current-sourced case, merely transposing current and voltage in the final expressions for the dependent quantities. The converter therefore exhibits the same property of reducing both dependent quantities to zero as the controlled pole reaches inversion end stop. This fact is very useful in dealing with another requirement, as discussed in Section 4.9.

For a polyphase single-quadrant phase-controlled voltage-sourced converter, each phase must be served by a full bridge. Therefore, the individual ac-phase currents must be available, all isolated. Not surprisingly, these combined requirements, coupled with the general lack of ac current sources, have severely inhibited the application of such converters. They are found so rarely that most practitioners of power electronics do not even realize they exist. Moreover, they are only used when circumstances absolutely demand it—i.e., when a compound conversion function is mandated and the ac-to-dc converter portion thereof must be single-quadrant, fully controlled, and voltage-sourced.

For the current-sourced converters, it was found that a single-quadrant converter operating in the inversion quadrant only was not possible. The reason was, of course, that the switch required is not realizable. Using a bidirectional current device in the current-sourced converter violates KLV, and if reverse blocking capability is present, then full two-quadrant operation is possible.

The switch requirement for a single-quadrant inverting voltage-sourced converter is also not realizable—the same switch, forward current-carrying, forward voltage-blocking only, would be needed. However, if a switch that also has reverse-blocking capability is used, neither of Kirchoff's laws will be violated. Thus, a voltage-sourced converter operating only at its inversion end stop is possible. Such single-quadrant converters are quite widely used in a variety of high-frequency dc-to-ac converter applications.

A fully controlled version can, of course, be created by combining a fully controlled pole with an uncontrolled pole in a bridge, or in each of a set of polyphase bridges. To the author's knowledge, such an arrangement has never actually been applied.

Apart from the uncontrolled dc-to-ac converter, single-quadrant voltage-sourced converters are of very little importance to power converter applications. It is sufficient to mention here that their interfacing and commutation requirements exactly parallel those of the two-quadrant versions. Hence, the discussions of these subjects in Chapters 6 and 8 are also germane to this much rarer artifact.

4.9 Dependent AC Voltage Control in Voltage-Sourced Converters

As mentioned in Section 4.7, practical voltage-sourced converters are not really phase-angle controlled, but phase-angle influenced. This is because of actual source considerations and converter usage. Voltage-sourced converters are used primarily as dc-to-ac converters. As such, their dependent ac voltages feed passive or dynamic (ac machine) loads or interface via a coupling network with a set of defined ac voltages (typically, the utility supply). The interfacing techniques used to make these real loads behave as reasonable approximations to current sources are discussed in Chapter 6. The concomitant problem, however, is that virtually all the applications require control to be exerted on the magnitude of the wanted components of the converters' dependent ac voltages.

Phase-angle control (variation of α) does not accomplish this. In fact, in most applications α is not a control variable but simply a consequence of the converters' ac voltages and the currents they cause to flow in the converters' loads. In the simplest case, passive load, α becomes simply the phase angle of the load-interfacing network combination. With machine loads, α is for the most part similarly determined, but is subject to transient disturbances if the exciting converters' voltage sets are suddenly phase-shifted with respect to the machines' frame of reference. In coupling to defined voltage sets, the α's of converters are determined by the magnitudes and phases of the wanted components of their voltage sets relative to those of the sets they interface.

Control of the magnitude of the wanted component in the dependent ac voltage of a voltage-sourced converter thus becomes a primary concern. There are a number of ways to accomplish the control; two of the most important are now described and analyzed.

As seen in Section 4.8, controlled single-quadrant operation is achieved by using an uncontrolled pole and a controlled pole as elements of the same bridge. As the controlled pole is shifted to inversion (or rectification) end stop, the dependent quantities go to zero. What if both poles in the bridge are fully controlled, i.e., if the bridge retains full two-quadrant capability? Obviously, nothing prevents the phase shift of the dependent voltage of either pole with respect to the other. The resultant bridge output voltage (line voltage for a polyphase bridge) is the phasor difference between the two voltages produced by the poles. In analyzing this case, it is convenient to presume that an arbitrary phase reference exists and that one pole is phase-advanced with respect thereto by some angle ϕ, while the other is delayed by the same angle. From Equation 4.63, ignoring α, the two voltages are then

$$v_{D1} = (2V_S/\pi) \sum_{n=1}^{n=\infty} [\sin(n\pi/2)/n]\cos[n(\omega_s t+\phi)]$$

$$v_{D2} = (2V_S/\pi) \sum_{n=1}^{n=\infty} [\sin(n\pi/2)/n]\cos(n\pi)\cos[n(\omega_s t-\phi)]$$

The resulting line voltage is $v_{D1}-v_{D2}$; the multiplier $\cos(n\pi)$ simply becomes -1 for all nonzero terms in v_{DL} and v_{D2}, so that (with n odd only):

$$\begin{aligned} v_D &= (2V_S/\pi) \sum_{n=1}^{n=\infty} [\sin(n\pi/2)/n]\{\cos[n(\omega_s t+\phi)] + \cos[n(\omega_s t-\phi)]\} \\ &= (4V_S/\pi) \sum_{n=1}^{n=\infty} [\sin(n\pi/2)/n]\cos(n\phi)\cos(n\omega_s t) \end{aligned} \tag{4.78}$$

The angle ϕ is thus a control variable, since the multiplier $\cos \phi$ appears in the amplitude of the wanted component. The useful range of ϕ is obviously 0 to $\pi/2$ rad, for at $\phi = \pi/2$, all $\cos(n\phi)$, n odd, are zero, and $v_D = 0$.

There are two disadvantages attached to this method of introducing ac voltage control to a voltage-sourced converter. As mentioned in connection with the single-quadrant converter, a full bridge must be used for each phase and the individual phase currents at a converter's ac terminals must be isolated. Moreover, the distortion index of v_D increases with ϕ, and individual low-order harmonics reassume their maximum absolute amplitudes ($4V_S/n\pi$) at some value of ϕ within the control range, when the wanted component has been reduced in amplitude.

This latter phenomenon arises because $\cos(n\phi)$ has the range -1 to $+1$; if ϕ assumes the value $m\pi/n$, m an integer less than n, then $\cos n\phi$ is -1 or $+1$ while $|\cos\phi|$ is less than 1. The distortion index is

$$\mu_V = [(\pi^2 - 2\phi\pi)/8\cos^2\phi - 1]^{1/2} \tag{4.79}$$

The limit of μ_V, as ϕ tends to $\pi/2$, is ∞. Thus, the problems of interfacing this type of converter with its ac loads become acute if this method of control is employed, and the range of ϕ required is a substantial fraction of the total available.

Other techniques for controlling the dependent ac voltages have been sought. In particular, techniques that dispense with the need for full bridges at every phase are, obviously, highly desirable. One such is PWM, which is discussed in Section 4.12. The second technique discussed here might be termed a *hybrid*, for it combines some of the properties of both PWM and the phase-displacement control just described. To date, there is no established nomenclature for this technique. Here it will be termed *conduction angle modulation* (CAM).

Consider the half-bridge in Figure 4-24, which is representative of any pole in a polyphase bridge. In establishing criteria for the switch-existence functions therein, it was stipulated that $H_2 = 1 - H_1$. Also, the requirement that no unwanted dc voltage component should exist led to the postulate that $A_1 = A_2 = 1/2$ provided the simplest pair of existence functions, H_1 and H_2, for the switches S_1 and S_2. However, a multitude of existence functions will satisfy both these conditions—any pair with equal unit- and zero-value periods will do so. In addition, it is possible to create existence functions with multiple unit- and zero-value periods within each cycle of the operating frequency of the converter. (PWM, as will be seen, is the ultimate extension and exploitation of this possibility.) Suppose then H_1 and H_2 take the general form depicted in Figure 4-30a. They still satisfy the requirements of the converter, but will obviously yield rather different expressions for dependent quantities than do the existence functions of Equations 4.61 and 4.62. Each existence function may now be considered to be the result of combining three simple existence functions, as depicted in Figure 4-30b. The basal existence functions, H_{1A} and H_{2A}, are those defined by Equations 4.61 and 4.62. Each has subtracted from it an existence function (H_{1B} and H_{2B}) with $A_B > 2$ and a relative phase shift, positive or negative, of θ (θ is shown positive in Figure 4-30). Each has added to it a simple existence function (H_{1C} and H_{2C}) identical to that subtracted but phase-shifted by a further $\pm\pi$ rad. Thus, these auxiliary existence functions may be written as

$$H_{1B} = 1/A_B + (2/\pi)\sum_{n=1}^{n=\infty} [\sin(n\pi/A_B)/n]\cos[n(\omega_s t + \theta)]$$

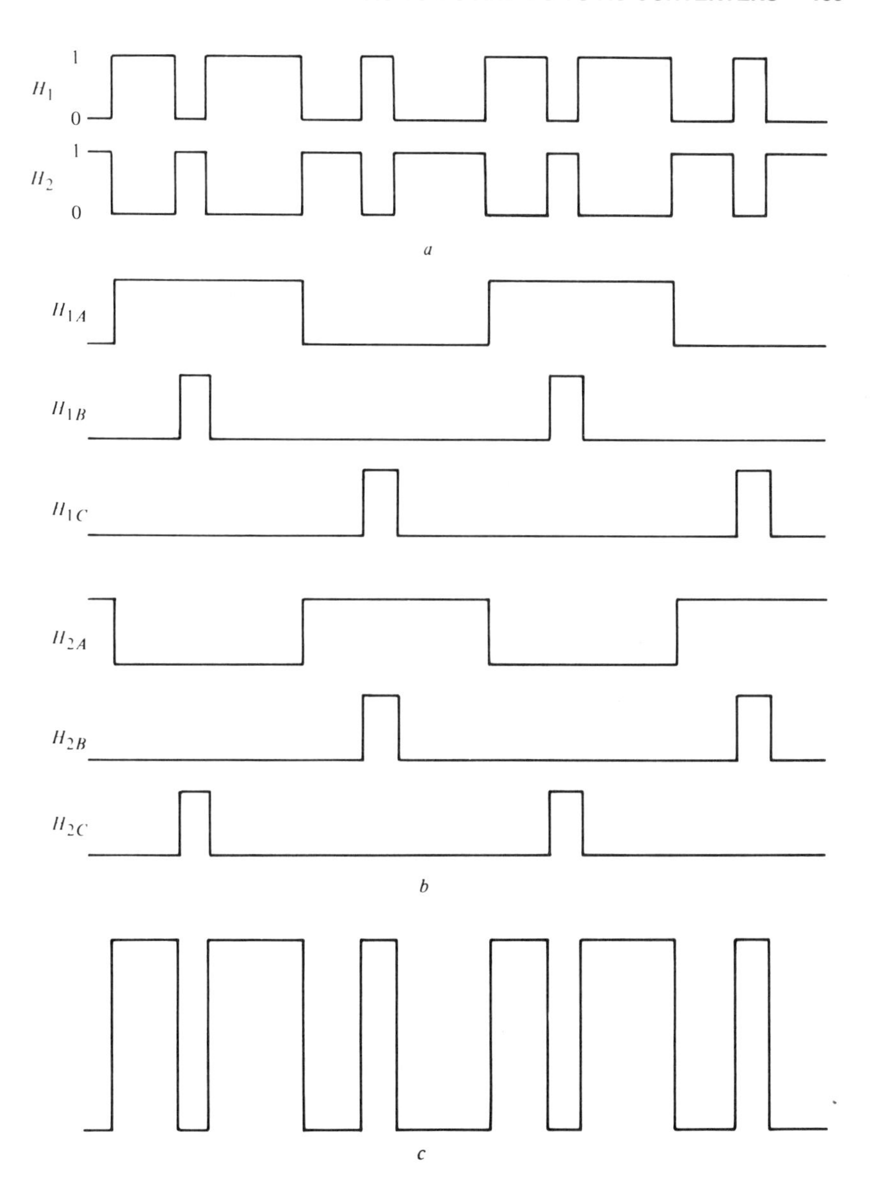

Figure 4-30. *a*, CAM existence functions; *b*, elementary constituent functions; *c*, pole voltage wave.

$$H_{2B} = 1/A_B + (2/\pi) \sum_{n=1}^{n=\infty} [\sin(n\pi/A_B)/n]\cos[n(\omega_s t + \theta + \pi)]$$

$$H_{1C} = H_{2B}$$

$$H_{2C} = H_{1B}$$

Note that α is ignored because, if it exists, it will be present in the arguments of all oscillatory terms and will not affect the manipulations that follow. H_1 and H_2 are then given by

$$\begin{aligned} H_1 &= H_{1A} - H_{1B} + H_{1C} \\ &= 1/2 + (2/\pi) \sum_{n=1}^{n=\infty} [\sin(n\pi/2)/n]\cos(n\omega_s t) \\ &\quad - (2/\pi) \sum_{n=1}^{n=\infty} [\sin(n\pi/A_B)/n]\cos(n\omega_s t+\theta) \\ &\quad + (2/\pi) \sum_{n=1}^{n=\infty} [\sin(n\pi/A_B)/n]\cos(n\pi)\cos(n\omega_s t+\theta) \end{aligned} \tag{4.80}$$

$$\begin{aligned} H_2 &= H_{2A} - H_{2B} + H_{2C} \\ &= 1/2 + (2/\pi) \sum_{n=1}^{n=\infty} [\sin(n\pi/2)/n]\cos(n\pi)\cos(n\omega_s t) \\ &\quad - (2/\pi) \sum_{n=1}^{n=\infty} [\sin(n\pi/A_B)/n]\cos(n\pi)\cos[n(\omega_s t+\theta)] \\ &\quad + (2/\pi) \sum_{n=1}^{n=\infty} [\sin(n\pi/A_B)/n]\cos[n(\omega_s t+\theta)] \end{aligned} \tag{4.81}$$

They are still clearly complementary and also give no dc component when the first part of Equation 4.63 is used to derive v_D. Now observe that the oscillatory terms in H_{1C} and H_{2C} have the same form as those of H_{1B} and H_{2B} but are multiplied by $-\cos(n\pi)$. Thus, the added oscillatory terms in H_1 and H_2 contain no even-order components, and the odd-order components can be combined so that H_1 and H_2 may be rewritten as

$$\begin{aligned} H_1 &= 1/2 + (2/\pi) \sum_{n=1}^{n=\infty} [\sin(n\pi/2)/n]\cos(n\omega_s t) \\ &\quad - (4/\pi) \sum_{n=1}^{n=\infty} [\sin(n\pi/A_B)/n]\cos[n(\omega_s t+\theta)] \end{aligned} \tag{4.82}$$

$$H_2 = 1/2 + (2/\pi) \sum_{n=1}^{n=\infty} [\sin(n\pi/2)/n]\cos(n\pi)\cos(n\omega_s t)$$
$$- (4/\pi) \sum_{n=1}^{n=\infty} [\sin(n\pi/A_B)/n]\cos(n\pi)\cos[n(\omega_s t+\theta)] \quad (4.83)$$

where n is restricted to be odd only.

It is obvious that the extra oscillatory terms in these existence functions will modify the converter's dependent quantities. The resultant pole voltage wave is shown in Figure 4-30c. It has the expression

$$v_D = H_1 V_S/2 - H_2 V_S/2$$
$$= (2V_S/\pi) \sum_{n=1}^{n=\infty} [\sin(n\pi/2)/n]\cos(n\omega_s t) \quad (4.84)$$
$$- (4V_S/\pi) \sum_{n=1}^{n=\infty} [\sin(n\pi/A_B)/n]\cos[n(\omega_s t+\theta)]$$

where n is odd only.
The wanted component of v_D is given by

$$v_{DD} = (2V_S/\pi)[\cos(\omega_s t) - 2\sin(\pi/A_B)\cos(\omega_s t+\theta)] \quad (4.85)$$

Its magnitude is controllable by varying A_B, θ, or both, for it is given by

$$A_V = (2V_S/\pi)[1+4\sin^2(\pi/A_B) - 4\sin(\pi/A_B)\cos\theta]^{1/2} \quad (4.86)$$

The control variables are not totally independent, being related by

$$\theta \not> \pi\ (1/2 - 1/A_B) \quad (4.87)$$

The pole voltage wave of Figure 4-30c may be regarded as the basal square wave from which a *notch wave* has been subtracted. The notch wave derives from a square wave that has twice the amplitude of the basal square wave, and the consequences in terms of harmonic amplitudes and distortion index can be most unfortunate. From Equation 4.86, it can be deduced that if $\theta = 0$, $A_V = 0$ for $A_B = 6$; the useful range of $2\pi/A_B$, the notch width, is 0 to $\pi/3$. For any other value of θ, A_V cannot be reduced to zero; its minimum value becomes $(2V_S/\pi)\ (1-\cos^2\theta)^{1/2}$, and the useful range of $2\pi/A_B$ is from 0 to arc $\sin(\cos\theta/2)$.

Individual harmonic amplitudes are given by

$$A_n = (2V_S/n\pi)[1+4\sin^2(n\pi/A_B)-4\sin(n\pi/2)\sin(n\pi/A_B)\cos(n\theta)]^{1/2} \quad (4.88)$$

From this expression, it can be deduced that A_n will become zero for $\theta = 0$ and $A_B = 6n$; it will also become zero for $\theta = k\pi/n$, where k is any integer, and $A_B = 6n$. However, A_n becomes a maximum of $6V_S/\pi$, three times its absolute maximum amplitude in the basal square wave, for $\theta = (2k-1)\pi/n$ and $2\pi/A_B = (4p-3)\pi/n$, where p is any integer, or for $\theta = 2k\pi/n$ and $2\pi/A_B = (4p-1)\pi/n$. From Figure 4-30c, the rms value of the pole wave is $V_S/2$ no matter what values of θ and A_B exist. This invariance is also of some significance to PWM waves, as will be seen. The fundamental amplitude is given by Equation 4.86, whence the distortion index becomes

$$\mu_V = \left(\pi^2 \Big/ 8[1+4\sin^2(\pi/A_B)-4\sin(\pi/A_B)\cos\theta]\} - 1 \right)^{1/2} \quad (4.89)$$

At $\theta = 0$, μ_V increases rapidly with $2\pi/A_B$ and becomes ∞ for $A_B = 6$, since the dependent voltage then consists solely of unwanted components. For θ not zero, μ_V still increases as $2\pi/A_B$ is increased from zero to the limiting useful value.

Thus, the advantage of CAM lies in its ability to effect control of any pole's wanted component of dependent ac voltage regardless of the presence or absence of other poles. Its disadvantage is in the distortion (i.e., larger amplitude unwanted components) it creates. In a polyphase converter—particularly in that most important case of the three-phase bridge—this effect can be mitigated somewhat by controlling θ and A_B such that the zero-sequence harmonics of the polyphase set tend to maximize. These harmonics, of order kM, where M is the number of phases, do not appear in the line voltages of the bridge, and thus their presence, with large amplitude, in the phase voltages (pole voltages) is of little concern.

A further disadvantage of CAM lies in the increased rate of operation of the switches. For the basic control described, the switching frequency is three times the converter operating frequency; thus, switching losses in a converter using given devices will increase threefold (or more, depending on the V-I conditions seen by the switches as they open and close). Despite all this, CAM has been a fairly popular way of implementing voltage control in voltage-sourced converters. It is easy to implement and, if executed with due concern for harmonic considerations, can give overall performance as good as that obtained with phase-displacement control.

There is yet another disadvantage to the simple CAM scheme so far described. Unless θ is set at zero, which is one of the least desirable options, the wanted component is shifted with respect to that of the basal square wave. If θ is not held constant, the phase of the wanted component varies with the control. This can be deleterious when voltage-sourced converters are used for machine drives or to interface with defined ac voltages. However, it may be avoided by using multiple CAM. If two complementary sets of auxiliary existence functions are combined with the basic ones, at $+\theta$ and $-\theta$ displacements, then the two notches appear in the pole-voltage wave with those displacements. In this event, no phase displacement of the wanted component occurs. Unfortunately, harmonic amplitudes can now reach five times the amplitudes they possess in the basal square wave. As a result, the distortion index increases, and the switching frequency is now five times the wanted component frequency.

Obviously, the use of multiple CAM can be further extended to produce more notches in the pole wave. Ultimately, a PWM wave will develop as this process is followed. Each additional pair of notches, created by adding four more auxiliary existence functions, adds four to the multiplier on switching frequency and the potential multiplying factor on harmonic amplitudes.

Regardless of which technique is used to control the wanted component of dependent voltage, phase displacement of pole voltages, or CAM of pole voltage, modifications to the dependent current expressions and waveforms will occur. The expressions can be developed easily enough. However, they are, very cumbersome and are of little use in developing insight into the possible effects on converter interfacing at the dc terminals. More useful information is obtained from some simple basic considerations in each case.

For phase-displacement control, consider that the two converters operating at $+\phi$ and $-\phi$ with respect to an arbitrary reference correspond to the complementary current-sourced converters at $+\alpha$ and $-\alpha$ postulated in Section 4.7 for operation at unity displacement factor. Then the ripple-current behavior of the voltage-sourced converters with phase-displacement control will correspond to the ripple-voltage behavior of complementary-phase-controlled current-sourced converters. The spectra will be the same as that for a simple uncontrolled bridge, insofar as constituent frequencies are concerned. However, the amplitudes of individual components can be twice those they would assume for the uncontrolled converters. This is because of the presence of the displacements $n\phi$ and $-n\phi$ in individual component arguments, and a resulting amplitude multiplier of $2\cos n\phi$, just as in the case of the dependent voltage.

In the case of CAM, the waveform modification of the dependent current is even more extensive. The exact waveshape depends on θ, A_B, and α. It can be constructed for any particular combination of these variables by the technique established for constructing any wave. The most important point, however, is that individual harmonic amplitudes can be tripled as compared to those of the uncontrolled converters, just as they can in the dependent voltage.

4.10 Switch Stresses in Voltage-Sourced Converters

Since the voltage-sourced converters are so clearly reciprocals of the current-sourced converters, and the switch requirements exhibit reciprocity too, it is to be expected that switch voltage and current waveforms will also show this characteristic. That they do is evident from the current waveforms of Figure 4-27 and the accompanying description of the voltage waves.

A switch in a voltage-sourced converter never has to support forward voltage (as viewed from the ac terminals). It does have to sustain, regardless of the phase angle α, instantaneous transition to reverse voltage just following conduction of reverse current. In other words, a switch with a very high reapplied $\partial V/\partial t$ capability is needed.

High $\partial I/\partial t$ is also evident for current in the reverse direction (if α is positive) and in the forward direction (if α is negative). The transition from reverse-to-forward (α positive) or forward-to-reverse (α negative) conduction is smooth, following the defined current sinusoid.

The peak value of current in a switch, reverse or forward, is I, the peak of the defined ac current. The average and rms values depend on α; because no truly single practical switch conducts current in both directions, two values of rms and average current are stipulated anyway for any α. Maximum values are identical—I/π for the average, $I/2$ for the rms in either direction. At any α, positive or negative, the values are $I(1+\cos\alpha)/2\pi$ for the forward average, $I(1-\cos\alpha)/2\pi$ for the reverse average, $I[(\pi-\alpha)/4\pi+\sin 2\alpha/8\pi]^{1/2}$ for forward rms, and $I[\alpha/4\pi-\sin 2\alpha/8\pi]^{1/2}$ for reverse rms.

The peak reverse voltage stress is invariably V_S. While reverse voltage in the converter's frame of reference, this is forward voltage to the switch that carries the converter's reverse current, which is the controlled portion of an actual switch.

The addition of phase-angle displacement control to voltage-sourced converters does not change the maximum device stresses; CAM, however, creates a considerable change. Peak voltage stress cannot change, of course, but the reapplied voltage after conduction occurs more often. Peak current stress does not change either, but average and rms values in the two directions become functions of θ and A_B as well as α. Maximum values do not change, however. Notably, the device exposure to high $\partial I/\partial t$ increases. High $\partial I/\partial t$ (instantaneous change in the idealized converters) occurs for both directions of conduction regardless of α; for extended CAM, multiple high $\partial I/\partial t$ exposure occurs in each cycle.

4.11 Displacement Factor Control in Current-Sourced Converters

The description of CAM so far has pertained solely to its use to control wanted-component magnitude in voltage-sourced converters. It can also be used to correct phase displacement in current-sourced converters. Until recently, such applications were rare; the quadrature current demands of current-sourced converters were accepted in almost all uses. This situation has now changed, and there has been a growing demand that current-sourced converters be operated without quadrature current. The complementary converter arrangement, by now clearly the reciprocal of phase-displacement-controlled voltage-sourced converters, is one way of achieving this goal. Since it requires two complete converters, and hence twice the number of switching devices that are required for power transfer, other ways of reducing or eliminating quadrature current demand are being sought.

One is to provide compensating current by using passive-reactive components. Another is the use of CAM; its use in this case differs somewhat in detail, but not in principle, from its application to voltage-sourced converters.

Consider first the single-quadrant, controlled, current-sourced midpoint converters. As the existence functions of the active switches are delayed by α, in operation with α negative, quadrature current demand develops. If, however, the existence-function unit-value period terminations are simultaneously advanced by α, there will be no quadrature current demand. In effect, the converter is made to be its own complementary converter. KLC will not be violated, of course, since the autocomplementary switch will assume conduction whenever the conduction period of an active switch is terminated. The existence function of the kth active switch becomes

$$H_k = 1/M - \alpha/\pi + (2/\pi) \sum_{n=1}^{n=\infty} \{\sin[n(\pi/M - \alpha)]/n\}\cos(n\omega_s t) \qquad (4.90)$$

Thus, the amplitude modifier $\sin(\pi/M - \alpha)$ applies to the wanted components of both the dependent voltage and the dependent current, and the useful range of α is from 0 to π/M.

An alternative technique for implementing CAM in a current-sourced converter is not to delay the existence function at all, but to subtract from it one that is in phase ($\theta = 0$) and of duration 2α. The amplitude multiplier which then develops for the wanted components is $\sin(\pi/M) - \sin\alpha$, so that the transfer function is changed. The dependent quantity waves will be different too, of course, but in general the effects on distortion index and individual harmonic amplitudes are similar.

To operate a single-quadrant bridge with CAM, it is necessary to add an autocomplementary freewheeling switch in shunt with the bridge's dc terminals, so that KLC is not violated when the unit-value periods of the active switches' existence functions are shortened or interrupted. In the case of a bridge, when CAM is employed, the useful range of α is 0 to π/P, where P is the pulse number. This is because curtailing or interrupting the active switches' existence functions interrupts or curtails those for the mid-point group comprised of autocomplementary switches. The amplitude multiplier thus becomes a composite, being $\sin(\pi/P-\alpha)$ for one of the midpoint groups and $\sin(\pi/P)-\sin\alpha$ for the other when M, the number of phases, is odd. When M is even, the multiplier takes the same form for both midpoint groups.

CAM may also be applied to the two-quadrant midpoint and bridge current-sourced converters. To do so, the midpoint must be equipped with a controlled, not autocomplementary, freewheeling switch. For the bridge, the controlled freewheeling path can be accomplished using the switches already present. Control becomes a little more involved than for the single-quadrant converters, since α has a useful range of 0 to π/P (in a midpoint, $P=M$) for two distinct reference positions of the existence functions: namely, 0 and π. For $\alpha' = 0$, these converters are controlled through the rectification quadrant as α is varied from 0 to π/P. For $\alpha' = \pi$, it is controlled through the inversion quadrant. The amplitude multipliers developing for the wanted components of the dependent quantities are the same as those for the single-quadrant converters.

From this variation of the conduction angles of the active and freewheeling switches in current-sourced converters, the term *CAM* is derived. While not quite so obvious, CAM also applies to the voltage-sourced case, for there it adjusts the conduction angles for which current is drawn from or injected into the defined dc voltage; during the periods when the basal existence functions are influenced by the auxiliary ones, current "circulates" within the converter. As for the voltage-sourced converters, CAM can be extended in current-sourced converters by adding more pairs of auxiliary existence functions to the complex forming the active switches' existence functions, and the ultimate result will be PWM. (Any CAM existence function can be regarded as a combination of two or more simpler functions.) The penalties paid for the use of CAM in current-sourced converters are the same as those paid in voltage-sourced converters—increased switching rate, higher distortion indices, and potentially much greater amplitudes of individual harmonic components. In addition, however, there are the penalties of the added switches needed for single-quadrant bridge and two-quadrant midpoint converters.

4.12 Harmonic Neutralization in Voltage-Sourced Converters

Since the dependent quantities of these converters are reciprocals of those of the current-sourced converters, it is to be expected that harmonic neutralization can be employed with them and will be based on the same principles and mechanisms. This is indeed so. The only difference lies in the manner of combination at the dc and ac terminals.

At their dc terminals, voltage-sourced converters are ripple-current sources. Hence, a number of them operated with appropriate phase displacements to achieve harmonic neutralization may be directly parallel-connected to a common defined dc voltage. If it is desired to connect them in series, independent and isolated dc source realizations must be used, and a good many of the benefits of harmonic neutralization are lost at the dc interface. Unfortunately, there is no passive component capable of achieving the series combination of phase-staggered ripple-current sources in the same way that an interphase reactor achieves the parallel combination of phase-staggered ripple-voltage sources.

At their ac terminals, the voltage-sourced converters are ''ripple''-voltage sources. (*Ripple* is not usually applied to unwanted oscillatory components accompanying a wanted ac component; however, they have precisely the same characteristics as those accompanying a wanted dc component.) Thus, voltage-sourced converters cannot be directly connected to phase-shifting transformers with a common primary supply, or worse yet to phase-shifting secondary windings on a common core. Either connection provides the implicit direct-parallel connection of the phase-staggered ripple voltages, and therefore violates KLV. Thus, interphase reactors must be used. This further evidence of reciprocity should be noted: interphase reactors at the dc terminals for parallel-connecting harmonic-neutralized current-sourced converters, at the ac terminals for achieving parallel connection of harmonic-neutralized voltage-sourced converters. The simplest approach is to use an individual transformer for each converter, providing appropriate phase shift, of course, and then to use the interphase reactors between the transformer primaries and the common set of defined ac currents as depicted in Figure 4-31. This is sometimes considered uneconomical, because of the multiplicity of transformers involved, each with a single phase shifting secondary. Humphrey and Mokrytzki have shown that interphase reactors can be constructed to connect between converter ac terminals and transformer secondaries.[3] However, when used to combine more than two converters, these

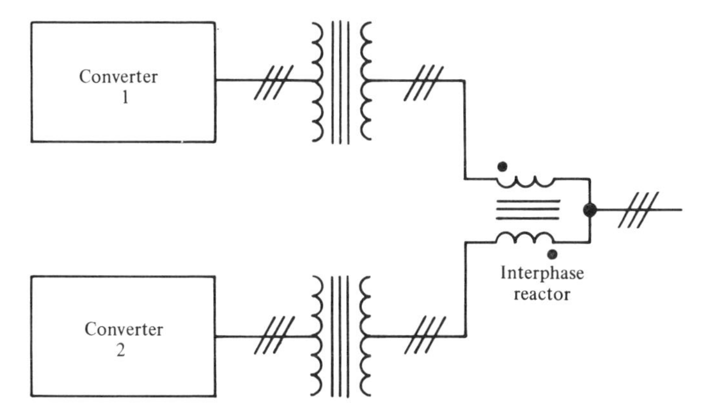

Figure 4-31. Application of interphase reactor to parallel combination of harmonic neutralized voltage-sourced converters.

become so complicated in winding arrangement that they are not generally recommended.

Series connection of the individual transformer primaries can be accomplished readily. However, it places on the transformers the restriction that all primaries be of identical configuration and consist of isolated windings for all phases. Moreover, the series connection of windings on different transformers of equal ratings is usually considered bad practice because of the transient voltage distribution problems that may arise. In the present case, this cautionary observation may not apply, since the transformer winding voltages are supposedly defined by the converters, not by any load or source to which they may be connected. In fact, this series connection has been fairly extensively used with two and sometimes three converters.

At this point, it should be observed that the use of harmonic neutralization is not limited to simple phase-angle-influenced converters. Phase-displacement-controlled and CAM-controlled converters (including the current-sourced versions thereof) may be combined in exactly the same fashion. Harmonic neutralization is a truly universal technique.

The harmonic neutralization of single-phase voltage-sourced converters is possible. If such converters are phase-staggered at π/ℓ rad consecutive displacement ($P = 2$ for any single-phase converter of this type), then their square-wave, phase-displacement-controlled, or CAM-controlled, dependent voltages may be combined by series-connecting appropriately proportioned transformer primaries, with the individual transformer secondaries being fed by the converters. Kernick et al. were the first to discuss harmonic neutralization as applied to voltage-sourced converters; they dealt with a single-phase approach, with a three-phase system derived by using three such single-phase systems.[4] This early misdirection obscured the relationships of the converters and the universality of harmonic neutralization for a considerable time.

The single-phase approach is not to be recommended unless it is absolutely unavoidable. The converters used (in a single-phase system) all operate at different α's and supply different fractions of the total loading. In a three-phase system derived by using three such single-phase systems, these problems disappear. However, the zero-sequence harmonics are unnecessarily neutralized in the individual phases, and the transformer arrangement is cumbersome, bulky, and fragmented.

4.13 Pulse-Width Modulation in Voltage- and Current-Sourced AC-to-DC and DC-to-AC Converters

It was seen in Chapter 2 that there is an alternate method of formulating existence functions to create ac-to-dc and dc-to-ac converters. In Sections 4.9 and 4.11, the use of CAM was described, and it was stated that in its ultimate extension, it becomes PWM, the alternate technique of Chapter 2. Recall that it was postulated that a basic existence function

$$H = 1/A + (2/\pi) \sum_{n=1}^{n=\infty} [\sin(n\pi/A)/n]\cos(n\omega t)$$

can be modulated so that the average term, $1/A$, assumes the form $(1/A_o)(1 \pm x\cos\omega_o t)$. This postulate is now verified, thus clarifying the relationships of the resulting existence functions and converter performance to those arising from extended CAM.

To create an existence function for which $1/A$ takes the form postulated, it is necessary to vary the duration of the unit value periods in a cyclic manner. In modulation theory, the technique used is called *natural sampling*. Black (1951) gives an excellent abstract of Bennett's technique for performing the analysis of a duration-modulated pulse train, which is what a pulse-width modulated existence function is.[5] The difficulties, which Bennett surmounted, arise because the normal evaluation of Fourier coefficients for a pulse train is no longer valid when the pulse durations become functions of time. When this happens, as it does in PWM, the limits of the integrals normally used to evaluate the Fourier coefficients become functions of the variable of integration. Remembering from Chapter 2 that, for a fixed-duration pulse train,

$$C_n = (1/\pi) \int_{-\pi/A}^{\pi/A} \cos(n\omega t)d(\omega t)$$

it can be seen that if $1/A$ is made to assume the postulated form, $(1/A_o)[1 \pm x\cos(\omega_o t)]$, the integration is no longer possible. Bennett circumvented this mathematical difficulty by creating a three-dimensional derivation of the pulse train in which ωt and $\omega_o t$ become independent variables on the orthogonal axes of a plane. Thus, the Fourier coefficients of the double Fourier series, which is the general form of the pulse train's expansion, were obtainable by a double integration in those two variables over a unit area. The mathematics of these integrations is a little complex, involving certain of the Bessel integrals. For that reason, only the specific case pertinent to power converters will be treated here.

First, consider Figure 4.32. The modulating function, $A_o\cos\omega_o t$, is plotted vertically in complementary pairs, with the plots repeated at intervals of 2π rad for the pulse-repetition frequency, f, along the horizontal axis. If a line of slope ω_o/ω is drawn through the origin, its intersections with the modulating function curves define the switching instants of the modulated pulse train. This type of modulation is called *natural sampling* because the duration of each pulse depends on the value of the modulating function at the instants of pulse transitions.

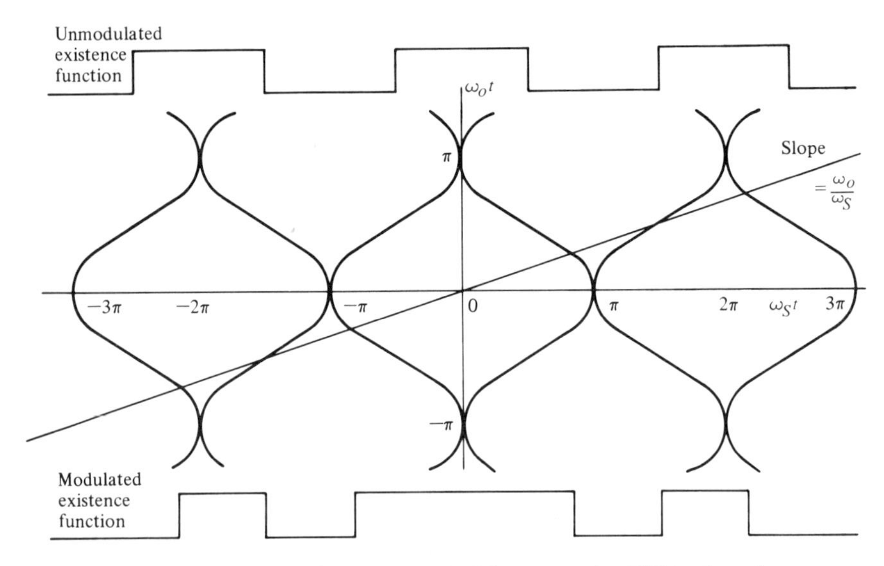

Figure 4-32. Adaptation of Bennett's method for constructing PWM pulse trains.

Modulation theorists recognize at least one other type of PWM, uniform sampling, in which the pulse durations depend on modulating function values at uniform time intervals. Black gives the analyses for this case also, but it is of no concern here.[4]

The third dimension in Bennett's formulation, of course, is the pulse height—in the case of existence functions, unity. Bennett visualized the modulating function as a series of "walls" built on the plane. The walls were intersected by an orthogonal plane (the line of slope ω_o/ω in Figure 4-32) and the projection of the intersections of that plane and the walls on a plane parallel to the ωt axis but orthogonal both to it and the $\omega_0 t$ axis became the modulated pulse train. This suggested that the Fourier expansion for the pulse train, its "height," could be expressed as a double series in the variables ωt and $\omega_0 t$, because, if the plane in which those axes lie is sectioned into squares with sides of length 2π, the portions of the "walls" contained within each such square are identical.

The double Fourier series may be developed as follows. Suppose a line is drawn on Figure 4-32 parallel to the ωt axis. Its intersections with the modulating functions ("walls") will define the switching instants of a set of unmodulated pulses that can be expressed as the Fourier series,

$$F(\omega t, \omega_o t_1) = a_o(\omega_o t_1)/2 + \sum_{n=1}^{n=\infty} [a_n(\omega_o t_1)\cos(n\omega t) + b_n(\omega_o t_1)\sin(n\omega t)] \quad (4.91)$$

The coefficients, $a_n(\omega_o t_1)$ and $b_n(\omega_o t_1)$, depend on $\omega_o t_1$, the point of intersection of the line with the $\omega_o t$ axis. Therefore, these coefficients are themselves functions of $\omega_o t$; since the modulating functions vary periodically with time along the direction of that axis, they themselves may be represented as the Fourier series,

$$a_n(\omega_o t) = c_{on}/2 + \sum_{m=1}^{m=\infty} [c_{mn}\cos(m\omega_o t) + d_{mn}\sin(m\omega_o t)] \quad (4.92)$$

$$b_n(\omega_o t) = e_{on}/2 + \sum_{m=1}^{m=\infty} [e_{mn}\cos(m\omega_o t) + f_{mn}\sin(m\omega_o t)] \quad (4.93)$$

Now the coefficients of Equation 4.91 are given by the integrals

$$a_n(\omega_o t_1) = (1/\pi)\int_{-\pi}^{\pi} F(\omega t, \omega_o t_1)\cos(n\omega t)d(\omega t) \quad (4.94)$$

$$b_n(\omega_o t_1) = (1/\pi)\int_{-\pi}^{\pi} F(\omega t, \omega_o t_1)\sin(n\omega t)d(\omega t) \quad (4.95)$$

while those of Equations 4.92 and 4.93 are in turn given by

$$c_{mn} = (1/\pi)\int_{-\pi}^{\pi} a_n(\omega_o t)\cos(m\omega_o t)d(\omega_o t) \quad (4.96)$$

$$d_{mn} = (1/\pi)\int_{-\pi}^{\pi} a_n(\omega_o t)\sin(m\omega_o t)d(\omega_o t) \quad (4.97)$$

$$e_{mn} = (1/\pi)\int_{-\pi}^{\pi} b_n(\omega_o t)\cos(m\omega_o t)d(\omega_o t) \quad (4.98)$$

$$f_{mn} = (1/\pi)\int_{-\pi}^{\pi} b_n(\omega_o t)\sin(m\omega_o t)d(\omega_o t) \quad (4.99)$$

Using Equations 4.94 and 4.95, Equations 4.96 through 4.99 may be expanded as follows:

$$c_{mn} = (1/\pi)\int_{-\pi}^{\pi}[(1/\pi)\int_{-\pi}^{\pi}F(\omega t,\omega_o t)\cos(n\omega t)d(\omega t)]\cos(m\omega_o t)d(\omega_o t)$$
$$= (1/\pi^2)\int_{-\pi}^{\pi}\int_{-\pi}^{\pi}F(\omega t,\omega_o t)\cos(n\omega t)\cos(m\omega_o t)d(\omega t)d(\omega_o t)$$
$$= (1/2\pi^2)\int_{-\pi}^{\pi}\int_{-\pi}^{\pi}F(\omega t,\omega_o t)\cos(n\omega t+m\omega_o t)d(\omega t)d(\omega_o t)$$
$$+ (1/2\pi^2)\int_{-\pi}^{\pi}\int_{-\pi}^{\pi}F(\omega t,\omega_o t)\cos(n\omega t-m\omega_o t)d(\omega t)d(\omega_o t) \quad (4.100)$$

$$d_{mn} = (1/2\pi^2)\int_{-\pi}^{\pi}\int_{-\pi}^{\pi}F(\omega t,\omega_o t)\sin(n\omega t+m\omega_o t)d(\omega t)d(\omega_o t)$$
$$- (1/2\pi^2)\int_{-\pi}^{\pi}\int_{-\pi}^{\pi}F(\omega t,\omega_o t)\sin(n\omega t-m\omega_o t)d(\omega t)d(\omega_o t) \quad (4.101)$$

$$e_{mn} = (1/2\pi^2)\int_{-\pi}^{\pi}\int_{-\pi}^{\pi}F(\omega t,\omega_o t)\sin(n\omega t+m\omega_o t)d(\omega t)d(\omega_o t)$$
$$+ (1/2\pi^2)\int_{-\pi}^{\pi}\int_{-\pi}^{\pi}F(\omega t,\omega_o t)\sin(n\omega t-m\omega_o t)d(\omega t)d(\omega_o t) \quad (4.102)$$

$$f_{mn} = (1/2\pi^2)\int_{-\pi}^{\pi}\int_{-\pi}^{\pi}F(\omega t,\omega_o t)\cos(n\omega t-m\omega_o t)d(\omega t)d(\omega_o t)$$
$$- (1/2\pi^2)\int_{-\pi}^{\pi}\int_{-\pi}^{\pi}F(\omega t,\omega_o t)\cos(n\omega t+m\omega_o t)d(\omega t)d(\omega_o t) \quad (4.103)$$

Substituting Equations 4.92 and 4.93 into Equation 4.91 yields

$$F(\omega t,\omega_o t) = (1/2)\left\{c_{oo}/2+\sum_{m=1}^{m=\infty}[c_{mo}\cos(m\omega_o t)+d_{mo}\sin(m\omega_o t)]\right\}$$
$$+\sum_{n=1}^{n=\infty}\left(\cos(n\omega t)\cdot\left\{c_{on}/2+\sum_{m=1}^{m=\infty}[c_{mn}\cos(m\omega_o t)+d_{mn}\sin(m\omega_o t)\right\}\right)$$

$$+\sum_{n=1}^{n=\infty}\left(\sin(n\omega t)\left\{e_{on}/2+\sum_{m=1}^{m=\infty}[e_{mn}\cos(m\omega_o t)+f_{mn}\sin(m\omega_o t)]\right\}\right)$$

$$= c_{oo}/4+(1/2)\sum_{m=1}^{m=\infty}[c_{mo}\cos(m\omega_o t)+d_{mo}\sin(m\omega_o t)]$$

$$+\ (1/2)\sum_{n=1}^{n=\infty}[c_{on}\cos(n\omega t)+d_{on}\sin(n\omega t)]$$

$$+\ \sum_{n=1}^{n=\infty}\sum_{m=1}^{m=\infty}[c_{mn}\cos(n\omega t)\cos(m\omega_o t)+d_{mn}\cos(n\omega t)\sin(m\omega_o t)]$$

$$+\ \sum_{n=1}^{n=\infty}\sum_{m=1}^{m=\infty}[e_{mn}\sin(n\omega t)\cos(m\omega_o t)+f_{mn}\sin(n\omega t)\sin(m\omega_o t)]$$

$$= c_{oo}/4+(1/2)\sum_{m=1}^{m=\infty}[c_{mo}\cos(m\omega_o t)+d_{mo}\sin(m\omega_o t)]$$

$$+\ (1/2)\sum_{n=1}^{n=\infty}[c_{on}\cos(n\omega t)+d_{on}\sin(n\omega t)]$$

$$+\ (1/2)\sum_{n=1}^{n=\infty}\sum_{m=1}^{m=\infty}(c_{mn}-f_{mn})\cos(n\omega t+m\omega_o t)$$

$$+\ (1/2)\sum_{n=1}^{n=\infty}\sum_{m=1}^{m=\infty}(c_{mn}+f_{mn})\cos(n\omega t-m\omega_o t)$$

$$+\ (1/2)\sum_{n=1}^{n=\infty}\sum_{m=1}^{m=\infty}(d_{mn}+e_{mn})\sin(n\omega t+m\omega_o t)$$

$$+\ (1/2)\sum_{n=1}^{n=\infty}\sum_{m=1}^{m=\infty}(e_{mn}-d_{mn})\sin(n\omega t-m\omega_o t) \qquad (4.104)$$

Now to the evaluation of coefficients. From Figure 4-32, it can be seen that $A_o = 2$, and the function $F(\omega t,\omega_o t)$ has the value 1 from $-\pi/2 - (x\pi/2)\cos(\omega_o t)$ to $\pi/2 + (x\pi/2)\cos(\omega_o t)$ along the ωt axis; it has the value zero elsewhere between $-\pi$ and π along that axis. Along the direction of the $\omega_o t$ axis, $F(\omega t,\omega_o t)$ is everywhere of unit value. Thus, from Equation 4.100, putting $\beta = \pi/2 + (x\pi/2)\cos(\omega_o t)$, c_{oo} is expressed as

$$c_{oo} = (1/\pi^2)\int_{-\pi}^{\pi}\int_{-\beta}^{\beta} 1\ d(\omega t)d(\omega_o t)$$

$$= (2/\pi^2)\int_{-\pi}^{\pi} \beta \, d(\omega_o t)$$

$$= 2.$$

$$c_{mo} = (1/\pi^2)\int_{-\pi}^{\pi}\int_{-\beta}^{\beta} 1 \cos(m\omega_o t)d(\omega t)d(\omega_o t)$$

$$= (2/\pi^2)\int_{-\pi}^{\pi} \beta \cos(m\omega_o t)d(\omega_o t)$$

$$= x \text{ for } m = 1, \quad 0 \text{ for all other } m.$$

$$d_{mo} = (1/\pi^2)\int_{-\pi}^{\pi}\int_{-\beta}^{\beta} \sin m\omega_o t \, d(\omega t)d(\omega_o t)$$

$$= 0 \text{ for all } m.$$

The remaining coefficients are not quite as simple to evaluate. They involve certain of the Bessel integrals. Consider first the Jacobi series:

$$\cos(y\sin\theta) = J_o(y) + 2J_2(y)\cos 2\theta + J_4(y)\cos 4\theta \ldots$$

$$\sin(y\sin\theta) = 2J_1(y)\sin\theta + 2J_3(y)\sin 3\theta + 2J_5(y)\sin 5\theta \ldots$$

$$\cos(y\cos\theta) = J_o(y) - 2J_2(y)\cos 2\theta + 2J_4(y)\cos 4\theta \ldots$$

$$\sin(y\cos\theta) = 2J_1(y)\sin\theta - 2J_3(y)\cos 3\theta + 2J_5(y)\cos 5\theta \ldots$$

where $J_p(y)$ is the Bessel function of the first kind of order p and argument y. These may be regarded as Fourier series, and on formulating coefficients, for them, it is found that

$$2\pi J_o(y) = \int_{-\pi}^{\pi} \cos(y\sin\theta)d\theta = \int_{-\pi}^{\pi} \cos(y\cos\theta)d\theta$$

and

$$0 = \int_{-\pi}^{\pi} \sin(y\sin\theta)d\theta = \int_{-\pi}^{\pi} \sin(y\cos\theta)d\theta$$

for $p = 2q$,

$$2\pi J_{2q}(y) = \int_{-\pi}^{\pi} \cos(y\sin\theta)\cos 2q\theta d\theta$$

$$= (-1)^{q-1} \int_{-\pi}^{\pi} \cos(y\cos\theta)\cos 2q\theta d\theta$$

for $p = 2q-1$,

$$2\pi J_{2q-1}(y) = \int_{-\pi}^{\pi} \sin(y\sin\theta)\sin(2q-1)\theta d\theta$$

$$= (-1)^{q-1} \int_{-\pi}^{\pi} \sin(y\cos\theta)\cos(2q-1)\theta d\theta$$

Also, for all p,

$$0 = \int_{-\pi}^{\pi} \cos(y\sin\theta)\sin p\theta d\theta$$

$$= \int_{-\pi}^{\pi} \sin(y\sin\theta)\cos p\theta d\theta$$

$$= \int_{-\pi}^{\pi} \cos(y\cos\theta)\sin p\theta d\theta$$

$$= \int_{-\pi}^{\pi} \sin(y\cos\theta)\sin p\theta d\theta$$

The definite integrals just defined are Bessel integrals; using them to evaluate the coefficients of the expansion of a PWM existence function leads to

$$c_{on} = (1/\pi^2) \int_{-\pi}^{\pi} \int_{-\beta}^{\beta} \cos(n\omega t) d(\omega t) d(\omega_o t)$$

$$= (2/n\pi^2) \int_{-\pi}^{\pi} \sin\beta \; d(\omega_o t)$$

Now $\sin\beta = \sin(n\pi/2)\cos[(xn\pi/2)\cos(\omega_o t)] + \cos(n\pi/2)\sin[(xn\pi/2)\cos(\omega_o t)]$. Therefore, for n odd, the integral has the value $\sin(n\pi/2) \cdot 2\pi J_o(xn\pi/2)$; for n even, the integral is zero. Hence,

$$c_{on} = (4/n\pi)J_o(xn\pi/2),\ n \text{ odd}$$

$$= 0,\ n \text{ even}$$

Now,

$$e_{on} = (1/\pi^2)\int_{-\pi}^{\pi}\int_{-\beta}^{\beta} \sin(n\omega t)d(\omega t)d(\omega_o t)$$

$$= 0 \text{ for all } n.$$

To evaluate the coefficients for the double summations, observe that

$$c_{mn} - f_{mn} = (1/\pi^2)\int_{-\pi}^{\pi}\int_{-\beta}^{\beta} \cos(n\omega t + m\omega_o t)d(\omega t)d(\omega_o t)$$

$$c_{mn} + f_{mn} = (1/\pi^2)\int_{-\pi}^{\pi}\int_{-\beta}^{\beta} \cos(n\omega t - m\omega_o t)d(\omega t)d(\omega_o t)$$

$$e_{mn} + d_{mn} = (1/\pi^2)\int_{-\pi}^{\pi}\int_{-\beta}^{\beta} \sin(n\omega t + m\omega_o t)d(\omega t)d(\omega_o t)$$

$$e_{mn} - d_{mn} = (1/\pi^2)\int_{-\pi}^{\pi}\int_{-\beta}^{\beta} \sin(n\omega t - m\omega_o t)d(\omega t)d(\omega_o t)$$

when terms in Equations 4.100 through 4.103 are collected, and the limits are inserted.

Now

$$\int \sin(n\omega t \pm m\omega_o t)d(\omega t) = -(1/n)\cos(n\omega t \pm m\omega_o t)$$

Taken over the limits β, $-\beta$ this becomes $\pm(2/n)\sin(n\beta)\sin(m\omega_o t)$. Expanding $\sin(n\beta)$, the trigonometric product becomes

$$\sin(n\pi/2)\cos[(xn\pi/2)\cos(\omega_o t)]\sin(m\omega_o t)$$

$$+ \cos(n\pi/2)(\sin[(xn\pi/2)\cos(\omega_o t)]\sin(m\omega_o t)$$

The integrals of both of these terms over the limits π and $-\pi$ have been seen to be zero, and hence,

$$e_{mn} + d_{mn} = e_{mn} - d_{mn} = 0$$

However,

$$\int \cos(n\omega t \pm m\omega_o t)d(\omega t) = (1/n)\sin(n\omega t \pm m\omega_o t)$$

Taken over the limits β and $-\beta$, this becomes $(2/n)\sin(n\beta)\cos(m\omega_o t)$. Expanding $\sin(n\beta)$, the trigonometric product becomes

$$\sin(n\pi/2)\cos[(xn\pi/2)\cos(\omega_o t)]\cos(m\omega_o t) + \cos(n\pi/2)\sin[(xn\pi/2)\cos(\omega_o t)]\cos(m\omega_o t)$$

The integral of the first term over the limits π to $-\pi$ has the value $\sin(n\pi/2)\cdot(-1)^q J_{2q}(xn\pi/2)$ for $m = 2q$, and zero for $m = 2q-1$. The integral of the second term over these same limits has the value $\cos(n\pi/2)\cdot(-1)^{q-1} J_{2q-1}(xn\pi/2)$ for $m = 2q-1$, and zero for $m = 2q$. However, $\sin(n\pi/2)$ is zero for n even, $\cos(n\pi/2)$ is zero for n odd. Hence, the only combinations for which the coefficients $c_{mn} \mp f_{mn}$ are nonzero are n odd, m even, and n even, m odd. In either case, the coefficients both have the value $(4/n\pi)\cdot(-1)^{(n+m-1)/2}J_m(xn\pi/2)$. Thus, with all coefficients evaluated, the Fourier expansion of the PWM existence function can now be written as

$$\begin{aligned} H(\omega t,\omega_o t) = {} & 1/2+(x/2)\cos \omega_o t \\ & -(2/\pi)\sum_{p=1}^{p=\infty} \cos(p\pi)\{J_o[x(2p-1)\pi/2]/(2p-1)\}\cos[(2p-1)\omega t] \\ & -(2/\pi)\sum_{p=1}^{p=\infty}\sum_{q=1}^{q=\infty} \cos[(p+q)\pi]\{J_{2q}[x(2p-1)\pi/2]/(2p-1)\} \\ & \qquad \cdot\cos[(2p-1)\omega t \pm 2q\ \omega_o t] \\ & -(2/\pi)\sum_{p=1}^{p=\infty}\sum_{q=1}^{q=\infty} \cos[(p+q)\pi][J_{2q-1}(xp\pi)/2p] \\ & \qquad \cdot\cos[2p\omega t \pm (2q-1)\omega_o t] \end{aligned}$$

where both signs, + and −, are to be taken in the arguments of the oscillatory terms.

Now suppose S_1 of the half-bridge of Figure 4-24 is to be assigned this existence function, now called H_1. Then to develop no dc component of output at $x = 0$ and to avoid violation of KLC and KLV, the function assigned S_2 must be complementary for $x = 0$. At $x = 0$, all $J_m(xn\pi/2) = 0$ and all J_o are unity.

Also, in order to avoid violation of Kirchoff's laws, the modulating function applied to S_2's existence function, H_2, must be phase-displaced by π rad from that of H_1; thus, H_2 may be written as

$$\begin{aligned}
H_2 = {} & 1/2-(x/2)\cos(\omega_o t) \\
& +(2/\pi)\sum_{p=1}^{p=\infty}\cos(p\pi)\{J_o[x(2p-1)\pi/2]/(2p-1)\}\cos[(2p-1)\omega t] \\
& +(2/\pi)\sum_{p=1}^{p=\infty}\sum_{q=1}^{q=\infty}\cos[(p+q)\pi]\{J_{2q}[x(2p-1)\pi/2]/(2p-1)\} \\
& \qquad \cdot\cos[(2p-1)\omega t\pm 2q\omega_o t] \\
& +(2/\pi)\sum_{p=1}^{p=\infty}\sum_{q=1}^{q=\infty}\cos[(p+q)\pi][J_{2q-1}(xp\pi)/2p] \\
& \qquad \cdot\cos[2p\omega t\pm(2q-1)\omega_o t]
\end{aligned}$$

From equation 4.63, the pole-dependent voltage is then given by,

$$\begin{aligned}
v_D = {} & H_1V_S/2 - H_2V_S/2 \\
= {} & xV_S\cos(\omega_o t)/2 \\
& -(2V_S/\pi)\sum_{p=1}^{p=\infty}\cos(p\pi)\{J_1[x(2p-1)\pi/2]/(2p-1)\} \\
& \qquad \cdot\cos[(2p-1)\omega t] \\
& -(2V_S/\pi)\sum_{p=1}^{p=\infty}\sum_{q=1}^{q=\infty}\cos[(p+q)\pi]\{J_{2q}[x(2p-1)\pi/2]/(2p-1)\} \\
& \qquad \cdot\cos[(2p-1)\omega t\pm 2q\omega_o t] \\
& -(2V_S/\pi)\sum_{p=1}^{p=\infty}\sum_{q=1}^{q=\infty}\cos[(p+q)\pi][J_{2q-1}(xp\pi)/2p] \\
& \qquad \cdot\cos[2p\omega t\pm(2q-1)\omega_o t] \qquad (4.105)
\end{aligned}$$

The wanted component is $xV_S\cos(\omega_o t)/2$, has peak amplitude $V_S/2$, and is controllable by varying x, the *modulation index*. The second term of the expansion comprises the *carrier frequency* and its harmonics. It is noteworthy that as x is increased from 0 to 1, the amplitudes of these components decrease. At x

= 0, of course, the wave is simply a square wave at the carrier frequency, of amplitude $V_S/2$, with a fundamental amplitude $2V_S/\pi$; at $x = 1$, the carrier frequency component is reduced to ~ 0.47 times its amplitude at $x = 0$, the fifth harmonic to ~ 0.21 times, and the seventh harmonic to about 0.16 times their amplitudes at $x = 0$.

They are "replaced" by the sideband components of the third and fourth terms of the Fourier expansion, which in general increase in amplitude as x is increased (but are zero amplitude at $x = 0$, of course). Observing these components, it is clear that if ω is not an integer multiple of ω_o, then some components will have lower frequency than the wanted output frequency. This is because, in the series $(2p-1)\omega t - 2q\omega_o t$ and $2p\omega t-(2q-1)\omega_o t$, it is possible to find a q for any p that makes the resultant component frequency $< f_o$ ($\omega_o = 2\pi f_o$). The largest amplitude components of this nature will be those for $p = 1$, since the Bessel coefficients in general diminish with increasing order and the demoninator $2p-1$ increases linearly with p. Suppose then $\omega = y\omega_o$, where y is not an integer. Then the two frequencies in question derive from $y\omega_o - 2q\omega_o$ and $2y\omega_o-(2q-1)\omega_o$. For the first of these, the resulting frequency is less that f_o if q lies in the range $(y-1)/2$ to $(y+1)/2$. The Bessel coefficient is of order $2q$, i.e., of the integer lying between $y-1$ and $y+1$, while the denominator is unity. Similarly, for the second case, the range of q is from y to $y+1$, and thus the Bessel coefficient is of the order of the integer between y and $y+1$, while the denominator is 2. In either case, the amplitude reduces for increasing y, and at large values of y becomes negligible. For low carrier-to- wanted output frequency ratios, however, these "subharmonic" components can have appreciable amplitude. The following list illustrates this point, giving the values of the Bessel coefficients of the term in $(2p-1)\omega t-2q\omega_o t$ for $y = 5.5$, 10.5, 20.5, and 30.5 with $x = 1$. In all these cases the corresponding subharmonic has a frequency one-half of f_o.

f/f_o	$J_{2q}(\pi/2)$
5.5	6.9×10^{-2}
10.5	2.2×10^{-4}
20.5	2×10^{-8}
30.5	$<10^{-10}$

It is highly desirable to have ω an integer multiple of ω_o if ω/ω_0 is to be kept low. It is desirable to do so, if possible, because $f(\omega = 2\pi f)$ is the switching rate.

Examining the sideband terms, if $\omega = k\omega_o$ and k is an integer, the resulting angular frequencies are $[(2p-1)k\pm 2q]\omega_o$ and $[2pk\pm(2q-1)]\omega_o$. The second of these is an odd multiple of ω_0 regardless of whether k is even or odd. The first is an even multiple if k is even, an odd multiple if k is odd. Even order harmonics are very undesirable in the dependent voltage, giving nonidentical half cycles. Also, there will be a value of q for every p that yields a dc component in the dependent voltage, which is even more undesirable.

Thus, it can be deduced that ω should always be an odd multiple of ω_o. The pole wave that arises for $k = 9$ and $x = 0.8$ is shown in Figure 4-33; its similarity to an extended CAM wave of particular θ's and A_B's is obvious.

Not so obvious is the extent of the distortion in such a wave. The process of PWM wave construction has led many workers to conclude erroneously that the PWM wave is a closer approximation to the wanted sinusoid than the basal square wave produced by the pole when operating uncontrolled at ω_o. That this is not so can be readily demonstrated. The rms value of the pole wave is clearly $V_S/2$ whether PWM or not. If it is PWM, then the maximum peak value of the wanted component is $V_S/2$, as has been seen, and the distortion index is given by

$$\mu_{VPWM} = (2-1)^{1/2} = 1$$

while that for the simple square wave is

$$\mu_{VS} = (\pi^2/8-1)^{1/2} = 0.483$$

What then is the benefit of PWM, apart from its giving the ability to control wanted component magnitude? The answer comes in two parts. First, if k (or y) is large, then the amplitudes of the low-order harmonics (sidebands) are very low, so that the higher distortion content is offset by the upward shifting in frequency of the high-amplitude components of the spectrum, making them easier to filter. If $k = 5$, for example, the highest amplitude harmonic will be the third, but if $k = 21$, it will be the nineteenth.

Note that the degeneracy of the spectrum makes it difficult to calculate the amplitudes of individual components when ω/ω_o is an integer. These amplitudes are then the sums of a number of contributions, each having a Bessel coefficient. There seem to be no closed-form solutions for such summations, so a digital computer must be called upon for accurate evaluations. Naturally, there is a slight perturbation of the fundamental, since there exists at least one q for every p in $(2p-1)k-2q$ and $2pk-(2q-1)$, which will make the result 1 or -1. However, the lowest orders of the resulting Bessel coefficients are high, and the variance of wanted component is very small unless k is small.

An appreciation of the distortion consequences of low ω to ω_o ratios can be gained by considering only the highest-amplitude sideband term. For $x = 1$, this has a Bessel coefficient $J_2(n\pi/2) \sim 0.25$. When the contributions of the next two or three terms are added, this rises to $\simeq 0.41$, and the amplitude relative

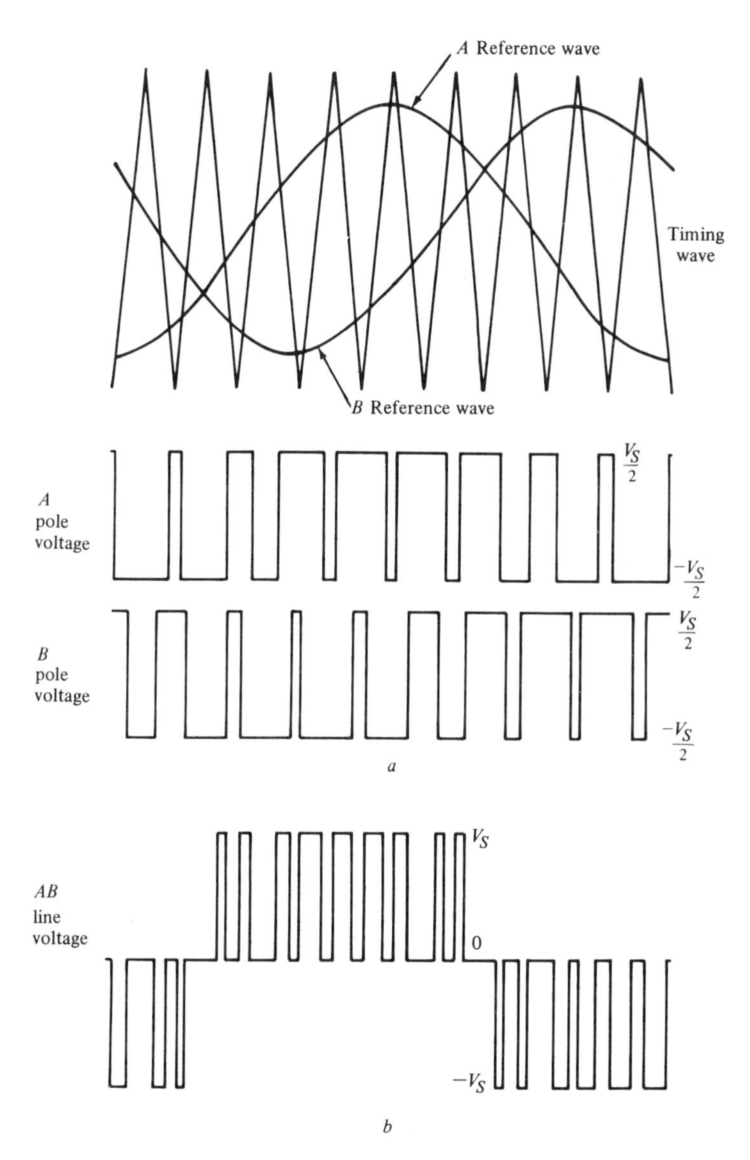

Figure 4-33. PWM by natural sampling.

to wanted component is obtained by multiplying by $4/\pi$. This can be disastrous, and reflects similar behavior to that exhibited by CAM.

The second reason for using PWM is that only in the single-phase case (and sometimes not even then) are the characteristics of a pole-wave germane. In polyphase bridges, it is the difference between two pole waves that provides the line voltage used. The specific case of interest is, of course, three-phase, and the wave produced is depicted in Figure 4-33b, again for $k = 9$ and $x = 0.8$. Two things are immediately obvious. First, the rms of this wave is less than the rms of the six-pulse wave that would be present if PWM were not being used. This leads to the expectation that the distortion index for this line-voltage wave will be considerably better than that of the pole wave, and may be little worse, or perhaps even better, then that of the six-pulse wave. Second, whereas the pole wave alternates between $V_S/2$ and $-V_S/2$, this difference-wave alternates between V_S and zero in the positive half-cycle and zero and $-V_S$ in the negative half-cycle. These behaviors have given rise to the terms *two-level PWM* and *three-level PWM* in much of the literature. As long as it is recognized that *three-level* is related to and derives from *two-level* and is not some radically different waveform construction, there is perhaps nothing wrong in the terminology. However, the general term *multi-level* has also been used to describe the waves of switching power converters that combine pulse-amplitude modulation (PAM) with PWM. Such waves are related to, but not derived from, those of simple PWM converters, and there is a grave danger of confusion if the precise nature of the wave is not specified more accurately than by the number of levels. Multilevel waves also occur when harmonic neutralization is used in conjunction with PWM.

The technique of combining PAM and PWM can yield spectral improvements in the dependent quantities as compared to simple PWM at the same switching rate. Nonetheless, it is, an asinine endeavor. The same spectral results can be obtained by using PWM at a higher switching rate. Using PAM requires additional switches and multlple dc levels or transformer windings. Either way, it is neither an economical nor efficient method of achieving the desired results.

Returning to the line voltage of the three-phase converter, there are two reasons for its improved distortion index over the pole wave. First, the carriers of the two poles—and all carrier harmonics—are in phase. As a result, none of these components appear in the line voltage—the first summation of Equation 4.105 disappears from v_L. Second, any zero-sequence components in the two double summations also disappear. Since the wanted component of the second pole is displaced by $-2\pi/3$ rad from that of the first, this displacement will apply to all $\omega_o t$ in the arguments of the sideband terms. Those with arguments $(2p-1)\,t \pm 2q\omega_o t$ will thus have displacements $\mp 4q\pi/3$, and are zero sequence if q is a multiple of 3. If desired, they can be removed from the summation by introducing the multiplier $(2/\sqrt{3})\sin(q\pi/3)$ and properly accounting for the sign, or by using the multipliers $(2/\sqrt{3})$ and $\sin(\pi/3)$, $\sin[(3m\pm1)\pi/3]$ with the term for $q=1$ segregated and the second variable of summation changed to m.

The terms with arguments $2p\omega t \pm (2q-1)\omega_o t$ will clearly be zero sequence when, and only when, $(2q-1)$ is an integer multiple of 3, which occurs for all $q = (3m+1)/2$, m any odd integer, 1 to ∞.

It is not possible to deduce, either from the expressions or the waves, the extent of the spectral improvement obtained in v_L. However, the frequency ratios, for which the highest-amplitude low-order harmonic component present in the pole wave is eliminated, are easy to establish. They are the series 9, 11, 13, etc., or $2m-1$, $m=5$ to ∞. For $k < 9$, the highest-amplitude low-order component is not eliminated in v_L. This is important, for it provides a very basic design criterion.

To determine the distortion index, the wave developed must be treated as a CAM wave and evaluated with the aid of a digital computer. To calculate individual harmonic amplitudes, the computer's aid must again be invoked. The procedures are straightforward and will not be discussed here.

The rather disappointing spectral performance of PWM, particularly with low k and especially for $k < 9$ in a three-phase bridge, has led to a search for a better modulation technique. Uniform sampling, which produces results very similar to the natural sampling technique so far discussed, is not the answer. However, properly applied, CAM is. It was seen in Section 4.9 that CAM can have disastrous results, because it can produce harmonic amplitudes multiplied by the ratio of ω to ω_o, the switching-frequency-to-wanted-output-frequency ratio. It was also seen, however, that it can produce very good results, for at certain values of θ and A_B, a given harmonic amplitude becomes zero. This suggests that extended CAM can make a multiplicity of harmonics have zero amplitude—in fact, for every notch introduced into the pole wave, a harmonic may be reduced to zero. The requirement that the wanted component magnitude be controlled modifies this postulate, so the requirement is for one notch more than the number of harmonics to be "neutralized."

Determining the θ's and A_B's needed to produce a given level of fundamental and optimum spectral results presents a formidable multivariate analytic problem. It is solved with the aid of a digital computer, using established and sophisticated numerical techniques for minimizing an *objective function* of the θs and A_Bs. The objective function need not necessarily be cancellation of the maximum possible number of harmonics; it may be minimum distortion index, minimum rms content of some number of harmonics greater than the number that can be cancelled, or any other collection of criteria that strikes the designer's fancy.[6] Once the θ's and A_B's are found, a digital control may be made to implement them in the converter.

This technique is termed *programmed modulation* or *programmed waveform* because the converter control is preprogrammed, in a read-only memory (ROM), with the pulse patterns needed for the existence functions to produce optimum performance with a given set of criteria. It has been found that dependent-quantity spectra can indeed be very much better than for natural or uniform sampling PWM at the same carrier (switching frequency). Depending on the criteria of the objective function, programmed waves can match the performance of PWM waves with 3 to 10 times higher switching rates. Moreover, the optimization process can take into account the practical imperfections of switching devices, passive components, and control systems. As a result, simple PWM has fallen into disuse except for low-power applications where very high values of k can be used without significant penalty.

So far, the dependent voltage of a PWM voltage-sourced converter has been addressed. Since the basic PWM existence function was formulated, the dependent current expression is easy enough to produce. Since it is even more cumbersome than that for the voltage, it will not be reproduced here. A few comments will suffice.

The first term, 1/2, gives rise to a component of frequency f; in full and polyphase bridges, such a component is a complete phasor set summing to zero. The second term, $(x/2)\cos(\omega_o t)$, gives rise to the wanted dc component. Its magnitude is clearly controlled by x. The first summation gives rise to a series of sidebands of frequencies $nf \pm f_o$; as x is increased, their amplitudes diminish. In single-phase and polyphase bridges, they form complete phasor sets and sum to zero. The double summations produce sideband terms of frequencies $(2p-1)f \pm (2q+1)f_o$, $(2p-1)f \pm (2q-1)f_o$, $2pf \pm 2qf_o$, and $2pf \pm 2(q-1)f_o$. If $f = kf_o$ and k is odd, only even harmonics of f_o are present. In polyphase bridges, only those for which $2q+1 = mM$, $2q-1 = mM$, $2q = mM$, and $2(q-1) = mM$ are present, where M is the number of phases. The spectrum is obviously as complex and difficult to evaluate as that of the voltage, but like the voltage it tends to be much worse than might be anticipated by a casual approach to the subject. Programmed waveform techniques effect improvements of the same order as they do for the dependent voltage.

So far, we have treated PWM and programmed waveform techniques as they apply to voltage-sourced converters. Clearly, they are not restricted to that application; they may equally well be implemented in current-sourced converters. Spectral results for the dependent quantities will be the same—those for the current-sourced converters can be derived from those for the voltage-sourced converters via the reciprocity relationship. Obviously, the switch additions needed to implement CAM in current-sourced converters are also needed to implement PWM or programmed waveforms. Until very recently, PWM was never used with current-sourced converters. Some usage has begun to develop and is discussed in Chapter 11.

The choice among PWM, programmed waveform techniques, and harmonic neutralization is another difficult multivariate problem. However, some generalizations are possible. Programmed waveform is much better than PWM except in low-power applications, where the simplicity of a natural sampling PWM control (see Chapter 10) and lack of great penalty for high switching rates make PWM attractive. Programmed waveform is better than harmonic neutralization for single-phase converters and for polyphase applications where isolating transformers are not required. Harmonic neutralization is almost invariably superior if a transformer is required. These comparisons are based on economic factors, efficiencies, sizes, weights, and reliabilities—not on attainable performance. Performance can be made equivalent for all types of converter implementation.

References

1. Wagner, C. F. Parallel inverter with resistance load. *AIEE Transactions* 1227–1235 (November 1935).
2. Wagner, C. F. Parallel inverter with inductive load. *AIEE Transactions* 970–980 (September 1936).
3. Humphrey, A. J., and B. Mokrytzki. Inverter paralleling reactors. *IEEE Power Converter Conference Record* 2-4-1-2-4-6 (May 1972).
4. Kernick, A., et al. Static inverter with neutralization of harmonics. *Transactions AIEE 81* (2) 56–68 (1962).
5. Black, F. *Modulation Theory*. New York: D. Van Nostrand Co., 1951, pp. 264–280.
6. Pitel, I., et al. Programmed waveform pulse width modulation. *IEEE/IAS Conference Record* (October 1979).

Problems

1. Sketch a three-phase midpoint current-sourced ac-to-dc/dc-to-ac converter. If the ac supply has a line-to-line voltage of 208 V rms, calculate:

a. The maximum and minimum dc voltage output of the converter

b. The maximum and minimum amplitudes of the first three components of the ripple voltage

c. The maximum *peak* amplitude of the ripple voltage

2. If the converter of Problem 1 has a defined current of 100 amp dc, calculate

a. The maximum and minimum amplitudes of the in-phase and quadrature fundamental currents

b. The amplitudes and harmonic orders of the first four unwanted components in the line currents

3. For the converter of Problems 1 and 2, calculate the displacement factors and power factors for α's of 0, $-\pi/3$, $-\pi/2$, $-2\pi/3$, and $-\pi$ rad.

4. Sketch a three-phase bridge current-sourced ac-to-dc/dc-to-ac converter. What line voltage is the supply required to have to produce a maximum dc output voltage of 600 V? Calculate the peak ripple voltage at $\alpha = 0$, $\alpha = \pi/2$.

5. Two six-pulse bridges are to be combined to form a 12-pulse current-sourced converter. Sketch a transformer arrangement to develop the required ac supplies from a single three-phase source, and state the phase displacement between the supplies developed. Sketch the dc terminal configurations for both series and parallel connection of the bridges, and describe the purpose of any additional components introduced.

6. An available dc voltage supply has a range of 105 V to 140 V. If it is used to feed a three-phase bridge voltage-sourced dc-to-ac converter, what is the range of the fundamental line-voltage component at the converter's output? Calculate the total rms distortion of the phase and line voltages, and explain the reason for the difference.

7. Given the dc supply of Problem 6, it is desired to produce a regulated three-phase 208 V fundamental line ac output, using phase-displacement control. Sketch the arrangement of converters and transformers needed; calculate the transformer ratio and the range of phase displacement.

8. It is desired to produce the ac supply of Problem 7 with the dc source of Problems 6 and 7 using a single bridge with CAM control. If θ is set to zero, and the minimum possible notch width ($2\pi/A_B$) is $2\pi/72$ rad, calculate the output transformer ratio needed, the maximum notch width required, and the maximum amplitudes of the first two unwanted components in the line voltages.

9. Repeat Problem 8 using two-notch CAM with minimum notch widths of $2\pi/144$ rad and θ's of $\pm\ \pi/3$ rad.

10. Calculate the range of fundamental line voltage produced by a PWM (natural sampling) voltage-sourced converter operated from the dc supply of Problem 6. If the output is regulated to a constant value, calculate the maximum total rms distortion that will be observed in the phase voltages, and compare it with that produced by the phase-displacement control of Problem 7 and the single-notch CAM control of Problem 8.

5
AC-to-AC Converters

5.1 Introduction

For all ac to ac converters, reciprocity is no longer a consideration, since a converter and its reciprocal are merely different versions of the same type of switching matrix.

It was shown in Chapter 1 that ac-to-dc/dc-to-ac switching matrices arose when one defined quantity was made unidirectional and switch capabilities were restricted to suit. If either of the switches used in ac-to-dc/dc-to-ac converters has its capabilities extended, a fully bilateral switch results. The only extension possible for a unidirectional current-carrying, bidirectional voltage-blocking switch is to make it bidirectional current carrying. For a bidirectional current-carrying, unidirectional voltage-blocking switch, the only possible extension is to make it bidirectional voltage blocking. In both cases, the result is a bidirectional current-carrying, bidirectional voltage-blocking switch.

If such switches are placed in the switching matrix of any two-quadrant ac-to-dc/dc-to-ac converter, a four-quadrant converter results. These converters, often used in machine-drive applications, share the properties of the two-quadrant versions. In fact, the limitations of actual switching devices generally cause realizations to take the form of two two-quadrant converters in appropriate combination. Therefore, four-quadrant converters will not be discussed separately.

As mentioned in Chapter 2, there are three possible forms of modulation that will permit realization of ac-to-ac converters when applied to the existence functions of bilateral switches in a four-quadrant converter matrix. A variety of interesting properties arise, varying with the modulating and control functions used. Of course, the ac regulators must also be considered. They will be dealt with last, since their relationship to the more sophisticated converters is tenuous and they are quite simple artifacts.

5.2 Converters with Linear Modulating Functions

Consider the midpoint converter in Figure 4-1. If its switches are made fully bilateral, it becomes a phase-controlled four-quadrant converter, capable of producing either polarity of dc voltage with either polarity of defined dc current.

The phase-control angle, α, will control the magnitude of the wanted dc component of the dependent voltage. If the converter is to be used as an ac-to-ac converter, this component must be set to zero, for a dc component in a dependent ac quantity is either totally impermissible or extremely undesirable. Thus, the basal existence function for the converter at zero output becomes

$$H_0 = 1/M+(2/\pi) \sum_{n=1}^{n=\infty} [\sin(n\pi/M)/n]\cos[n(\omega_s t\pm\pi/2)] \qquad (5.1)$$

With a set of existence functions of this form, the converter will produce no wanted component in the dependent voltage. If $\pi/2$ delay is used, it will produce a positive-type ripple wave regardless of the direction of the defined current, whereas, if $\pi/2$ advance is used, the unwanted dependent voltage will be a negative-type wave. These voltages are shown in Figure 5-1 for a three-phase converter.

In Chapter 2, the simplest modulating functions capable of producing an ac wanted component in the dependent voltage were shown to be $+\omega_o t$ and $-\omega_o t$. Inserting these into the existence function of Equation 5.1 yields

$$H_{\pm\omega_o t} = 1/M+(2/\pi) \sum_{n=1}^{n=\infty} [\sin(n\pi/M)/n]\cos[n(\underline{\omega_s t\pm\pi/2\pm\omega_o t})] \qquad (5.2)$$

where the form $\underline{\omega_s t\pm\pi/2\pm\omega_o t}$ indicates that the two sign combinations are to be taken independently. From Equations 4.2 and 4.14, the wanted components are given by

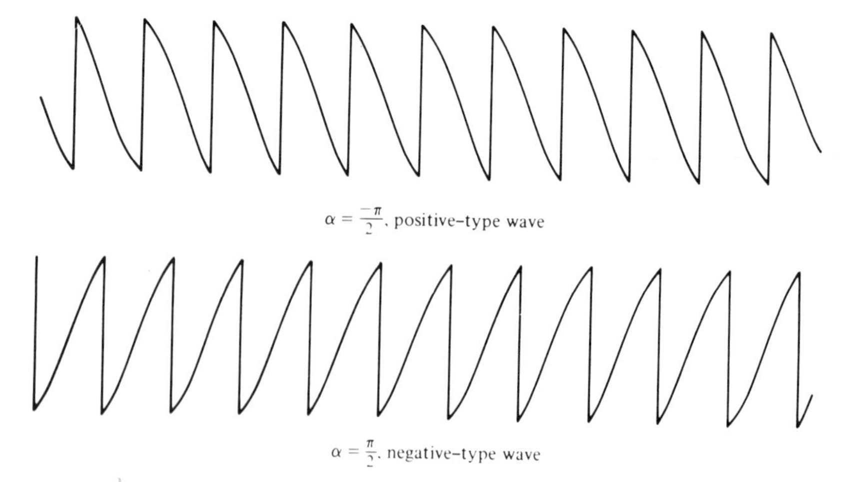

Figure 5-1. Dependent-voltage waves of ac-to-ac converters producing zero wanted components.

$$v_{DD} = (M/\pi)\sin(\pi/M)V\cos(\underline{\pm\omega_o t \pm \pi/2})$$

which for the choice $+\omega_o t$ results in

$$v_{DD+} = \mp(M/\pi)\sin(\pi/M)V\sin(\omega_o t) \tag{5.3}$$

and for the choice $-\omega_o t$ results in

$$v_{DD-} = \pm(M/\pi)\sin(\pi/M)V\sin(\omega_o t) \tag{5.4}$$

Thus, both choices result in the same wanted component, except for phase displacement. Note that there is no amplitude-control variable in these expressions; the amplitude of the wanted component with these modulating functions is fixed at $(M/\pi)\sin(\pi/M)V$. Methods of introducing amplitude control are discussed in Section 5.4.

The unwanted components of converters with these modulating functions are, from Equations 5.2 and 4.15, given by the expressions

$$v_{DU} = (M/\pi)\sin(\pi/M)V \sum_{p=1}^{p=\infty} \cos(p\pi)\{\cos[pM\omega_s t + (pM+1)(\omega_o t \pm \pi/2)]/(pM+1)$$
$$- \cos[pM\omega_s t + (pM-1)(\omega_o t \pm \pi/2)]/(pM-1)\} \tag{5.5}$$

if the modulating function is $\omega_o t$ and

$$v_{DU} = (M/\pi)\sin(\pi/M)V \sum_{p=1}^{p=\infty} \cos(p\pi)\{\cos[pM\omega_s t - (pM+1)(\omega_o t \mp \pi/2)]/(pM+1)$$
$$- \cos[pM\omega_s t - (pM-1)(\omega_o t \mp \pi/2)]/(pM-1)\} \tag{5.6}$$

if the modulating function is $-\omega_o t$.

Examining these expressions, it is clear from Equation 5.5 that the unwanted components when $\omega_o t$ is the modulating function have angular frequencies $pM\omega_s + (pM+1)\omega_o$ and $pM\omega_s + (pM-1)\omega_o$. This represents quite a good spectrum, with no unwanted components of angular frequency less than $pM\omega_s$ or $(pM-1)\omega_o$. Note that these oscillatory components are not, in general, harmonics of either $f_s(=\omega_s/2\pi)$ or $f_o(=\omega_o/2\pi)$. They will only be harmonics of f_o if $f_s/f_o = k/M$, where k is any integer, and can only be harmonics of f_s if f_o is an integer multiple of f_s.

If the modulating function is $-\omega_o t$, the angular frequencies of the unwanted components are $pM\omega_s - (pM+1)\omega_o$ and $pM\omega_s - (pM-1)\omega_o$. Again, in general, they are not harmonics of either f_o or f_s. They all are harmonics of f_o if $f_o/f_s =$

k/M, where k is any integer, and some are harmonics of f_o when the relationship $f_s/f_o = 1 + (k \pm 1)/pM$ is satisfied. They can never be harmonics of f_s. Thus, the spectrum for $-\omega_o t$ as the modulating function is much worse than that for $\omega_o t$. Instead of only sums of multiples of the two frequencies, only differences exist. This is a marked disadvantage for the choice $-\omega_o t$, for it obviously can make for much greater difficulty in interfacing the converter at the dependent voltage terminals.

Observe also that if the choice is $\omega_o t$, Equation 5.2 allows ω_o to have any value from zero to infinity without the existence functions becoming unrealizable. This lack of restriction on ω_o—it may be lower than, equal to or higher than ω_s—led to use of the name *Unrestricted Frequency Changer* (UFC) for an ac-to-ac converter with $\omega_o t$ as the modulating function.[1] However, if $-\omega_o t$ is the choice, then Equation 5.2 shows that the existence functions become unrealizable at $\omega_o = \omega_s$, and thus for this modulating function, the restriction $\omega_o < \omega_s$ must be invoked. A converter with $-\omega_o t$ as the modulating function is called a *Slow Switching Frequency Changer* (SSFC) because the switching rate, $f_s - f_o$, is the lowest that can be used to produce a wanted component of frequency f_o from an ac-to-ac converter.[1]

As Equations 5.3 and 5.4 show, either modulating function yields a wanted ac component of frequency f_o. The phase of this component can be made anything desired with respect to an arbitrary reference by adding the appropriate phase displacement to the modulating function. Thus, if $\omega_o t + \phi$ is used, the wanted component becomes $\pm(M/\pi)\sin(\pi/M)V\sin(\omega_o t + \phi)$, and ϕ appears, multiplied by $pM \pm 1$, in all arguments of the unwanted oscillatory components. With $-\omega_o t - \phi$ as the modulating function, the wanted component is $\mp(M/\pi)\sin(\pi/M)V\sin(\omega_o t + \phi \cdot)$, and $\phi \cdot$ again appears in all arguments of the unwanted oscillatory components. Thus, to create an M-phase-to-N-phase midpoint converter, it is merely necessary to use N M-phase midpoint converters with phase delays (or advances) in the modulating functions of 0, $2\pi/N$, $4\pi/N$. . . $2(N-1)\pi/N$ rad to match the phase displacements of the defined current phasor set.

Waveforms for both modulating functions at $f_o/f_s = 1/3$ are shown in Figure 5-2. The waves vary with the combination of initial conditions, $\pm\pi/2$, and ϕ, of course, because of variations in the phase angles of the unwanted components relative to each other and the wanted component. Waveforms for the Slow-Switching Frequency Changer operating at $f_o/f_s = 1/2$ and $2/3$ are shown in Figure 5-3, while waveforms for the Unrestricted Frequency Changer operating at $f_o/f_s = 1$ and 3 are shown in Figure 5-4. Examining these waveforms carefully, it can be seen that the UFC produces a positive-type wave whenever the slope of the wanted component is positive, a negative-type wave whenever the wanted component's slope is negative. The SSFC does the converse, producing a negative-type wave whenever the slope of its wanted component is positive, and a

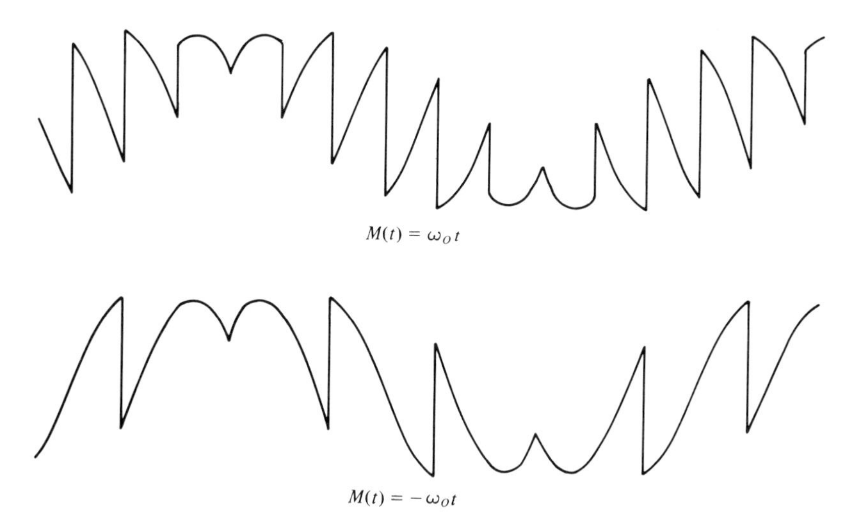

Figure 5-2. Dependent-voltage waves produced by linear modulating functions at $f_o/f_s = 1/3$.

positive-type wave whenever that slope is negative. In Chapter 4, it was observed that positive-type wave and positive current or negative-type wave and negative current are associated with a lagging displacement factor. Conversely, positive-type wave and negative current or negative-type wave and positive current are associated with a leading displacement factor. If these associations are assumed to apply on an instantaneous basis, rather than merely in the steady

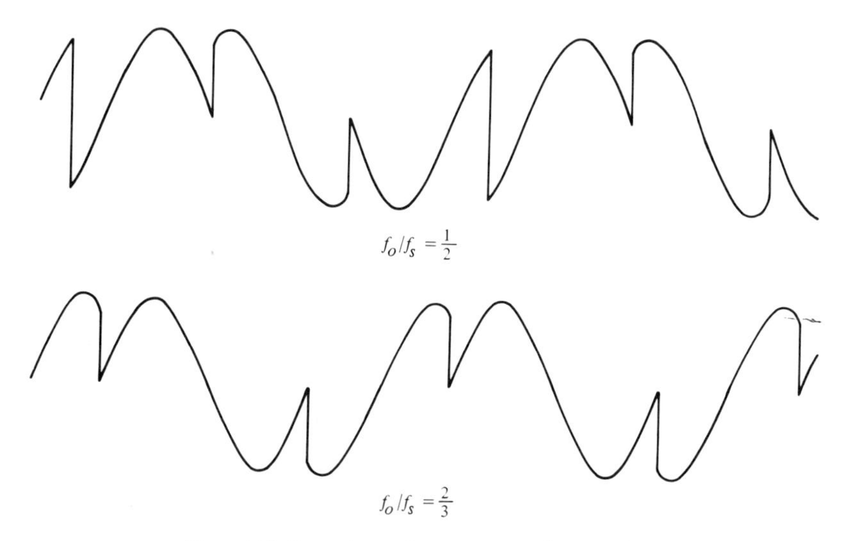

Figure 5-3. Dependent-voltage waves produced by SSFC.

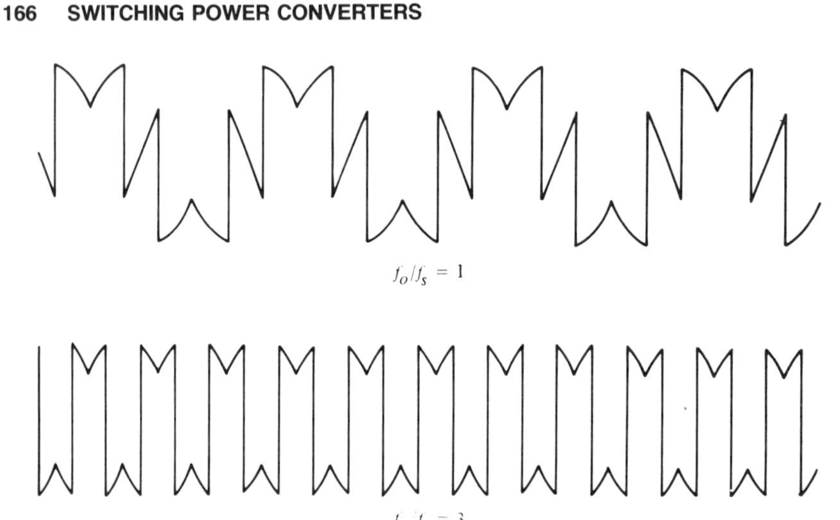

Figure 5-4. Dependent-voltage waves produced by UFC.

state, some interesting deductions can be made about the dependent-line current-displacement factors of UFCs and SSFCs.

Consider the UFC with a defined ac current in phase with the wanted component of its dependent voltage. Then there will be equal time periods of positive-wave, positive-current and negative wave, positive current in each positive-current half-cycle. Similarly, there will be equal time periods of negative wave, negative current and positive wave, negative current in each negative-current half-cycle. It can be concluded that the UFC should show unity-displacement factor (i.e., on the average, have neither a lagging nor a leading dependent-line current-displacement factor) for unity-displacement factor loading at the dependent-voltage terminals. Now suppose the defined current is made to lag the wanted component of the dependent voltage. Then the positive-current half-cycle will coexist with longer periods of negative-type wave, and the negative-current half-cycle will coexist with longer periods of positive-type wave. Thus, the converter should exhibit, on the average, a leading displacement factor of its dependent currents. If the defined current leads the wanted component of dependent voltage, then the current-wave-type relationships become the opposites of those cited for lagging defined current, and the converter can be expected to show a lagging displacement factor of its dependent currents. This is indeed a curious property—apparently, a UFC will "image" its output displacement angle at its input lines, where output is stipulated to be the defined currents connected to the dependent voltages, and input is the defined voltage set wherein flow the dependent currents. This property will shortly be verified by analysis of the dependent currents; however, it has just been seen that the behavior can be deduced without performing the analysis.

Following the same line of reasoning and argument with regard to the SSFC leads to the conclusion that it will reproduce its output's displacement angle at its input terminals. Again, this will shortly be verified by analysis, but it is deducible without performing the analysis. This ability to deduce displacement factor through current- and wave-type associations is of considerable significance. In approaching any switching converter, an examination of its current- and wave-type relationships will reveal its displacement factor behavior. This fact can often save valuable time in comparative performance assessments of switching power converters.

It should be quite obvious that the spectral improvements achieved in ac-to-dc/dc-to-ac converters by using the bridge connection and harmonic neutralization are also achieved for both the UFC and the SSFC, as far as their dependent voltages are concerned. The phase displacements required are identical—$2\pi/lP$ rad for successive P pulse converters of l used in a harmonic neutralized set and, of course, π rad displacement for the existence functions of the complementary converter used to form a bridge.

Attention turns now to the dependent currents of ac-to-ac converters with linear modulating functions. Suppose the modulating functions are defined as $\omega_0 t \mp \pi/2$ or $-\omega_0 t \mp \pi/2$ so that the basal existence functions become

$$H_{UFC} = 1/M + (2/\pi) \sum_{n=1}^{n=\infty} [\sin(n\pi/M)/n] \cos[n(\omega_s + \omega_o)t]$$

and

$$H_{SSFC} = 1/M + (2/\pi) \sum_{n=1}^{n=\infty} [\sin(n\pi/M)/n] \cos[n(\omega_s - \omega_o)t]$$

From Equations 5.3 and 5.4, the wanted components become:

$$v_{DDUFC} = v_{DDSSFC} = (M/\pi)\sin(\pi/M)V\cos(\omega_o t)$$

The defined current for either converter will take the form $I\cos(\omega_0 t + \psi)$, where ψ is an arbitrary phase angle between this current and the wanted component of the dependent voltage. Clearly, all three line currents will have identical form, differing only by phase displacement, and hence it is only necessary to evaluate one.

Consider first that of the UFC; the dependent current is given by:

$$i_{DUFC} = I\cos(\omega_o t + \psi)/M + (2I/\pi) \sum_{n=1}^{n=\infty} [\sin(n\pi/M)/n] \cos[n(\omega_s + \omega_o)t] \cos(\omega_0 t + \psi) \quad (5.7)$$

The first term of this expression is a component at the output frequency. It will clearly vanish for the bridge configuration, since the defined current for the complementary group is $-I\cos(\omega_o t+\psi)$. The remaining terms of Equation 5.7 expand to

$$i'_{DUFC} = (I/\pi)\sum_{n=1}^{n=\infty}[\sin(n\pi/M)/n]\{\cos[n\omega_s t+(n+1)\omega_o t+\psi] + \cos[n\omega_s t+(n-1)\omega_o t-\psi]\} \quad (5.8)$$

The wanted component is that at frequency f_s, from the term for $n = 1$; it is given by

$$i_{DDUFC} = (I/\pi)\sin(\pi/M)\cos(\omega_s t-\psi) \quad (5.9)$$

Observe that the phase angle ψ appears as $-\psi$ in this current; the UFC does indeed image the displacement angle of its output in presenting it to its input, as was deduced from the current-wave-type associations.

The unwanted components are all oscillatory terms of Equation 5.8 other than the wanted component. The frequencies present are: for all $n = 1$ to ∞, $nf_s+(n+1)f_o$; for $n = 2$ to ∞, $nf_s+(n-1)f_o$. The lowest unwanted frequencies in these series are f_s+2f_o and $2f_s+f_o$. Like the unwanted components of the dependent voltages, none of these unwanted components are, in general, harmonics of either f_s or f_o. In fact, they can only be harmonics of f_s if $f_o \geqslant f_s$ and are all harmonics of f_s only if f_o/f_s is an integer. The lowest frequency components can be quite a problem, for if $f_o < f_s$ then $f_s + 2f_o$ will be close to f_s and very difficult to filter. If $f_o << f_s$, not only will $f_s + 2f_o$ be very close to f_s, but $2f_s + f_o$ is very close to being a second harmonic.

Note that this M-phase-to-single-phase ac-to-ac converter presents a balanced burden to its input set of M-defined voltages. This is an inherent characteristic of ac-to-ac converters—the unbalance inherent in the M-phase-to-single-phase conversion is reflected in the unwanted components of the dependent currents, not in unbalanced wanted components. If N converters are used to feed a complete set of N-defined current phasors, then the most objectionable unwanted components disappear, as is shown below. In this case, N switches, one from each of the N converters, feed currents into each of the input (defined-voltage) lines. These N currents all have the form of Equation 5.7, but $\omega_o t$ is replaced by $\omega_o t-2k\pi/N$ for the kth contribution of N, $k = 0$ to $N - 1$.

It is clear that the output (defined current) frequency components will sum to zero. Equally clearly, the wanted component, of frequency f_s, defined by Equation 5.9, simply has its amplitude multiplied by N. However, the unwanted components of Equation 5.8 become complete phasor sets, the key elements in their arguments being $(n+1)(\omega_o t - 2k\pi/N)$ and $(n-1)(\omega_o t - 2k\pi/N)$. Hence, they will sum to zero except for the conditions $n+1 = pN$ and $n-1 = pN$; for these conditions, their amplitudes are multiplied by N. Thus, for N-phase output, the unwanted components of the input current (in the A, or reference line, the 0th) become:

$$i_{DUUFCN} = (NI/\pi) \sum_{p=1}^{p=\infty} \{\sin[(pN+1)\pi/M] \cdot \cos[(pN+1)\omega_s t + \psi + pN\omega_o t]/(pN+1)$$

$$+ \sin[(pN-1)\pi/M] \cdot \cos[(pN-1)\omega_s t - \psi + pN\omega_o t]/(pN-1)\} \quad (5.10)$$

The most important practical case is $M = N = 3$; in this event, Equation 5.10 reduces to

$$i_{DUUFC3} = (3\sqrt{3}I/2\pi) \sum_{p=1}^{p=\infty} \cos(p\pi)\{\cos[(3p+1)\omega_s t + \psi + 3p\omega_o t]/(3p+1)$$

$$- \cos[(3p-1)\omega_s t - \psi + 3p\omega_o t]/(3p-1)\}$$

The complementary natures of this and the dependent-voltage spectrum of Equation 5.6 are quite evident. Typical waveforms for three-phase-to-single-phase and three-phase-to-three-phase currents are shown in Figure 5-5, with $f_o/f_s = 1/3$ and $\psi = 0$.

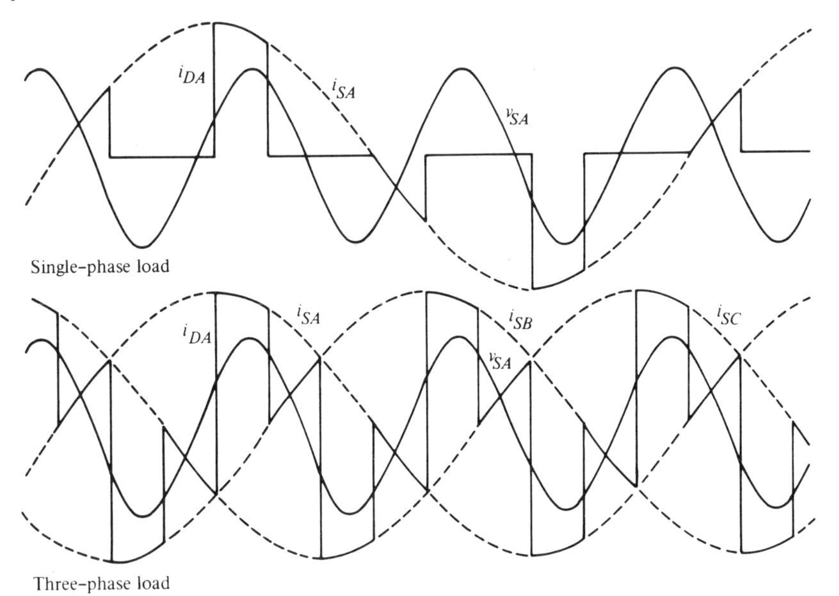

Figure 5-5. Dependent-current waves produced by UFC at $f_o/f_s = 1/3$.

The analysis of the line currents for an SSFC follows the same lines. The total current for single-phase output is given by

$$i_{DSSFC} = I\cos(\omega_o t+\psi)/M + (2I/\pi)\sum_{n=1}^{n=\infty}[\sin(n\pi/M)/n] \cdot \cos[n(\omega_s-\omega_o)t]\cos(\omega_o t+\psi)$$

The first term is again at the output frequency f_o, and again it clearly disappears if the bridge configuration is used or if N converters burdened by a complete set of N-defined current phasors are fed from the same defined-voltage set. The remaining terms expand to

$$i'_{DSSFC} = (I/\pi)\sum_{n=1}^{n=\infty}[\sin(n\pi/M)/n]\{\cos[n\omega_s t-(n+1)\omega_s t-\psi] + \cos[n\omega_s t-(n-1)\omega_o t+\psi]\} \quad (5.11)$$

From Equation 5.11, the wanted component is

$$i_{DDSSFC} = (I/\pi)\sin(\pi/M]\cos(\omega_s t+\psi)$$

and the SSFC is seen to reproduce its output displacement angle at its input, as was deduced from current-wave-type associations. The unwanted components are the remaining terms of Equation 5.11 and contain frequencies $nf_s-(n+1)f_o$ and $nf_s-(n-1)f_o$. Just like the dependent voltage of the SSFC, its dependent current has a poor spectrum, consisting of differences of multiples of the input and output frequencies. The largest amplitude terms have frequencies f_s-2f_o and $2f_s-f_o$. The former is clearly "subharmonic" and can be very distressing. Observe, however, that if N converters are involved (with a complete set of N output current phasors), these components disappear, and the unwanted components become

$$i_{DUSSFCN} = (NI/\pi)\sum_{p=1}^{p=\infty}\{\sin[(pN+1)\pi/M] \cdot \cos[(pN+1)\omega_s t-\psi-pN\omega_o t]/(pN+1) + \sin[(pN-1)\pi/M] \cdot \cos[(pN-1)\omega_s t+\psi-pN\omega_o t]/(pN-1)\} \quad (5.12)$$

and in the three-phase-to-three-phase case, the frequencies are $(3p\pm1)f_s-3pf_o$; the largest amplitude components of the three-phase-to-single-phase currents vanish. Again, the unbalance inherent in the M-phase-to-single-phase conversion is reflected in increased and more obnoxious unwanted components in the input line currents, not unbalanced fundamental currents. Typical waveforms, single- and three-phase output, are shown in Figure 5-6, with $f_o/f_s = 1/3$ and $\psi = 0$.

At this point, observe that in the UFC, with modulating function $+\omega_o t$, all unwanted components in the dependent quantities have frequencies that are sums of multiples of the input and output frequencies. In the SSFC, with modulating function $-\omega_o t$, all unwanted components in the dependent quantities have frequencies that are differences of multiples of the input and output frequencies. These statements may be paraphrased to read that modulating functions of positive slope produce only the upper sidebands; those of negative slope produce only the lower sidebands. This conclusion can be used to make deductions about the probable spectra produced by nonlinear modulating functions, which are discussed in Section 5.5.

From Equations 5.10 and 5.12, it should be clear that harmonic neutralization as applied at the ac terminals of dc-to-ac/ac-to-dc converters also works, in exactly the same manner, at the defined-voltage terminals of these ac-to-ac converters with linear modulating functions, just as it worked on their dependent voltages. It will work on the dependent currents because the unwanted components contain the multipliers $3p \pm 1$ (midpoint) and $6p \pm 1$ (bridge) on f_s just as do the harmonics produced by the converter of Chapter 4. Hence, all manipulations with phase shifts and sequences that were valid for those harmonics are also valid for these unwanted components. The results are essentially the same,

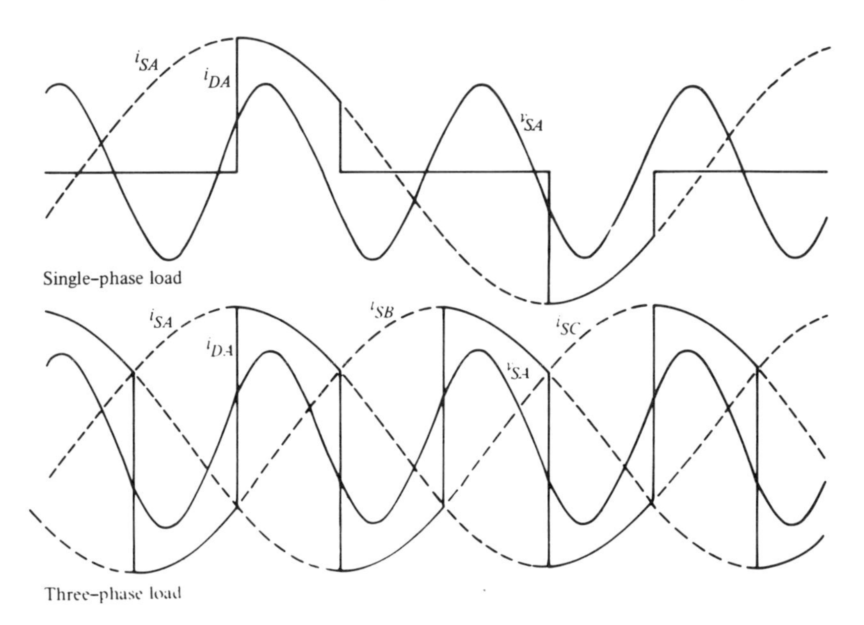

Figure 5-6. Dependent-current waves produced by SSFC at $f_o/f_s = 1/3$.

too. If ℓ three-phase midpoint groups are combined, the pulse number becomes $3l$; the dependent voltage(s) than contain unwanted components with frequencies $3lf_s+(3l\pm1)f_o$ *(UFC) or* $3lf_s-(3l\pm1)f_o$ (SSFC). The dependent currents contain unwanted components with frequencies $(3l\pm1)f_s+3lf_o$ (UFC) or $(3l\pm1)f_s-3lf_o$ (SSFC). If $3l$ in these expressions is replaced by $6l$, the statements apply equally well to harmonic neutralized UFC and SSFC three-phase bridges. As before, the same principles apply to M phase to M-phase systems too. The M-phase-to-N-phase case is rather different, but its complexity is beyond the scope of this book.

5.3 Switch Stresses in Converters with Linear Modulating Functions

From Figures 5-5 and 5-6, which show the line current waves for the converters, it is clear that the peak current in any switch equals the peak of the defined current. Also, the switches must tolerate high $\partial I/\partial t$ for both directions of current when turning on and are subjected to high reverse $\partial I/\partial t$ ($\partial I_R/\partial t$) when turning off. These statements apply to both the UFC and the SSFC. Establishing switch average (average of the absolute value of the current) and rms currents is very difficult. Both quantities vary with f_o/f_s. For integer ratios of f_o/f_s in the UFC or f_s/f_o in either converter, the average currents in the several switches are not identical. Fortunately, neither average nor rms current is very meaningful in terms of switch rating, since in general a switch conducts a train of current pulses of equal time durations but varying amplitudes. Practical switches are temperature-limited (see p. 312) and thus dissipation-limited, but their thermal time constants are short, comparable to or shorter than the conduction periods of the converters, and internal temperature fluctuates considerably. Hence, establishing a rating for a switch in one of these ac-to-ac converters (and most others) is quite difficult. Methods of attacking this problem and related rating problems are described in Chapter 7.

A very pessimistic approach is to consider that each current pulse has a constant amplitude equal to the peak amplitude of the defined current. This will result in a safe but overly conservative design. Less pessimistic, but somewhat risky, is considering that each current pulse in a switch has a constant amplitude equal to the average of the absolute values of the defined current. This approach will yield fairly accurate results for most f_s/f_o(or f_o/f_s) that are incommensurable, but is dangerous for integer and rational quotients, since it then can give very optimistic results for the switches under the highest current stresses.

From the voltage waves for the switches in UFCs and SSFCs in Figure 5-7, it can be seen that the maximum voltage to be withstood is the peak line voltage of the input set. It is also clear that in both converters the switches operate in two "regimes" as regards voltage blocking. Whenever positive-current, positive-type wave or negative-current, negative-type wave conditions exist, the

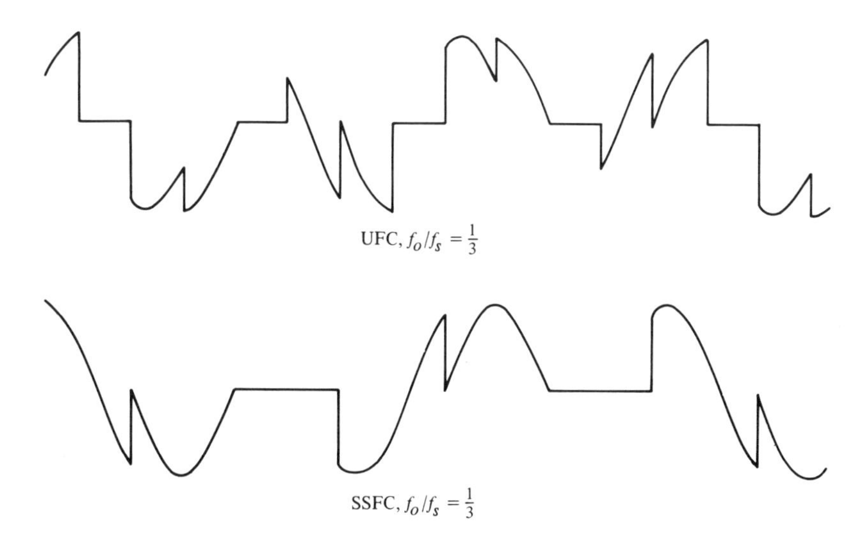

Figure 5-7. Switch-voltage waves for ac-to-ac converters operating with linear modulating functions.

switch voltage stresses are similar to those of ac-to-dc/dc-to-ac converters operating with negative α. That is, immediately following conduction, they are subjected to the sudden application of voltage of opposite polarity (to the current just quenched), and subsequent reapplied voltage of the same polarity as the current is gentle, following an input sinusoid. Also, immediately prior to closing, the switches are withstanding a voltage of the same polarity as the current they will carry subsequent to closure.

When positive-current, negative-type wave and negative-current, positive-type wave conditions exist, the switch voltages reverse characters and become like those of switches in ac-to-dc/dc-to ac converters with positive α. That is, immediately upon ceasing to conduct current they are subjected to a high reapplied $\partial V/\partial t$, the voltage being of the same polarity as the current just quenched; immediately prior to turning on the switches, they are withstanding voltage of opposite polarity to the current that is to flow.

Thus, the switches for these ac-to-ac converters with linear modulating functions are required to withstand the worst possible combination of stresses. In this light, it is perhaps not surprising that no combination of switching devices, let alone any one device, has yet been developed to meet the demands of these applications efficiently and economically.

5.4 Dependent Voltage Control in UFCs and SSFCs

It was observed in Section 5.2, upon the analyses for the dependent voltages, that no control variables exist in the expressions for the wanted components

(Equations 5.3 and 5.4). Control of the amplitude of these components is essential to successful application of these converters. The means for implementing such control are now explored.

The quantity to be controlled is a dependent voltage, of which the magnitude is a function of that of the defined voltage set. This strongly suggests that the control techniques for voltage-sourced dc-to-ac converters (discussed in Section 4.9) will work for these converters too. It is a trivial matter to observe that this is true for phase-displacement control. If two converters, UFC or SSFCs, are operated from the same input in order to produce the same wanted component of dependent voltage, with one component at a phase angle ϕ with respect to an arbitrary reference and the other at $-\phi$, then the multiplier $2\cos\phi$ will appear in the expression for their combined wanted component.

For harmonic neutralized, bridge, and simple midpoint UFCs and SSFCs, combination may be effected by series connection of isolating transformer windings. It may also be accomplished by parallel connection of such windings, or of the converters directly, via an interphase reactor. However, the interphase reactor must withstand the fundamental (wanted) voltage difference that develops with ϕ, of $2\sin\phi$ times the magnitude of either individual component, and thus tends to be large, expensive, and inefficient. Hence, parallel connection is not generally favored.

In the case of midpoint converters only, direct series connection can be used. The converter so created is, of course, analogous to a controlled current-sourced single-quadrant ac-to-dc converter. However, to prevent phase displacement of the variable wanted component of the combined output with respect to an external reference, the two constituent midpoint groups are simultaneously displacement-controlled, one by $+\phi$ and the other by $-\phi$ with respect to that reference.

All these phase-displacement-controlled UFCs and SSFCs exhibit control of the magnitude of the wanted components of their input currents as a consequence of their control of the wanted component of their combined outputs. None, however, modify their inherent displacement factor behavior. This is easy to prove, for if the combined output phase angle is ψ, then one converter of a phase-displacement-controlled pair sees an output displacement angle of $\phi + \psi$, while the other sees one of $-\phi + \psi$. If they are UFCs, then individual input-displacement angles will be $-\phi-\psi$ and $\phi-\psi$. Then

$$I_{DD}[\cos(\omega_s t-\psi-\phi) + \cos(\omega_s t-\psi+\phi)] = 2I_{DD}\cos\phi\cos(\omega_s t-\psi)$$

gives the combined wanted component of input current. Replacing $-\psi$ by $+\psi$ in the individual input currents of an SSFC obviously leads to the same result with $-\psi$ replaced by $+\psi$. However, suppose one converter is a UFC and the other an SSFC. Then the input displacement angle of the UFC will be $\pm\phi-\psi$ and that of the SSFC $\mp\phi + \psi$, yielding a combined input current given by

$$I_{DD}[\cos(\omega_s t \pm \phi - \psi) + \cos(\omega_s t \mp \phi + \psi)] = 2I_{DD}\cos\psi\cos(\omega_s t \pm \phi)$$

Such a combination will clearly have unity displacement factor regardless of the load displacement, ψ, if $\phi = 0$. It is then a form, although not a preferred one, of the Unity Displacement Factor Frequency Changer (UDFFC)[1] discussed in Section 5.5. Unfortunately, in general, the dependent component spectra of such a combination will contain both upper and lower sidebands—frequencies $pMf_s \pm (pM+1)f_o$ and $pMf_s \pm (pM-1)f_o$ in the combined output voltage, $(pM+1)f_s \pm pNf_o$ and $(pN-1)f_s \pm pNf_o$ in the combined input current. As will be seen, this is true of all ac-to-ac converters using alternative modulating functions and is not the primary reason for eschewing this implementation of a UDFFC. That reason is the difference in controls between the two constituent converters. A second obviously unwelcome consequence of this combination is the restriction $\omega_o < \omega_s$, imposed by the SSFC. This is also a restriction applied to other UDFFC implementations.

For $\phi \neq 0$, this peculiar converter produces an input current wanted component with a displacement factor that is a function of ϕ but not ψ, i.e., its input displacement factor is controllable, by ϕ, regardless of output displacement factor. As such, it is a form, although not the preferred one, of the Controlled Displacement Factor Frequency Changer (CDFFC).[1]

The use of two converters, UFCs or SSFCs, to produce a controlled wanted component by phase-displacement control is not generally regarded as a desirable configuration. Obviously, as with using the same type of control for dc-to-ac converters, it requires that twice as many switches as are needed to accomplish the basic conversion function be employed. Thus, the question arises as to the suitability of CAM as a control technique for the ac-to-ac converters. Obviously, CAM cannot be implemented with the midpoint circuit (Figure 4-1 with bilateral switches) without violating KLC. However, if a freewheeling switch is added to the converter, as in Figure 4-13 with all switches fully bilateral, then CAM may be employed. The existence function of the kth switch is modified by subtracting from it a control existence function

$$H'_k = 1/A_B + (2/\pi) \sum_{n=1}^{n=\infty} [\sin(n\pi/A_B)/n]\cos\{n[(\omega_s \pm \omega_o)t - 2k\pi/M + \theta]\}$$

to become

$$H_k = 1/M - 1/A_B + (2/\pi) \sum_{n=1}^{n=\infty} \Big([\sin(n\pi/M)/n]\cos\{n[(\omega_s \pm \omega_o)t - 2k\pi/M]\} - [\sin(n\pi/A_B)/n]\cos\{n[(\omega_s \pm \omega_o)t - 2k\pi/M + \theta]\} \Big)$$

with ϕ, the phase angle of the wanted component, set at $\mp \pi/2$.

Obviously, $A_B > M$ and $\theta + \pi/A_B < \pi/M$ are necessary restrictions. The existence function for the freewheeling switch becomes the sum of all the control existence functions, i.e.,

$$H_F = M/A_B + (2/\pi) \sum_{k=0}^{k=M-1} \sum_{n=1}^{n=\infty} [\sin(n\pi/A_B)/n] \cos\{n[(\omega_s \pm \omega_o)t - 2k\pi/M + \theta]\}$$

$$= M/A_B + (2M/\pi) \sum_{p=1}^{p=\infty} [\sin(pM\pi/A_B)/pM] \cos\{pM[(\omega_s \pm \omega_o)t + \theta]\}$$

The dependent voltage of this converter is given by

$$v_D = \sum_{k=1}^{k=M-1} H_k \cdot v_k$$

Using superposition, one element of this voltage is the uncontrolled voltage. The other element is similar in form, but is substractive and dependent on both A_B and θ. The wanted component then becomes

$$v_{DDC} = (M/\pi)\sin(\pi/\mathrm{M})\mathrm{V}\cos(\omega_o t) - (M/\pi)\sin(\pi/A_B)\cos(\omega_o t + \theta) \quad (5.13)$$

which is clearly controlled in magnitude by both A_B and θ. Its phase angle is also dependent on both these variables; this is undesirable, and can be avoided by putting $\theta = 0$, when the magnitude becomes simply $(M/\pi)V[\sin(\pi/M) - \sin(\pi/A_B)]$. Clearly, the unwanted components produced have the same frequencies as those of the uncontrolled voltage. Their amplitudes are modified, with the multiplier $\sin[(pM \pm 1)\pi/M]$ implicit in Equation 5.5 changing to $\sin[(pM \pm 1)\pi/M] - \sin[(pM \pm 1)\pi/A_B]$. This expression maximizes for $\sin[(pM \pm 1)\pi/A_B] = \pm 1$, becoming $\pm 1 + \sin[(pM \pm 1)\pi/M]$. The ratio of maximum amplitude when controlled, to amplitude when uncontrolled, is thus $\pm 1 + 1/\sin[(pM \pm 1)\pi/M]$; the absolute maximum value of this ratio is $1 + 1/\sin(\pi/M)$. For the common case, $M = 3$, this is $1 + 2/\sqrt{3}$, somewhat less than the ratio of 3 observed with CAM in the dc-to-ac converters but still a severe increase in individual unwanted components for the appropriate values of the control variable A_B.

As might be expected, the amplitudes of the wanted and unwanted components in the CAM-controlled ac-to-ac converters' dependent currents are affected in very similar fashion to those in the dependent voltages. The wanted component of current acquires the same amplitude multiplier, and the converters' displacement-factor properties are not affected. The multiplier for unwanted components with single-phase output becomes $\sin(n\pi/M) - \sin(n\pi/A_B)$, and with N-phase output, the multiplier becomes $\sin[(pN \pm 1)\pi/M] - \sin[(pN \pm 1)\pi/A_B]$.

This obviously behaves similarly to the expression for the voltage amplitude—indeed, when $N = M$, and $N = M = 3$ is the common case, it is the same expression—and maximizes when $\sin[(pN \pm 1)\pi/A_B] = \pm 1$.

When CAM control for the wanted components is applied to a bridge configuration, there is clearly no need to add a freewheeling switch. The freewheel paths can be created, using the switches already present in the bridge, by adding the control existence function displaced by π rad to each switch existence function as well as subtracting it at zero displacement.

Obviously, there is no restriction on the degree of CAM that may be used. Provided the control existence functions are symmetrically disposed, any number of them can be introduced for either the midpoint or bridge configuration. The penalties—increased maximum unwanted component amplitudes—will escalate in similar fashion to those suffered in extended CAM dc-to-ac/ac-to-dc converters. In the limit, a UFC or SSFC with PWM superposed will result. Such systems have not been constructed, and it is difficult to perceive their usefulness, particularly in the case of the UFC. Simple CAM control gives adequate performance for the majority of possible applications.

5.5 AC-to-AC Converters with the Nonlinear Modulating Function

As seen in Chapter 2, there exists an alternate modulating function, $M(t) = \arccos[\cos(\omega_o t)]$, which will yield a wanted component containing $\cos(\omega_o t)$ when used in ac-to-ac converters. This choice offers one advantage: if it can be realized in the form given, then $M(t,x) = \arccos[x\cos(\omega_o t)]$ can also be realized where $0 \leq x \leq 1$. This function will result in a wanted component containing $x\cos(\omega_o t)$, and x, the modulation index (or *modulation depth*), is a control variable for the wanted component amplitude.

Figure 5-8 shows this function, $M(t,x)$, for values of x ranging from 0.2 to 1. It is clearly periodic and therefore subject to expression by Fourier expansion.

Now consider the midpoint ac-to-ac converter operating with $\alpha = -\pi/2$, producing zero wanted components in its dependent voltage. Its switches' existence functions contain the oscillatory elements, $\cos[n(\omega_s t - \pi/2)]$. If the modulating function $\pi/2 - M(t,x)$ is introduced into the arguments of the existence functions, then the resulting wanted component of its dependent voltage is obviously

$$v_{DD} = (M/\pi)\sin(\pi/M)V\cos[M(t,x)]$$

$$= (M/\pi)\sin(\pi/M)V\, x\cos(\omega_o t)$$

There is obviously nothing to prevent the realization of $M(t,x,\phi) = \arccos[x\cos(\omega_o t + \phi)]$, so that a phase displacement ϕ will appear in this wanted

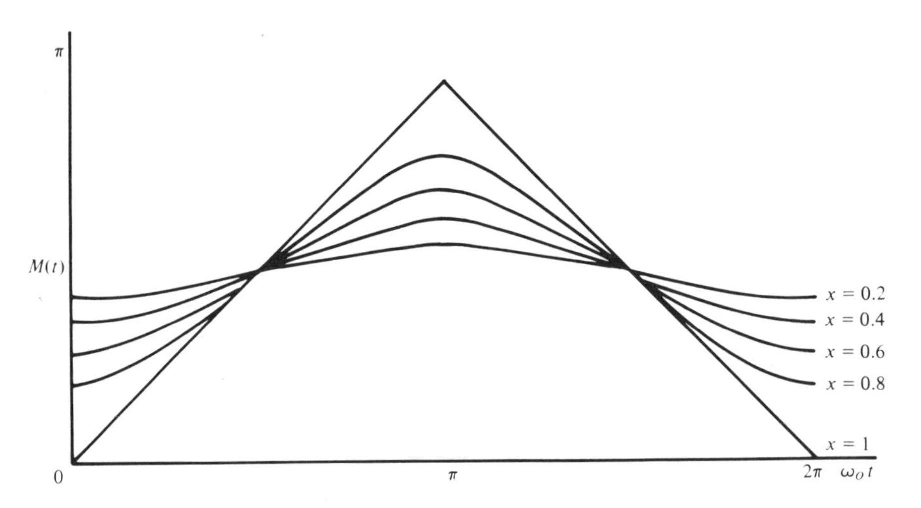

Figure 5-8. Nonlinear modulating function, $M(t) = \arccos(x\cos\omega_o t)$.

component in order to allow the creation of polyphase sets using multiple converters.

The manner in which this converter operates is quite interesting. Observe the modulating function $\pi/2 - M(t, 1)$ in Figure 5-9a. At the origin, the effective α of the converter is clearly 0. As the function traverses the first π rad of $\omega_o t$, its slope is negative and the converter's effective α is clearly being swung from 0 to $-\pi$. However, during the traverse from π to 2π rad of $\omega_o t$, the function has a positive slope, and α is being swung back from $-\pi$ to 0. Hence, at all times the converter operates with negative α and thus *always produces a positive-type dependent-voltage wave.* Since this wave coexists with equal time periods of positive and negative current, the defined current taking the form $I\cos(\omega_o t + \psi)$, this converter can be expected to exhibit unity input displacement factor *regardless of its output displacement factor.*

This behavior remains the same, of course, for any modulating function $\pi/2 - M(t,x)$ introduced to the converter's existence functions when $\alpha_o = \pi/2$. The effect of $x < 1$ is simply to restrict the maximum excursions of $\alpha - \alpha_{max} < \pi$ and $\alpha_{min} > -\pi$ and to make the excursions nonlinear. Consider now the same converter operated with $\alpha_o = \pi/2$, so that it produces a negative-type wave when the wanted component of dependent voltage is zero. If the modulating function $-\pi/2 + M(t,x,\phi)$ is then introduced to its existence functions, it will produce a wanted component of dependent voltage given by

$$v_{DD} = (M/\pi)\sin(\pi/M)V\,x\cos(\omega_o t + \phi)$$

just as did the first converter considered. The modulating function now used is

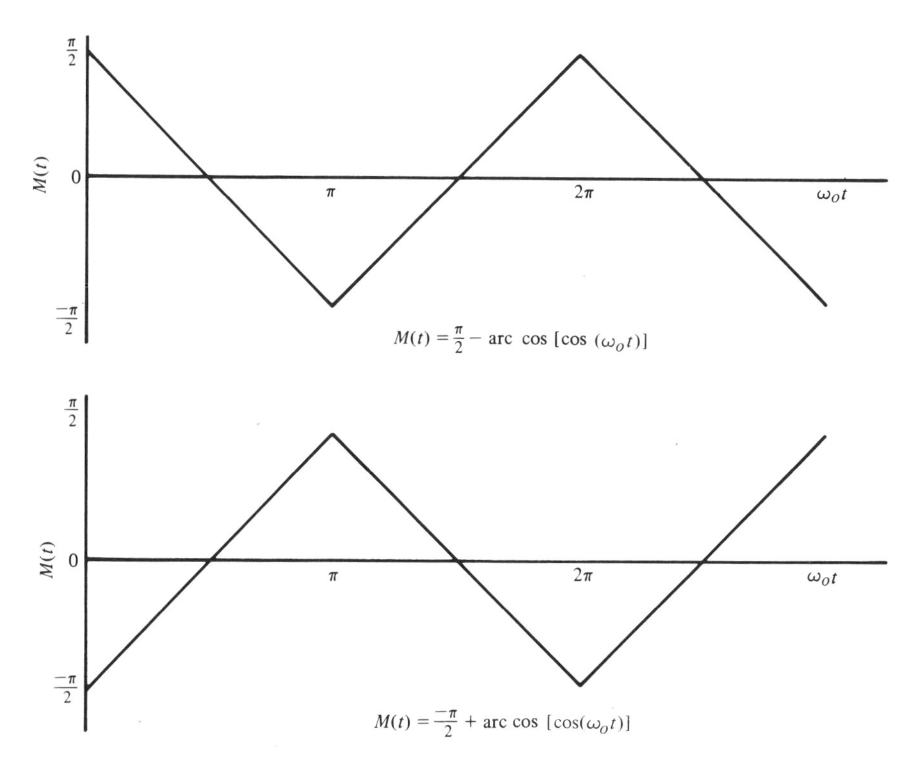

Figure 5-9. Nonlinear modulating functions.

shown in Figure 5-9b (for $x = 1$ and $\phi = 0$). Observe that the converter is operating at $\alpha = 0$ at the origin again, but that over the first π rad of $\omega_o t$, the function has a positive slope and α is advanced to $+\pi$ rad. From π to 2π rad of $\omega_o t$, the function assumes a negative slope and α is retarded from π rad back to 0.

Thus, this converter always produces a negative-type wave, since α is ever positive. It too, then, can be expected to have an input displacement factor of unity regardless of that of its output.

These two converters are evidently complementary. One always produces a positive-type wave and uses $-M(t,x,\phi)$ as its modulating function; the other always produces a positive-type wave and uses $+M(t,x,\phi)$ as its modulating function. Henceforth, they will be called simply the *positive converter* and the *negative converter,* nomenclature first used by Gyugyi (1970).[2] They are both forms of the UDFFC; moreover, they have the ability to control the amplitudes of the wanted voltage components they produce. However, individually, they are not the preferred form for the UDFFC, for reasons which will become apparent very shortly.

Now we will examine the unwanted components produced by these converters in their dependent voltages. Since the modulating functions contain periods of both positive and negative slope, it might be deduced that both upper- and lower-sideband unwanted components will be present. That is, the spectra can be expected to contain components of frequencies $pMf_s + (pM\pm1)f_o$ and $pMf_s - (pM\pm1)f_o$. In addition, it might be expected that the singularities of the derivatives of these modulating functions would have some effects on the spectra, which is indeed found to be the case.

For the positive converter (setting $\phi = 0$ for convenience), the unwanted components are given by

$$v_{DUP} = (M/\pi)\sin(\pi/M)V \sum_{p=1}^{p=\infty} \cos(p\pi)\{\cos[pM\omega_s t-(pM+1)M(t,x)]/(pM+1) - \cos[pM\omega_s t-(pM-1)M(t,x)]/(pM-1)\} \quad (5.14)$$

Those of the negative converter are given by a similar expression, with $-M(t,x)$ replaced by $+M(t,x)$.

The spectra cannot be evaluated unless the trigonometric terms representing the oscillatory components in Equation 5.14 are expanded. Using $\cos(A\pm B) = \cos A \cos B \mp \sin A \sin B$ gives

$$\cos[pM\omega_s t\mp(pM\pm1)M(t,x)] = \cos(pM\omega_s t)\cos[(\underline{pM\pm1})M(t,x)] \pm \sin(pM\omega_s t)\sin[(\underline{pM\pm1})M(t,x)]$$

where the form $(\underline{pM\pm1})$ indicates that both signs therein are to be taken regardless of whether "+" or "−" is taken in the other double-sign combination. Now it can be seen that the unwanted components of these dependent voltages are of the form

$$v_{DUP} = A_U + B_U$$

$$v_{DUN} = A_U - B_U$$

where A_U contains the oscillatory components of the product $\cos(pM\omega_s t)\cos[(pM\pm1)M(t,x)]$, and B_U contains those of $\sin(pM\omega_s t)$ $\sin[(pM\pm1)M(t,x)]$. Thus, if a positive and a negative converter are combined by series connection of their outputs or by parallel connection via an interphase reactor, the unwanted components in B_U will disappear in their combined output. Therefore, the preferred form for a unity displacement factor frequency changer (UDFFC) is such a combination, since it eliminates what will be seen to be a rather nasty set of unwanted components from the dependent voltage.

Expanding further, the oscillatory components of A_U come from

$$\cos(pM\omega_s t)\cos[(pM\pm 1)\arccos(x\cos\omega_o t)]$$

and those of B_U from

$$\sin(pM\omega_s t)\sin[(pM\pm 1)\arccos(x\cos\omega_o t)]$$

For the case $x = 1$, the further development of the oscillatory components of A_U is trivial, for it is clear from Figure 5-8 that $\cos[N\arccos(\cos\omega_o t)] = \cos(N\omega_o t)$ and A_{U1} is seen to contain the oscillatory terms

$$\cos[pM\omega_s t+(pM\pm 1)\omega_o t],\ \cos[pM\omega_s t-(pM\pm 1)\omega_o t]$$

just as was forecast by observing that the modulating functions have both positive- and negative-slope periods. For $x < 1$, the evaluation is more complex.

Observe that $\cos(N\theta)$, where N is any integer, can be expressed as a finite series of R terms if $N = 2R-1$, and $R = 1$ to ∞, of the form

$$\cos[(2R-1)\theta] = K_{R,2R-1}(\cos\theta)^{2R-1} + K_{R,2R-3}(\cos\theta)^{2R-3} + \ldots K_{R,1}\cos\theta$$

whereas if $N = 2R$, the finite series will contain $R+1$ terms and be of the form

$$\cos(2R\theta) = K_{R,2R}(\cos\theta)^{2R} + K_{R,2R-2}(\cos\theta)^{2R-2} + \ldots K_{R,0}$$

Thus, $\cos[(pM\pm 1)\text{arc}\cos(x\cos\omega_o t)]$ can be expressed as one of these finite series, with $(\cos\theta)^Q$ becoming $x^Q[\cos(\omega_o t)]^Q$. Then, observe that involuting the finite series for $\cos(N\theta)$ gives finite series for $\cos(\phi)^Q$ in the form

$$\cos(\phi)^{2R-1} = B_{R,2R-1}\cos[(2R-1)\phi]+B_{R,25R-3}\cdot\cos[(2R-3)\phi]$$
$$+\ldots B_{R,1}\cos\phi$$

if $Q = 2R-1$ and

$$\cos(\phi)^{2R} = B_{R,2R}\cos(2R\phi)+B_{R,2R-2}\cos[(2R-2)\phi] + \ldots B_{R,0}$$

if $Q = 2R$. Using these expansions, $\cos[(pM+1)\text{arc}\cos(x\cos\omega_o t)]$ can always be expressed as a finite series of terms involving powers of x and the cosines of multiples of $\omega_o t$, where the general term has the form $C_n x^n\cos(n\omega_o t)$ with n ranging from 0 to $pM\pm 1$ if $pM\pm 1$ are even and from 1 to $pM\pm 1$ if $pM\pm 1$ are odd. Thus, these unwanted components expand each sideband into a finite series of sidebands, when $x < 1$, so that the frequencies become pMf_s, $pMf_s \pm 2f_o$, $pMf_s \pm 4f_o$ . . . $pMf_s \pm (pM \pm 1)f_o$ if pM is odd and $pMf_s \pm f_o$, $\pm 3f_o$, $\pm 5f_o$ . . . $\pm(pM \pm 1)f_o$ is pM is even. Calculation of the coefficients C_n is extremely laborious and is not presented here. Pelly (1971)[3] and Gyugyi and Pelly (1975)[1] both tabulate values for the important three-phase case.

The evaluation of the unwanted components contained in B_U requires yet more Herculean labors. Even with $x = 1$, this is not a completely trivial problem, for $\sin[N \text{arc} \cos(\cos\omega_o t)]$ is not a simple function. However, it can be expressed as a Fourier expansion, since it is obviously periodic, in the general form

$$\sin[N\arccos(\cos\omega_o t)] = A_{N,0}/2 + \sum_{m=1}^{m=\infty} (A_{N,m}\cos m\omega_o t + B_{N,m}\sin m\omega_o t)$$

Observing Figure 5-8, it can be concluded that the function $\sin[N\arccos(\cos\omega_o t)]$ is equal to $\sin N\omega_o t$ during all intervals for which $2q\pi \leqslant \omega_o t \leqslant (2q+1)\pi$, where $q = -\infty$ to ∞, and is equal to $-\sin N\omega_o t$ for the intervals $(2q+1)\pi \leqslant \omega_o t \leqslant 2q\pi$. This permits evaluation of the Fourier coefficients in the expansion of this function as

$$A_{N,m} = 1/\pi \int_{\pi\,2\pi}^{\pi\,\pi} \sin(N\omega_o t)\cos(m\omega_o t)d(\omega_o t)$$

and

$$B_{N,m} = 1/\pi \int_{\pi\,2\pi}^{\pi\,\pi} \sin(N\omega_o t)\sin(m\omega_o t)d(\omega_o t)$$

where the double limits indicate that each integral is to be evaluated over both sets. Evaluating these coefficients yields

$$B_{N,m} = 0 \text{ for all } N, m.$$

$$A_{N,0} = 0 \text{ if } N \text{ is even}$$

$$= \frac{2}{N\pi} \text{ if } N \text{ is odd.}$$

$$A_{N,m} = 0 \text{ if } N \text{ and } m \text{ are both even}$$

$$= 0 \text{ if } N \text{ and } m \text{ are both odd}$$

$$= \frac{2}{\pi}[1/(N+m)+1/(N-m)] \text{ if } N \text{ is odd and } m \text{ even or } N \text{ even and } m \text{ odd.}$$

Thus, there are in effect two different expansions, given by

$$\sin[(2n-1)\arccos(\cos\omega_o t)] = A_{2n-1,0}/2 + \sum_{m=1}^{m=\infty} A_{2n-1,2m}\cos(2m\omega_o t)$$

and

$$\sin[2n\arccos(\cos\omega_o t)] = \sum_{m=1}^{m=\infty} A_{2n,2m-1}\cos[(2m-1)\omega_o t]$$

The terms in B_U are of the form

$$\sin(pM\omega_s t)\sin[(pM \pm 1)\cos(x\cos\omega_o t)].$$

From the foregoing, it is clear that each term in B_U separates into an infinite series of sidebands even for $x = 1$, with frequencies pMf_s, $pMf_s \pm 2f_o$, $pMf_s \pm 4f_o$. . . if pM is even and $pMf_s \pm f_o$, $\pm 3f_o$, $\pm 5f_o$. . . if pM is odd. Now putting $x<1$ clearly does not affect the periodicity or character of sin[Narc cos($x\cos\omega_o t$)]. Therefore, the effect of a fractional x is merely a modification of the Fourier coefficients $A_{2n-1,2m}$ and $A_{2n,2n-1}$. *Merely* is perhaps inappropriate in this context, for the evaluation of these coefficients becomes very laborious. Again, Gyugyi and Pelly (1975)[1] and Pelly (1971)[3] contain such evaluations, and they are not pursued further here.

The dependent-voltage waveforms produced by three-phase midpoint positive and negative converters for $f_o/f_s = 1/3$ are illustrated in Figure 5-10. Numerous other illustrations can be found in References 1 and 3 for midpoint, bridge, and harmonic neutralized converters. It should be clear that the spectral benefits observed in previous converters for bridge and harmonic neutralized configurations apply also to these converters and, hence, to all combinations thereof.

The dependent currents of the positive and negative converters are given by [with $\phi = \pm\pi/2$ in $M(t,x,\phi)$]:

$$i_{DP,N} = I\cos(\omega_o t+\psi)/M + (2I/\pi)\sum_{n=1}^{n=\infty} [\sin(n\pi/M)/n]\cos[n(\omega_s t \mp M(t,x)]\cos(\omega_o t+\psi)$$

for single-phase output, $I_S = I\cos(\omega_o t+\psi)$, with only the reference line ($k = 0$, of $k = 0$ to $M-1$) expressed. The first term, at f_o, disappears for a bridge configuration or when the output is polyphase. The remaining terms all contain the products $\cos(\omega_o t+\psi)\cos\{n[\omega_s t \mp \arccos(x\cos\omega_o t)]\}$ which expand to

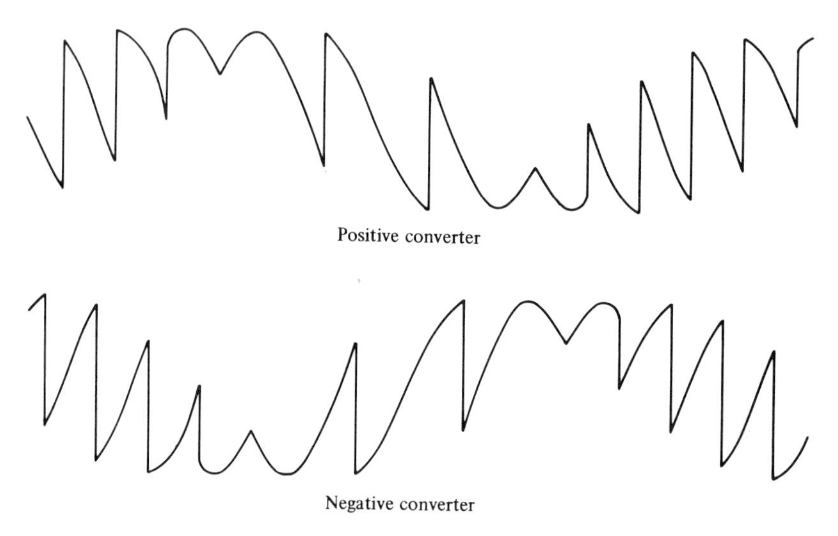

Figure 5-10. Dependent-voltage waves produced by ac-to-ac converters operating with nonlinear modulating functions at $f_o/f_s = 1/3$.

$$\cos(\omega_o t+\psi)\{\cos(n\omega_s t)\cos[n\arccos(x\cos\omega_o t)]\pm\sin(n\omega_s t)\sin[n\arccos(x\cos\omega_o t)]\}$$

To evaluate these expressions, recall that during the analysis of the dependent voltages it was established that $\cos[n\arccos(x\cos\omega_o t)]$ expands into a finite series of terms, each of the form $C_{n,m}x^m\cos(m\omega_o t)$. For n odd, all m are odd and for n even, all m are even in these expansions. It is also established that $\sin[n\arccos(x\cos\omega_o t)]$ expands into an infinite series, with terms of the form $A_{n,m}\cos(m\omega_o t)$, where, if n is odd all m are even (including $m = 0$) and if n is even all m are odd. Thus, the only term giving a wanted component in the dependent current is that for $n = 1$, and this fundamental frequency component is

$$i_{DDP,N} = (I/\pi)\sin(\pi/M)x\cos\psi\cos\omega_s t$$

This clearly has unity displacement factor regardless of the magnitude or sign of ψ, as was predicted from wave-type current associations.

The term having $n = 1$ also results in unwanted components in the dependent current, as do the terms for all other n. From the first part of the term for $n = 1$,

$$\cos(\omega_o t+\psi)\cos(\omega_s t)\cos[\arccos(x\cos\omega_o t)] = \cos(\omega_o t+\psi)\cos(\omega_s t).x\cos(\omega_o t)$$

$$= (1/2)\cos(\omega_s t)[x\cos\psi+x\cos(2\omega_o t+\psi)]$$

There arise two unwanted components:

$$(I/2\pi)\sin(\pi/M)x\cos(\omega_s t+2\omega_o t+\psi)$$

and

$$(I/2\pi)\sin(\pi/M)x\cos(\omega_s t-2\omega_o t-\psi)$$

Clearly, these are analogous to the unwanted components of frequencies f_s+2f_o and f_s-2f_o produced, respectively, by positive and negative linear modulating functions. Like those, they will disappear when the converters now under consideration are operated with polyphase output. From the expansion of $\cos[n\arccos(x\cos\omega_o t)]$, if n is odd, a finite series of terms exists that contains $\cos(m\omega_o t)$, where m is odd ($m = 2q-1$). The cross products of these with $\cos(\omega_o t+\psi)$ result in a finite series of terms containing $\cos(2q\omega_o t)$ and $\cos[2(q-1)\omega_o t]$, and in the case of single-phase output, all frequencies $(2p-1)f_s\pm2qf_o$ ($n = 2p-1$) exist in the unwanted components, with q ranging from 0 to p (maximum value of $2q-1$ is n in the finite series). For N phase output, only those components for which $2q = mN$ will exist.

If n is even, the expansion of $\cos[n\arccos(x\omega_o t)]$ produces a finite series of terms containing $\cos m\omega_o t$ where $m = 2q$, and $q = 0$ to ∞. Their cross products with $\cos(\omega_o t+\psi)$ produce terms containing $\cos[(2q-1)\omega_o t]$, where $q = 1$ to ∞. Thus, for single-phase output, all frequencies $2pf_s\pm(2q-1)f_o$ exist, $p = 1$ to ∞ and $q = 1$ to p. For N-phase output, only those terms for $(2q-1) = mN$ will exist. For $x = 1$, the unwanted frequencies present reduce to those comprising the sets $(pN\pm1)f_s\pm pNf_o$; for $x<1$, more components appear. The set $(2pN\pm1)f_s$ is then present, and the sets $(pN\pm1)f_s\pm pNf_o$ expand to become $(pN\pm1)f_s\pm Nf_o$, $\pm2Nf_o, \ldots \pm pNf_o$. Unwanted frequencies of the forms $2qNf_s\pm Nf_s$, $\pm3Nf_s, \ldots \pm(2q-1)Nf_s$ and $(2q-1)Nf_s$, $(2q+1)Nf_s\pm2Nf_s$, $\pm4Nf_s, \ldots \pm2qNf_s$, with a range of q from 1 to ∞ in both cases, also appear for fractional x.

Coefficients for all these components in the important three-phase-to-single-phase and three-phase-to-three-phase cases have been evaluated and are tabulated in Pelly (1971)[1] and Gyugyi and Pelly (1975)[3]. They are not repeated herein. It should be observed, however, that in the three-phase-to-three-phase case (in fact, in any M-phase-to M-phase case), the last two sets of frequencies are not present because $\sin(kM\pi/M) = 0$.

The second part of the expansion is observed to have alternate signs—positive for the positive converter, negative for the negative converter. Thus, like those of the dependent voltages, the dependent currents' unwanted components take the forms

$$i_{UP} = I_U + J_U$$

$$i_{UN} = I_U - J_U$$

where the components in I_U are those just discussed and those in J_U are those arising from the second part of the expansion. Again, if positive and negative converters are combined to form a UDFFC, the unwanted components in J_U vanish, just as did those of B_U for the dependent voltage. Again, too, the components in J_U are extensive. They arise from the products

$$\cos(\omega_o t+\psi)\sin(n\omega_s t)\sin[n\arccos(x\cos\omega_o t)]$$

When n is even, $\sin[n\arccos(x\cos\omega_o t)]$ expands to an infinite series of terms of the form $A_{n,2q,2p+1}\cos[(2p+1)\omega_o t]$ giving rise to unwanted frequencies $2qf_s \pm 2pf_o$. A double summation exists, of course, over $q = 1$ to ∞ and over $p = 0$ to ∞. When n is odd, $\sin[n\arccos(x\omega_o t)]$ expands into an infinite series having terms $A_{n,2q-1,2p}\cos(2p\omega_o t)$, where $p = 0$ to ∞, resulting in unwanted components of frequencies $(2q-1)f_s \pm (2p+1)f_o$. The double summation is again involved, of course, with $q = 1$ to ∞ and $p = 0$ to ∞. All these components will be present with single-phase output. Some disappear for polyphase output, because they form complete phasor sets summing to zero. For the first group, only those for $2pf_o = mN$ will remain, while for the second group only those for $(2p+1)f_o = mN$ will remain.

Examining the frequencies of the components in these cases, it can be seen that the even-order harmonics of the fundamental, $2qf_s$, where $q = 1$ to ∞, will be present. If N, the number of output phases, is even, then so will be all frequencies $2qf_s \pm mNf_o$; if N is odd, then this series of components reduces to those of frequencies $2qf_s \pm 2mNf_o$, $m = 0$ to ∞ in both cases. The second series $(2q-1)f_s \pm (2p+1)f_o$, will not exist if N is even since $2p+1$ cannot then be a multiple of N. If N is odd, it can be so only as an odd multiple of N, resulting in unwanted component frequencies of the form $(2q-1)f_s \pm (2m+1)Nf_o$. Double summations, over $q = 1$ to ∞ and $m = 0$ to ∞, are, of course, involved for both these sets. Again, References 1 and 3 contain evaluations and tabulations of coefficients, and they will not be reported here.

It is noteworthy that some of these components vanish in the three-phase-to-three-phase case, since $\sin(2q\pi/M) = 0$ for all $q = 3l$ ($l = 1$ to ∞) and $\sin[(2q-1)\pi/M]$ vanishes for all $2q = 3l+1$. However, the even-order harmonics of f_s refuse to vanish for any case.

As for the dependent voltages, bridge configurations, and harmonic neutralization bring spectral benefits to both the individual positive and negative converters' dependent currents and, in the same manner, to those of the combination used to form a UDFFC; unwanted components in the dependent voltages and dependent currents are formed into complete phasor sets that sum to zero except when their multipliers are multiples of the number of converters involved. For the bridge configuration, all dependent-current components with even multiples of f_o vanish; only those with odd multiples remain. Harmonic neutralization, on the other hand, attacks the multiples of f_s, removing all components whereof the m of mN in $(mN \pm 1)f_s$ is not an integer multiple of l, the number of converters used.

Waveforms for the dependent currents of the positive and negative converters at $f_o f_s = 1/3$ and for $\psi = 0$ are shown in Figures 5-11 and 5-12. It should be noted that for both these converters, and all combinations thereof, the restriction $f_o < f_s$ applies. This is not because the existence functions become unrealizable, as they do for the SSFC, but because the "information" content of the modulating function becomes irrecoverable as a wanted component in the dependent voltage.

5.6 Combinations of the Positive and Negative Converters

One combination of positive and negative converters, the UDFFC, has already been discussed. In that ac-to-ac converter, the constituent positive and negative converters are both in continuous operation. Gyugyi (1970) calls the resultant dependent voltage wave a "concurrent composite" wave—"concurrent" because both constituent converters operate simultaneously and continuously, and "composite" because the resulting dependent quantities are composites of those of the positive and negative converters.[2]

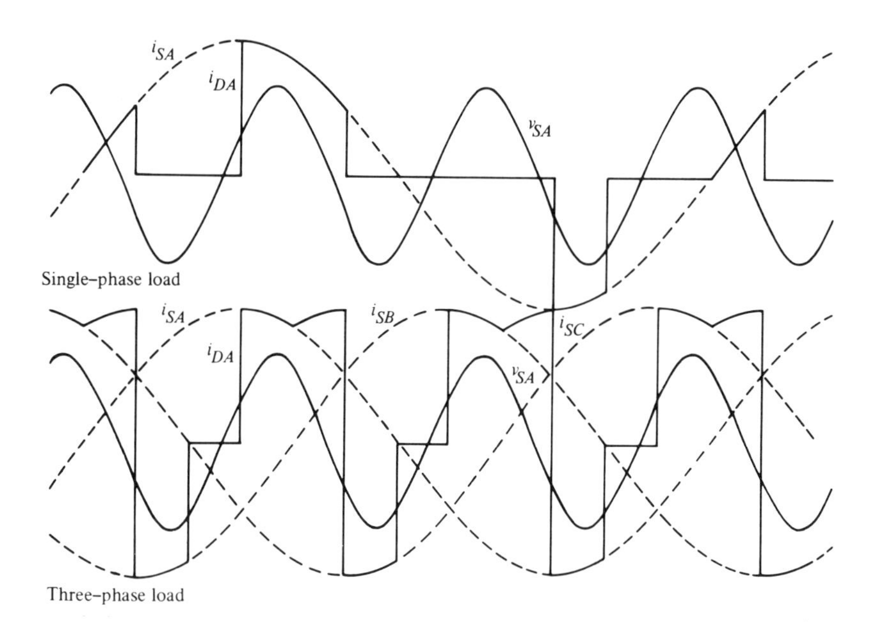

Figure 5-11. Dependent-current waves produced by positive converter at $f_o/f_s = 1/3$.

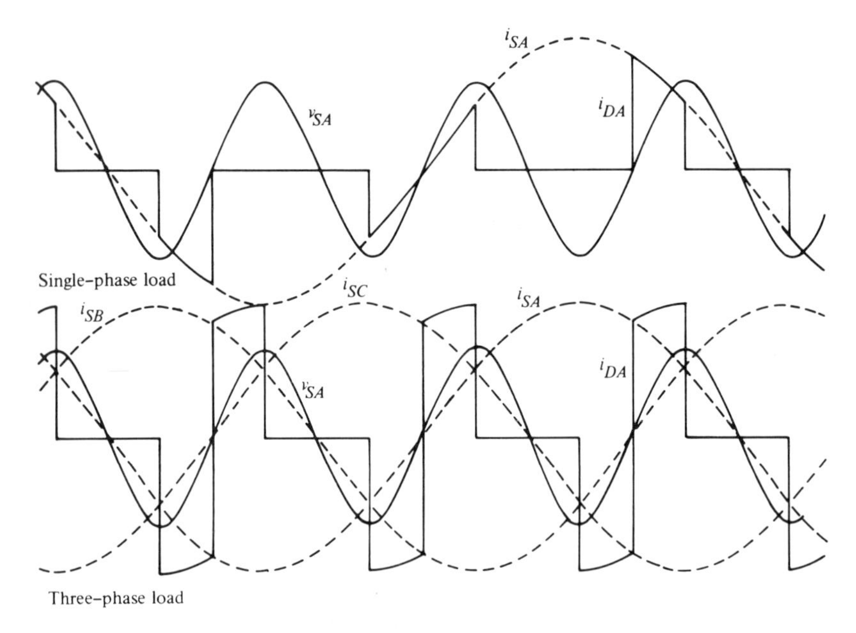

Figure 5-12. Dependent-current waves produced by negative converter at $f_o/f_s = 1/3$.

The two converters can also be combined by forming a "consecutive composite" wave. As the terminology implies, this technique involves nonsimultaneous operation of the positive and negative constituent converters of a combination. To accomplish this, the two converters are governed by supervisory existence functions that will be designated H_P and H_N. The sum of these is maintained at unity, so that KLC is not violated with respect to the common defined output current, nor is KLV with respect to the defined input voltage. H_P and H_N take the forms

$$H_P = 1/2+(2/\pi) \sum_{n=1}^{n=\infty} [\sin(n\pi/2)/n]\cos[n(\omega_o t+\rho)]$$

$$H_N = (1/2)-(2/\pi) \sum_{n=1}^{n=\infty} [\sin(n\pi/2)/n]\cos[n(\omega_o t+\rho)]$$

where ρ is an arbitrary phase angle.

H_N, obviously, is simply H_P phase-shifted by π rad. These two supervisory existence functions are identical in form to the simple existence functions for the switches in a pole of a voltage-sourced ac-to-dc/dc-to-ac converter.

The converters' outputs may be directly connected for parallel combination, since the converter which is inoperative at any given time has all its switches in the blocking state at that time. If it is desired to connect the converter outputs in series, then bypass switches must be provided for midpoint converters. The bypass switch for the positive converter is assigned H_N; that for the negative converter is assigned H_P. For bridges, the bypass needed for series connection can be effected by assigning H_N and H_P to pairs of switches in the bridges, which is analogous to the procedure used in creating CAM-controlled UFCs and SSFCs.

The dependent voltage of the combination is given by

$$v_D = H_P v_P + H_N v_N$$

It was seen that the wanted components of both v_P and v_N are given by

$$v_{DDP,N} = (M/\pi)\sin(\pi/M)x\cos(\omega_o t + \phi)$$

where x is the modulation index and ϕ the prescribed phase angle of this wanted component, usually $2k\pi/N$, where $k = 0$ to $N-1$ and N is the number of output phases. It is then obvious that the operation of H_P and H_N on these components to produce s_D for the combination reproduces this component; i.e., the wanted component of the dependent voltage of the combination is identical to the wanted component of the dependent voltage of either of the constituent converters. Now the unwanted components of the constituent positive and negative converters were seen to be expressible in the forms

$$v_{DUP} = A_U + B_U$$

$$v_{DUN} = A_U - B_U$$

Thus, the unwanted components in the combined dependent voltage also contains those of A_U for the constituent converters, with no modification. However, they also contain contributions from the unwanted components in B_U, and these are

$$v_{UB} = (4/\pi)B_U \sum_{n=1}^{n=\infty} [\sin(n\pi/2)/n]\cos[n(\omega_o t + \rho)] \qquad (5.15)$$

B_U was seen to contain components of frequencies $pMf_s \pm 0$, $\pm 2f_o$, $\pm 4f_o$, . . . $\pm \infty$ when pM is even, and $pMf_s \pm f_o$, $\pm 3f_o$,. . . $\pm \infty$ when pM is odd, and thus to consist of two double summations with infinite ranges. When modified as indicated by Equation 5.15, each term of each of these series becomes an infinite series. This twofold triple summation, each summation with infinite range, modifies the spectrum so that components of frequencies $pMf_s \pm 0$, $\pm f_o$, $\pm 2f_o$, $3f_o$,

∞ are generated regardless of whether pM is even or odd. The complication is that the amplitude of each component is the sum of an infinite number of terms. Pelly (1971) and Gyugyi and Pelly (1975), with the aid of a digital computer, have waded through the evaluation of these unwanted components.[1,3]

There is obviously no rationale for employing the converters in this consecutive composite mode as far as the dependent voltage is concerned. It is in the dependent current behavior, specifically that of the wanted components, and in the possible switch operating conditions, that compelling motivations are discovered. Clearly, the total dependent current is given by

$$i_D = H_P i_{DP} + H_N i_{DN}$$

and again the wanted component of both i_{DP} and i_{DN}, $i_{DDP,N} = (I/\pi)\sin(\pi/M) \cdot x\cos\psi\cos\omega_s t$ (where ψ is the output phase angle) is reproduced unchanged. Consider, however, the unwanted component in i_{DN} and i_{DP} coming from J_U of the expressions for those currents and having the form

$$\pm(I/\pi)\sin(\pi/M)\ A_{x,1,0}\{\sin[(\omega_s+\omega_o)t+\psi)+\sin[(\omega_s-\omega_o)t-\psi]\}$$

On processing by H_P and H_N, these components will yield in the combined dependent current the components

$$(2I/\pi^2)\sin(\pi/M)\ A_{x,1,0}[\sin(\omega_s t+\psi-\rho)+\sin(\omega_s t-\psi+\rho)]$$

as a part of the resultant of the product

$$(4/\pi)J_U \cdot \sum_{n=1}^{n=\infty} [\sin(n\pi/2)/n]\cos[n(\omega_o t+\rho)]$$

for the term in $n = 1$.

Simple trigonometric manipulation condenses this interesting component into

$$i_Q = (4I/\pi^2)\sin(\pi/M)A_{x,1,0}\cos(\psi-\rho)\sin\omega_s t$$

This is a quadrature component of fundamental current whereof the amplitude is a function of $\psi-\rho$. However, ρ is an arbitrarily selectable phase angle, and thus the input displacement factor of this converter can be controlled independently of the phase angle, ψ, at the output. If this were the only such quadrature component of fundamental current created, this Controlled Displacement Factor Frequency Changer (CDFFC)[1,2] would have a very wide range of input displacement factor, limited only by the relative amplitudes of the controllable quadrature current and the in-phase current. However, higher-order cross products of J_U and H_P, H_N also result in fundamental quadrature currents. Half of these have amplitudes depending on $\psi - n\rho$; the other half have amplitudes depending on $\psi + n\rho$. Their effect is, typically, to limit the degree to which controllable leading input displacement factors can be produced, particularly with lagging output displacement factors. Nonetheless, this property of the CDFFC is both interesting and potentially useful.

The remainder of the unwanted components in the combined input current wave comprise those of I_U unmodified and those of J_U, with each component thereof expanding into an infinite series of components. Again, the result is to produce all possible sidebands with the amplitude of each being the sum of an infinite series of terms, and again References 1 and 3 provide tabulations for the three-phase-to-single-phase and three-phase-to-three-phase cases.

Of particular interest is the case where ρ is made to equal ψ. In this event, the positive converter will only carry the positive half-cycles of the defined output current; the negative converter will only carry the negative half-cycles. Also, *there will be a lagging quadrature component of fundamental input current regardless of the load displacement factor*. As a consequence of the restrictions on positive and negative current polarity, the switches in the constituent converters have their capabilities reduced to unidirectional current-carrying, bidirectional voltage-blocking.

This most special and restricted of ac-to-ac converters with the nonlinear modulating function is of great historical and practical importance. Called the *Naturally Commutated Cycloconverter* (NCC), it was the first ac-to-ac converter of any type ever to find practical application. Because of practical switch limitations, it is still the only ac-to-ac converter ever implemented outside the laboratory. Until very recently, its lagging quadrature current demand was thought to originate because its constituent converters always operate with α negative; this has been seen not to be so—the fundamental reason lies much deeper in converter theory.

Finally, on the subject of converters with nonlinear modulating functions, the function $M(t,x,\phi) = \text{arc}\cos[x\cos(\omega_0 t + \phi)]$ is very easy to realize and to insert in the arguments of oscillatory terms in existence functions of converter switches. References 1 and 3 both give excellent discussions on this topic, and the modulating techniques used are briefly described in Chapter 10.

5.7 Inverter-Type AC-to-AC Converters

Gyugyi and Pelly state that "inverter type" ac-to-ac converters are more closely related to dc-to-ac converter technology than to the family of ac-to-ac converters they treat.[1] There is some validity to their view, inasmuch as ac-to-ac converters producing dependent-voltage waves of the types depicted in Figure 5-13 certainly are not based on the introduction of simple modulating functions to the

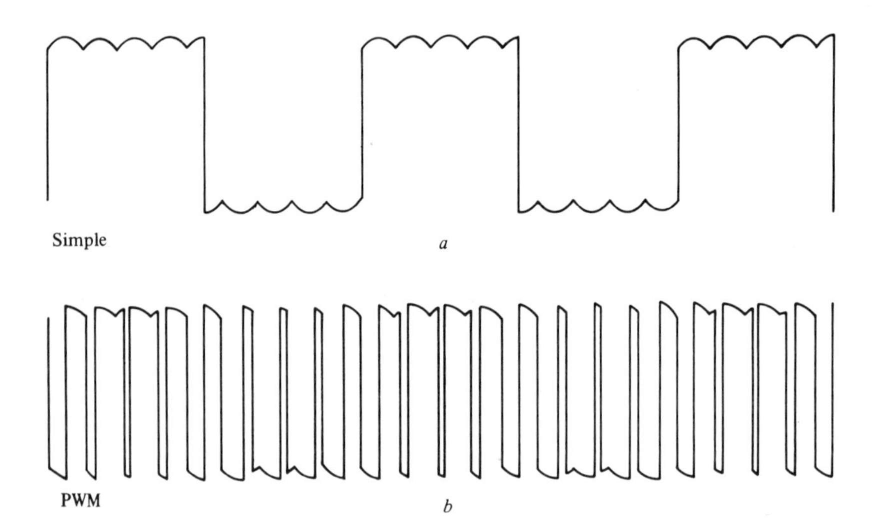

Figure 5-13. Dependent-voltage waves produced by inverter-type ac-to-ac converters.

switch existence functions. They perform by creating a consecutive composite wave from a single converter; the waves used in the composite are, from examination of Figure 5-13, the dependent voltages produced by the converter at $\alpha = 0$ and $\alpha = \pm\pi$.

In the simple manifestation producing the first waveform depicted in Figure 5-13, two supervisory existence functions

$$H_+ = 1/2+(2/\pi) \sum_{n=1}^{n=\infty} [\sin(n\pi/2)/n]\cos[n(\omega_o t+\rho)]$$

and

$$H_- = 1/2-(2/\pi) \sum_{n=1}^{n=\infty} [\sin(n\pi/2)/n]\cos[n(\omega_o t+\rho)]$$

are assigned in the converter so that the existence functions of converter switches 0, 1, 2, 3. . .$M-1$ are modified to become

$$H_0 = H_+ \, H + H_- \cdot \, H_\pi$$

where

$$H = 1/M+(2/\pi) \sum_{m=1}^{m=\infty} [\sin(m\pi/M)/m]\cos(m\omega_s t)$$

is the normal unmodulated and unmodified existence function of the 0th switch and H_π the same function displaced by $\pm\pi$ rad. It is known from Equations 4.7 and 4.11 that the converter-dependent voltage at $\alpha = 0$ is given by

$$v_+ = (M/\pi)\sin(\pi/M)V\{1+\sum_{p=1}^{p=\infty} \cos p\pi[\cos(pM\omega_s t)/(pM+1) - \cos(pM\omega_s t)/(pM-1)]\}$$

whereas the voltage at $\alpha = \pm\pi$ is, from Equations 4.14 and 4.15,

$$v_- = -(M/\pi)\sin(\pi/M)V\{1+\sum_{p=1}^{p=\infty} \cos p\pi \cos pM\pi[\cos(pM\omega_s t)/(pM+1) - \cos(pM\omega_s t)/(pM-1)]\}$$

The dependent voltage of a converter with fully bilateral switches governed by the supervisory existence functions H_+ and H_- is simply

$$v_D = H_+ \, v_+ + H_- \, v_-$$

regardless of the polarity of the defined current. Clearly, this voltage has no unwanted dc component arising from the dc components of v_+ and v_-. However, the dc terms in v_+ and v_- do give rise to the first part of v_D, v_{D1}, which is given by

$$v_{D1} = (4M/\pi^2)\sin(\pi/M)V \sum_{n=1}^{n=\infty} [\sin(n\pi/2)/n]\cos[n(\omega_o t+\rho)] \quad (5.16)$$

Equation 5.16 clearly defines a square wave of amplitude $(M/\pi)\sin(\pi/M)V$ with a wanted component of frequency f_o, completely independent of f_s. Note that none of the expressions place any restriction on f_o with respect to f_s—this converter is a form of the UFC insofar as lack of restriction on the ratio f_o/f_s is concerned. It produces a wanted component of higher amplitude than those produced by any of the ac-to-ac converters so far considered (see Equations 5.3 and 5.4 and the expression for v_{DD} on p. 177). Why then is it not preferred? The answer lies in part in the unwanted components in the dependent quantities, and in part in other properties (or the lack thereof).

Clearly, from Equation 5.16, one set of unwanted components in the dependent voltage are all the odd harmonics of the wanted frequency, each with relative amplitude $1/Q$ where Q is the harmonic order.

A second set of unwanted components arise from the expression

$$v_{D2} = (M/2\pi)\sin(\pi/M)V \sum_{p=1}^{p=\infty} \mp(1-\cos pM\pi)\ \cos p\pi\cos(pM\omega_s t)/(pM\pm 1) \quad (5.17)$$

If M, the number of input phases, is even, then these components are of zero amplitude, whereas if M is odd, then they exist for all odd values of p. They then represent the residue of the dc-terminal voltage ripple generated by the converter operating as a dc-to-ac/ac-to-dc converter. The remainder of the unwanted components in v_D arise from the third part thereof, v_{D3}, given by

$$v_{D3} = (2M/\pi^2)\sin(\pi/M)V \sum_{n=1}^{n=\infty} \sum_{p=1}^{p=\infty} [\sin(n\pi/2)\cdot\cos p\pi(1+\cos pM\pi)/n]$$

$$\cdot\ [\cos(pM\omega_s t)/(pM+1)-\cos(pM\omega_s t)/(pM-1)]\ \cos[n(\omega_o t+\rho)] \quad (5.18)$$

Expanding this will obviously give rise to terms having frequencies $pMf_s \pm (2q-1)f_o$, where $q = 1$ to ∞, $p = 1$ to ∞. This series of unwanted components are very similar to those produced by the B_U terms for the positive and negative converters of Section 5.6, but their amplitudes are smaller, being $(2M/\pi^2)\sin(\pi/M)/\{(2q-1)[2/(p^2M^2-1)]\}$, or $1/[(2q-1)\ (p^2M^2-1)]$ times the wanted component amplitude. Note that if M is even, then these components exist for all p, whereas if M is odd, they exist only for even values of p. This pulse number increase is analogous to that obtained from going to the bridge configuration in ac-to-dc/dc-to-ac converters. It arises, of course, because in this ac-to-ac converter, the midpoint group is used both as the normal and complementary group in constructing the consecutive composite voltage wave.

Whether the dependent-voltage spectrum thus created is better or worse than the spectra of the ac-to-ac converters previously described is not an easy question to answer. Clearly, it is much worse than that of the UFC under any circumstances. However, its sideband content is "comparable" to the contents of the other converters in frequencies, but lower in magnitude. It does contain the direct odd-order harmonics, of which the lower orders are very objectionable. In general, if f_o/f_s is low, then this spectrum is worse than that of the other converters. As f_o/f_s increases, its relative quality improves, and as f_o approaches f_s, it becomes superior to all except the UFC. For $f_o \geqslant f_s$, it is, of course, only to be compared with the UFC.

It should be noted that ρ is not an arbitrarily selectable phase angle for this converter, at least in regard to the output conditions. The wanted-component phase angle, ρ, can be selected for multiple converters in order to give the progressive displacements needed to form a complete phasor set (usually three-phase) of wanted output voltages. However, ρ is not selectable with regard to the defined current at any output, since, as in the case of voltage-sourced ac-to-dc/dc-to-ac converters, the output phase angle is a consequence of the converter's output network behavior. In view of this, ρ will be set to zero (or $-2k\pi/N$, where N is the number of output phases) in all further manipulations.

If N-phase output is created using N converters, then the output line voltages become of interest. As in any such case, all unwanted components that are zero-sequence sets of the output-phasor sets vanish from these dependent voltages. Thus, all terms of Equation 5.16 for which n is an integer multiple of N disappear, as do all terms of Equation 5.17 for which n (or $2q-1$, as it is later expressed) is an integer multiple of N. In the case of a three-phase output, the case of greatest practical importance, the terms of Equation 5.16 form a six-pulse wave.

Obviously, the spectral benefits of harmonic neutralization apply to the dependent voltages of such a converter, provided a polyphase version is being used and appropriate phase-shifting transformers are used to effect the direct series combination or parallel combination via an interphase reactor.

To analyze the dependent current behavior of such converters, it is first necessary to expand the expression for a switch existence function,

$$H_0 = H_+ \cdot H + H_- \cdot H_\pi$$

The expansion of this expression is

$$H_o = 1/M + (1/\pi) \sum_{m=1}^{m=\infty} [\sin(m\pi/M)/m](1+\cos m\pi)\cos(m\omega_s t)$$

$$+ (4/\pi^2) \sum_{m=1}^{m=\infty} \sum_{n=1}^{n=\infty} [\sin(m\pi/M)\sin(n\pi/2)/mn](1-\cos m\pi)\cos(m\omega_s t)\cos(n\omega_o t)$$

The first term of this expansion is simply $1/M$, the unchanged average value. When the dependent input current is evaluated, it produces a term at the output frequency—$I\cos(\omega_o t+\psi)/M$—which vanishes when polyphase output is produced.

The second term produces a series of unwanted components from

$$i_{D2} = (I/\pi) \sum_{m=1}^{m=\infty} [\sin(m\pi/M)/m](1+\cos m\pi)\cos(m\omega_s t)\cos(\omega_o t+\psi)$$

Since $1+\cos m\pi = 0$ if m is odd and $= 2$ if m is even, this reduces to

$$i_{D2} = (I/\pi) \sum_{m=1}^{m=\infty} [\sin(2m\pi/M)/2m][\cos(2m\omega_s t+\omega_o t-\psi)+\cos(2m\omega_s t-\omega_o t-\psi)]$$

containing components of all frequencies $2mf_s \pm f_o$. These components also disappear when polyphase output is being generated.

The double summation terms have zero value if m is even. They are multiplied by 2 when m is odd and thus reduce to

$$i_{D3} = (4I/\pi^2) \sum_{m=1}^{m=\infty} \sum_{n=1}^{n=\infty} [\sin(2m-1)\pi/M)\sin(n\pi/2)/(2m-1)n]$$
$$\cdot \cos[(2m-1)\omega_s t]\{\cos[(n-1)\omega_o t+\psi]+\cos[(n-1)\omega_o t-\psi]\}$$

The wanted component, at f_s, comes from the term for $n = 1$ and $m = 1$ and is

$$i_{DD} = (4I/\pi^2)\sin(\pi/M)\cos\psi\cos\omega_s t \tag{5.19}$$

Thus, this converter is a form of the UDFFC with regard in its input displacement factor. The remaining, unwanted components of i_{D3} have all frequencies $(2m-1)f_s \pm 2pf_o$, where $p = 1$ to ∞, $m = 1$ to ∞.

The amplitude of the component for m, p is

$$A_{m,p} = (4I/\pi^2)\sin[(2m-1)\pi/M][1/(2m-1)][2/(4m^2p^2-1)]$$

For polyphase output, only components for which $2p = qN$, where N is the number of output phases, will exist.

Spectrally, these input line currents are rather similar to those of the converters with nonlinear modulating functions insofar as frequency distribution is concerned. However, unwanted component amplitudes are generally higher for this converter than in the cases previously discussed. Harmonic neutralization brings benefits, of course, by increasing the basal multiplier on f_s. It is curious, however, that the dependent-voltage spectrum is not affected at all by adopting the bridge configuration. The dependent-current spectrum improves due to the elimination of the output-frequency term and the unwanted components of i_{D2}, but none of the components of i_{D3} are affected.

Obviously, no control variable exists in the expression for the wanted component of dependent voltage deriving from Equation 5.16. Control of the amplitude of the wanted component may be introduced in a number of ways. The first, and most obvious, is to introduce α to the basic switch existence functions so that the dependent voltages, which are processed by H_+ and H_-, are $v_{D,\alpha}$ and $v_{D,\pi-\alpha}$ given by

$$v_{D,\alpha} = (M/\pi)\sin(\pi/M)V\{\cos\alpha + \sum_{p=1}^{p=\infty} \cos p\pi \cos[pM\omega_s t + (pM \pm 1)\alpha]/(pM \pm 1)\}$$

$$v_{D,\pi-\alpha} = -(M/\pi)\sin(\pi/M)V\{\cos\alpha + \sum_{p=1}^{p=\infty} \cos p\pi \cdot \cos pM\pi$$

$$\cdot \cos[pM\omega_s t - (pM \pm 1)\alpha]/(pM \pm 1)\}$$

The resulting wanted component is

$$v_{DD,\alpha} = (4MV/\pi^2)\sin(\pi/M)\cos\alpha\cos\omega_o t$$

Of the wanted components, the harmonics of the square wave of equation 5.16 are also all multiplied by $\cos\alpha$, and so they diminish at the same rate as does the wanted component as α is increased over the range 0 to $\pi/2$. At $\pi/2$, the amplitude of the square wave is zero, as is that of its wanted component. Note that if α is taken beyond $\pi/2$, the square wave reappears with reversed sign, indicating phase reversal of all its components.

The unwanted components of Equation 5.17 behave in a rather complex fashion, as do those of Equation 5.18. For α not equal to zero, the components of Equation 5.17 are no longer zero when pM is even. This is because the expression

$$(1 - \cos pM\pi)\cos(pM\omega_s t)/(pM \pm 1)$$

expands to become

$$\mp\cos[pM\omega_s t + (pM \pm 1)\alpha]/(pM \pm 1) \pm \cos pM\pi \cos[pM\omega_s t - (pM \pm 1)\alpha]/(pM \pm 1)$$

Thus, all frequencies pMf_s are present, and their amplitudes increase rapidly with $|\alpha|$. At $|\alpha| = \pi/2$, these components are identical to the ripple produced by the converter operating as an ac-to-dc/dc-to-ac converter. The sideband components of Equation 5.18 also change character. No longer are those for odd values of pM absent—the sidebands exist for all values of p regardless of whether M is even or odd. As $|\alpha|$ approaches $\pi/2$, their amplitudes diminish, since they vanish at those operating points.

Over the range of $|\alpha|$, 0 to $\pi/2$, these amplitudes can be shown to follow the square root of the very cumbersome expression

$$[1/(pM+1)^2+1/(pM-1)^2-2\cos 2\alpha/(p^2M^2-1)]$$
$$+\cos pM\pi\{\cos[2(pM+1)\alpha]/(pM+1)^2+\cos[2(pM-1)\alpha]/(pM-1)^2$$
$$-2\cos pM\alpha/(p^2M^2-1)\}$$

Since little work has been done on these converters, the behavior of this function has not been fully tabulated to date. However, it is clearly oscillatory in nature, and for pM odd, it can be shown to exhibit maxima (of its absolute value) for all $k\pi/pM<\pi/2$, with zeros interspersed, and with the largest maximum occuring when $k = (pM-1)/2$. For pM even, similar absolute-value oscillations occur with the maxima occurring for all $(2k-1)\pi/2pM<\pi/2$, with the largest at $k = pM/2$.

The introduction of α to effect wanted-component magnitude control also affects the dependent currents. With the introduction of α, the switch existence functions become

$$H = 1/M+(1/\pi)\sum_{m=1}^{m=\infty}[\sin(m\pi/M)/m][\cos\{m(\omega_s t+\alpha)+\cos m\pi\cos[m(\omega_s t-\alpha)]\}$$
$$+(4/\pi^2)\sum_{m=1}^{m=\infty}\sum_{n=1}^{n=\infty}[\sin(m\pi/M)\sin(n\pi/2)/mn]\cos(n\omega_o t)\{\cos[m(\omega_s t+\alpha)]$$
$$-\cos\pi\cos[m(\omega_s t\alpha)]\}$$

The dependent current for single-phase output is given by

$$i_{D,\alpha} = H_\alpha I\cos(\omega_o t+\psi)$$

Examine this expression term by term, as before; the first, at output frequency, vanishes for the bridge configuration or polyphase output. The next group of terms now contains all frequencies $mf_s \pm f_o$ except those for which m is an integer multiple of M, with amplitudes $(2I/\pi)\sin(m\pi/M)\cos m\alpha$ for those with m even and $(2I/\pi)\sin(m\pi/M)\sin m\alpha$ for those with m odd. As for the uncontrolled case ($\alpha \doteq 0$), these components also vanish for the bridge configuration or polyphase output. It is again from the term in the double summation for $n = 1$ and $m = 1$ that the wanted component, of frequency f_s, comes. This component is

$$i_{DD,\alpha} = (4I/\pi^2)\sin(\pi/M)\cos\alpha\cos\psi\cos\omega_s t$$

Thus, the converter remains a unity displacement factor device despite the introduction of phase delay or advance (α) as the means for controlling wanted component amplitude. The remaining terms of the double summation contain unwanted oscillatory components of frequencies $mf_s \pm (n \pm 1)f_o$. Not all these combinations are present; those values of m that are integer multiples of M are not, nor are those for even values of n. Hence, the combinations reduce to $mf_s \pm 2pf_o$; the amplitude of such a component depends on whether m is odd or even, since $\cos m\pi$ alternates sign. These amplitude expressions are very cumbersome because of the cosine and sine terms involved. There is a basal amplitude multiplier $(2I/\pi^2)\sin(m\pi/M)/m$ for all m. For m odd, the remainder of the term becomes

$$-[\cos p\pi/(2p-1)][\cos(m\omega_s t \pm 2p\omega_o t + m\alpha \pm \psi) + \cos(m\omega_s t \pm 2p\omega_o t - m\alpha \pm \psi)]$$

$$+[\cos p\pi/(2p+1)][\cos(m\omega_s t \pm 2p\omega_o t + m\alpha \mp \psi) + \cos(m\omega_s t \pm 2p\omega_o t - m\alpha \mp \psi)]$$

while for m even the differences of the two oscillatory components are taken rather than the sums. The absolute amplitudes $A_{2q-1,2p}$ and $A_{2q,2p}$ can be derived by expanding these expressions. None of these sideband components vanish for the bridge configuration, but many do for polyphase outputs, provided that N, the number of output phases, is greater than 2. Then, all components for which $2p$ is not an integer multiple of N will form complete phasor sets summing to zero, and the resulting sideband spectra contains only frequencies $mf_s \pm 2pNf_o$.

The introduction of α is not the only way to effect control of wanted-component magnitude in these converters. Clearly, two converters with phase displacement control can be used. In such a case, the supervisory existence functions H_+ and H_- are phase-displaced in the two converters, and the usual multiple $2\cos\phi$. where ϕ is the phase displacement of each converter from a reference, appears in the wanted component amplitude expression. The consequences to unwanted component magnitudes are, as might be expected, very similar to those observed for voltage-sourced ac-to-dc/dc-to-ac converters using the same control technique.

CAM can also be introduced as the means for wanted-component magnitude control. With these converters, it is not necessary to add a freewheel switch to the midpoint configuration since CAM can be introduced to the supervisory existence functions. Again, the consequences to unwanted-component magnitudes are similar to those observed for CAM-controlled voltage-sourced ac-to-dc/dc-to-ac converters, as they are also when extended CAM and PWM are introduced. The second waveform of Figure 5-13 illustrates the introduction of PWM to an inverter-type ac-to-ac-converter. Obviously, all these control techniques can be used individually or in combination. For example, PWM or CAM might be combined with phase delay or advance. Not all the various possibilities arising will be treated in detail here, since the analytic technique used in this book makes it possible to evaluate any desired combination.

Observing the waveforms of Figure 5-13, it is seen that these inverter-type ac-to-ac converters operate with an "implicit dc link." Without α control, their waveforms are fabricated essentially from the maximum dc dependent voltages the converters could produce; with α control, their waveforms are fabricated from lower magnitude dc dependent voltages. Therefore, it is concluded that when PWM or programmed waveform techniques are invoked, there may be a better approach. It can be observed, in the second waveform of Figure 5-13, that for each switching an input voltage generally exists, which is closer to the wanted-output-component magnitude at that time than is the input voltage actually selected. It is then natural to ask whether better converter performance can be achieved by always selecting the input voltage nearest in value to the instantaneous wanted-component value. If this technique is used with a switching frequency of $\omega_s+\omega_o$, then a UFC results. If the frequency of switching is $\omega_s-\omega_o$, an SSFC results. For other switching frequencies, particularly those higher than $\omega_s+\omega_o$, it might be expected that better converter performance, in terms of dependent-quantity spectra, could be achieved. To the author's knowledge, there has been very little, if any, investigation of this possibility. The analysis could be approached using the universal technique expanded here, but its complexity is such as to render it beyond the scope of this book.

5.8 AC Regulators

So far, the ac-to-ac converters discussed in this chapter have been of the full switching-matrix variety. As indicated in Chapter 1, a class of ac-to-ac converters exists that uses degraded switching matrices. These converters are generally called *ac regulators* or *ac voltage controllers,* since their prime function is usually simply to regulate or control the rms value of an ac quanity.

The single-phase version of such a regulator, using one switch, is depicted in Figure 5-14. There are four possible three-phase versions, shown in Figure 5-15; each uses three switches. Clearly, these converters cannot operate with a defined sinusoidal output current. They are the first cases where detailed analysis will be performed for discontinuous dependent quantities. If discontinuous current is permitted at their outputs, then these converters are seen to operate in a manner analogous to a buck dc-to-dc converter with the same postulate. Whenever the switch of Figure 5-14 is closed, the source voltage v_S is applied to the load. Whenever that switch is open, the output current is zero. The output voltage is then determined by the characteristics of the load network at zero current, and may be nonzero. Similar statements apply to the three-phase versions, and would apply to polyphase versions with $M>3$ if they were encountered.

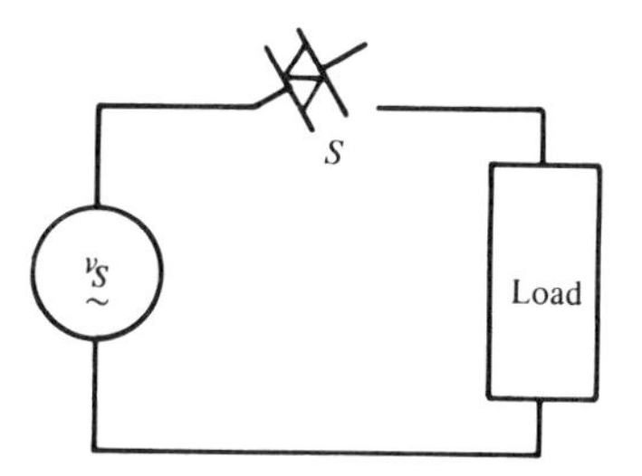

Figure 5-14. Single-phase ac regulator.

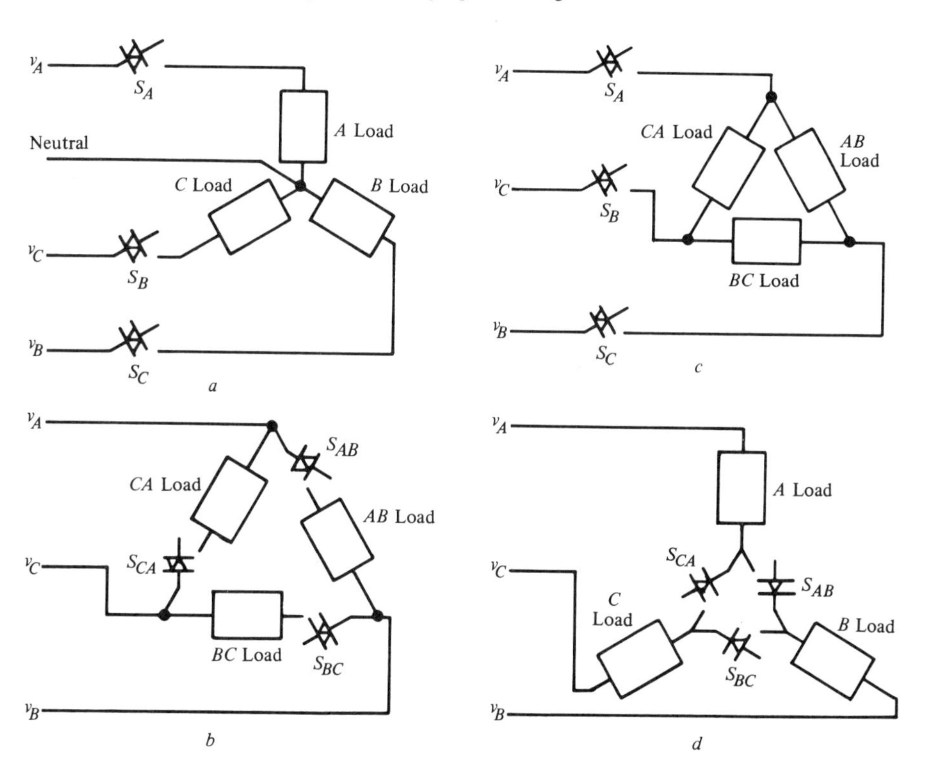

Figure 5-15. Three-phase ac regulators.

Were the switches used fully bilateral and controllable, the analysis of these converters would be quite trivial. However, the situation is complicated by a universal switch property—the switches used automatically open when the current flowing in them goes to zero. This means that the switch existence functions become load-dependent; while control can determine the inception of the unit-value periods of the existence functions of the switches, the subsequent interaction of source voltage and load determines their terminations. In fact, this is an almost universal feature for converters of any type operating with discontinuous dependent currents. Thus, the analytic approach used in this section, which consists of an extension of the general approach used previously, is typical of the approaches used in all cases where such discontinuous currents exist.

Examining Figures 5-14 and 5-15, it is seen that, insofar as individual load currents and voltages are concerned, the single-phase configuration of Figure 5-14 will produce the same results as the three-phase four-wire configuration of Figure 5-15a and the three-phase delta-switching configuration of Figure 5-15b. The two three-phase versions are simply threefold reproductions of the single-phase version, and except for unbalanced loading or regulator settings, can be treated as single-phase problems. The three-phase wye-switching regulator of Figure 5-15c is a little more complex, for it requires that at least two of the switches be closed before current will flow between source and load. Shown with a delta-connected load network, it can be used equally well with a wye-connected load without neutral return. The configuration of Figure 5-15d is known as the "British delta"; it is unique because the switches need only unidirectional current-carrying capability.

Consider now the single-phase regulator of Figure 5-14, implemented with a switch that automatically opens every time the current therein goes to zero. Assume first a purely resistive load. Then, in order to maintain the switch closed at all times, a control must reclose it at each current zero, i.e., at each voltage zero. Under these circumstances, the source voltage v_S is permanently connected to the load resistor R_O, and the dependent current flowing in both R_O and v_S is simply v_S/R_O. The objective of the converter, used as a regulator, is to control the voltage applied to R_O and hence the power delivered to R_O. Clearly, it can only apply a voltage lower than v_S when doing so, and will be able to operate only by creating periods of time when v_S is not connected to R_O, i.e., when the switch is open. There arise two basic possibilities for the creation of such periods.

First, suppose that the control allows the switch to remain open when it opens at one of the voltage zero crossings, but recloses it after some number of cycles of the supply voltage, v_S have elapsed. If this procedure is executed repeatedly at a fixed rate, with the switch closed for m cycles and open for n cycles, the control mode known as *integral cycle control* exists. The rms voltage applied to R_O is $[m/(m+n)]^{1/2}$ times the rms voltage of v_S, and hence the power delivered to R_O is $[m/(m+n)]$ times the maximum power delivered when the switch is always closed. Now suppose that instead of waiting for a number of cycles of the supply before reclosing the switch, the control recloses the switch at some time in the succeeding half-cycle of the supply voltage. This method of control is called *phase-delay control,* since switch closures are phase-delayed with respect to the source voltage.

The resulting load voltage waves for both these techniques are shown in Figure 5-16. The dependent-current waves, flowing in both load and source as Kirchoff's law demands, will simply be reproductions of these waves scaled by the factor $1/R_O$, or G_O, the load conductance. As might be expected, there are advantages and disadvantages attending both methods of control. However, one claim often made for integral cycle control—that it gives unity power factor loading on the source, v_S—will now be shown to be preposterous.

Whatever means of control are used, the load rms voltage will be lower than or equal to that of v_S. Designate that load rms voltage V_{OE} and the rms voltage of v_S as V_{SE}. Then the power delivered to the load is V^2_{OE}/R_O. This must also be the power extracted from the source, since the system must obey the laws of thermodynamics. Now the rms current flowing in R_O, V_{OE}/R_O, must also flow in v_S to satisfy KLC. Hence, the volt-ampere product in v_S is $V_{SE}\ V_{OE}/R_O$. By definition, the power factor of the burden on v_S is power-divided by volt-amperes, or V_{OE}/V_{SE}. This is true, from the above, *no matter what means of control is employed.*

Analysis of the performance of this converter with resistive load is quite straightforward. For both integral-cycle and phase-delay controls, switch existence functions are readily defined as shown in Figures 5-17 and 5-18. For integral-cycle control, the switch existence function has a repetition frequency $f_s/(m+n)$ and a unit-value angular period of $2\pi m/(m+n)$, where m is the number of cycles for which the switch is closed and n the number for which it is open. The existence function can thus be written as

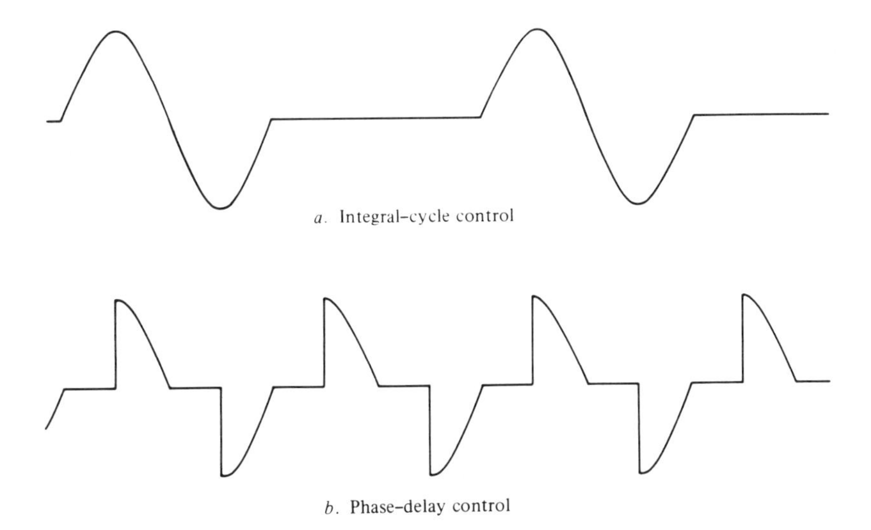

Figure 5-16. AC regulator dependent-voltage waves (resistive load).

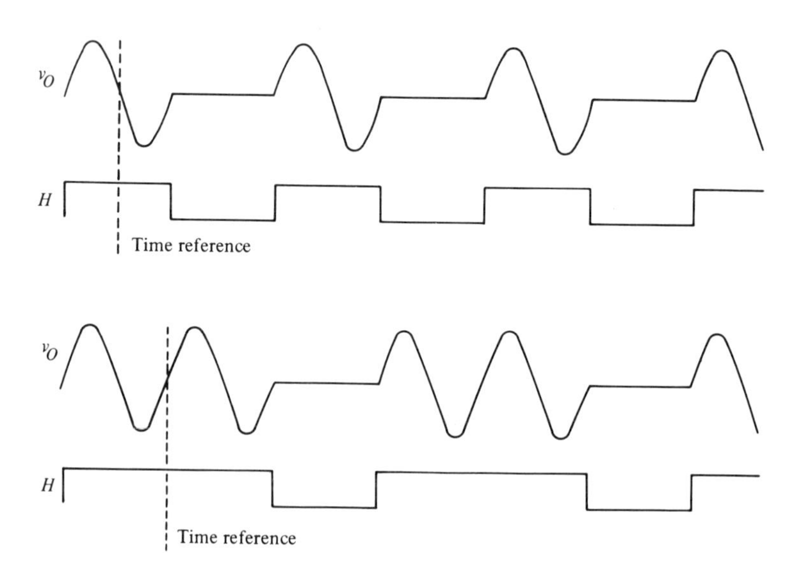

Figure 5-17. Switch existence functions for integral cycle control.

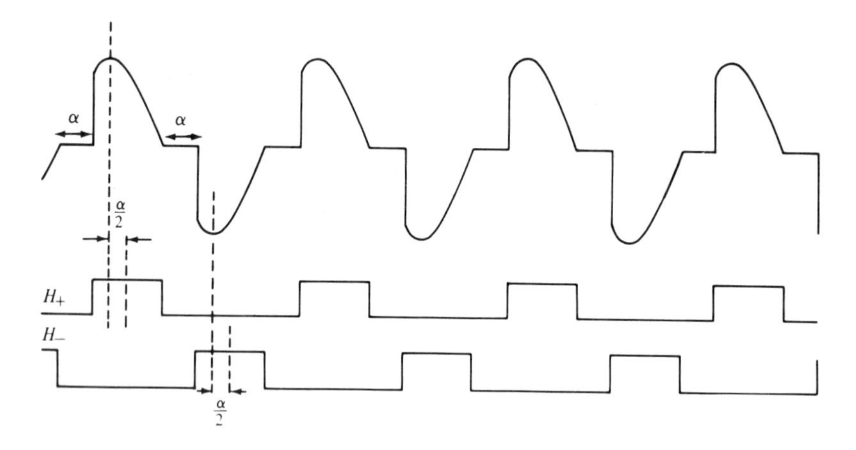

Figure 5-18. Switch existence functions for phase-delay control.

$$H = m/(m+n) + (2/\pi) \sum_{p=2}^{p=\infty} \{\sin[pm\pi/(m+n)]/p\}\cos[p\omega_s t/(m+n)]$$

v_S is defined as $-V\sin\omega_s t$ if m is odd and $+V\sin\omega_s t$ if m is even. The dependent voltage, applied to R_O, is thus

$$v_D = mV\sin(\omega_s t)/(m+n)+(2V/\pi)\sum_{p=1}^{p=\infty}\{\sin[pm\pi/(m+n)]/p\}$$
$$\cdot\sin(\omega_s t)\cos[p\omega_s t/(m+n)]$$

for m even and the same with $\sin(\omega_s t)$ replaced by $-\sin(\omega_s t)$ if m is odd.

All components of v_D might be termed *wanted,* since all contribute to the power delivered to R_O. However, in the dependent current, obtained by dividing v_D by R_O, only the component $G_o mV\sin(\omega_s t)/(m+n)$ can be considered wanted. The remaining unwanted components in i_D will have frequencies $f_s \pm pf_s/(m+n)$, except those for which $pm/(m+n)$ is an integer. This is clearly not a desirable spectrum, for many of the components with frequencies $f_s - pf_s/(m+n)$ are subharmonic. However, it can be seen that the *displacement factor* of the converter is unity, although its power factor is not. One reason for the common misconception regarding the power factor of this converter is that if $m+n$ is large, as it usually is, then the sideband components of largest amplitude are typically close to the fundamental component and are not resolved by common measuring equipment. Nonetheless, any transformers and reactors in the source network are "aware" of the presence of these sidebands, particularly the subfundamental frequency components, as are the source's generators and any inductance or synchronous machine loads connected to the same source.

As shown in Figure 5-18, it is convenient to split the existence function into two complementary parts when phase-delay control is used. With the source voltage defined as $V\cos(\omega_s t)$ and a switch-closing delay angle α, these existence functions become

$$H_+ = 1/2-\alpha/2\pi+(2/\pi)\sum_{p=1}^{p=\infty}\{\sin[p(\pi-\alpha)/2]/p\}\cos[p(\omega_s t-\alpha/2)]$$

and

$$H_- = 1/2-\alpha/2\pi+(2/\pi)\sum_{p=1}^{p=\infty}\{\sin[p(\pi-\alpha)/2]/p\}\cos p\pi\cos[p(\omega_s t-\alpha/2)]$$

giving

$$H = H_+ + H_-$$
$$= 1-\alpha/\pi+(2/\pi)\sum_{p=1}^{p=\infty}\{\sin[p(\pi-\alpha)/2]/p\}(1+\cos p\pi)\cos[p(\omega_s t-\alpha/2)]$$

$$= 1-\alpha/\pi+(4/\pi)\sum_{q=1}^{q=\infty}\{\sin[q(\pi-\alpha)]/2q\}\cos(2q\omega_s t-q\alpha)$$

since $1+\cos p\pi = 0$ if p is odd. The dependent voltage is then given by

$$v_D = (1-\alpha/\pi)V\cos\omega_s t+(4V/\pi)\sum_{q=1}^{q=\infty}\{\sin[q(\pi-\alpha)]/2q\}\cos(2q\omega_s t-q\alpha)\cos(\omega_s t)$$

and it is seen that a fundamental component in v_D, and hence in i_D, comes from the summation term for $q = 1$ in addition to that deriving from the average value term of H. Thus

$$v_{DD} = (1-\alpha/\pi)V\cos\omega_s t+(V/\pi)\sin(\pi-\alpha)\cos(\omega_s t-\alpha)$$

and $i_{DD} = v_{DD}/R_O$ have both in phase (cosine) and lagging quadrature (sine) components. The in-phase component of v_{DD} is

$$v_{DDC} = (1-\alpha/\pi)V\cos\omega_s t+(V/\pi)\sin(\pi-\alpha)\cos\alpha\cos(\omega_s t)$$

$$= [1-\alpha/\pi+\sin(2\alpha)/2\pi]V\cos(\omega_s t)$$

while the quadrature component is

$$v_{DDS} = (V/\pi)\sin^2\alpha\sin(\omega_s t)$$

The remaining terms of v_D and i_D are clearly all the odd harmonics of f_s. Their amplitudes are rather cumbersome expressions, since each is composed of two contributions,

$$v_{DUk1} = (V/\pi)\{\sin[k(\pi-\alpha)]/k\}\cos[(2k-1)\omega_s t-k\alpha]$$

and

$$v_{DUk2} = (V/\pi)\{\sin[(k-1)(\pi-\alpha)]/(k-1)\}\cos[(2k-1)\omega_s t-(k-1)\alpha]$$

The resulting amplitude function, like many encountered for unwanted components in switching power converters, is oscillatory. However, it has no zeros except at $\alpha = 0$ and $\alpha = \pi$; in between, it shows maxima and minima. For those harmonics of order $2.2l-1$ ($k = 2l$), such as the 3rd, 7th, 11th, etc., the largest maximum occurs at $\alpha = \pi/2$ when those harmonics have amplitudes $2V/l\pi$. Those of orders $2\,(2l-1)-1$ ($k = 2l-1$) exhibit their highest-value minima at $\alpha = \pi/2$, when their amplitudes are $2V/k\pi$. Their absolute maxima occur for two values of α symmetrically disposed about $\alpha = \pi/2$ and not simply expressible; the magnitudes are not simply expressible either, but are slightly greater than $[1/(2k-1)+1/2k](V/\pi)$. It is interesting to compare the maximum amplitudes of these unwanted components with those of the converter under integral cycle control. In that case, each pair of sideband components $f_s+pf_s/(m+n)$ has joint maximum amplitude $V/p\pi$ when $pm\pi/(m+n) = (2k-1)\pi/2$. Thus, if $m/(m+n) = 1/2$, those for p odd maximize, while those for even p are zero. The largest harmonic for phase-delay control is third, with maximum amplitude V/π at $\alpha = \pi/2$. This equality of largest components reflects the equality of total rms unwanted components in the two cases at any given level of power transfer. However, since the converter with phase delay generates a quadrature fundamental component in the dependent current, which is actually unwanted, its dependent-current wave has lower total rms distortion than that of the integral-cycle-controlled converter.

In the three-phase realization of Figures 5-15b, using delta-switching, these observations become yet more pertinent. For the integral-cycle-controlled converter, none of the sideband components of nonzero amplitude can be zero sequence. This is because, for a component to be zero sequence, it is necessary that the three switch-existence functions be phase-displaced with respect to each other and that $p/(m+n)$ be an integer. However, if $p/(m+n)$ is an integer, the component's amplitude is zero. In contrast, if the converter is phase-delay controlled, has balanced load, and is operated at equal delay angles in all three phases, then all triplen (order $3[2l-1]$) components in its load currents are zero sequence and flow only in the delta—they do not appear in the input lines. Thus, for this three-phase configuration, the phase-delay converter's input line currents have markedly less distortion than those of the integral cycle controller and the phase-delay controlled converter's input-power factor is better than that of the integral-cycle-converter despite the lagging displacement factor of the former.

With nonresistive loading, the situation becomes more complex for both types of control. If the load has an impedance Z_O, expressible as $A_O = R_O+jX_O$, the first problem is that X_O is a function of frequency and, thus, so is Z_O. In view of the unwanted component spectra produced, this dependence of impedance on frequency might present formidable problems to the analysis of the converters' behavior were it not for the existence of the Laplace transform technique for analyzing networks under transient conditions. Using this technique, an exciting voltage $V\cos(\omega_s t+\phi)$ is transformed into

$$\overline{V} = Vs\cos\phi/(s^2+\omega_s{}^2)-V\omega\sin\phi/(s^2+\omega_s{}^2)$$

where s is the complex variable of the Laplace transformation. An impedance expressed as $R_O+j\omega L_O$ in the frequency domain is transformed to

$$\bar{Z}_o = R_O + sL_O$$

Thus, current defined as $V\cos(\omega_s t+\phi)/Z_O$ becomes

$$\bar{i} = (Vs\cos\phi - V\omega_s\sin\phi)/[(s^2+\omega_s{}^2)(R_O+sL_O)]$$

Putting $R_O/L_O = \beta$ results in

$$\bar{i} = (V/L_O)(s\cos\phi - \omega_s\sin\phi)/[(s^2+\omega_s^2)(s+\beta)]$$

$$= (V/L_O)[(\beta\cos\phi + \omega_s\sin\phi)s/(s^2+\omega_s^2) + (\omega_s^2\cos\phi - \beta\omega_s\sin\phi)/(s^2+\omega_s^2)$$

$$-(\beta\cos\phi+\omega_s\sin\phi)/(s+\beta)]/(\beta^2+\omega_s^2)$$

The inverse transform of this expression is

$$i = (V/L_O)[(\beta\cos\phi + \omega\sin\phi)\cos(\omega_s t) + (\omega\cos\phi - \beta\sin\phi)\sin(\omega_s t)$$

$$-(\beta\cos\phi+\omega\sin\phi)\epsilon^{-\beta t}]/(\beta^2+\omega_s^2)$$

$$= [VL_O/(R_O^2+\omega_s^2 L_O^2)][\beta\cos(\omega_s t+\phi) + \omega_s\sin(\omega_s t+\phi) - (\beta\cos\phi + \omega\sin\phi)\epsilon^{-\beta t}]$$

$$= [V/(R_O^2+\omega_s^2 L_O^2)^{1/2}][\cos(\omega_s t+\phi-\psi) - \epsilon^{-\beta t}\cos(\psi-\phi)]$$

where $\psi = \arctan(\omega_s L_O/R_O)$.

This expression consists of two parts. The first, the oscillatory term, is the steady-state current resulting from the application of the source voltage to the impedance. The exponential term represents a decaying transient current, the magnitude of which depends on both ϕ and ψ.

Now consider the effect of this on a converter with integral cycle control. With resistive load, switch closure is at a voltage zero, i.e., with $\phi = -\pi/2$. If this condition is maintained for nonresistive load, then the current subsequent to closure will be given by

$$i_Z = (V/Z_O)[\sin(\omega_s t - \psi) + \epsilon^{-\beta t}\sin\psi]$$

and contains the transient term. Subsequent current zeroes do not occur at $\omega_s t = \pi$, 2π, etc.; they gradually converge toward occurrence at $n\pi+\psi$, with the rate of convergence dependent on β. The smaller β, the slower this process, and so converter analysis becomes extremely complex, highly dependent on β and m, the number of cycles for which the switch is supposed to be on. In fact, the switch will not be closed for an integral number of cycles under any circumstances—at best, if m is very large compared to $2\pi\beta/\omega_s$, m cycles + ψ rad is the time of closure.

All this complexity can be avoided if ϕ is put equal to $-\pi/2+\psi$, i.e., if switch closures are delayed by an amount equal to the phase angle of the load. The transient term disappears, and the dependent quantities have the same expressions as in the resistive-load case with Z_0 substituted for R_0 and the angle ψ introduced into oscillatory-term-arguments. Although it is doubtful that many (if any) actual integral-cycle-controllers use this maneuver, the analysis for the case where it is not used is so laborious that it will not be pursued further. Similar results can be obtained by applying the same analytic technique for capacitive loads, but they are generally of little importance in the actual applications of such ac regulators.

In the case of phase-delay control, the transient term must be considered in determining current zero subsequent to switch closure and hence the end of the switch conduction period. Putting $\phi = -\pi/2+\psi$, the expression for the current becomes

$$i = (V/Z_0)[\sin(\omega_s t+\alpha-\psi)+\epsilon^{-\beta t}\sin(\psi-\alpha)]$$

Now for $\alpha<\psi$, the switch will remain permanently closed since the succeeding current zero will be beyond the time at which the control recloses the switch. Thus, the control range becomes from $\alpha = \psi$ to π in this case. The current zero crossing occurs at an angle θ_o beyond the zero crossing, which is the solution of

$$\sin(\psi-\theta_o)-\epsilon^{-(\beta/\omega_s)(\pi-\alpha+\theta_o)}\sin(\alpha-\psi)=0$$

There is no general closed-form solution to this; a digital computer should be used to perform iterative solutions for the ranges of ψ and α under consideration in a particular application. Given such a solution, the two complementary existence functions for the switch can be redefined as

$$H_+ = (\pi-\alpha+\theta_o)/2\pi+(2/\pi)\sum_{p=1}^{p=\infty}\{\sin[p(\pi-\alpha+\theta_o)/2]/p\}\cos\{p[\omega_s t-(\alpha+\theta_o)/2]\}$$

and H_-, which is similar, except that it has $\cos p\pi$ multiplier on the oscillatory terms. The analysis can be performed as for resistive load; the results are obviously identical insofar as the frequencies present in the dependent quantities are concerned. An additional delay of $\theta_o/2$ appears in oscillatory components, translating into θ_o because only the even-order components of $H_+ + H_-$ exist.

A particularly interesting case, of great practical importance, occurs when $\psi = \pi/2$, i.e., for purely inductive load. In this event, $\beta = 0$, and the equation for θ_o reduces to

$$\sin(\pi/2-\theta_o) - \sin(\alpha-\pi/2) = 0$$

to which the solution is $\theta_o = \pi-\alpha$. In this case, α must be greater than $\pi/2$. Let $\alpha' = \alpha-\pi/2$, α' being the delay from the current zero crossing. Then $\theta_o = \pi/2-\alpha'$, and the existence functions become

$$H_+ = (\pi-2\alpha')/2\pi+(2/\pi)\sum_{p=1}^{p=\infty}\{\sin[p(\pi/2-\alpha')]/p\}\cos[p(\omega_s t-\pi/2)]$$

and

$$H_- = (\pi-2\alpha')/2\pi+(2/\pi)\sum_{p=1}^{p=\infty}\{\sin[p(\pi/2-\alpha')]/p\}\cdot\cos p\pi\cdot\cos[p(\omega_s t-\pi/2)]$$

As before, when combining these to create H, the multiplier $1+\cos p\pi$ is zero for p odd so that

$$H = 1-2\alpha'/\pi+(4/\pi)\sum_{q=1}^{q=\infty}\{\sin[q(\pi-2\alpha')]/2q\}\cos(2q\omega_s t-q\pi)$$

$$= 1-2\alpha'/\pi-(2/\pi)\sum_{q=1}^{q=\infty}[\sin(2q\alpha')/q]\cos(2q\omega_s t)$$

The dependent voltage is then given by

$$v_D = (1-2\alpha'/\pi)V\cos(\omega_s t)-(2V/\pi)\sum_{q=1}^{q=\infty}[\sin(2q\alpha')/q]\cos(2q\omega_s t)\cos(\omega_s t)$$

and all components are seen to be cosine, i.e., in phase. The fundamental component comes from the average term and the first term ($q=1$) of the summation and is

$$v_{DD} = (1-2\alpha'/\pi-\sin 2\alpha'/\pi)V\cos(\omega_s t)$$

All unwanted components are odd-order harmonics of f_s again. The amplitude expression is considerably simpler than for resistive load, reducing to

$$|V_{DU,2k+1}| = (V/\pi)\{\sin[2(k+1)\alpha']/(k+1)+\sin(2k\alpha')/k\}$$

for the harmonic of order $2k+1$, $k = 1$ to ∞. Putting $n = 2k+1$, this transposes to

$$|V_{DU,n}| = (2V/\pi)\{\sin[(n+1)\alpha']/(n+1)+\sin[(n-1)\alpha']/(n-1)\}$$

This amplitude function is oscillatory with zeroes. Its zeroes occur for

$$\tan(n\alpha') = (1/n)\tan\alpha'$$

with α' in the range 0 to $\pi/2$, and there are also two trivial zeroes at $\alpha' = 0$ and $\alpha' = \pi/2$. The maxima (which are amplitude maxima, of course) and minima occur when

$$\alpha' = (2l-1)\pi/2n$$

with l any integer from 1 to $(n+1)/2$. There are k-1 (n=$2k+1$) nontrivial zeroes in the amplitude function and, clearly, $k+1$ amplitude maxima. The amplitude at these maxima is given by

$$|V_{DU,n}|_{\max} = \{4nV/[\pi(n^2-2)]\cdot\cos[(2l-1)\pi/2n]$$

which obviously has its absolute maximum for $l = 1$, $\alpha' = \pi/2$, and

$$|V_{DU,n}|_{\text{absmax}} = \{4nV/[\pi(n^2-1)]\}\cdot\cos(\pi/2n)$$

With a purely inductive load, the dependent voltage has been seen to consist entirely of cosine components. The dependent current, therefore, will consist entirely of sine components whereof the absolute maximum amplitudes are

$$|I_{DU,n}|_{\text{absmax}} = \{4V/[\pi\omega_s Ln(n^2-1)]\}\cdot\cos(\pi/2n)$$

Expressed as percentages of the maximum fundamental, these values are

n	*% Harmonic, Absolute Maximum*
3	13.8
5	5.05
7	2.59
9	1.57
11	1.05
13	.752

In three-phase operation, as in Figure 5-15b, this is quite a respectable dependent-current spectrum in the input lines since the triplen harmonics, of which the third and the ninth are members, are not present for balanced operation.

Before proceeding to consider the wye-switched three-phase regulator, it should be observed that harmonic neutralization can be applied to improve the dependent-quantity spectra of any of the three-phase (or M-phase, $M > 3$) regulators. Until recently, this was not done because the applications served by these converters sought the cheapest possible realizations and the power levels did not usually demand multiple switching devices. The three-phase inductively loaded regulator (delta-switched) has recently been quite widely used in an application where power levels are high enough and the benefits of harmonic neutralization are welcome, so in some instances it has been employed or proposed. The rules for its implementation are exactly the same as in the cases of voltage- and current-sourced ac-to-dc/dc-to-ac converters; the regulator is treated as a current-sourced converter, so that the constituent converters of a harmonic neutralized set are connected implicitly in parallel at their ac-source terminals. However, at their load terminals, they are kept isolated; there is no possibility, with this configuration, of series or parallel (via interphase reactor) connection.

The three-phase wye-switched regulator, as mentioned earlier, must be considered separately because switch existence functions are not independent in that case. From an examination of Figure 5-15c, this is because two switches must be closed simultaneously for current to flow in the load circuit. In the case of the integral cycle controller, this fact gives rise to rather minor but very annoying perturbations of behavior. For the phase-controlled converter, there are some substantial changes in operation.

The situation is further complicated by practical considerations surrounding the switches. The bidirectional current-carrying, bidirectional voltage-blocking switches needed do exist, in the form of triacs, thyristor relatives which will be discussed in Chapter 7. However, high-power regulators usually use two inverse parallel-connected unidirectional current-carrying devices to achieve bilateral current capability. Those applications seeking the most economical implementations are wont to substitute unidirectional voltage-blocking devices for one of each such pair of devices. Provided that they are placed with the same polarity in each line, they may only conduct at the pleasure of their controlled, bidirectional voltage-blocking partners. Thus, in effect, three versions of the wye-switched three-phase regulator exist: the three-triac, six-thyristor, and three-thyristor three-diode configurations. The first two exhibit essentially identical behavior; the third is quite different in the phase-delay control mode. Analyses will be carried out for the resistive load cases only. It is rare to find significantly nonresistive loads used with these regulators, and the analyses for such cases are tedious in the extreme. They may, if desired, be pursued by the unified technique used throughout this book.

Consider first the three-triac or six-thyristor version with integral cycle control. It would seem logical that all three switches should be turned on and off simultaneously, but there are serious problems with such an approach. First, since at no time are the three phase or line voltages simultaneously zero, there are inevitably transient currents in two of the lines at turn-on. For the same reason, simultaneous turn-off of the switches, which open at current zeros, is not possible. The resulting load voltage waveforms are illustrated, for nominal one-cycle closures, in Figures 5-19 and 5-20 for wye- and delta-connected loads,

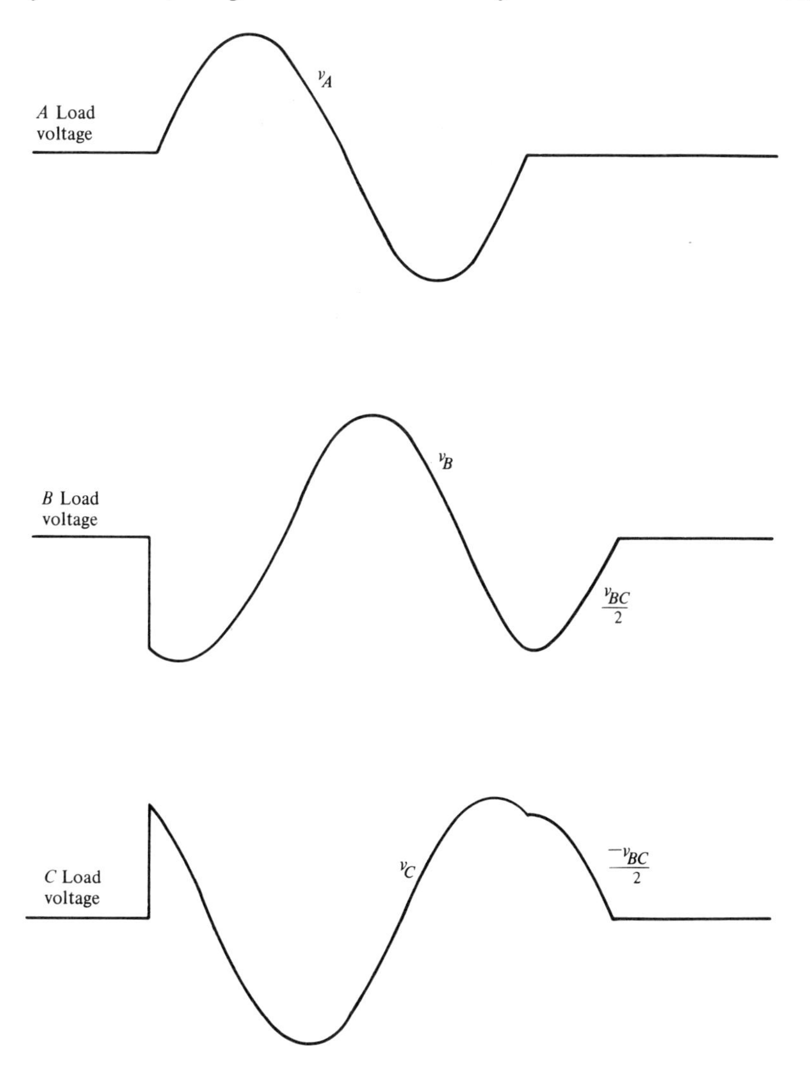

Figure 5-19. Wye-switched regulator voltage waves with integral cycle control (wye-loaded).

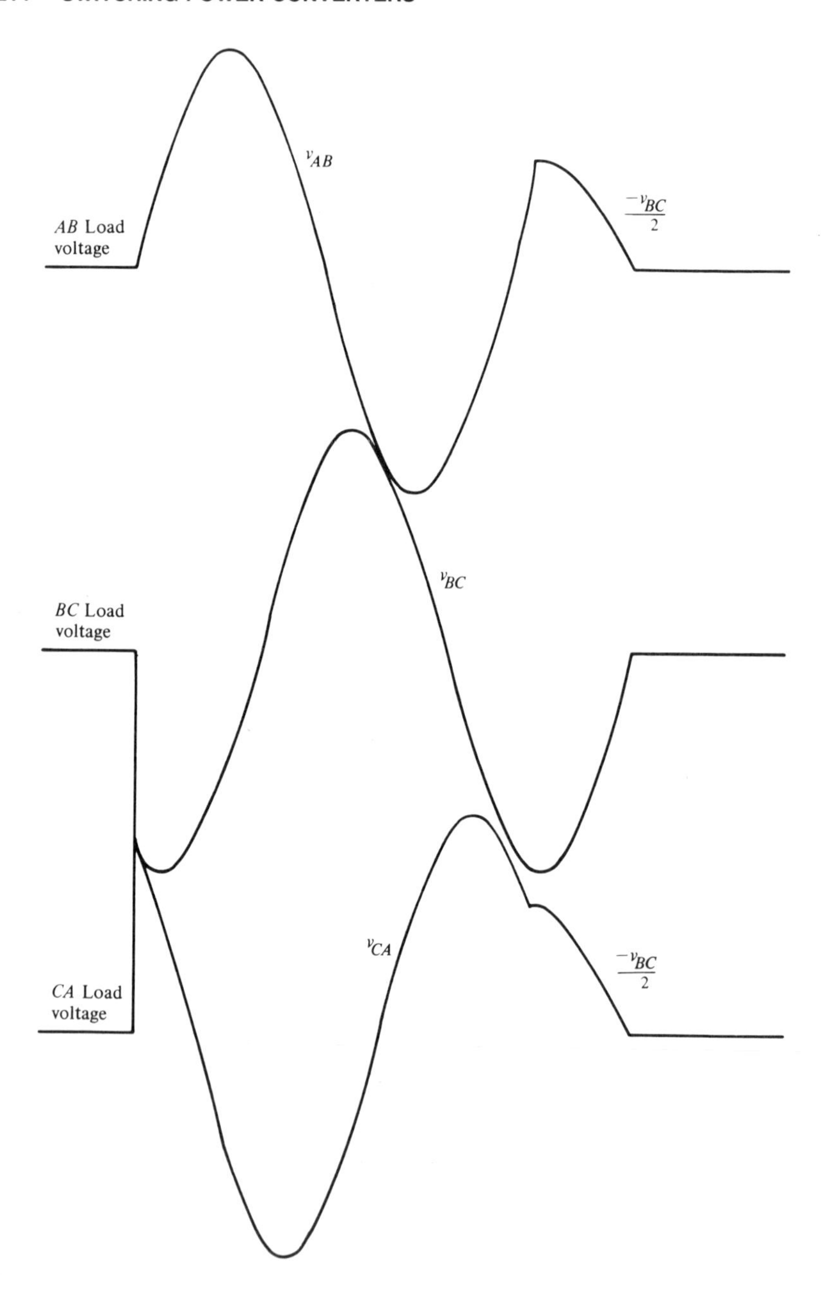

Figure 5-20. Wye-switched regulator voltage waves with integral cycle control (delta-loaded).

respectively. Load currents have the same appearance, of course, since the loads are assumed resistive.

It can be seen from these waves that transient currents and nonsimultaneous switch openings are not the only problems entailed when this approach is used. The load voltages and currents are evidently unbalanced and have dc components. The degrees of unbalance and magnitudes of dc components obviously decrease as the number of cycles for which the switches are supposed to be closed increases, but both phenomena are most undesirable. They are also inevitable unless a more sophisticated control strategy is employed.

The transient currents can be avoided by adopting the strategy of closing two switches at a line voltage zero and having the third's closure delayed by $\pi/2$ rad, as depicted in Figures 5-21 and 5-22. However, the problems of unbalance and dc components remain. They can be removed by "circulating" the orders of switch closure from cycle to cycle of the control frequency, $f_s/(m+n)$. In this case, the sequence is:

First cycle: Close switches A and B at voltage zero of line voltage v_{AB}; close switch C $\pi/2$ rad later at voltage zero of phase voltage v_C.

Second cycle: Close switches B and C at voltage zero of line voltage v_{BC}; close switch A $\pi/2$ rad later at voltage zero of phase voltage v_A.

Third cycle: Close switches C and A at voltage zero of line voltage v_{CA}; close switch B $\pi/2$ rad later at voltage zero of phase voltage v_B.

The corresponding sequence of terminations is:

First cycle: Allow switch A to remain open at current zero in A line; switches B and C will open at subsequent current zeros in B and C lines $\pi/2$ rad later.

Second cycle: Same with B first, A and C later.

Third cycle: C first, A and B later.

Such a procedure introduces unwanted components of frequencies $f_s \pm f_s/[3(m+n)]$ into the dependent quantities. The amplitudes of these components are dependent on m and $m+n$ and diminish as both m and n increase. Generally, they are of significantly lower amplitude than the sidebands of frequencies $f_s \pm f_s/(m+n)$; in view of this and because the exact analysis is very laborious, it will be omitted here.

A further, and perhaps not so obvious, consequence of using the wye-switching regulator is that the power delivered to the load is not exactly $m/(m+n)$ times full power. This is because the power delivered in the nominally one-cycle closure periods of Figures 5-21 and 5-22 (and also in those of Figures 5-19 and 5-20) does not exactly equal $1/(1+n)$ times full power. From the wave-

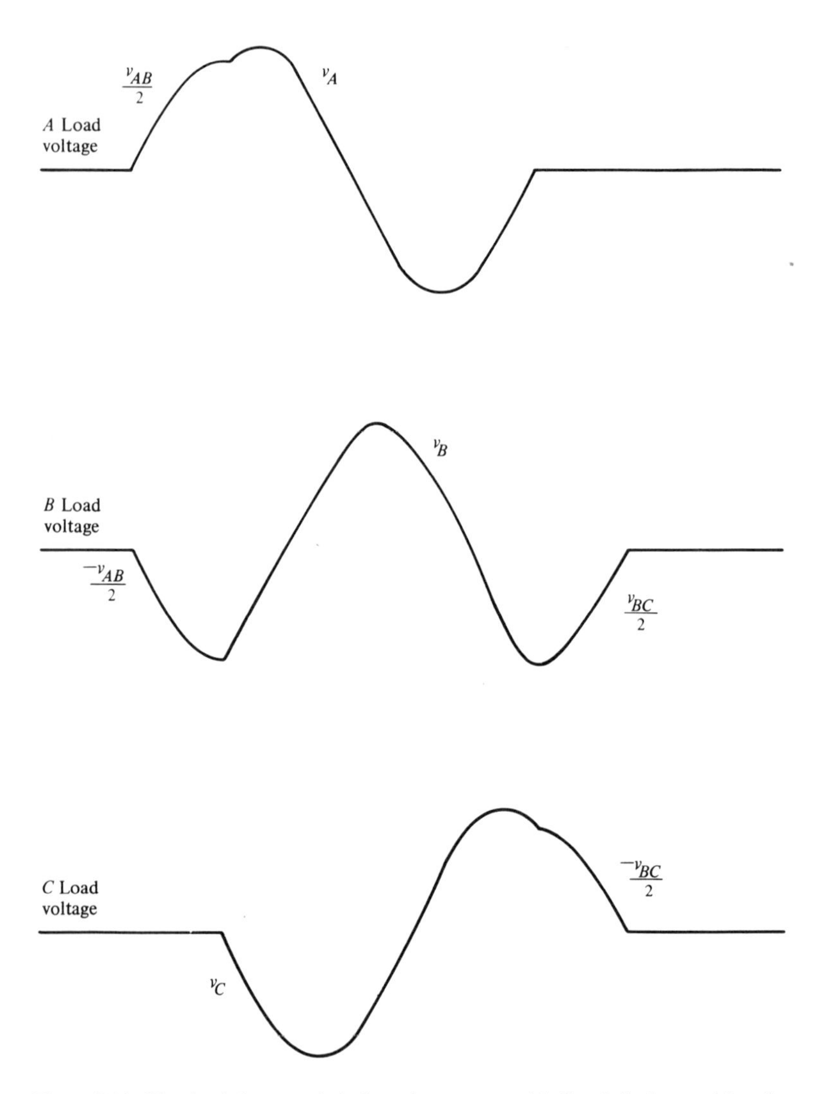

Figure 5-21. Wye-loaded wye-switched regulator waves with C-switch closure delayed.

forms of Figures 5-21 and 5-22, it is easy to show that 1.09 times the power that would be calculated on the idealized basis is delivered in such nominal one-cycle closure periods. Obviously, the error diminishes with increasing m, becoming $< 1\%$ at $m = 9$; however, it may need to be accounted for in some applications.

The three-thyristor three-diode regulator will give exactly the same waveforms for both wye and delta loads, if the same control sequences are employed. Only one point has to be noted—a positive-going zero crossing of the appropriate line

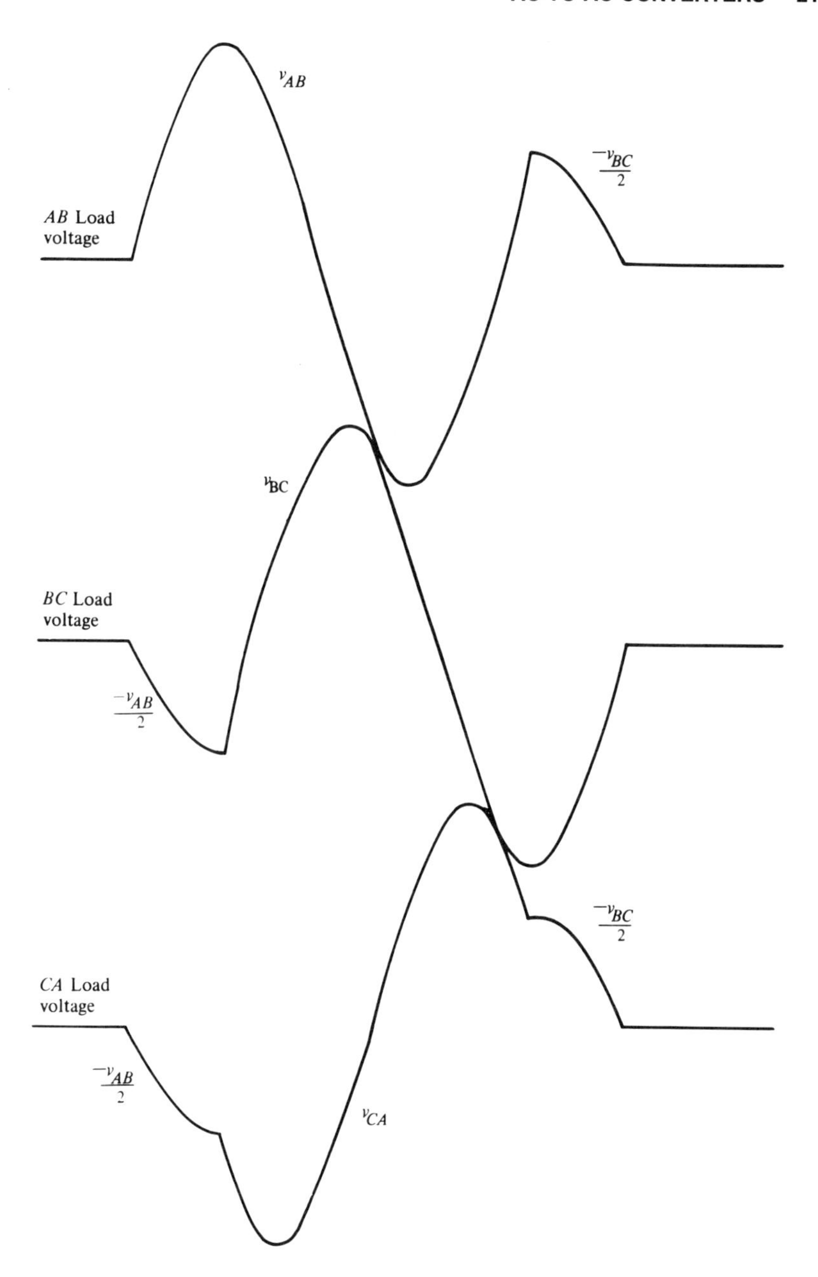

Figure 5-22. Delta-loaded wye-switched regulator waves with C-switch closure delayed.

voltage must be selected for initial turn-on; in the six-thyristor version, either positive- or negative-going voltage zeros may be chosen.

The British Delta of Figure 5-15d gives slightly different results. When on, in the steady state, the switch currents are as depicted in Figure 5-23, and two differences in conditions from those of the wye-switched regulator can be seen. First, only two of the three switches conduct at any given time; second, the switches open only at negative-going current zeros in the line currents. The *ab* switch opens as the *A* line current goes to zero, the *bc* switch opens as the *B* line current goes to zero, and the *ca* switch opens as the *C* line current goes to zero, but only at the negative-going zero crossings. Also, it is only necessary to close one switch at turn-on. If that switch is closed at the positive-going zero crossing of the corresponding line voltage, then the next switch in sequence is closed $\pi/2$ rad later. The third switch is subsequently closed (at its normal

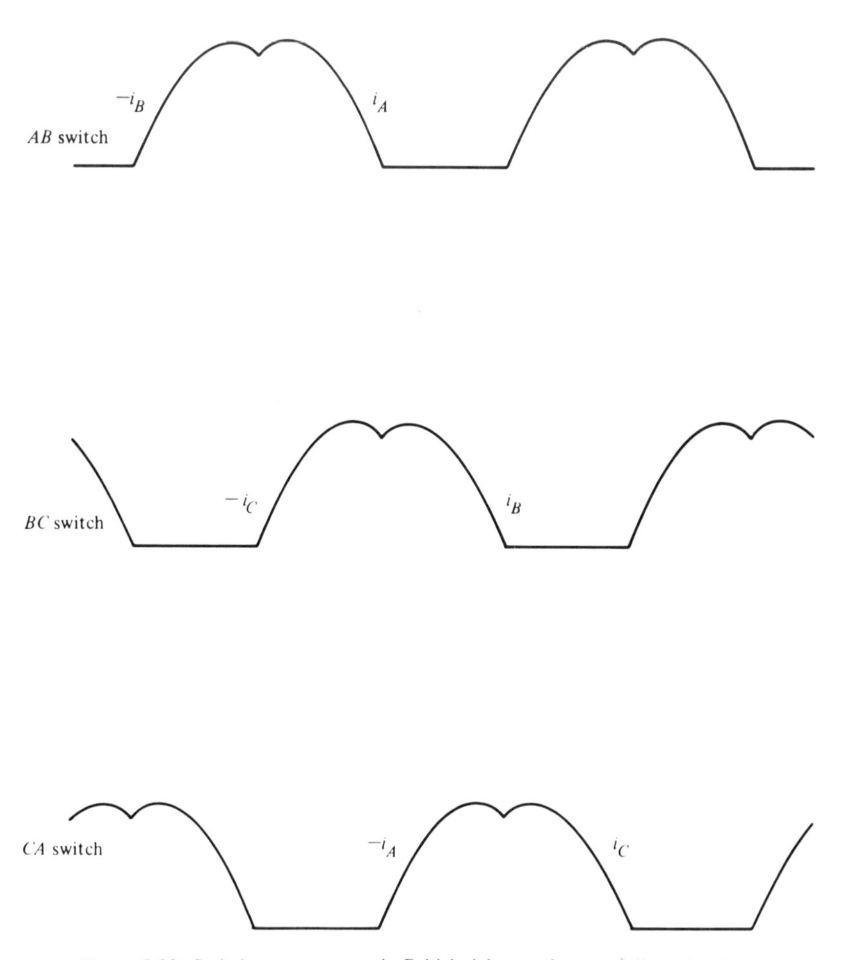

Figure 5-23. Switch current waves in British delta regulator at full conduction.

closing time), $2\pi/3$ rad later still, and a transient free turn-on of the same type as for the wye-switched regulator with wye-connected load results. However, the turn-off sequence differs significantly, as depicted in the waveforms of Figure 5-24. There, the *ab* switch was closed first, followed by the *bc* switch at $\pi/2$ rad delay and the *ca* switch $2\pi/3$ rad later still. The *ab* switch turns off after $7\pi/6$-rad conduction, when the *ca* switch is closed, but is reclosed $2\pi/3$ rad later as the *bc* switch opens. The *ca* switch is then allowed to open $2\pi/3$ rad

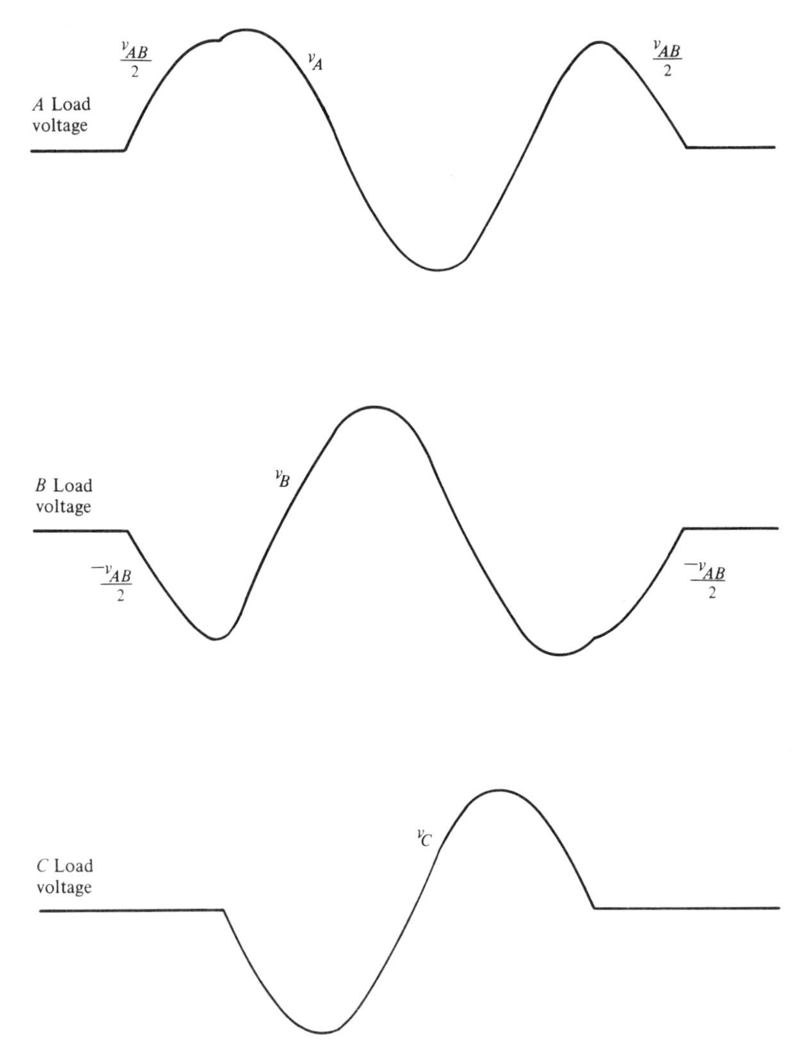

Figure 5-24. Voltage waves for British delta regulator with integral cycle control and BC-switch closure delayed.

later without reclosing the *bc* switch, and the *ab* switch then opens at the succeeding zero of the *AB* phase current. These waves have higher dc content, greater unbalance, and more power throughput than those of the wye-switched regulator. Thus, it is even more important to rotate the starting through the *ABC* sequence, and the correction factor for the power delivered by a nominal one-cycle closure is 1.25, which must definitely be accounted for in some applications.

Before considering the behavior of these regulators with phase-delay control, it is helpful to contemplate the basic rules of their operation from which the waveforms of Figures 5-19 through 5-24 are derived. For the wye-switched regulator with wye load, the phase voltages are impressed on the loads whenever all three switches are closed. When only two switches are closed, two of the loads share a line voltage while the other sees zero voltage, but if less than two switches are closed no load has any voltage applied. With delta loads, the line voltages are impressed on all three whenever all three switches are closed. With only two switches closed, one load is subjected to a full line voltage, while the other two share that line voltage. Again, with less than two switches closed, no voltage is applied to any load.

It is a trivial matter to demonstrate that these two configurations—wye and delta loading—are exactly equivalent in all three situations, provided that the loads are equivalent (i.e., if the resistance of each wye-load branch is R, then the resistance of each delta-load branch is $3R$). This being so, the line currents are identical for both cases regardless of control technique, and the delta-load voltages and currents are easily derived from those of the wye load, so that only the wye-load case need be treated.

The British delta's behavior is different due to the unidirectional current-carrying capacity of its switches and its conformation. With no switches closed, of course, no voltage is impressed on any load. With one switch closed, current will flow in two loads *as long as the line voltage to which they are connected is of appropriate polarity*. During such a period of conduction, closing the apposite switch of the other two establishes conduction in *all three loads,* and the imposition of the phase voltages thereon. During the first $\pi/2$ rad of the *ab* switch conduction period, closure of the *ca* switch is required to fulfill this condition; during the second $\pi/2$ rad, the *bc* switch must be closed. Moreover, as can be observed from the full on waveforms of Figure 5-23, closing both switches up to $\pi/6$ rad in advance of the positive-going zero crossing of the appropriate line voltage will produce this condition. This is because, for example, a positive current flowing from C to B via the *ca* and *ab* switches will allow a negative current flow from b to a via the *ab* switch.

If these simple situations are recognized and accounted, the behavior of both regulators under phase-delay control becomes easy to understand. Consider first the wye-switched regulator of the six-thyristor (or three-triac) variety. The waveform of Figure 5-25a shows the voltage applied to the A load (resistive) for α, the phase-delay angle with respect to the A phase voltage, lying between 0 and $\pi/3$ rad. This wave develops in the following manner. From 0 to α rad, there is no voltage—the A switch is open. At α, the A switch (positive thyristor) closes, and, since the B and C switches (negative and positive thyristors, respectively) are closed, the A-phase voltage appears. At $\pi/3$ rad (i.e., after

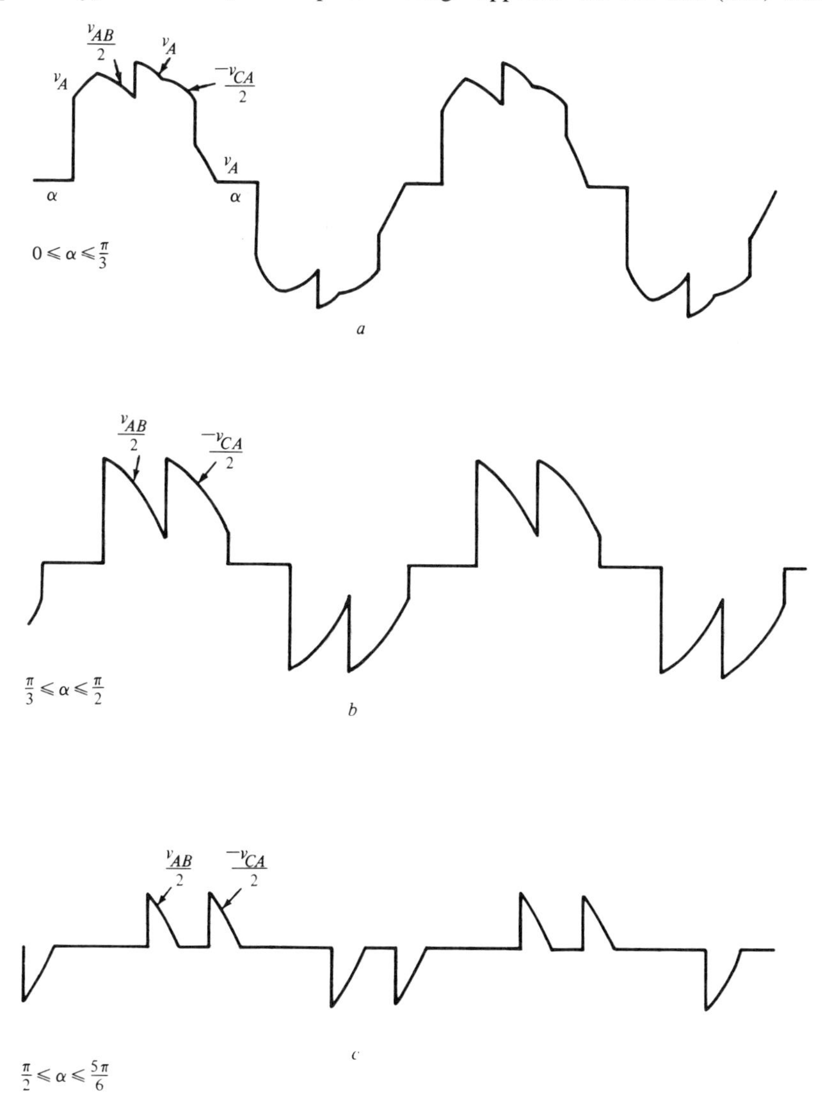

Figure 5-25. Dependent-voltage waves of three-thyristor–three-diode wye-switched regulator.

$\pi/3-\alpha$ rad of the A-phase voltage), the C switch opens and the voltage becomes one-half of the AB line voltage. After a further α rad delay, at $\pi/3+\alpha$ rad, the C switch (negative thyristor) closes and the A voltage is again present. At $2\pi/3$ rad, the B switch opens, and the voltage becomes one-half of the AC voltage, whereas, after a further α rad, at $2\pi/3+\alpha$ rad, the B switch (positive thyristor) recloses and the A voltage is present once more. The A switch opens at π rad, at A current zero, and the whole sequence of events then repeats for the negative half-cycle and beyond. Analysis of this wave is straightforward, if tedious, by defining existence functions for the various periods. Thus, $H = 0$ from 0 to α and from π to $\pi+\alpha$, quite obviously. $H = 1$ from α to $\pi/3$ while operating on v_A, and $H = 1$ from $\pi/3$ to $\pi/3+\alpha$ while operating on $v_{AB/2}$ and so on. Redefining as Fourier expansions:

$$H_{1A} = (1/6-\alpha/2\pi)+(2/\pi)\sum_{n=1}^{n=\infty}\{\sin[n(\pi/6-\alpha/2)]/n\}\cos[n(\omega_s t+\pi/3-\alpha/2)]$$

$$H_{1A\pi} = (1/6-\alpha/2\pi)+(2/\pi)\sum_{n=1}^{n=\infty}\{\sin[n(\pi/6-\alpha/2)]/n]\cos[n(\omega_s t-2\pi/3-\alpha/2)]$$

$$H_{AB} = \alpha/2\pi+(2/\pi)\sum_{n=1}^{n=\infty}[\sin(n\alpha/2)/n]\cos[(\omega_s t-\alpha/2+\pi/6)]$$

$$H_{AB\pi} = \alpha/2\pi+(2/\pi)\sum_{n=1}^{n=\infty}[\sin(n\alpha/2)/n]\cos[n(\omega_s t-\pi-\alpha/2-5\pi/6)]$$

$$H_{2A} = (1/6-\alpha/2\pi)+(2/\pi)\sum_{n=1}^{n=\infty}\{\sin[n(\pi/6-\alpha/2)]/n\}\cos[n(\omega_s t-\alpha/2)]$$

$$H_{2A\pi} = (1/6-\alpha/2\pi)+(2/\pi)\sum_{n=1}^{n=\infty}\{\sin[n(\pi/6-\alpha/2)]/n\}\cos[n(\omega_s t-\pi-\alpha/2)]$$

$$H_{AC} = \alpha/2\pi+(2/\pi)\sum_{n=1}^{n=\infty}[\sin(n\alpha/2)/n]\cos[n(\omega_s t-\pi/3-\alpha/2+5\pi/6)]$$

and $H_{AC\pi}$ the same except for an additional π rad delay.

$$H_{A3} = (1/6-\alpha/2\pi)+(2/\pi)\sum_{n=1}^{n=\infty}\{\sin[n(\pi/6-\alpha/2)]/n\}\cos[n(\omega_s t-\pi/3-\alpha/2)]$$

and $H_{A3\pi}$ the same except for an additional π rad delay. Note that these existence functions are not the existence functions of the switches, but logical combinations thereof. Using the standard notations $X \cdot Y$ to indicate "X and Y" and $X \cdot (Y \cdot Z)$ to indicate "X and Y and Z," the combinations are

$$H_{1A} + H_{1A\pi} + H_{2A} + H_{2A\pi} + H_{3A} + H_{3A\pi} = H_A \cdot (H_B \cdot H_C)$$

$$H_{AB} + H_{AB\pi} = \overline{H}_C \cdot (H_A \cdot H_B)$$

$$H_{AC} + H_{AC\pi} = \overline{H}_B \cdot (H_A \cdot H_C)$$

where $\overline{H}$ means the complement of X ("not X") and H_A, H_B, and H_C are the existence functions of the three switches.

The dependent voltage is given by

$$v_D = (H_{1A} + H_{2A} + H_{3A} + H_{1A\pi} + H_{2A\pi} + H_{3A\pi})V\cos(\omega_s t) + (\sqrt{3}/2)[(H_{AB} + H_{AB\pi}) \cdot V\cos(\omega_s t + \pi/6) + (H_{AC} + H_{AC\pi})V\cos(\omega_s t + 5\pi/6)]$$

When α reaches $\pi/3$ rad, a transition takes place, and the dependent-voltage wave takes on the appearance shown in Figure 5-25b. This exists until α reaches $\pi/2$ and is constructed as follows. For α, the voltage is zero because the A switch is off. At α, the voltage becomes $v_{AB}/2$ because only the B switch is then on. At $\pi/3+\alpha$ rad, the C switch is closed. This immediately forces the B switch to open by driving the B current to zero, since if all three switches were to remain closed, the B current would be positive. The voltage is now $v_{AC}/2$, but $\pi/3$ rad later, the B switch recloses and forces the A switch to open by attempting to produce a negative A current. Thus, the total conduction period remains $2\pi/3$ rad, $\pi/3$ rad on $v_{AB}/2$, and $\pi/3$ rad on $v_{AC}/2$, regardless of α in the range $\pi/3 \leqslant \alpha \leqslant \pi/2$. The existence functions are now defined as

$$H_{AB} = 1/6 + (2/\pi) \sum_{n=1}^{n=\infty} [\sin(n\pi/6)/n]\cos[n(\omega_s t + \pi/3 - \alpha)]$$

and $H_{AB\pi}$ the same except for an additional π rad delay;

$$H_{AC} = 1/6 + (2/\pi) \sum_{n=1}^{\infty} [\sin(n\pi/6)/n]\cos[n(\omega_s t + \pi/3 - \alpha)]$$

and $H_{AC\pi}$ the same except for an additional π rad delay. Thus,

$$v_D = (\sqrt{3}/2)[(H_{AB}+H_{AB\pi})V\cos(\omega_s t+\pi/6) + (H_{AC}+H_{AC\pi})V\cos(\omega_s t+5\pi/6)] \quad (5.20)$$

When α reaches $\pi/2$ rad, yet another transition occurs. At this point, the *AB* and *AC* currents just reach zero at the end of their $\pi/3$ rad conduction periods. Thus, the *A* switch is about to open as the *C* switch closes, and it is necessary that the *A* switch be reclosed with the *C* switch for normal operation of the regulator for $\alpha \geqslant \pi/2$. The dependent voltage is as depicted in Figure 5-25c, $v_{AB}/2$ from α to $5\pi/6$ rad and $v_{AC}/2$ from $\pi/3+\alpha$ to $7\pi/6$ rad. At $\alpha = 5\pi/6$, v_D becomes zero, since both conduction periods vanish simultaneously. The existence functions can now be defined as

$$H_{AB} = (5/12-\alpha/2\pi)+(2/\pi)\sum_{n=1}^{n=\infty}\{\sin[n(5\pi/12-\alpha/2)]/n\} \cdot \cos[n(\omega_s t+\pi/12-\alpha/2)]$$

$$H_{AC} = (5/12-\alpha/2\pi)+(2/\pi)\sum_{n=1}^{n=\infty}\{\sin[n(5\pi/12-\alpha/2)]/n\} \cdot \cos[n(\omega_s t+3\pi/4-\alpha/2)]$$

and $H_{AB\pi}$ and $H_{AC\pi}$ are the same except for an additional π rad delay. Of course, v_D is again defined by Equation 5.20 when these new definitions of H_{AB}, etc., are substituted therein.

The appearance of the waves indicates, and completion of the analysis outlined confirms, that these waves contain only odd harmonics of the supply frequency f_s. Moreover, since all three load waves are identical except for mutual phase displacement of $2\pi/3$ rad and represent currents flowing in the three lines of a three-phase system, they cannot contain any zero-sequence harmonics. Thus, the triplens are absent; only those harmonics of orders $6k\pm1$ are present. A word of caution here. It is commonly assumed that triplens are always zero sequence in a three-phase system. This is definitely not so. They *are* zero sequence if the three waves involved are identical in all respects except for progressive phase displacements of $2\pi/3$. If these conditions are not met for any reasons—unbalanced loads, unbalanced source voltages, or unbalanced α's to name the most common—then triplen harmonics will appear in the three-phase system as mixed positive- and negative-sequence components.

If the loads are inductive, the situation becomes more difficult to assess because the currents are no longer free to make instantaneous changes, even though the voltages are. Thus, those voltage transitions that depend on current zeros are delayed, and may be modified depending on the degree of delay, which in turn depends on the load time constant. Accurate analysis is possible using Laplace transform techniques to establish existence-function bounds, as was done for the single-phase regulator. However, as stated previously, the primary applications for these regulators are with resistive loads, and fortunately, such analyses for the inductive load case are rarely necessary.

The three-thyristor–three-diode regulator exhibits rather different behavior, as illustrated in Figure 5-26. It differs because the turn-on of the diodes, which

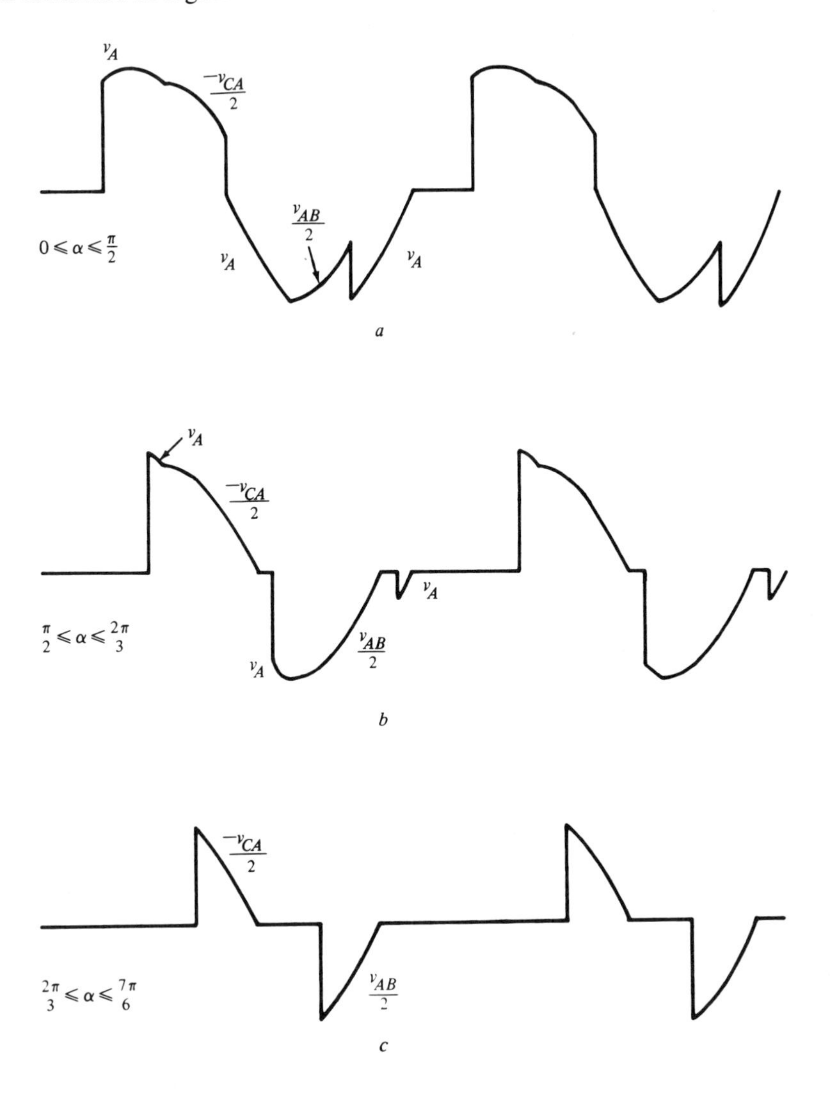

Figure 5-26. Dependent voltage waves of a three-thyristor–three-diode wye-switched regulator.

carry the negative portions of the line currents in this example, cannot be delayed. Thus, for $0 \leqslant \alpha \leqslant \pi/2$, the following sequence, giving the wave of Figure 5-26a, occurs in the A load voltage and line current. For α, the voltage is zero at the beginning of a positive half-cycle because the thyristor of the switch is off. From α to $2\pi/3$ rad, the A-phase voltage is impressed on the load because the C switch does not open at $\pi/3$ rad—its diode conducts immediately the C line current goes negative, and thus keeps the C switch closed. At $2\pi/3$ rad, the B switch opens; its diode ceases to conduct and its thyristor's turn-on is delayed, so that for a period of α, from $2\pi/3$ rad to $2\pi/3+\alpha$, the A load has one-half of the AC-line voltage impressed on it. When the B switch recloses as the B thyristor is turned on, the A phase voltage is reestablished and continues into its negative half-cycle until $4\pi/3$ rad when the C switch opens (diode off, thyristor delayed). There then appears an interval of α rad during which this load sees one-half of the AB-line voltage; as the C switch recloses, the A-phase voltage reappears and is present from $4\pi/3+\alpha$ to 2π. Note that the transition at $\alpha = \pi/3$ that was seen in the six-thyristor version does not occur in the three-thyristor–three-diode version. Also, note that in this sequence the A thyristor is forced off as the C thyristor turns on; this is because the resulting A-line current is negative at that time.

When α reaches $\pi/2$, a transition does occur, and waveforms of the type shown at Figure 5-26b arise. Now, the AC current goes to zero before the B thyristor turns on, so that the A thyristor extinguishes at current zero and the A and C switches open. Also, of course, the AB current goes to zero before the C switch closes, resulting in the B and A switches opening at that current zero but reclosing (diodes conduct) when the C thyristor is turned on. At $\alpha = 2\pi/3$, the periods of A-phase voltage application disappear entirely, and waveforms of the type shown at Figure 5-26c exist from $2\pi/3 \leqslant \alpha \leqslant 7\pi/6$. Observe that the voltages and currents do not become zero until $\alpha = 7\pi/6$; thus, the range of α is extended by $\pi/3$ rad as compared to the six-thyristor version.

The existence functions are simple to formulate, for resistive loads at least. For $0 \leqslant \alpha \leqslant \pi/2$, it is seen that H is 0 from 0 to α; from α to $2\pi/3$ H is 1 and operates on v_A; from $2\pi/3$ to $2\pi/3+\alpha$ $H = 1$ but operates on $-v_{CA}/2$; from $2\pi/3+\alpha$ to $4\pi/3$, $H = 1$ and operates on v_A again; from $4\pi/3$ to $4\pi/3+\alpha$ $H = 1$ and operates on $v_{AB}/2$. Finally, from $4\pi/3+\alpha$ to 2π, $H = 1$ and operates on v_A once more. Note that as before, this is a combination of switch existence functions; those periods operating on V_A represent $H_A \cdot (H_B \cdot H_C)$, where H_A, H_B, and H_C are the existence functions of switches A, B, and C. Those periods operating on $-v_{CA}/2$ represent $\overline{H}_B \cdot (H_A \cdot H_C)$, while those operating on $v_{AB}/2$ represent $\overline{H}_C \cdot (H_A \cdot H_B)$, as before.

Thus, for $0 \leqslant \alpha \leqslant \pi/2$, the existence functions operating on v_A are

$$H_{1A} = 1/3-\alpha/2\pi+(2/\pi) \sum_{n=1}^{n=\infty} \{\sin[n(\pi/3-\alpha/2)]/n\}\cos[n(\omega_s t+\pi/6-\alpha/2)]$$

$$H_{2A} = 1/3-\alpha/2\pi+(2/\pi)\sum_{n=1}^{n=\infty}\{\sin[n(\pi/3-\alpha/2)]/n\}\cos[n(\omega_s t-\pi/2-\alpha/2)]$$

$$H_{3A} = 1/3-\alpha/2\pi+(2/\pi)\sum_{n=1}^{n=\infty}\{\sin[n(\pi/3-\alpha/2)]/n\}\cos[n(\omega_s t-7\pi/6-\alpha/2)]$$

while those operating on $v_{AB}/2$ and $v_{AC}/2$ are

$$H_{AB} = \alpha/2\pi+(2/\pi)\sum_{n=1}^{n=\infty}[\sin(n\alpha/2)/n]\cos[n(\omega_s t-5\pi/6-\alpha/2)]$$

$$H_{AC} = \alpha/2\pi+2/\pi\sum_{n=1}^{n=\infty}[\sin(n\alpha/2)/n]\cos[n(\omega_s t+5\pi/6-\alpha/2)]$$

Giving

$$v_D = (H_{1A} + H_{2A} + H_{3A})\cdot V\cos(\omega_s t) + (\sqrt{3}/2)[H_{AB}V\cos(\omega_s+\pi/6) + H_{AC}V\cos(\omega_s t+5\pi/6)] \quad (5.21)$$

For $\pi/2 \leqslant \alpha \leqslant 2\pi/3$ H_{1A}, H_{2A}, and H_{3A} remain unchanged. H_{AB} and H_{AC} are modified to become

$$H_{AB} = 1/4+(2/\pi)\sum_{n=1}^{n=\infty}[\sin(n\pi/4)/n]\cos[n(\omega_s t+11\pi/12)]$$

$$H_{AC} = 1/4+(2/\pi)\sum_{n=1}^{n=\infty}[\sin(n\pi/4)/n]\cos[n(\omega_s t+7\pi/12)]$$

The dependent voltage is again given by Equation 5.21. For $2\pi/3 \leqslant \alpha \leqslant 7\pi/6$, H_{1A}, H_{2A}, and H_{3A} all become zero, whereas H_{AB} and H_{AC} are further modified to become

$$H_{AB} = 7/12-\alpha/2\pi+(2/\pi)\sum_{n=1}^{n=\infty}\{\sin[n(7\pi/12-\alpha/2)]/n\}\cdot\cos[n(\omega_s t-3\pi/4-\alpha/2)]$$

and H_{AC} the same except for an advance of $11\pi/12$ rather than a delay of $3\pi/4$ in the oscillatory term arguments. Equation 5.21 again gives v_D.

It can be observed that the waveforms of Figure 5-26 are highly asymmetric. This observation is confirmed by the lack of complementary pairs of existence functions and means that these waveforms contain even-order harmonics. They still do not contain triplens, of even or odd order, but do contain all harmonics of orders $3k\pm1$, not just those of order $6k\pm1$ as in the six-thyristor case. The lower-order even harmonics are of large amplitude and can be very distressing. The second harmonic in particular, since it is a negative-sequence component, can be a great problem to rotating machine loads that are either fed by the regulator or connected to the same supply as the regulator. Note, however, that there is no dc content in the waves of Figure 5-26.

As is the case for integral cycle control, the British delta regulator of Figure 5-15d behaves differently. It does not, of course, obey the same basic rules in regard to voltage imposition as a consequence of switch closure. This is shown quite clearly by the full-power switch-current waves of Figure 5-23, for only two switches are closed at any given time.

Waveforms under phase control are shown in Figures 5-27 and 5-28. Figure 5-27 shows the current waves in the *ab* switch, Figure 5-28 the voltage waves impressed on the *A* load. These voltages are seen to be identical to those produced by the three-thyristor–three-diode regulator except for reversal of signs (i.e., exchange of positive half-cycles and negative half-cycles). Reversing switching-device polarity in either regulator will, of course, cause them both to produce exactly the same waves. Hence, the only real difference lies in the switch-current waves shown in Figure 5-27. Switch-current waves for the wye-switched regulators, of course, are identical to the line currents, which in turn are identical in waveshape to the voltages impressed on the wye-connected loads in Figures 5-25 and 5-26.

Since the voltages impressed on the loads by a phase-controlled British delta regulator are identical to those for a three-thyristor–three-diode wye-switched regulator, the same analysis will serve. However, the existence functions derived do not have the same relationship to switch existence functions. This is reflected in the switch-current waves of Figure 5-27.

There now remains only the question of switch-voltage stresses in these ac regulators. For single-phase regulators and the three-phase wye regulator with neutral of Figure 5-15a, peak switch voltage stress obviously equals the ac-supply peak-phase voltage. For the delta-switched regulator of Figure 5-15b, it is equally obviously the ac-supply peak line voltage, as it is for the British delta. The wye-switched regulators impose slightly less peak voltage stress on their switches—one-and-one-half times the ac-supply peak phase voltage. Voltage transitions follow a supply sinusoid as unidirectional current-carrying switches have forward voltage reapplied, and impose sudden reverse voltages when such switches are turned off. The switch-voltage waves are easily constructed, if desired, from the load voltage waves given and the supply sinusoids.

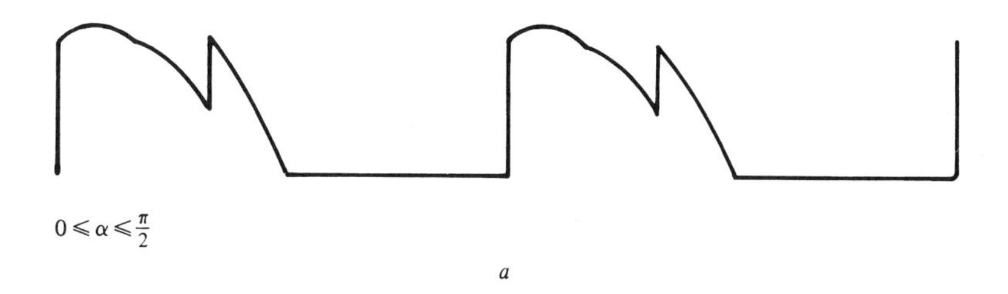

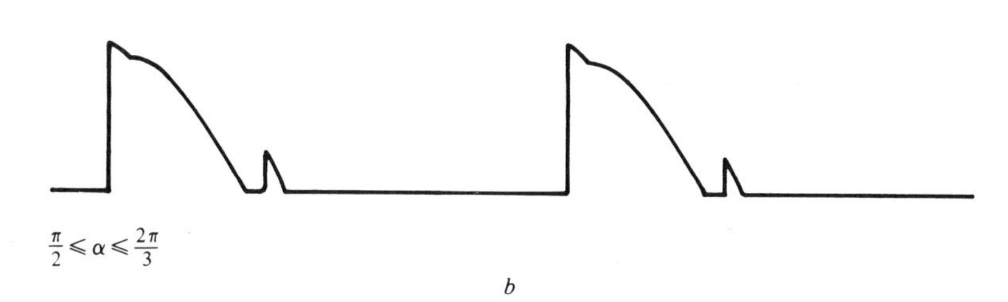

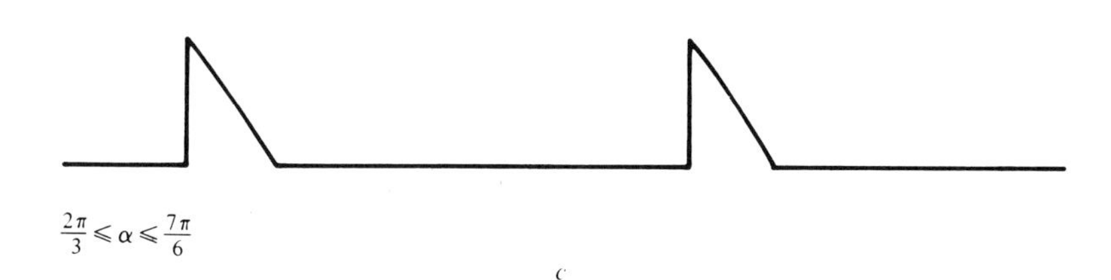

Figure 5-27. Switch current waves in British delta regulator.

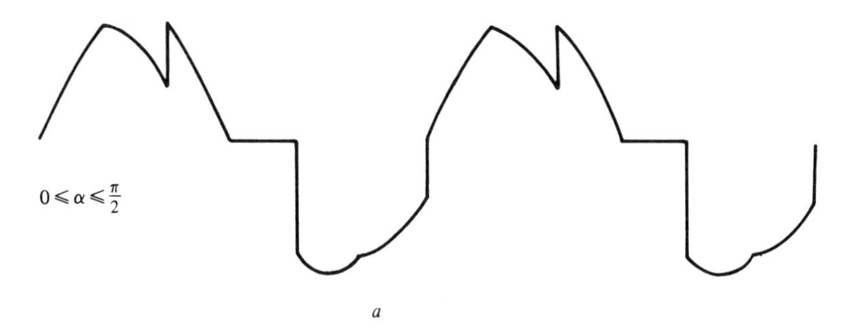

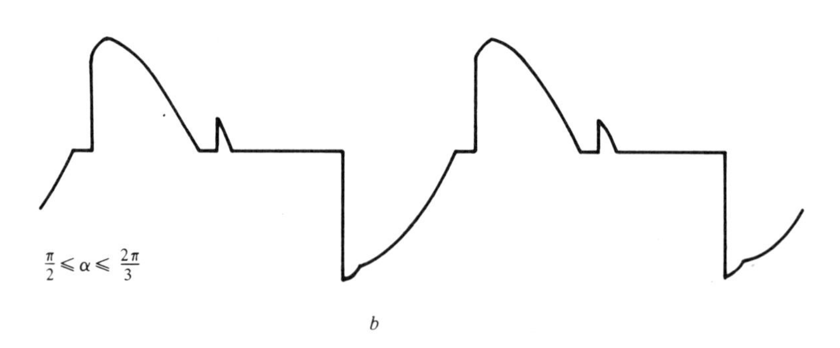

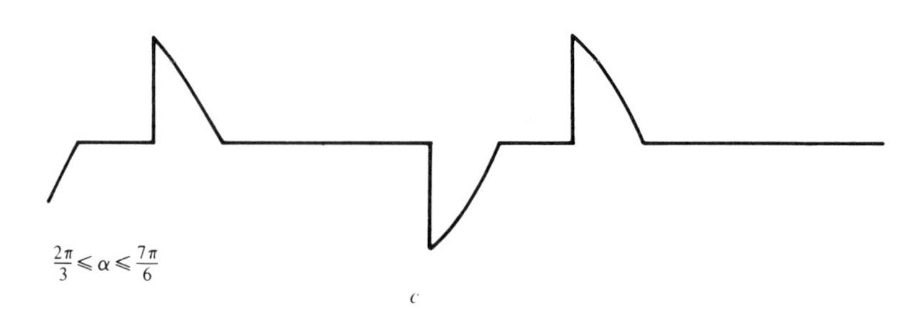

Figure 5-28. Dependent-voltage waves of British delta regulator.

References

1. Gyugyi, L., and B. R. Pelly. *Static Power Frequency Changers*. New York: John Wiley & Sons, 1975.
2. Gyugyi, L. "Generalized Theory of Static Power Frequency Changers." Ph.D. thesis, University of Salford, Salford, 1970.
3. Pelly, B. R. *Thyristor Phase Controlled Converters and Cyclo-converters*. New York: John Wiley & Sons, 1971.

Problems

1. A three-phase to three-phase midpoint UFC is used with 60-Hz input to provide wanted output frequencies ranging from 20 to 200 Hz. Calculate the frequencies and relative magnitudes of the unwanted components in the output voltage that are closest in frequency to the output frequency when operation is at 20, 40, 60—by increments of 20 to 200 Hz.

2. Repeat Problem 1 for the unwanted components of input current closest to the input (60-Hz) frequency.

3. Repeat Problem 1 for an SSFC with output frequency range of 2 to 20 Hz, at 2 Hz increments.

4. Repeat problem 2 for the SSFC of Problem 3.

5. A three-phase-to-three-phase UFC with voltage control has its input connected to a 60-Hz supply and is operated to produce 60-Hz wanted output. Its output is connected back to its input supply via a transformer and a small reactor. The wanted output voltage set is kept in phase with the input voltage set, but adjusted in magnitude so that

a. It is sufficiently less than the input set that full-rated current flows in the inductance between input and output.

b. It is sufficiently greater to cause full-rated current flow. For both (a) and (b), what are the burdens imposed on the 60^{ζ} supply? What happens if an SSFC is substituted for the UFC?

6. A three-phase-to-three-phase bridge (six-pulse) NCC is fed from a 60-Hz supply and produces wanted output frequencies ranging from 2 to 20 Hz. For output frequency increments of 2 Hz, calculate the frequencies of unwanted output voltage and input current components closest to the wanted output and input frequencies, respectively. Compare these with the unwanted component frequencies produced by a six-pulse UFC and a six-pulse SSFC used in the same circumstances, and explain why any differences occur.

7. Repeat problem 6 for a simple (non-PWM) inverter-type converter with $\alpha = 0$.

8. A three-phase delta-switched ac regulator is used with pure inductive load on a 60-Hz ac system. Prove that the triplen harmonic currents in the inductors do not flow in the ac lines if all three branches are operated at the same delay angle. What happens if the delay angles are not identical?

9. Calculate the total rms harmonic distortion in the output voltage and input current of a single-phase ac regulator (two thyristor or triac) producing 2/3, 1/2, and 1/3 of the rms voltage feeding it at its output. Calculate, and contrast with the total rms distortion produced when integral cycle control is used to produce the same results.

10. Explain why the six-thyristor wye-switched phase-controlled ac regulator has a better line-current spectrum than the three-thyristor–three-diode version. What does this imply regarding the displacement factors of these two converters when operated to produce the same total rms voltage across given resistive loads from a given supply?

6
Interfacing Converters with Real Sources and Loads

6.1 Introduction

So far, we have been concerned with the fundamental theoretical concepts and idealized analyses of switching power converters. These aspects are necessary to a full understanding of the nature, relationships, capabilities, and limitations of converters. In almost all analytic treatments, perfect voltage and current sources have been assumed to exist at the converters' terminals. In reality, of course, this is not so, especially not for current sources. For converters to operate as the analyses postulate, and indeed they do so operate in the vast majority of cases, they have to be interfaced with real sources and loads so that their electrical world appears to contain the ideal sources, or at least reasonably close approximations thereto.

Since the converters are power-processing artifacts, and efficiency is all important, the means for interfacing should be lossless. Thus, any networks connected between converter terminals and real sources and loads, or in shunt with converter-source (or load) terminals, must be nondissipative. The interfacing is generally accomplished by using passive component networks—filters—consisting of inductors and capacitors. For the most part, no resistance is introduced save for that which inevitably accompanies the use of practical components.

In this respect, filters for switching power converters differ markedly from many of their counterparts in the world of signal processing. There, insertion loss is an accepted penalty for filter usage, while in power conversion any significant loss is totally unacceptable.

The duty of the interfacing filters for power converters may be simply stated. In general, they must allow free passage of wanted input or output voltage or current components between a converter and its real sources and loads. At the same time, they must obstruct the passage of unwanted components that have been seen to be inevitable companions of the wanted components at the converters' terminals. They must also present to the converters impedances which, in conjunction with real load and source characteristics, create reasonable facsimiles of the ideal voltage and current sources that the analyses postulate. In most cases, meeting these requirements is not as difficult as it might seem.

This chapter deals with the electrical requirements of interfacing components and filters, and to some extent with the practical design considerations of the components. Also, of necessity, it deals with the characteristics of a wide variety of loads and sources, since interface design depends on those characteristics in addition to those of the converters.

6.2 Approximation of DC Current Sources

In view of the great diversity of dc loads and sources served by converters, it is perhaps surprising that there is a great deal of commonality of basic electrical characteristics. The actual loads served usually fall into one of three major categories:

1. dc machine armatures and fields
2. electrochemical loads and sources
3. electronic equipment

As presented to the converter, all exhibit a ''back emf'' together with some series resistance and inductance. For dc machine armatures and fields, these characteristics are obvious, as they are for electrochemical loads and sources. In these cases, an electro-chemical potential generally dominates the situation, with the resistance and inductance appearing as part of the source or load regulation behavior. In this category, loads may be exemplified by plating baths, metal refining-pot lines, and batteries under charge. Sources can be exemplified by fuel cells and batteries under discharge. Electronic loads take the form of varying resistors—usually cyclically varying. However, in order to assume these characteristics, they must be fed from a good approximation to a voltage source. In consequence, they are universally subject to a double interface when driven by current-sourced switching power converters. First, the load itself is provided with the means to approximate a voltage source, by the techniques described in Section 6.3, then the current-sourcing interface is applied between the converter and the modified load.

There is, of course, one other major application for current-sourcing interfaces—between converters. This is most apparent in the current-source transfer buck-boost dc-to-dc converter, but also occurs when ac-to-ac conversion is accomplished by cascade-connecting ac-to-dc and dc-to-ac converters. This latter situation is very common, since the switching devices needed to implement many of the direct ac converters are not yet available, and in any case switch utilization tends to be poorer in direct ac-to-ac converters than in compound converters. The following discussion on the dc current-sourcing interface applies as well to the requirements for inter-converter interfaces as to converter-source and -load interfaces.

In examining the characteristics of the dc-terminal dependent voltages of current-sourced ac-to-dc/dc-to-ac converters in Chapter 4 and of voltage-sourced and current-sourced dc-to-dc converters in Chapter 3, it is apparent that in all cases the dependent voltage consists of a wanted, perfectly smooth, dc component accompanied by an infinite series of unwanted oscillatory components. These ripple components are harmonics of the converter-switching frequency, which for dc-to-dc converters, is arbitrarily chosen. For ac-to-dc/dc-to-ac converters, the switching frequency equals the ac source (or load) frequency unless the converter is PWM, in which case the switching frequency will usually be an odd multiple of the ac source frequency. Now these statements hold true, the analyses tell, if the defined current at the converter terminals in question is perfectly smooth dc. It can be deduced, from the analytic manipulations performed, that the statements will also hold true as long as the dc terminal current remains unidirectional and greater than zero at all times. Under such conditions, the same switch existence functions will apply, and hence the same dependent-voltage expressions will arise.

With these facts established, it can be concluded that the ideal dc current-sourcing interface is a component offering zero impedance to the flow of dc current while offering infinite impedance to ac currents or, in the converse, supporting any finite-magnitude oscillatory voltage but permitting no oscillatory current flow. The component required is thus quite obviously an infinite inductance. Equally obviously, it cannot be realized in practice. However, if, as has been postulated, oscillatory components of current are allowed provided that the total current remains unidirectional and greater than zero, then a finite inductance will serve.

Thus, the basic current-sourcing interface consists simply of an inductance. There are two design questions regarding this component in a given situation:

1. What value of inductance is needed?
2. To what extent and in what manner does the ripple current permitted influence converter performance at the defined voltage terminals?

To answer the first question, it is clear that the minimum tolerable value of inductance is that which just permits the total current to reach zero instantaneously once in each cycle of the dc-terminal ripple voltage. This value of inductance—the *critical inductance* for the interface—is a function not only of the converter's ripple voltage but also of the dc (wanted-component) current level. The degree to which the current-sourcing inductor in a given application must exceed the critical inductance depends on the load's (or source's) tolerance for ripple current. Thus, in many instances, particularly with electronic loads, voltage-sourcing interfaces are used in conjunction with the finite-inductance current-sourcing interface.

Although the critical inductance for a given situation can be calculated using the spectral information generated by the analyses of Chapters 3 and 4, it is not usually so established. This is because such a calculation involves the summation of a large number of the current terms due to the infinite series of ripple voltage components. This can be done with the aid of a digital computer, but there is a simpler way to approach the determination of critical inductance if the ripple voltage waveshape can be defined accurately. As we have seen, this definition can be made by using the graphical representations of switch existence functions and defined voltage waves. Since the total change in current in a finite inductor is given by

$$I_{PPL} = \text{Impressed Volt-Seconds}/L$$

From

$$i(t) = (1/L)\int V dt$$

the peak-to-peak ripple current excursion in a finite current-sourcing inductor is easily calculated by evaluating the total volt-seconds in the ripple voltage wave over one-half cycle of that wave. Thus, if v_D is the dependent voltage wave and V_{DD} its wanted dc component, then the volt-second integral involved is

$$VSI = \int_{(v_D=V_{DD})_1}^{(v_D=V_{DD})_2} (v_D - V_{DD})dt$$

The limits, of course, represent two successive times when the ripple wave intersects the dc component. The resulting volt-second integral may be positive or negative, depending on the choices of upper and lower limits. It is positive if $(v_D = V_{DD})_1$ is the time of a positive-going intersection of v_D and V_{DD}; it is negative for the converse. This simply reflects the fact that in steady-state operation, the maximum-to-minimum current excursion equals, in magnitude, the minimum-to-maximum current excursion in the inductor.

Figure 6-1 shows the configurations, dependent-voltage waves, and quantity definitions for dc-to-dc converters. Beginning with the buck converter of Figure 6-1a, the average current through the current-sourcing inductor is I_S, the defined source current, and the voltage impressed thereon is $V_S - V_{DD}$, the difference between the input and output voltages, for a time $2\pi/\omega A$ sec. Thus, the volt-second integral can be expressed as

$$VSI = \int_0^{2\pi/\omega A} (V_S - V_{DD})dt$$

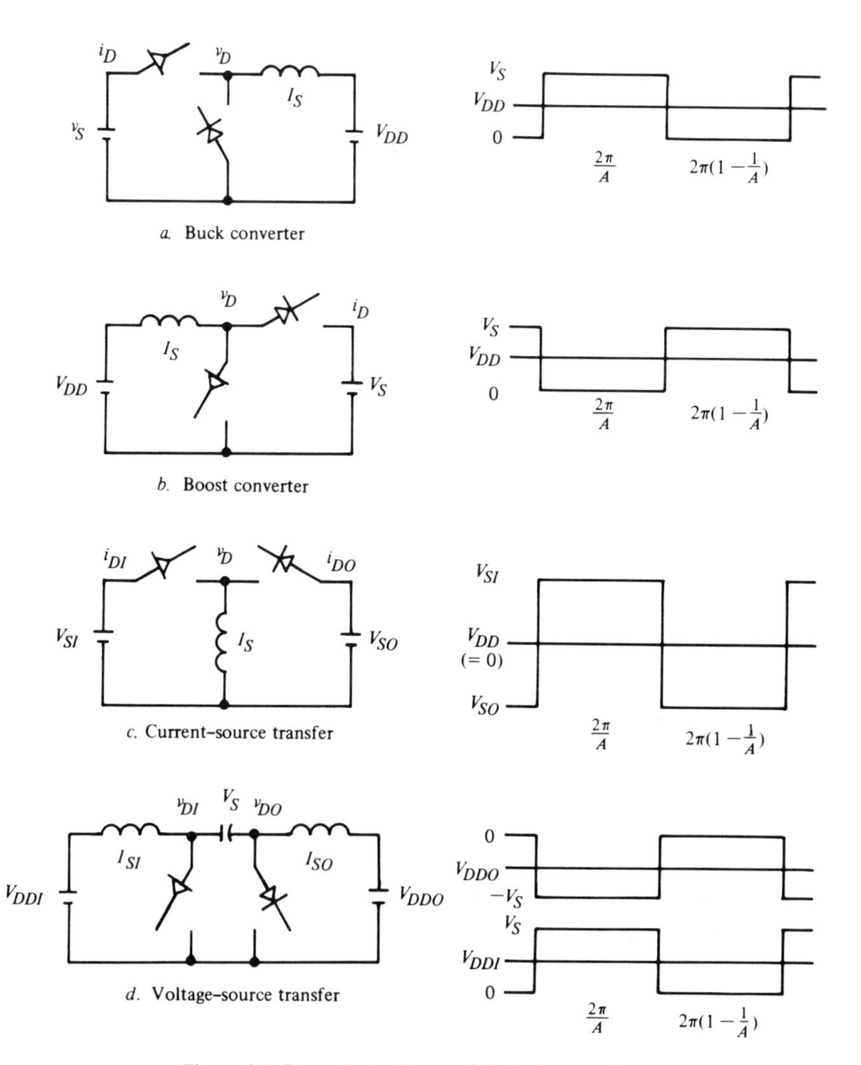

Figure 6-1. Dependent voltages of dc-to-dc converters.

$$= V_s \int_0^{2\pi/\omega A} (1-1/A)dt$$

$$= (V_S/f)[(1-1/A)(1/A)] \tag{6.1}$$

since $V_{DD} = V_S/A$.

Since the voltage impressed on the current-sourcing inductor is constant in both periods of the ripple cycle, the current changes are linear, and hence the peak-to-peak ripple current is exactly twice the peak-to-average current and average-to-minimum current excursions. The peak-to-peak ripple current is given by

$$I_{PP} = VSI/L = (V_S/fL)(1/A - 1/A^2)$$

and thus the average-to-minimum current excursion is given by

$$I_S - I_{MIN} = (V_S/2fL)(1/A - I/A^2)$$

The inductance is critical when $I_{MIN} = 0$ for a given I_{SMIN}, the minimum value at which the converter is expected to operate, so that the critical inductance is given by

$$L_C = (V_S/2fI_{SMIN})(1/A - 1/A^2) \quad (6.2)$$

Using $A = V_S/V_{DD}$, where V_{DD} is the wanted component of the dependent (output) voltage, yields

$$L_C = (1/2fI_{SMIN})(V_{DD} - V_{DD}{}^2/V_S)$$

The design value of the inductance is the value obtained under worst-case operating conditions. If V_{DD} is fixed, as it often is, the maximum value of L_C clearly occurs when V_S assumes its maximum value in the specified range. If both V_{DD} and V_S are variable, then the worst-case value of L_C can be found from

$$dL_C/dV_{DD} = (1/2fI_{SMIN})(1 - 2V_{DD}/V_S)$$

giving $dL_C/dV_{DD} = 0$ when $V_{DD} = V_S/2$. Thus, for this case, the maximum value of critical inductance occurs when V_S assumes its maximum value and $A = 2$ to give $V_{DD} = V_S/2$.

In many practical situations, the value of inductance used is well above the critical value. In these cases, the ripple current permitted, often implicitly specified as a consequence of a ripple voltage limit at a voltage-sourcing interface, forces the use of larger inductances.

The boost converter of Figure 6-1b can be approached in exactly the same manner, as can the buck-boost converters of Figures 6-1c and d. For the boost converter, a voltage $V_S - V_{DD}$ is applied to the inductor over the time period $2\pi(1 - 1/A)/\omega$. In this case, of course, V_S is the output voltage and V_{DD} the wanted component of the dependent input voltage. The relationship between V_S and V_{DD} for the converter is

$$V_{DD} = (1-1/A)V_S$$

and thus the volt-seconds applied to the inductor are given by Equation 6.1 and the peak-to-peak ripple current by

$$I_{PP} = (V_S/fL)(1/A-1/A^2)$$

The peak ripple, $I_P = I_{PP}/2$, is again

$$I_P = I_S - I_{MIN}$$

since the input current flows in the current-sourcing inductor of the boost converter, and thus the critical inductance is once more given by Equation 6.2. However, it is desirable to transpose this so that L_C is given in terms of the input and output voltages and the output current, since these are the quantities usually given the designer. For the boost converter,

$$I_{DD} = (1-1/A)I_S$$

where I_{DD} is the wanted (stipulated) component of output current and I_S the defined (analytically perfectly smooth) input current that results. Thus,

$$L_C = (1/2fI_{DDMIN})(V_{DD}^2/V_S - V_{DD}^3/V_S^2)$$

after substituting for I_S and A. When the output voltage (in this case, V_S), is fixed, L_C assumes its maximum value when $dL_C/dV_{DD} = 0$, which occurs when V_{DD} (input voltage) $= 2V_S/3$. If both V_S and V_{DD} vary, then L_C has its maximum value when V_S is maximum and $V_{DD} = V_S/2$.

The current-source transfer buck-boost converter of Figure 6-1c is rather different, since the average voltage (wanted component) impressed on the current source is zero. This allows an inductor to be used, of course, but the voltage impressed thereon is observed to be V_{SI} for a time $2\pi/\omega A$, where V_{SI} is the input voltage. Thus, the volt-seconds are

$$VSI = V_{SI}/fA$$

and the peak-to-peak ripple current is given by

$$I_{PP} = V_{SI}/fAL$$

and the critical inductance by

$$L_C = (V_{SI}/2fI_{SMIN})(1/A)$$

where I_{SMIN} is the minimum anticipated average inductor current.

The average quantity relationships for this converter are

$$V_{SO} = -(A-1)V_{SI}$$

giving the output voltage and

$$I_{DDI} = I_S/A$$

$$I_{DDO} = -(1-1/A)I_S$$

giving the input and output currents in terms of the source (inductor) current. Making appropriate substitutions, the expression for the critical inductance becomes

$$L_C = (1/2f\overline{V}_{DDOMIN})[\overline{V}_{SO}\overline{V}_{SI}^2/(\overline{V}_{SO}+\overline{V}_{SI})^2]$$

where $\overline{X}$ is used to mean "the absolute value of X." If $\overline{V}_{SO}$ is fixed, then L_C is again an increasing function of the input voltage, with maximum L_C occurring for the maximum value of $\overline{V}_{SI}$ in the design range. If both $\overline{V}_{SO}$ and $\overline{V}_{SI}$ are variable, then the maximum value of L_C occurs when $\overline{V}_{SI}$ has its maximum and $\overline{V}_{SO} = \overline{V}_{SI}(A = 2)$.

The voltage-source-transfer buck-boost converter differs in that two current sources, and hence two inductors, must be considered. The output inductor, L_O, is subjected to a voltage V_{DDO} (the wanted component of the output voltage) for a time $2\pi(1-1/A)/\omega$ and thus to a volt-second integral

$$VSIO = (V_{DDO}/f)(1-1/A)$$

The input inductor, L_I, is subjected to a voltage V_{DDI} (the wanted component of the input voltage) for a time $2\pi/\omega A$ and thus to a volt-second integral

$$VSII = V_{DDI}/fA$$

The average quantity relationships are

$$V_{DDO} = -V_{DDI}/(A-1)$$

and

$$I_{SI} = -I_{SO}/(A-1)$$

where I_{SI} and I_{SO} are the input and output currents, respectively. Using these expressions, together with the volt-second expressions, yields

$$L_{CO} = (1/2f\overline{I}_{SOMIN})[\overline{V}_{DDO}\overline{V}_{DDI}/(\overline{V}_{DDO}+\overline{V}_{DDI})]$$

and

$$L_{CI} = (1/2f\overline{I}_{SOMIN})[\overline{V}_{DDI}^{2}/(\overline{V}_{DDO}+\overline{V}_{DDI})]$$

L_{CO} is an increasing function of both $\overline{V}_{DDI}$ and $\overline{V}_{DDO}$; hence, with $\overline{V}_{DDO}$ fixed, the maximum value of L_{CO} occurs for the maximum $\overline{V}_{DDI}$. If both are variable, it occurs for maximum $\overline{V}_{DDO}$ in conjunction with maximum $\overline{V}_{DDI}$. L_{CI} is an increasing function of $\overline{V}_{DDI}$ but a decreasing function of $\overline{V}_{DDO}$. Hence, if $\overline{V}_{DDO}$ is fixed, the maximum value of L_{CI} occurs when $\overline{V}_{DDI}$ is at its maximum, but if both $\overline{V}_{DDO}$ and $\overline{V}_{DDI}$ are variable, then L_{CI} has its maximum value when $\overline{V}_{DDI}$ is at its maximum and $\overline{V}_{DDO}$ at its minimum.

Harmonic neutralized dc-to-dc converters present slightly more complex problems. When dependent voltages are connected in parallel via interphase reactors, as they usually are in such cases, ripple amplitude decreases and ripple frequency increases. The ripple waves are still rectangular, but now have repetition frequencies lf, where l is the number of converters used and f their common operating frequency. Concomitantly, the peak-to-peak ripple amplitude is decreased by the factor l. As A, the converters' control variable, covers the range 1 to ∞, the ripple vanishes for values of A for which l/A is an integer, as was shown in Chapter 3. Thus, there are a multiplicity of maxima in the volt-second integrals that are applied to inductors used for current sourcing. This is illustrated in Figure 6-2, showing the dependent-voltage waves for a three-phase buck converter. The general case of the buck, l harmonic neutralized converters, will now be analyzed in detail. Analyses for the other converters are similar, and may easily be derived therefrom.

It can be seen from Figure 6-2 that a ripple cycle of the combined (averaged) output voltage of the l converters extends from $2k\pi/l$ rad to $2(k+1)\pi/l$ rad, where k has the range 0 to $l-1$. The maximum and minimum voltages during the cycle depend on an integer p, being $V_{MAX} = (p+1)V_S/l$ and $V_{MIN} = pV_S/l$ while $A = l/(p+z)$, where $0 \leqslant p \leqslant l-1$ and $0 \leqslant z \leqslant 1$. From these definitions, it is easy to show that the volt-seconds applied to the inductor are given by

$$VSI(l) = (V_S/l^2f)(1-z)z$$

However, $V_{DD}/V_S = (p+z)/l$, which translates to

$$VSI(l) = (1/l^2f)\{[(p+1)V_S-lV_{DD}](lV_{DD}/V_S-p)\}$$

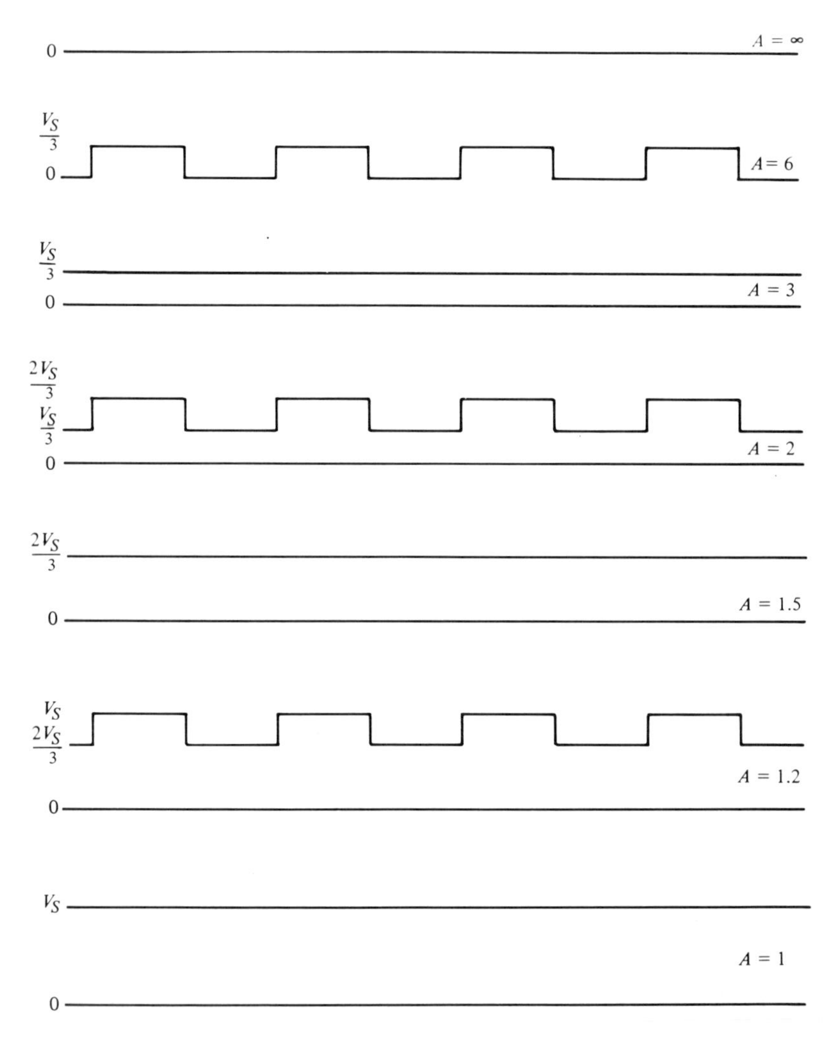

Figure 6-2. Dependent-voltage waves of three-phase harmonic neutralized buck converter.

giving the critical inductance as

$$L_C = (1/l^2 f I_{SMIN})[(p+1)(V_S - lV_{DD})(lV_{DD}/V_S - p)]$$

If V_{DD} is fixed, the maximum value for L_C is obtained when $z = \sqrt{p(p+1)} - p$. This is an increasing function of p, and thus the absolute maximum value of L_C occurs when $z = \sqrt{l(l-1)} - (l-1)$, $p = l-1$, and $V_S = lV_{DD}/\sqrt{l(l-1)}$ (provided that this condition exists in the converter's operating range). If both V_S and V_{DD}

are variable, then maxima of the volt-seconds (and hence for the critical inductance) occur whenever $z = 1/2$; clearly, the absolute maximum occurs when V_S is maximum and $z = 1/2$.

For current-sourced dc-to-ac/ac-to-dc converters, the situation is made more complex by the lack of constant voltages and the resultant nonlinearity of the ripple current excursions in a current-sourcing inductor. The waveform diagrams of Figure 6-3 (for a phase delay) show that there are two distinct ripple-voltage conditions for such a converter. In Figure 6-3a, the dependent voltage does not have a positive-going intersection with its average value until after the switching has occurred in the converter. In Figure 6-3b, the situation is that the positive-going intersection occurs at the instant of switching. Since the incoming phase voltage (that being switched to) is defined as

$$v_i = V\cos(\omega_s t - \pi/M)$$

and the average voltage is given by

$$V_{DD} = (M/\pi)\sin(\pi/M)V\cos\alpha$$

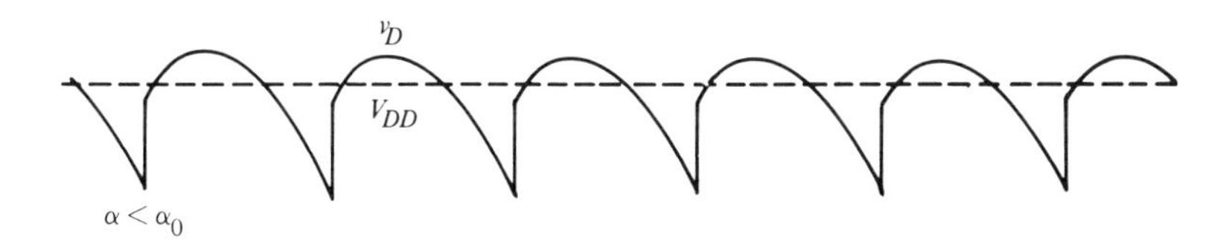

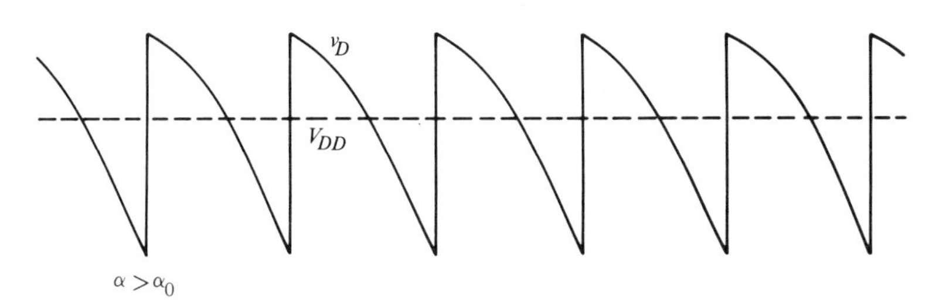

Figure 6-3. Dependent and average dc terminal voltages for three-phase midpoint current-sourced converter.

the boundary between the two ripple conditions is a delay angle α_o such that

$$\cos(\alpha_o - \pi/M) = (M/\pi)\sin(\pi/M)\cos\alpha_o$$

or

$$\alpha_o = \arctan[M/\pi - \cot(\pi/M)]$$

For the most important practical cases, $M = 2$ and $M = 3$, this gives $\alpha_{02} \simeq 32.5°$ and $\alpha_{03} \simeq 20.7°$. For an M-phase bridge with M odd, the expression is modified by the fact that the sinusoidal segments forming the dependent voltage are line voltages, not phase voltages. Hence, the incoming voltage becomes

$$v_i = 2V\sin(\pi/M)\cos(\omega_s t - \pi/2M)$$

with the average voltage being

$$V_{DD} = (2M/\pi)\sin(\pi/M)V\cos\alpha$$

and α_o given by

$$\alpha_o = \arctan[2M\operatorname{cosec}(\pi/2M)/\pi - \cot(\pi/2M)]$$

which for a three-phase, six-pulse bridge gives $\alpha_{06} \simeq 10.1°$.

In the vast majority of applications, these delays are much below those at which maximum ripple occurs and for which the critical inductance is of interest. Only for simple rectifiers (uncontrolled single-quadrant ac-to-dc converters) is this usually not so, and in such cases, the perturbation in dependent voltage caused by commutation overlap (discussed in Chapter 8) should be accounted for. The detailed analysis for $\alpha < \alpha_o$ will not be pursued; it is approached in the same way as that for $\alpha > \alpha_o$, should it prove necessary in any particular case, with the only difference being the definition of the angle of the-positive going intersection of the dependent voltage and its average value. For $\alpha > \alpha_o$, that angle is α in Figure 6-3b; the angle of the negative-going intersection, which is the end of the period of positive volt-seconds for the ripple wave applied to a current-sourcing inductor, is defined by

$$V\cos(\theta_o - \pi/M) = (M/\pi)\sin(\pi/M)V\cos\alpha$$

giving

$$\theta_o = \pi/M + \arccos[(M/\pi)\sin(\pi/M)\cos\alpha]$$

The volt-seconds applied to the current-sourcing inductor are then

$$VSI = (V/\omega_s) \int_{\alpha}^{\theta_0} [\cos(\omega_s t - \pi/M) - (M/\pi)\sin(\pi/M)\cos\alpha] d\omega_s t$$

$$= (V/\omega_s)[\sin(\theta_o - \pi/M) - \sin(\alpha - \pi/M) - (\theta_o - \alpha)(M/\pi)\sin(\pi/M)\cos\alpha]$$

The resultant peak-to-peak ripple current is obtained simply by dividing by L. Since differentiating this expression with respect to α and equating to zero produces an equation that has no closed-form solution and can therefore be solved only by iterative numerical techniques, it is fortunate that the case producing absolute maximum volt-seconds is clearly evident from a study of the waveforms. Maximum volt-seconds unquestionably occurs for $\alpha = \pi/2$, when $\theta_o = \pi/2 + \pi/M$, and is

$$VSI(\max) = (V/\omega_s)[(1 - \cos(\pi/M)]$$

If the design calls for a peak-to-peak (or rms) ripple-current limitation, evaluation of this expression (or the expression for volt-seconds at the value of α closest to $\pi/2$ if $\pi/2$ does not lie within the operating range) will give the required value of the current-sourcing inductor. If the critical inductance is of interest, however, this is not the case. The current ripple wave produced is invariably asymmetric, consisting of a series of sinusoidal segments, and hence the average-to-maximum and average-to-minimum current excursions are not of equal magnitude. The critical inductance is a function only of the average-to-minimum excursion. It may be evaluated by the following procedure.

The voltage impressed on the inductor is given by

$$e_L = V\cos(\omega_s t - \pi/M) - (M/\pi)\sin(\pi/M)V\cos\alpha$$

and therefore the current in the inductor may be expressed as

$$i_L = (1/L) \int e_L dt + C$$

$$= (1/\omega_s L)[V\sin(\omega_s t - \pi/M) - (M\omega_s t/\pi)\sin(\pi/M)V\cos\alpha] + C$$

where C is an arbitrary constant, a "constant of integration."

The average current (I_S) can be found by integrating i_L over the angular interval α to $\alpha + 2\pi/M$. Thus

$$I_S = (M/2\pi) \int_{\alpha}^{\alpha+2\pi/M} i_L d\omega_s t$$

$$= (M/2\pi)(V/\omega_s L)[2\sin\alpha\sin(\pi/M)$$
$$-(M/2\pi)\sin(\pi/M)\cos\alpha(4\alpha\pi/M+4\pi^2/M^2)]+C$$

Now the minimum current, with α a delay as shown in Figure 6-3, clearly occurs at the instant of switching since that terminates the period of negative volt-seconds. The maximum current occurs, just as clearly, at θ_o. Hence, the average-to-minimum and average-to-maximum excursions are given by

$$I_S - I_{\min} = (V/\omega_s L)\sin\alpha[(M/\pi)\sin(\pi/M) - \cos(\pi/M)] \tag{6.3}$$

and

$$I_{\max} - I_S = (V/\omega_s L)[\sin\theta_o' - (\theta_o - \alpha)\cos\theta_o' - (M/\pi)\sin(\pi/M)\sin\alpha + \sin(\pi/M)\cos\alpha]$$

where

$$\theta_o' = \theta_o - \pi/M = \arccos[(M/\pi)\sin(\pi/M)\cos\alpha]$$

It can be shown, with a great deal of labor that will not be executed here, that $I_S - I_{\min} > I_{\max} - I_S$ for $\alpha_o \leqslant \alpha \leqslant \pi - \alpha_o$. Thus, if the critical inductance were supposed to be the value obtained by setting $I_S - I_{\min}$ equal to one-half of $I_{\max} - I_{\min}$, with $I_{\min}$ put equal to zero, a conservative design would result. The degree of conservatism can be considerable; taking the worst-case ripple, i.e., $\alpha = \pi/2$, the inductance value given by this "simplified" procedure for $M = 2$ will be $\simeq$ 1.75 times greater than the actual critical inductance value; as M increases, this ratio increases toward a limit of 2. In practical converter systems, particularly high-power converter systmes, the cost, size, weight, and losses of current-sourcing inductors are significant enough to make this a serious matter.

The true value of critical inductance can be derived from Equation 6.3 when $I_{\min}$ is set to zero and I_S to the minimum value anticipated in the converter's operating range.

Since all voltage waves for phase-advance control are mirror images in time of those for phase delay, all equations derived for phase delay apply equally well to phase advance because

$$\int_A^B y\,d\psi = -\int_B^A y\,d\psi$$

and all functions integrated have the appropriate form, i.e., $y(-t) = -y(t)$ if an appropriate time reference is chosen. There is some difficulty in extending the results to apply to bridge and harmonic neutralized converters, because the sinusoidal segments appearing in the dependent-voltage waves no longer correspond to the phase voltages. For a bridge with M even, they do, except for an amplitude multiplier of 2, which also appears in the average dc voltage (V_{DD}). For bridges with M odd and all harmonic neutralized configurations, the situation is more involved, for the sinusoidal segments in question are from phasor sums of the phase voltages. However, all this means in practice is that the V in all expressions except that for V_{DD} must be modified by a factor that depends on the number of converters involved and the nature of their dc terminal connections—series (simply summed) or parallel (summed and averaged). For a bridge with M odd, this modifier is $2\cos(\pi/2M)$, while the multiplier for V in V_{DD} is simply 2. For l converters (or bridges) in series connection, the multipliers are simply l in V_{DD} (previously multiplied by 2 if bridges) and

$$Q = 1+2 \sum_{p=1}^{p=(l-1)/2} \cos(2p\pi/lM) \text{ if } l \text{ is odd}$$

$$Q = 2 \sum_{p=1}^{p=l/2} \cos[(2p-1)\pi/lM] \quad \text{if } l \text{ is even}$$

for V in all other expressions. For parallel connection, these multipliers are simply divided by l. In these cases, it is easier to calculate the peak-to-peak ripple by using Equation 6.3 and the corresponding expression for $I_{\max}-I_S$ since no new integrations are then necessary—merely modify V and reevaluate θ_o. In addition, π/M must be replaced by π/lM($\pi/2M$ for a bridge with M odd, $\pi/2lM$ for l such bridges) everywhere except in those parts of the expressions deriving from V_{DD}. Some of the resulting expressions from these translations are:

- θ_o:
  1. Bridge with M odd

     $$\pi/2M+\arccos[(2M/\pi)\sin(\pi/2M)\cos\alpha]$$

  2. l midpoint groups in parallel or in series

     $$\pi/lM+\arccos[(lM/Q\pi)\sin(\pi/M)\cos\alpha]$$

  3. l bridges in parallel or in series, M odd

     $$\pi/2lM+\arccos[(2lM/Q\pi)\sin(\pi/2M)\cos\alpha]$$

- Peak-to-peak ripple current
  1. Bridge with M odd:

$$(2V/\omega_s L)\{\cos(\pi/2M)[\sin(\theta_o-\pi/M)-\sin(\alpha-\pi/2M)] \\ -(\theta_o-\alpha)(M/\pi)\sin(\pi/M)\cos\alpha\}$$

2. l midpoint groups in series (divide by l for parallel connection)

$$(V/\omega_s L)\{Q[\sin(\theta_o-\pi/lM)-\sin(\alpha-\pi/lM)]-l(\theta_o-\alpha)(M/\pi)\sin(\pi/M)\cos\alpha\}$$

3. l bridges in series, M odd (divide by l for parallel connection)

$$(2V/\omega_s L)\{Q\cos(\pi/2M)[\sin(\theta_o-\pi/2lM)-\sin(\alpha-\pi/2lM)] \\ -(lM/\pi)(\theta_o-\alpha)\sin(\pi/M)\cos\alpha\}$$

- $I_S - I_{min}$:

1. Bridge with M odd:

$$(2V/\omega_s L)\sin\alpha[(M/\pi)\sin(\pi/M)-\cos^2(\pi/2M)]$$

2. l midpoint groups in series (divide by l for parallel connection)

$$(QV/\omega_s L)[(lM/\pi)\sin(\pi/lM)\sin\alpha-\sin(\alpha-\pi/lM)-\sin(\pi/M)\cos\alpha/Q]$$

3. l bridges in series, M odd (divide by l for parallel connection)

$$(2QV/\omega_s L)[(2lM/\pi)\sin(\pi/2lM)\cos(\pi/2M)\sin\alpha \\ -\cos(\pi/2M)\sin(\alpha-\pi/2lM)-\sin(\pi/M)\cos\alpha/2Q]$$

In the case of harmonic neutralization with parallel connection, the current-sourcing inductor is not the only component at the interface. For both dc-to-dc and ac-to-dc/dc-to-ac converters, interphase reactors are required. These components contribute significantly to the cost, size, weight, and losses at the interface, and introduce some practical design problems of a particularly unpleasant nature. Section 6.6 discusses the rating and design procedures for current-sourcing inductors and interphase reactors.

If CAM or PWM is introduced to a current-sourced converter, the dependent-voltage wave grows more complex, and, consequently, the determination of critical or otherwise adequate current-sourcing inductance becomes more difficult. The same analytic technique suffices to give closed-form expressions, but the derivation is so tedious in these cases that summation of terms in the spectral expressions, using a digital computer, is strongly recommended. Typically, amplitudes diminish so rapidly with increasing order that a 10-term summation is accurate enough for design purposes, and 100 terms will be within 0.5% or so of ideal values. Some care is needed, though, with the degenerate spectra for PWM when the carrier/wanted frequency ratio is an integer. It may then be necessary to sum very many terms, 1000 or more, to be sure of achieving sufficient accuracy because of continuing contributions to the lowest frequency terms. Using programmed-waveform techniques reduces this problem somewhat, because lower frequency ripple terms will generally be eliminated intentionally in the modulating process.

If the inductor placed at the dc current-source terminals of any converter is subcritical (i.e., has an inductance less than the critical inductance), discontinuous current results. Since converter switches almost invariably turn off automatically at current zeros, their existence functions become load-dependent in a manner similar to that investigated in the ac regulators of Section 5.8. Methods of analysis are similar to those in Section 6.9, where the performance of some converters operating with discontinuous current is discussed. For the most part, such operation is permitted only at very low power levels—either for very low power, part-load operation of moderate-to-high-power converters or over the full load range of low-power converters.

It should be apparent that the use of a finite inductance to approximate a dc current source influences converter performance. Provided the inductor used is supercritical (i.e., its inductance is greater than the critical inductance), the *dependent voltage* waveforms and spectra are *not affected*. The dependent currents are affected, however, and the nature of the changes that occur therein will now be discussed.

In ideal dc-dc converters, the defined currents are pure, smooth dc, and the dependent currents consist of pure, smooth dc wanted components accompanied by *ripple*. This is because the dependent currents result from multiplying the defined currents by switch existence functions that have the form

$$H = 1/A + (2/\pi) \sum_{n=1}^{n=\infty} [\sin(n\pi/A)/n]\cos(n\omega t)$$

If a finite inductor is used, then the ripple in the dependent voltage causes ripple current to flow therein. Thus, the defined current, ideally perfectly smooth, is in practice accompanied by an infinite series of unwanted oscillatory terms. These will also be multiplied (modulated) by the switch existence functions to create additional unwanted components in the dependent currents. They cannot create wanted components therein, since oscillatory currents flowing in an inductor produce reactive, not active (real) power, and any wanted current component in a dc-to-dc converter produces active power.

Qualitatively, the nature of the oscillatory components added to the dependent current is easy to establish. Since all oscillatory components in both the switch existence functions and the ripple current in the current-sourcing inductor are harmonics of the converter's switching frequency, so will be these additional components. Most designers assume that such effects can be ignored. If the current-sourcing inductance is much greater than critical, i.e., if the ripple current therein is small, this is a valid assumption; if not, and particularly if the inductor approaches criticality, it may not be. Quantitative analysis of the effect is not difficult; the buck converter will be used as an example.

The dependent voltage is

$$v_D = H\,V_S$$

The resulting ripple current in a finite inductor L is

$$i_L = (2V_S/\pi\omega L)\sum_{n=1}^{n=\infty}[\sin(n\pi/A)/n^2]\sin(n\omega t)$$

since the reactance offered by the inductor to a component of frequency nf is $n\omega L$ and the current in an inductor lags the impressed voltage by $\pi/2$ rad.

The total current at the converter output is I_S+i_L, where I_S is the perfectly smooth defined current, and thus the dependent current becomes

$$\begin{aligned} i_D &= H\,(I_S+i_L) \\ &= I_S/A+(2I_S/\pi)\sum_{m=1}^{m=\infty}[\sin(m\pi/A)/m]\cos(m\omega t) \\ &+ (2V_S/\pi\omega LA)\sum_{n=1}^{n=\infty}[\sin(n\pi/A)/n^2]\cos(n\omega t) \\ &+ (4V_S/\pi^2\omega L)\sum_{n=1}^{n=\infty}\sum_{m=1}^{m=\infty}[\sin(n\pi/A)/n^2][\sin(m\pi/A)/m]\sin(n\omega t)\cos(m\omega t) \end{aligned}$$

The double summation can be expanded and consists of terms containing $\sin[(m \pm n)\omega t]$. The spectrum is degenerate, and it is necessary to resort to computer summation for evaluation of the magnitudes of individual harmonic components. Waveform construction for this case is rarely, if ever, essayed. It can be done by constructing i_L (inductor current) from $(1/L)\int e_L dt$, knowing that the average component is I_S. However, this is such a laborious process, and the result provides so little enlightenment, that the effort is not worthwhile.

The same reasoning, for the most part, and the same analytic approach may be used to evaluate the effects of finite inductance on the dependent currents of ac-to-dc/dc-to-ac converters. There is one difference, however. The dependent-voltage ripple, and hence the current ripple in the inductor, contains only components having frequencies mPf_s, where P is the pulse number of the converter. The switch existence functions contain components having frequencies nf_s, where n is not an integer multiple of M, the number of supply phases. The additional unwanted components created by the $1/M$ term of an existence function are thus harmonics that are not normally present in a midpoint converter's dependent current spectrum; they disappear in bridge and harmonic neutralized converters because of phasor summation. The components created by the double summation in all cases have frequencies $(mP \pm n)f_s$. Since n can certainly assume the values $mP \pm 1$, an infinite series of components of frequency f_s are produced. Clearly, this composite fundamental frequency component must be a quadrature current, since the ripple components in the inductor that are responsible for its creation produce no active power contribution. Detailed analysis reveals that it is always a lagging quadrature current, regardless of the magnitude or sign of α (or of the use of PWM, CAM, or complementary converters trying to produce a unity displacement factor). Hence, in addition to producing a degenerate spectrum in regard to the harmonics in the dependent line current, an extra fundamental lagging quadrature current is produced. Other than for two pulse midpoint or bridge converters, its magnitude is usually small compared to the quadrature current produced by phase control, and neglecting it is not unjustified. In the two-pulse cases, it can be very significant. The very simple waveform construction of Figure 6-4 shows that with $\alpha = \pm \pi/2$ and critical inductance (not a likely practical operating condition), all harmonics are eliminated from the dependent line currents of two-pulse converters; such a current is then a pure sinusoid, wholly lagging if $\alpha = - \pi/2$ and wholly leading, albeit smaller in magnitude, if $\alpha = + \pi/2$.

Finally, in all cases evaluated, it has been found that the critical inductance, or inductance needed to meet a given ripple-current specification, is hyperbolically related to the defined current. The rating (which determines cost, size, weight, and loss) of an inductor is a function of its inductance and the maximum current it must carry, and the inductance value by simple reasoning is determined by the minimum operating average current.

Obviously it would be desirable for an inductor to obey the hyperbolic law, $L \alpha 1/I_{av}$, for its rating would then be a function only of the maximum I_{av}. Such inductors, called *swinging chokes*, can be, and in fact are, made. Design principles are discussed in Section 6.6.

6.3 Approximation of DC Voltage Sources

The actual dc voltage sources used in conjunction with switching power converters are legion, but like the loads discussed in Section 6.2, many of them have common characteristics. Almost all can be equivalenced by an emf in series with some inductance and resistance. This applies to batteries, fuel cells,

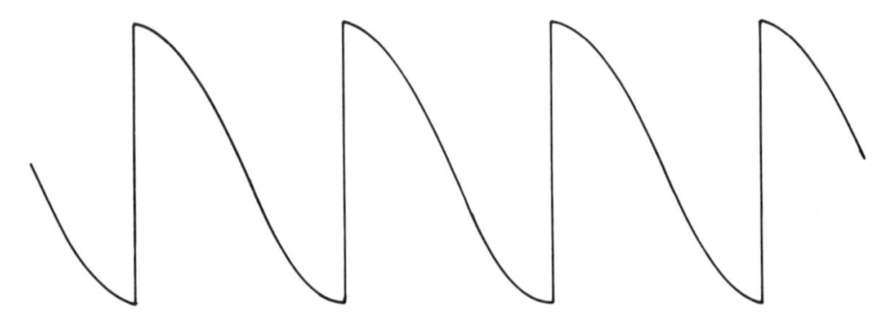

a. Dependent voltage at $\alpha = \frac{\pi}{2}$.

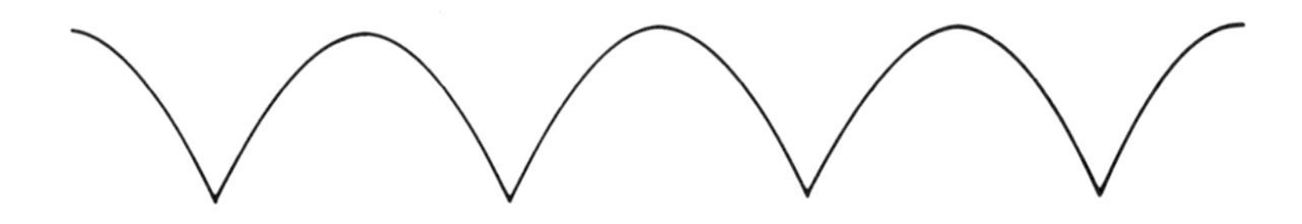

b. Defined current when interfacing inductance has critical value.

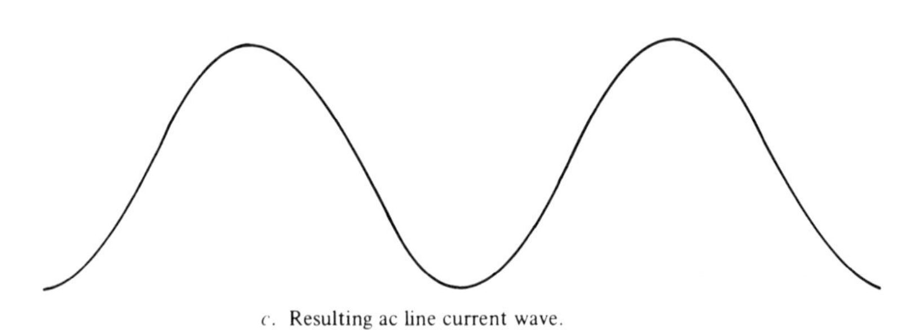

c. Resulting ac line current wave.

Figure 6-4. Waves for two-pulse converter.

rotary dc generators, and magneto-hydrodynamic generators. They share another characteristic too—an intolerance for the ripple in the dependent currents of switching power converters. For a variety of reasons—thermal, electrochemical, electromagnetic, and electromechanical—the oscillatory current that may be injected into these sources is limited. Even if it were not, the ripple voltage caused by converter dependent-current ripple flowing in the source impedance would usually be unacceptable in view of its effect on converter performance. Thus, when these sources are used, a voltage-sourcing interface is used at the converter terminals with a current-sourcing interface between the voltage source and the actual source.

Converters are also routinely supplied by other converters. This is, in fact, the most common configuration. Current-sourced ac-to-dc converters are used to supply voltage-sourced dc-to-dc and dc-to-ac converters and voltage-sourced dc-to-dc converters are used ahead of voltage-sourced dc-to-ac converters. In both instances, a voltage-sourcing interface is required to convert the current-source characteristic of the supply converter to a voltage-source characteristic acceptable to the supplied converter. Finally, of course, there are the electronic loads for converters, which must, for their own successful operation, be made to present themselves as voltage sinks.

Such an electronic load generally takes the form of a time-varying resistance (see Figure 6-5a). Were it fed directly by a converter with a current-source dc interface, the ripple current would flow in the load, developing ripple voltage in the process. Ripple voltage in excess of a few tenths of a percent of the average dc voltage is generally unacceptable for such loads, since it is likely to degrade or interfere with their internal functions. To achieve such low values of ripple voltage, very low levels of ripple current would be necessary in the current-sourcing inductor; therefore, very large inductance would be required. It can be seen that an entity is needed that, when connected in shunt with the actual load, will divert the ripple current without generating ripple voltage. It must also support the average dc voltage. This is, clearly, a dual of the component needed for a perfect current-source interface, and an infinite capacitor will fit the bill.

Like infinite inductors, infinite capacitors are not available, and a designer must be content with a finite-value component. As a result, ripple voltage will develop, in an amount proportional to the ripple current and the finite capacitive impedance. The configuration that emerges, which is depicted in Figure 6-5c, is a simple low-pass filter, comprising the current- and voltage-sourcing interfaces, between the converter and the actual load. The true attenuation factor (ratio of input voltage to load voltage) of this filter is

$$F = 1-\omega^2LC+j\omega L/R$$

where ω is the angular frequency of any oscillatory component at the converter terminals and $j = \sqrt{-1}$. Since R, in most practical cases, is widely variable, it is common to design so that F is dominated by its real term, $1-\omega^2LC$. This requires that $\omega^2LC>>1$, or $\omega_o<<\omega$, where $\omega_o{}^2 = 1/LC$, i.e., that the resonant frequency of the filter be well below the lowest unwanted component frequency in the converter's dependent voltage. Because of common design practices, it

a. Actual variable resistor load.

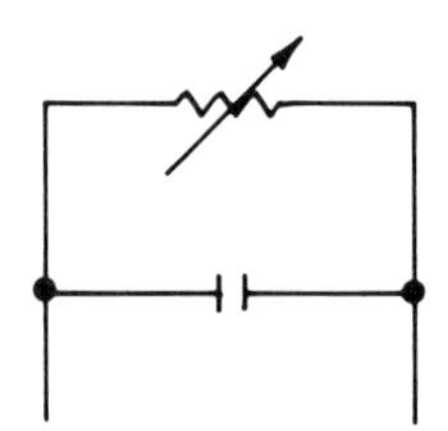

b. Shunted by capacitor, becomes dc voltage sink.

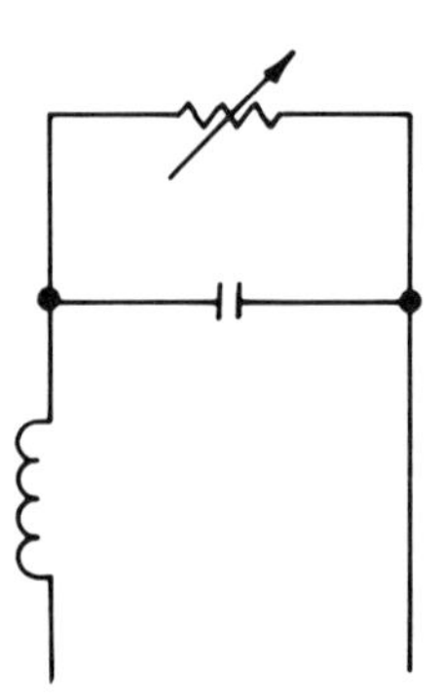

c. With current–source interface, becomes low–pass filter.

Figure 6-5. Development of typical dc interface.

is of interest to compare F with the attenuation factor that is presumed when the interfaces are considered separately. In that case, an oscillatory voltage component of amplitude V_ω is presumed to create an oscillatory current component of magnitude $V_\omega/\omega L$ in the inductor. This current is then assumed to flow in the capacitor and to create an oscillatory voltage component of magnitude $V_\omega/\omega^2 LC$ in so doing. The presumed attenuation factor is thus $\omega^2 LC$; if $\omega_o << \omega$, the error is not substantial, since for $R = \infty$ the true attenuation factor is $1-\omega^2 LC$; for $R < \infty$, the error becomes even less, so that the common design practice of separating the interfaces can be justified.

Given that a finite capacitor can satisfy, with limitations, the requirements of this dc voltage-source interface, it should be clear that it can also satisfy the requirements for all others. The general needs are, after all, the same—to pass ripple current, ideally without developing ripple voltage, while supporting the average dc voltage without consuming power. When applied between a primary dc source and a converter (dc-to-dc or dc-to-ac), a low-pass filter again results as depicted in Figure 6-5c. In this case, two related aspects of interface performance are of interest to the designer—the level of ripple current flowing in the real source and the ripple voltage developed at the converter terminals. Neglecting source impedance, the impedance seen from the converter terminals is

$$Z_\omega = -j\omega L/(\omega^2 LC - 1)$$

so that an oscillatory current of magnitude I_ω produces a ripple voltage of magnitude $\omega L I_\omega/(\omega^2 LC - 1)$. Again, for this ripple voltage component to be small, ω_o must be much less than ω. Consider for a moment that the switching-frequency component of ripple current at the input terminals of a buck dc-to-dc converter has, ideally, amplitude $(2I_S/\pi)\sin(\pi/A)$. The resulting ripple-voltage component will have amplitude $(2\omega L I_S/\pi)\sin(\pi/A)/(\omega^2 LC - 1)$ and a corresponding percentage amplitude of

$$(200\ \omega L I_S/\pi)\sin(\pi/A)/[V_S(\omega^2 LC - 1)]$$

For this to be small, either $\omega^2 LC >> 1$ or ωL must be small. However, the design is also constrained by the need for limiting ripple-current injection into the source; the amplitude of the component of current flowing therein will be

$$I_{D\omega} = I_\omega/(1 - \omega^2 LC)$$

and for this to be small, ω_o must be much less than ω. If the allowable percentages of voltage and current ripple for this component are P_V and P_I, respectively, the design must simultaneously satisfy

$$P_I = 100\ I_\omega/[I_{DD}(\omega^2 LC - 1)]$$
$$= 100\ A I_\omega/[I_S(\omega^2 LC - 1)]$$

and

$$P_V = (200\ \omega L I_S/\pi)\sin(\pi/A)/[V_S(\omega^2 LC - 1)]$$
$$= (200\ \omega L I_S/\pi)\sin(\pi/A)/[A V_{DD}\cdot(\omega^2 LC - 1)]$$

Manipulating appropriately yields

$$\omega^2 LC - 1 = 100\ AI_\omega / P_I I_S$$

and hence

$$\omega L = \pi A^2 V_{DD} I_\omega P_V / [2 I_S{}^2 \sin(\pi / A) P_I]$$

and

$$1/\omega C = \omega L / (1 + 100\ AI_\omega / P_I I_S)$$

Quite often, the source impedance cannot be neglected, but must be incorporated into the expressions if accurate analysis is required. However, ignoring it invariably produces conservative designs. If the source impedance is largely inductive, and source resistance can be neglected, the expressions change only by substituting $L + L_S$ for L; if the source resistance is significant, the damping it produces makes for considerable complications.

Obviously, similar considerations apply to converter-converter interfaces. Capacitors are used for voltage-sourcing, and the ripple voltage and current limitations are the primary design constraints for the low-pass filter that invariably is created. Fortunately, the effects of ripple voltage on converter performance are so deleterious in most cases that the phenomenon of "critical capacitance" never arises in practical designs. This condition and the resultant "discontinuous voltage operation" are clearly a possibility at all dc voltage-source interfaces if the capacitance is sufficiently reduced. The only practical case where it may be encountered is in the voltage-source-transfer buck-boost dc-to-dc converter.

Since voltage-ripple levels are usually kept low, it is rarely considered necessary to evaluate the effects they might have on converter performance. This can be done, if necessary, by using exactly the same technique that was delineated for evaluating ripple-current effects. When both are evaluated simultaneously, equations without closed-form solutions result, and a computer is needed to arrive at interpretable results.

The characteristics required of and found in real capacitors used for voltage-source interfaces are discussed in Section 6.7. One point is worth mentioning here: practical capacitors themselves are limited, thermally, as regards the ripple current they can absorb. Thus, a designer must often use far more capacitance than is needed simply to meet ripple voltage and current specifications at a converter-source, converter-load, or converter-converter interface.

6.4 Approximation of AC Current Sources

If dc current sources are uncommon in the electrical world at large, ac current sources may be said to be nonexistent. Since they are a supposition of the analyses for both ac-to-ac converters and voltage-sourced ac-to-dc/dc-to-ac converters, reasonable approximations thereto must be available for these converters to operate properly. Again, consideration of the converter-load or converter-source relationships leads to a simple realization.

With the exception of purely passive load networks, all real ac loads and sources may be equivalenced as voltage sources in series with impedance. This is obviously true of the public utility supply at all levels, from primary transmission to residential distribution. It is also true of both induction and synchronous machine loads. It is easy to conclude, and for the most part true, that the ac current-source interface should reduce to the same problem in all cases, by providing any simple passive load networks with an ac voltage-source interface. When this is done, the problem universally becomes that of interfacing two ac voltage sources. The load, or source, consists of a sinusoidal voltage or voltage set with series impedance, the converters produce sinusoidal wanted components of voltage accompanied by unwanted components that are harmonics for the ac-to-dc/dc-to-ac converter but that in general are not harmonics for the ac-to-ac converters. What is needed, then, is a component or network, connected between converter and loand or source, which offers no impedance to the wanted current components and infinite impedance to all unwanted current components. As in the dc current-source interface, it must support the unwanted oscillatory voltage components without allowing current flow and must be transparent to the wanted components while absorbing no real power.

No such simple component exists, nor can any passive component network be synthesized to totally meet these requirements. The closest "ideal" approximation is a series-tuned circuit resonant at the wanted frequency and of infinite Q. This exhibits zero impedance at the wanted frequency, absorbs no power, and can be made to exhibit an arbitrarily high impedance at all unwanted frequencies since

$$Z = \omega L - 1/\omega C$$

and if L is made large enough, Z is very large for any ω other than $\omega_o = 1/\sqrt{LC}$. Since ideal components do not exist, and a designer must therefore use L and C with resistance present, infinite Q is not attainable. As a result, power is absorbed, and the higher the inductance used, the greater the power lost. Thus, in practical cases, the attainable ratio of impedance at unwanted frequencies to impedance at the wanted frequency is limited by power-loss considerations.

In the early days of power converters using sold-state switching devices, such interfaces were in fact used. They later fell into disuse, however, for a number of reasons. First, it is difficult, in most situations, to get good utilization of the capacitor dielectric, and capacitor costs tend to be high. Second, the transient behavior of series-tuned circuits leaves a great deal to be desired—converter performance under load switching and other application-related transient phenomena was often unacceptable. Third, and probably most important, designers found that reactive impedance at the wanted frequency was generally desirable to facilitate converter control. Thus, a simple inductor interface could be predicated, eliminating the capacitor entirely.

A simple inductor interface provides a higher impedance at unwanted frequencies higher than the wanted frequency than at the wanted frequency, and hence is not an unreasonable approximation to the requirements. More often than not, designers ignore the true constraints of the interface, selecting the inductance value on the basis of control-related criteria or permissible converter short-circuit current. There is some risk in much approaches, since ''discontinuous'' ac current operation can arise, which has influence on converter characteristics similar to that observed in the dc case. In the ac case, a discontinuous current has more than two zero crossings per cycle of the wanted frequency. If such a condition exists, then converter switch existence functions are affected because switches turn off automatically at current zeros; in the case of the unidirectional voltage-blocking, bidirectional current-carrying switches of the ac-to-dc/dc-to-ac converters, they can also turn on automatically in the event of current reversal.

It is not a simple matter to determine the value of interfacing inductance needed to avoid this condition, and this may perhaps be a major reason why designers have traditionally avoided the issue. The fundamental, wanted, current magnitude depends on both the inductive reactance and the magnitude and phase relationships of the converter and load/source fundamental voltages. The unwanted current magnitudes depend solely (presuming purely sinusoidal, or at least negligibly distorted, load/source voltage) on the reactance. Whether multiple zero crossings are produced depends not only on the magnitudes of the unwanted components, but also on their phase relationships to each other and to the fundamental. The situation is further complicated because, while there exist closed-form expressions for summations of the type

$$I_1\cos(\omega_s t+\phi_1) + I_3\cos(3\omega_s t+\phi_3) + I_5\cos(5\omega_s t+\phi_5)\ . . .$$

they are so unwieldy, and not subject to direct interpretation, as to be of little use in assessing the situation. When the unwanted components are not harmonics, but sidebands, no closed-form expressions exist. Thus, any attempt to formulate completely general expressions for the critical inductance at an ac current-source interface is doomed to failure; even if such an expression can be derived, computer assistance will be needed for evaluation, and the labor expended in its derivation is largely wasted.

In this case, in fact, all the analyst's wiles are of no avail. Each design must be taken to the computer for numerical evaluation if an accurate reading of the situation is desired. As mentioned, designers have traditionally evaded the issue. As a result, the inductor values in ac current-sourcing interfaces are normally selected using criteria other than the avoidance of discontinuous current. It is a happy coincidence that, for the most part, designs executed thus produce inductance values above critical for most of a converter's operating range.

The most common criterion is simply the rate of rise of fault current into a short circuit (zero voltage) load or source. This happens to be a critical parameter in the design of commutating circuits (discussed in Chapters 8 and 9) and is thus a valid concern. A secondary criterion sometimes invoked relates to converters' control. With a simple inductor between converter and source voltage, the real component of the fundamental current is determined by the quadrature component of converter fundamental voltage acting on the reactance, and the quadrature component of fundamental current is determined by the difference between the in-phase component of fundamental converter voltage and the source voltage acting on the reactance. Thus, the reactance determines the fundamental current transients developing in response to amplitude and phase transients of the converter voltage, and dictates converter closed-loop response behavior, including behavior in current-limiting operation if such is employed. In consequence, the reactance is sometimes selected in order to tailor closed-loop response to some desired characteristic.

6.5 Approximation of AC Voltage Sources

In a great many instances, the need for an ac voltage-sourcing interface is not perceived. A converter is to be supplied from the public utility supply or some other ac generating system, and is simply connected thereto. The majority of designers blithely ignore the facts that ac supplies have finite (largely reactive) impedances, may possess resonances that can be excited by converter harmonics, and serve other loads that may be adversely affected by converter behavior. A converter requiring an ac voltage source is, of course, a current source or sink as far as the ac supply is concerned. The two most important aspects of its behavior are harmonic or other unwanted component injection and fundamental quadrature current demand or supply. Since the means provided to combat the former influences the latter, it will be dealt with first.

One way of attacking the harmonic problem—using programmed-waveform techniques—was discussed in Chapter 4. When introduced, such techniques also provide a means for implementing control of displacement factor, thus also avoiding any quadrature current problems. However, because of switching device limitations, they are not yet prevalent, particularly in the higher powered converters where problems are most likely to be severe, (see Chapter 7). The other means available is to resort to passive components at the interface, as in all other cases. A component or network is needed that will support the fundamental sinusoidal ac voltage without absorbing real power, as well as be a short circuit to all unwanted sinusoidal current components. As for the dc case, the simplest realization is an infinite capacitor. Not only is such a component unobtainable, it is totally unacceptable in this case, for it will create an infinite leading quadrature-current supply to the ac source. Even a finite capacitor may be unacceptable in view of the leading quadrature-current supply it generates, although in many instances it will be offset to some extent by lagging quadrature-current demand from the converter.

Consider a two-pulse phase-delay controlled current-sourced ac-to-dc/dc-to-ac converter interfaced by the arrangement of Figure 6-6. Its dependent current contains a third harmonic component that has an amplitude one-third that of the fundamental. Suppose the source reactance is x per unit relative to the converter rating and that it is desired to limit the third harmonic current flowing in the source to 5% of the converter fundamental current. If the magnitude of the fundamental is I, that of the third harmonic in the converter current is $I/3$; this splits between the capacitor and the source reactance according to the relationships

$$I_L + I_C = I/3$$

$$j3\omega_s L I_L = -jI_C/3\omega_s C$$

giving

$$I_L = I/[3(1-9\omega^2 LC)]$$

putting

$$I_L = 0.05\, I \text{ and } \omega_s L = xV/I \textit{ yields}$$

$$I_C = 0.85\, I/(1.35x)$$

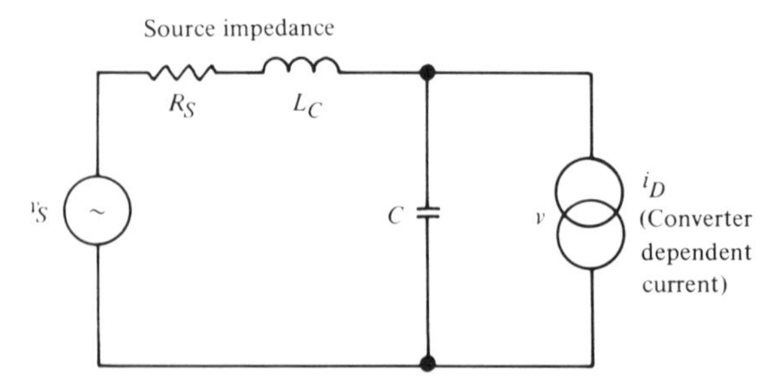

Figure 6-6. Simple capacitor ac voltage interface.

where $I_C = V/X_C$ is the leading quadrature-current supply produced by the capacitor. Allowing $x = 0.1$, which would be a relatively high value of source reactance, the capacitive current is over six times the converter current for this rather modest restriction on third harmonic injection.

A six-pulse converter produces a somewhat better result, since the lowest order harmonic is the fifth, with amplitude one-fifth that of the fundamental. Even so, for the same criterion, 5% fifth harmonic in the source current, the analysis yields

$$I_C = 0.85\, I/(6.25x)$$

and the capacitor current is still 1⅓ times the converter current if $x = 0.1$.

Such large leading quadrature-current supplies are most undesirable—in fact, they are intolerable. To begin with, the capacitors are large and expensive; beyond this, the impact on voltage regulation at the tie point is appalling. In the two pulse case, the open circuit voltage would be 2.7 times the source voltage; for the six-pulse case, it would be 1.16 times the source voltage. The former is clearly quite intolerable, the latter most undesirable. Thus, the use of simple capacitive interfaces is not to be recommended, unless offsetting lagging quadrature current demand is incorporated by connecting shunt reactance. Shunt inductors, of course, introduce additional costs and losses.

This unfortunate situation can be alleviated by using tuned harmonic filters, as shown in Figure 6-7a. It is customary, when such an interface is designed, to use from two to six high-Q (greatly underdamped) branches for the lowest order harmonics and a damped "broadband" arm for the remaining higher order harmonics. Such an arrangement greatly reduces leading quadrature-current supply, as will now be shown, with consequent economies and a reduction in voltage regulation problems.

Suppose the six-pulse converter case is considered. The Q of the "tuned traps" at harmonic frequency will be assumed as 50, which is in fact readily attainable in practice; the fifth harmonic will be limited not to 5% but to 1%. The harmonic current now divides between the filter resistance, $r = 5\omega_s L_5/Q = 1/5\omega_s C_5 Q$, according to the relationships

$$j\omega_s L I_L = r I_R$$

and

$$I_L + I_R = I/5$$

giving, after appropriate manipulation and setting $I_L = 0.01\, I$,

$$r = \omega_s L/4$$

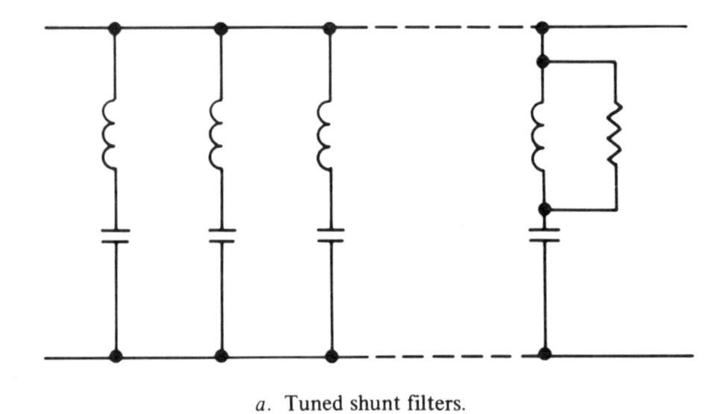

a. Tuned shunt filters.

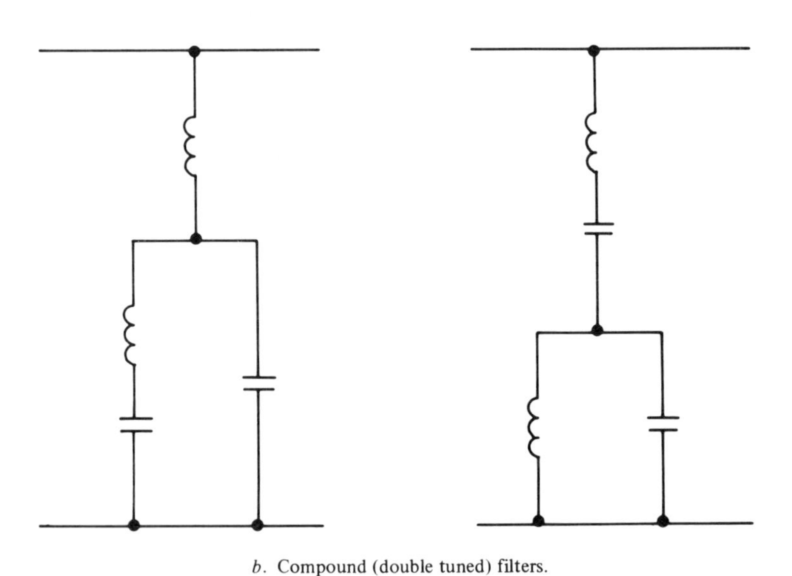

b. Compound (double tuned) filters.

Figure 6-7. AC voltage-source interfaces.

and thus

$$\omega_s L_f = 1/25\ \omega_s C_f = 2.5\ \omega_s L$$

The resultant leading quadrature-current supply, determined by the impedance $\omega_s L_f - 1/\omega_s C_f$, is only $I/60x$, or one-sixth of the converter current if $x = 0.1$. Even when the other traps and the broadband filter are added, quite tolerable leading quadrature-current supplies and quite reasonable component sizes (and, consequently, costs) accrue.

The design procedures for tuned traps and the damped broadband arm are somewhat complicated because component tolerances and temperature drifts, coupled with supply frequency variations, make it necessary to tailor the Q to guarantee performance. Although knowledge of the source impedance is desirable, it is not essential if harmonic voltage levels at the tie point are specified, as is often the case. Ainsworth[1] gives an excellent exposition of the procedure for tuned traps, including compound traps of the type shown in Figure 6-7b and damped broadband filters. However, he does not pay much heed to a danger inherent in the use of such an interface arrangement. A network such as that of Figure 6-7a obviously possesses a number of admittance poles (impedance zeros) equal to the number of tuned traps present. Consequently, it possesses admittance zeros interlacing the poles. Since the traps are tuned to odd harmonics of the supply frequency, these admittance zeros will occur close to—perhaps at—even-order harmonics of the supply frequency. Practical converters, and other components and loads on a supply system, can generate small amplitude "non-canonical" even-order harmonics, and the results can be quite distressing unless the admittance zeros are both displaced from exact coincidence with such unwanted (and theoretically unexpected) components and sufficiently well damped to prevent severe resonance effects.

Such filter arrangements are not useful with ac-to-ac converters, since the unwanted components in the converters' dependent currents are not then harmonics of the supply frequency, but sidebands. In such cases, the designer must fall back on a simple capacitive interface, and suffer its attendant problems. This is a strong incentive for two courses of action as regards ac-to-ac converter design: (1) to use a high pulse number and (2) to institute control measures (discussed briefly in Chapter 10) that eliminate or greatly reduce the lower sideband components of ac-to-ac converter dependent currents.

Where an actual voltage source is not present (i.e., when the load is purely passive), some means of fabricating a sinusoidal voltage is needed. This type of voltage-sourcing interface is encountered in induction-heating applications, in compound converter (ac-to-ac converter feeding ac-to-ac converter) interfaces, and with voltage-sourced dc-to-ac converters feeding general loads. The basic need is fulfilled by a parallel resonant circuit, as depicted in Figure 6-8a and b. This consumes little real power, and, if tuned to the wanted frequency, it can provide a suitably low impedance for the unwanted components of the dependent currents. Its performance, as regards harmonic distortion of the voltage supply it produces, depends on the accuracy of tuning and its loaded Q. Of course, the accuracy of tuning is influenced by reactive components of load current, which poses a particular problem in the converter-converter interface of Figure 6-8b. This problem is addressed in either of two ways: (1) by allowing the operating frequency to vary with converter loading and control in order to maintain resonance, or (2) by forcing the converters to consume appropriate amounts of quadrature current regardless of their operating conditions.

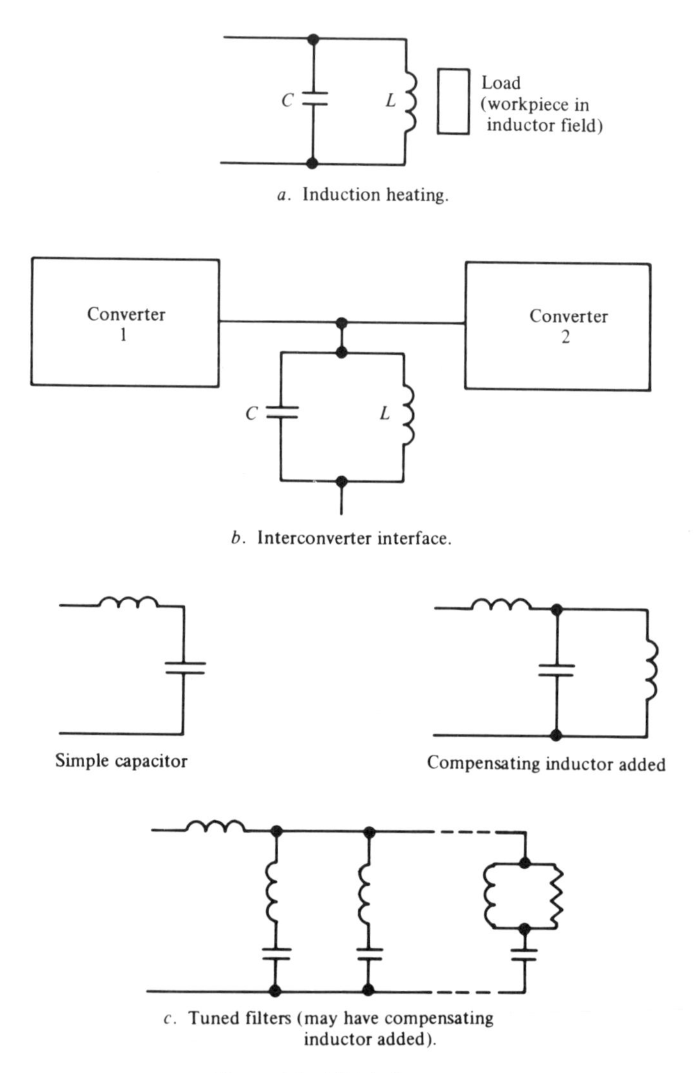

a. Induction heating.

b. Interconverter interface.

c. Tuned filters (may have compensating inductor added).

Figure 6-8. AC interfaces.

In the case of the voltage-sourced converter, where the voltage-source interface is preceded by a current-source interface, a simple capacitive interface is often used, as depicted in Figure 6-8c. Just as for the dc case, the result is a low-pass filter, but there may be acute problems. To achieve good attenuation of the lowest order harmonic, the filter resonance must be well below the frequency thereof, as discussed in Section 6.3. If there is not good separation between the wanted (fundamental) frequency and the lowest order harmonic, the result will be excessive regulation at the fundamental, akin to that seen with a simple capacitive interface at an ac supply. It is avoided by the same techniques—compensating shunt reactors or harmonic trap arrangements. The compensating shunt reactor, of course, produces the parallel tuned circuit interface. Where ac-to-ac converters are used to feed general loads, the harmonic trap arrangement is again not useful because of the sideband nature of unwanted components.

In low-power converters (5 to 10 kVA rating or below), neither tuned traps nor compensating reactors are considered economic. Moreover, such converters are often single-phase output, and the third harmonic can hardly be said to be well separated from the first. Thus, programmed waveform modulation is often employed to modify the spectrum, or simple PWM with a very high carrier frequency is employed by those ignorant of the better way to accomplish the desired objectives. Where simple converters are used, ferroresonant regulating and waveshaping transformers are often found as the voltage-sourcing interface in low-power single-phase applications. Despite their bulk, cost, and inefficiency, they have been popular because they relieve the converter-system designer of two burdens—the voltage-sourcing interface is satisfactorily accomplished and regulation of converter output voltage is achieved also.

6.6 Inductor, Transformer, and Interphase Reactor Rating and Design

It cannot be emphasized too strongly that the design of magnetic components is a specialized branch of engineering that should be left to the specialists whenever possible. However, a designer of power converters must develop an understanding of the factors that influence magnetic component design and thus influence the cost, size, weight, and losses of the components used in conjunction with converters. Many design decisions in the converter field affect the electrical specifications of magnetic components, and the characteristics of those components can have profound effects on converter systems.

Transformers are used for isolation at ac interfaces and to provide the phase-shifted sources needed by harmonic neutralized converters. A transformer's nameplate rating is the volt-ampere product of its burden. (More usually, for higher power units, the rating is the kilovolt-ampere product.) The transformer's windings must have at least that rating for both primary and secondary winding if full isolation is required, so that the winding rating of an isolation transformer is at least twice the transformer rating. As shown in Chapter 4, phase-shifting transformers often have secondary winding ratings considerably greater than their burden.

Both conductor and core material used in a converter transformer are selected to conform with application requirements. The key characteristics of core materials are saturation flux density and losses, and losses in a given material are a function of both operating flux density and frequency. Conductor losses are also a function of frequency, since skin and proximity effects can significantly increase the resistance of a given conductor as the frequency of oscillation of the current therein increases.

The two basic design equations for a transformer with sinusoidal exciting voltage are

$$B_{max} = 10^8 \, V/(4 \cdot 44 \, fNA_i)$$

which defines the peak flux in gauss for an rms exciting voltage V of frequency f applied to N turns on a core area of A_i square centimeters and

$$A_W = NA_C/F_W$$

where A_W is the window area occupied by one winding of N turns; F_W a winding space factor ($0<F_W<1$) depending on the shape of the conductor and the insulation and mechanical support requirements therefor; A_C is the area of the conductor. When the exciting voltage is not sinusoidal, the more general form of the flux expression is

$$B_{max} = (1/2) \text{ (Volt-seconds in one-half-cycle) } 10^8/NA_i$$

The use of cgs units is still prevalent in magnetic component design because magnetic material manufacturers still characterize their products in those units. For a double-wound transformer, the total window area is given by

$$A_W = N_P A_{CP}/F_{WP} + N_S A_{CS}/F_{WS}$$

where the subscripts P and S define primary and secondary quantities, respectively. Now for a simple isolating transformer $N_P/N_S = V_P/V_S = I_S/I_P$; if the conductors are operated at a current density ρ_I amperes/square centimeter, then $A_{CP} = I_P/\rho_I$ and $A_{CS} = I_S/\rho_I$. Substituting and manipulating, with the assumption $F_{WP} = F_{WS}$ yields

$$N_P = F_W \rho_I A_W / 2I_P$$

(In practice, $F_{WP} = F_{WS}$ is often not true.)
Substituting in the flux equation and transposing yields

$$A_i A_W = 2 \, x \, 10^8 \, V_P I_P/(4 \cdot 44 \, f \rho_I F_W B_{max})$$

The product A_iA_W is a measure of the physical size of the transformer. Since it is proportional to a linear dimension to the fourth power (for a given geometry), the volume and weight will be proportional to $(A_iA_W)^{0.75}$; for given materials, cost and losses are (to a first order) proportional to weight. From this expression, it is easily deduced that transformer size (and hence, to a first order, cost and losses) is proportional to the three-quarter power of rating and inversely proportional to the three-quarter powers of operating frequency, conductor current density, winding space factor, and peak operating flux density.

For frequencies up to 200 Hz or so, the core material usually used is grain-oriented silicon steel in 0.012-to-0.015-in.-thick laminations. This material has a high (~ 18 kG) saturation flux density and relatively low losses, ~1.3 W/lb at 17 kG with 60-Hz excitation. From about 200 through 600 to 800 Hz, the same material is employed in 0.004-in.thick laminations, retaining high flux capability and reducing eddy current loss so that losses remain reasonable, ~ 9 W/lb at 15 kG with 400-Hz excitation. Above 600 to 800 Hz, a wide variety of core materials are encountered. Grain-oriented silicon steel is still used, up to 2000 Hz or so, in 0.002- and 0.001-in.-thick tape in ring cores and C cores. It is limited, however, to peak operating flux densities of 2 to 5 kG because of high losses, ~ 15 W/lb in 0.002-in. tape at 4 kG with 5000-Hz excitation. Other materials such as Hi-Mu80 and Supermalloy are also used in tape-wound ring cores. Their major advantage is much higher permeability, due to the nickel and transition metals alloyed with steel; their loss/flux relationships are similar to those of grain-oriented silicon steel. For lower losses, and in almost all applications above 2000 to 5000 Hz, ferrite core materials are used. These ceramiclike materials, made from mixtures of metallic oxides, are pressed into cores of various shapes and styles; the most popular for low-power converter transformers is the pot core, while U cores are more common for larger transformers. The saturation flux densities of ferrites are quite low, from 3 to 5 kG, but their high-frequency losses are low. There is, because of the nature of these materials and manufacturing techniques used to produce transformer and inductor cores made from them, a distinct size limitation on ferrite cores. Cross sections greater than about 1 in.^{2} are not found in low-cost cores; larger cross sections, if needed, are fabricated by machining the core shape from solid blocks, a very expensive process.

Where the excitation is from a bona fide ac source, transformer cores usually have no intentional gap in their magnetic circuit(s). Only the unavoidable residual gaps resulting from the various types of core construction are present. When converters provide the excitation, this is no longer tenable. A converter's ac dependent voltage is wholly ac only under ideal conditions. In most practical cases, imperfections in the existence functions and differences in device conducting drops result in the presence of a small unwanted dc voltage component.

A dc unwanted current component results, and its magnitude is limited only by parasitic resistances, including transformer winding resistances, in the loop where the dc voltage component is generated. The dc magnetization equation for a magnetic circuit is, in cgs units,

$$(4\pi/10)NI = H\ell_i + B\ell_a$$

where N is the number of turns carrying I, the dc current, and H is the magnetizing (coercive) force in Oersteds, B the flux density in gauss, and ℓ_i and ℓ_a the lengths in centimeters of the magnetic and air gap paths, respectively. If ℓ_a is very small, then H tends to be large; the coercivity of the core materials used ranges from low to very low, in no case exceeding 1 Oe or so. Consequently, dc saturation results in ungapped cores with quite small dc currents. To avoid this, transformer cores used with converter excitation are usually provided with air gaps of sufficient length for the $B\ell_a$ term to dominate the dc magnetization, with B a small fraction of the saturation flux density of the core material.

Conductor materials commonly used for transformer windings are copper and aluminum. Conductor formats vary widely; round wires, square and rectangular wires, flat strips, and the stranded conductors known as "Litzendraht" or "Litz" are all used. Current densities employed typically range from 500 to 2000 A/in.^{2}. The larger the transformer, the lower the current density is likely to be, for reasons that will shortly be made plain. At these current densities, copper loss for dc current ranges from ~ 0.09 to ~ 1.39 W/lb at 150°C, a common winding temperature limit. Aluminum loss is about 1.75 times higher per unit volume, about one-half that of copper per unit weight.

With conductor losses the effect of most interest and concern is the increase in effective resistance with frequency. This arises because of two related phenomena known as *skin effect* and *proximity effect*. Skin effect occurs because the magnetic field created by a current flowing in an isolated conductor exerts a force on the current and drives it away from the center of the conductor. At high frequencies, or with very large conductors at lower frequencies, the current crowds into the "skin" of the conductor. In other words, current flow is restricted to a region near the surface of the conductor, and the resistance offered to that current can be very much higher than the dc resistance of the conductor. When conductors are wound into a coil, the magnetic fields due to currents in adjacent turns also exert forces on the current in a given turn, causing further current crowding and further increase in ac resistance. This is proximity effect. Terman gives an excellent summary of the known quantitative relationships pertaining to these effects.[2]

The increase in resistance is, as has been noted, a function of both frequency and conductor size. Thus, although it is most serious at high frequencies, it becomes a problem at low frequencies, including the 50- or 60-Hz utility frequencies, if currents (and in consequence conductor sizes) are large enough. Transformer and reactor designers use two techniques to alleviate these problems when they become severe. The first might be considered begging the question—windings are made from tubing, rather than solid conductor, so that the material in which current will not flow is not present. The second, more effective technique is to use Litz conductor. Such a conductor is made up of a great many insulated strands of thin round wire, so that skin effect is not a problem in individual strands; each strand is laid up in such a way that in a given length—the *pitch length*—it occupies every position in the bundle once and once only. Very large Litz conductors are made by taking several smaller ones and laying them up in similar fashion.

This method of conductor construction minimizes proximity effects among strands with the conductor; when the conductor is wound into a coil, proximity effects are much less than for solid (or tubular) conductors because the current in an isolated conductor is much more uniformly distributed through its total cross section. Note that flat strip, while very good as far as skin effect is concerned, is subject to severe proximity effect since adjacent turns have large contiguous surfaces.

There is one final aspect of transformer design to consider, which also applies to interphase reactors and reactors. It was observed that the size of a transformer is proportional to the three-fourths power of its rating. Transformers, like most electrical components, are temperature-limited. The heat generated by core and conductor losses escapes, by radiation and convection, from the transformer's surface; heat-transfer considerations dictate the surface area needed to reject a given amount of power to ambient for a stipulated temperature rise.

Now the volume (size) is proportional to the cube of a linear dimension, whereas the surface area is proportional to the square of a linear dimension. Thus, the area is proportional to the two-thirds power of the volume, i.e., to the square root of the rating. To maintain a constant temperature rise, the losses must be reduced in proportion to the square root of a rating increase.

While this is a somewhat simplistic argument, taking no account of more efficient cooling techniques, the point should be clear. Despite all measures available, the percentage losses of magnetic components go down as ratings increase *because they must*. However, the fundamental causes making this so tend to force designers to use lower conductor current densities and, to some extent, lower flux densities as component ratings increase.

Interphase reactors are very similar to transformers insofar as design and construction are concerned. They always endure a harmonic rich, nonsinusoidal excitation, which usually makes core and conductor losses higher than for transformers of equivalent rating. They are much more susceptible to saturation due to converter imperfections than are transformers, because they rely on balanced wanted current components from multiple converters to avoid magnetization by those current components. Even with gapped cores, unbalances among harmonic neutralized converters connected to an interphase reactor must be limited by closed-loop control.

Because of harmonic rich excitations, core losses and acoustic noise are usually considerably higher in interphase reactors and converter-fed transformers than in sinusoidally excited components operating at comparable flux densities. These effects can only be ameliorated by reducing flux density, thus increasing the size and cost of the components. The design problems for interphase reactors are severe enough to discourage their use in many instances.

The rating of an interphase reactor is fairly easy to establish for sizing purposes. The winding assigned to one converter (dc-to-dc) or one converter phase (ac-to-dc/dc-to-ac and ac-to-ac) is assumed to carry only the wanted component of current and to be excited by the total ripple or harmonic voltage involved. The product of their rms values gives the winding rating, and the equivalent transformer rating is taken to be one-half of the sum of all winding ratings.

Inductors present a somewhat different problem. All the ampere turns on an inductor are uncompensated, so that the total magnetizing force exerted on the inductor's magnetic circuit is large. Thus, the core operating point, where there is a core, invariably has to be controlled by an air gap or a series of air gaps. As a result, in many instances, any magnetic material does not really contribute to the inductance, but merely acts to constrain the flux to the immediate vicinity of the reactor. There comes a point, highly dependent on reactor size and function, when the designer decides the benefit of a magnetic core no longer justifies its cost, weight, and losses; inductors without magnetic cores are widely used, and are almost universally called *air-cored* inductors (or reactors).

The basic equation defining the inductance of a coil with a magnetic core is

$$L = 4\pi[10^{-9}N^2A_i/(\ell_a + \ell_i/\mu_i)]$$

where

N is the number of turns

A_i is the core cross section in square centimeters

ℓ_a is the effective length of the air gap, which is generally less than its physical length

ℓ_i is the length of the magnetic path

μ_i is the incremental permeability of the magnetic material under the operating conditions of the inductor

This deceptively simple equation is made enormously complex, insofar as calculation of an optimum core size (A_iA_w as for a transformer) for the inductor is concerned, because μ_i is a function of both H, the magnetizing force, and B_{ac}, the flux density due to ac excitation. Although much has been written on the subject of inductor design, most authors stubbornly try to force-fit magnetic material characteristics into a common relationship. Since the materials vary widely in the functional relationship of μ_i to H and B_{ac}, and in the functional relationship of H to B (i.e., their dc magnetization curves), such efforts are invariably not fruitful. The correct approach is to accept the dependence on particular material characteristics and use modified material characteristic curves.

Where an inductor is used for ac applications, the dependence of μ_i on H is of no concern. Moreover, in most instances, the reluctance term ℓ_i/μ_i will be very much smaller than the reluctance term ℓ_a. A first-order design can be accomplished by using the flux equation.

$$B_{ac} = 10^8 \times \text{Volt-seconds}/2NA_i$$

or, for sinusoidal excitation, $B_{ac} = 10^8V/4.44fNA_i$, the window area equation

$$F_W A_W = NI/\rho_I$$

and the simplified inductance equation

$$L = 4\pi(10^{-9}N^2A_i/\ell_a)$$

As for a transformer, the flux and window area equations are manipulated to yield

$$A_iA_w = 10^8VI/(4.44f\rho_I F_w B_{ac})$$

for sinusoidal excitation, or

$$A_iA_w = 10^8(\text{Volt-seconds})[I/(2\rho_I F_w B_{ac})]$$

for nonsinusoidal excitation.

This permits selection of a core; N can then be determined from either equation and ℓ_a established from the simplified inductance equation. The physical length of the air gap almost invariably needs to be greater than ℓ_a. This is due to the phenomenon known as *fringing flux*, illustrated in Figure 6-9. The proclivity of the flux to follow paths beyond the core's boundaries increases the effective area of the gap and, in consequence, reduces its effective length. For large gaps (greater than a few thousandths of an inch), the effective length, ℓ_a, may be approximated by

$$\ell_a = ab\ell/(a+\ell)(b+\ell)$$

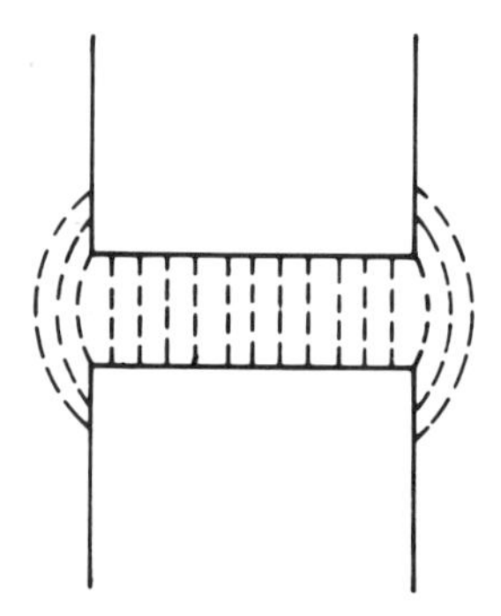

Figure 6-9. Fringing flux at air gap in magnetic circuit.

where a and b are the dimensions of the core cross section and ℓ the gap's physical length. From this, it can be established that ℓ_a increases with ℓ only so long as $\ell < \sqrt{ab}$. Literally interpreted, this would indicate that the winding has a minimum inductance at some finite physical gap length $\ell = \sqrt{ab}$, and the inductance increases again as ℓ is further increased. This is not the case, since the relationship given is an approximation. What happens in practice is that the inductance rapidly approaches the air core inductance of the winding as ℓ approaches $\sqrt{ab}$, and does not reduce farther for ℓ in excess thereof.

Long before ℓ reaches such a value, other effects of fringing flux have become of considerable concern. As ℓ, the physical length of the air gap, is increased, more of the total flux becomes fringing flux (i.e., strays beyond the confines of the core cross section). As Figure 6-9 indicates, this flux leaves and reenters the core material at high incidence angles—it is not parallel to the core laminations or tape. It can therefore produce high local eddy-current losses in the core. Moreover, it will also intersect conductors in the vicinity of the gap at high-incidence angles, and induce locally high eddy-current losses there too. These can be particularly distressing if a rectangular conductor with a low aspect ratio is used for the winding; strip conductors should be avoided in ac reactors with gapped cores. Any core supports that are conductive and that bridge or are near the gap will also suffer. As the gap gets larger, these stray and highly localized losses increase, and serious "hot spot" problems can arise.

The problems that arise are sometimes evaded by having a series of shorter gaps (interspersed by sections of the core) make up the total gap length required. Such an approach increases core-assembly costs and is not generally favored. In most cases, designers retreat to air-cored designs when required gap lengths become large in magnetic cored designs.

In the case where dc current flows in the inductor, the dependence of μ_i on H must be considered if the most economical design is to be achieved. The defining equations are the inductance equation in its full form, the window area equation, and the induction equation:

$$(4\pi/10)NI = B\ell_a + H\ell_i$$

The situation is somewhat simplified because, in the vast majority of cases, the ac excitation, and hence ac flux, is small in dc reactors, but optimum design depends on the empirically determined characteristics of the core material. The equations can be transposed to give

$$A_iA_w = 10^8LI^2(1+\ell_i/\mu_i\ell_a)/[(B+H\ell_i/\ell_a)\rho_I F_w]$$

Using manufacturers' published curves for B versus H and μ_i versus H and B_{ac}, optimum (minimum A_iA_w) designs can be derived. For the grain-oriented silicon steels, A_iA_w is not quite proportional to LI^2, as shown in Figure 6-10 where the value of K in the expression

$$A_iA_w = 10^3KLI^2/\rho_I F_w$$

is plotted against LI^2 for 0.012 and 0.004-in. steel laminations operating at $2\ kG\ B_{ac}$. In all cases, optimum designs arise when the quantity $(1+\ell_i/\mu_i\ell_a)/(B+H\ell_i/\ell_a)$ is minimized. This is a function of material characteristics and core size. Defining the gap ratio, r_g, as ℓ_a/ℓ_i, it becomes $(1+1/\mu_i r_g)/(B+H/r_g)$. Unfortunately, r_g and H are *not* independent variables, so the minima can only be

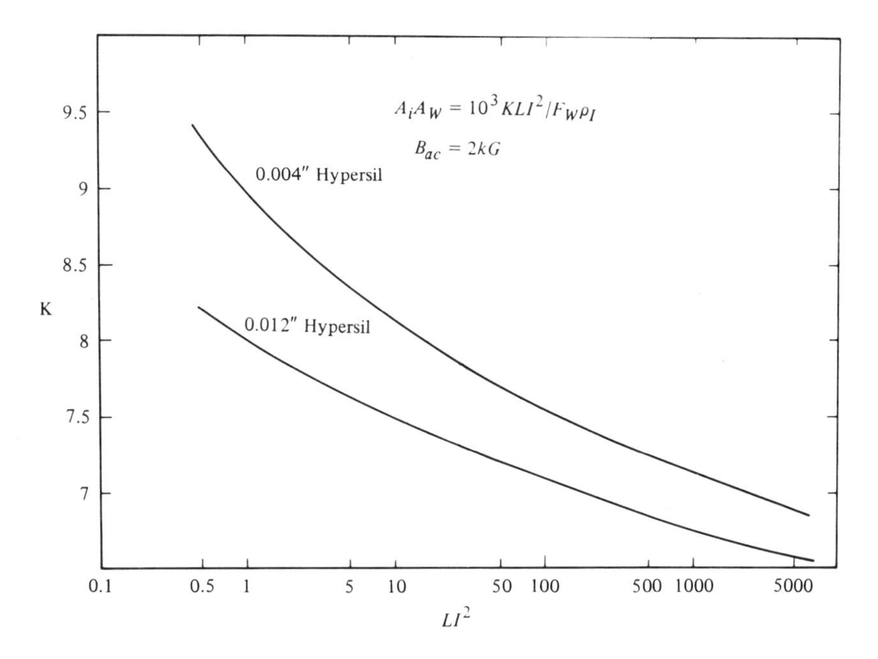

Figure 6-10. Design curves for inductors with gapped cores.

found by executing iterative searches. These minima are also weak functions of core geometry—i.e., of the ratio of A_w to A_i and of the aspect ratios of both the window and the core cross section. However, this dependence is slight, with no more than 1% variation over a very wide range of core geometries, and is usually ignored.

The designs giving the curves of Figure 6-10 have gap ratios ranging from about 0.003 for low LI^2 to perhaps 0.02 for high LI^2. The dc inductions range from ~0.7 Oe for LI^2 in the range 0.5 to 10 to 1.6 Oe for LI^2 in the range 500 to 2000. With large gaps, fringing flux reduces the effective length (and any ac flux produces locally high losses); the gap ratio at which this phenomenon finally becomes unmanageable is ~ 0.04 for most core geometries encountered. As this is approached, designers abandon the use of magnetic cores and use air-cored reactors.

Of particular interest is the behavior of an iron-cored inductor as the dc current therein varies. For a given design, μ_i—and hence the reluctance term ℓ_i/μ_i—will vary and thus so will the inductance. Figure 6-11 shows typical curves of the inductance variation with dc current for various gap ratios. Now, in designing reactors as dc current-sourcing interfaces, a problem arises. The critical inductance is determined by the lowest average current expected, while the inductor's

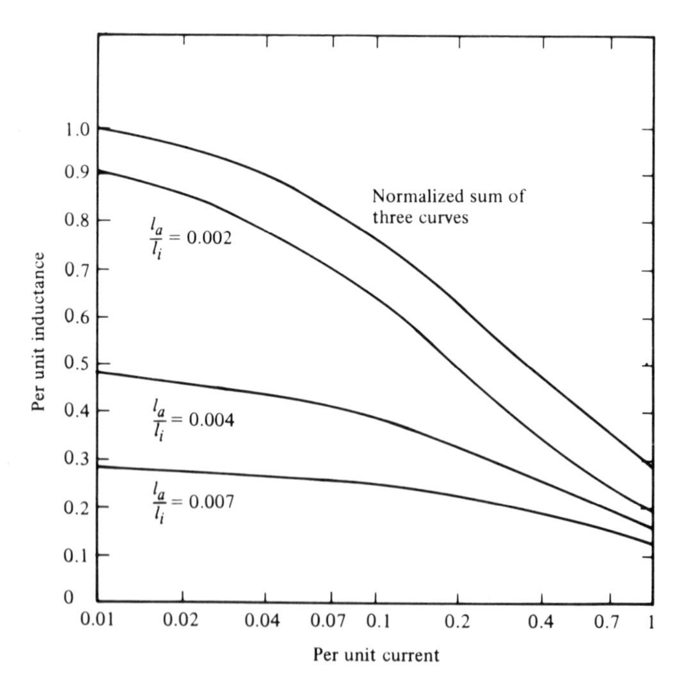

Figure 6-11. L versus I for various gap ratios.

size and rating (LI^2) are determined by the highest average current. However, at that current, a much lower inductance is tolerable. It would obviously be desirable to have such a reactor's inductance be inversely proportional to the dc current flowing. Figure 6-11 shows that this is not the case for any fixed-gap ratio, but suggests that such a characteristic could be obtained by series-connecting several reactors with different gap ratios. Now, since a single winding on several parallel magnetic circuits is essentially equivalent to the series connection of such windings on the individual magnetic circuits, it follows that a reactor with a stepped gap configuration as shown in Figure 6-12b can approximate the desired hyperbolic relationship of L and I over a fairly wide range of

a. Series connection of reactors with different gap ratios.

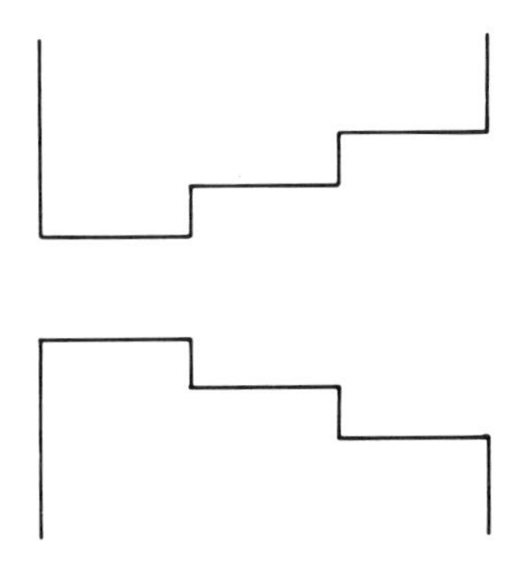

b. Achieving equivalent to (*a*) from single winding by stepping gap in core.

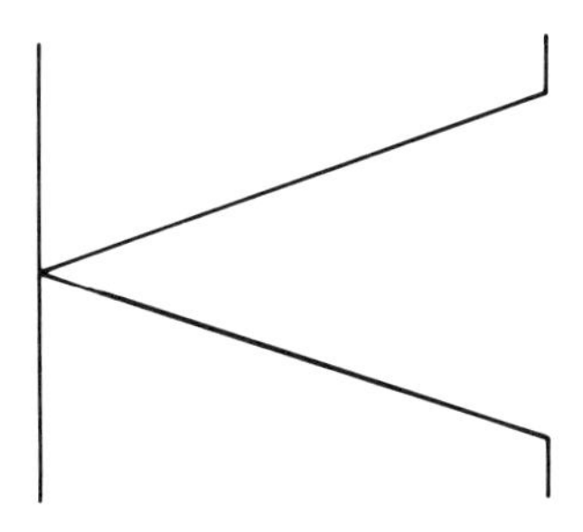

c. Refining to tapered gap.

Figure 6-12. Swinging choke gap profiles.

current. Carrying this procedure to the limit results in the tapered gap shown in Figure 6-12c. These swinging chokes are quite often encountered. Usually, they can maintain $L \alpha 1/I$ over a dc current range of 10:1 to 15:1, which is adequate for most applications, and they are considerably smaller than a design that maintains the inductance at or above the critical value for the lowest current over the whole current range.

Some reasons for abandoning the use of magnetic cores in inductors have already been discussed. There is one other, also suggested by the curves of Figure 6-11. In many cases, it is required that the inductance remain at its nominal value for currents very much larger than normal steady state. These requirements arise because of either an overload specification on the converter or the need to have the reactor play a significant part in fault current, or fault current rate of rise, limiting. Since all magnetic cores are saturable, iron-cored reactors are unable to fulfill such requirements unless they are designed, insofar as core conditions are concerned, for continuous service at the worst-case transient conditions. Designing them in that fashion typically makes for very large, heavy, costly and inefficient components. An air-cored reactor is not subject to saturation effects, and hence only conductor-related thermal considerations influence its transient and overload current capacity. In most instances, it can carry the stipulated fault or overload currents without needing to be designed for other than steady-state stresses, except perhaps for the need to survive mechanical forces generated by supranormal currents.

Although fairly simple in principle, the actual practical design of air-cored inductors is complicated by considerations regarding insulation, cooling, and the aforesaid mechanical forces. A converter designer can, however, gain an appreciation of the probable approximate size, weight, and losses in a given reactor by using and manipulating certain empirically derived formulas. The two most useful basic formulas in this respect are

$$L = r^2N^2/(9r+10\ell)$$

for the inductance of a single layer solenoid, in microhenries, and

$$L = 0.8a^2N^2/(6a+9b+10c)$$

for the inductance of a multilayer solenoid. In both expressions, N is the number of turns. For the single-layer solenoid, r is the mean radius of the winding and ℓ its overall length, both in inches. In the expression for the multilayer solenoid, a is the mean radius, b the overall length, and c the depth or "build" of the winding, all in inches. Obviously, if dimensions are in centimeters, the multiplier 2.54 must be applied in both cases.

In the case of a single-layer solenoid, the overall length is $N \times$ Conductor diameter (or width)+$(N-1)\times$ Conductor spacing. If these dimensions are d_c and s_c, respectively, the equation can be rewritten as

$$(r^2N^2) - 10L[Nd_c + (N-1)s_c] - 9rL = 0,$$

a quadratic in r if N is assigned. The total length of conductor used is $2\pi rN$ (plus an allowance for terminations or fly leads), and a simple iteration of solving for r with a given N and evaluating rN will produce an "optimum" design for a given L and conductor diameter. The latter is, of course, determined by the current to be carried by the inductor.

Similarly, both b and c in the expression for a multilayer solenoid can be expressed as functions of d_c and s_c. If there are m layers of n turns each, so that $N = mn$, then

$$b = nd_c + (n-1)s_{cn}$$

and

$$c = md_c + (m-1)s_{cm}$$

Note that conductor spacing in a layer may be different from that between layers. A quadratic in a results if m and n are assumed, and the total conductor length is $2\pi amn$; a fairly rapid iteration will produce an "optimum" design. The conductor spacings, s_c, s_{cn}, and s_{cm} are dictated by thermal considerations in practical designs. They range from $(0.1 \times d_c)$ to $(2 \times d_c)$, typically increasing as the current in the conductor and the number of layers increases.

Configurations other than simple solenoids, single or multilayer, are rarely used. Approximate formulas for the inductances of various geometries can be found in Terman[3]; more accurate formulas for all cases can be found in Grover[4], but their use is only necessary if actual designs are to be undertaken.

A major problem with the use of air-core reactors, particularly solenoids, is their external field. All the flux they produce must close in paths, mainly air paths, external to the reactor. Since the field intensity close to the reactor will be high, any magnetic or conductive material in this vicinity will influence both the inductance and the losses of the component because of currents induced by ac flux. To avoid serious problems in this regard, it is advisable to allow clearance of about one reactor diameter between the surfaces of the component and any substantial aggregation of magnetic or conducting material. Where possible, all such material, even small pieces, should be kept well away from the reactor.

Designers occasionally attempt to mitigate these problems by using toroidal windings. Winding an air-core inductor as a toroid reduces its external field essentially to that produced by a single turn having the toroid's mean radius and carrying the same current as the inductor. However, this approach is not popular because toroids are difficult to wind with large conductors and, for a given L at a given I, a toroid occupies very nearly the same volume as a solenoid, including the spacing from the solenoid necessitated by its external field.

It should be noted in passing that the *Brooks coil* configuration, so beloved by elementary textbooks on inductor design because it produces the maximum inductance from a given length of wire, is rarely possible for high-current inductors. A Brooks coil is a multilayer solenoid having $b = c$ and $c \simeq 2a/3$. Thermal considerations usually prohibit the use of a winding with so many layers when currents and conductors are large. In addition, the inner radius required is often below the minimum winding radius of a large conductor. This latter point is of some significance in all configurations. Any conductor has a minimum winding radius, from purely mechanical considerations. For solid round copper conductors, it is usually taken to be about four to six times the conductor diameter, and it sets a fundamental limit to the geometry of a winding of that conductor.

6.7 Capacitors Used in Converter Interfaces

It has been seen that capacitors are the primary means to accomplish voltage-source interfacing. A converter designer must have an appreciation of the characteristics of various types of capacitors in order to select those that are appropriate for the various interfaces that must be created.

For dc voltage-sourced interfacing, the requirement is generally for a capacitor that will support the dc voltage without problems while passing substantial ripple current. For low-to-medium voltages, electrolytic capacitors are almost universal, not because of any superiority in characteristics over other types, but because of low cost and small size and weight. For low dc voltages, particularly the 5-to-15-V supplies so prevalent for solid-state electronic loads, tantalum electrolytics are often used. Up to 500 V or so, etched foil aluminum electrolytics are used; above this voltage, they are still employed, but it is then necessary to series-connect capacitors.

Electrolytic capacitors suffer from a number of characteristic deficiencies, which makes application somewhat difficult. To begin, all electrolytics possess appreciable leakage resistance, which results in unwanted losses from the imposition of continuous dc stress. While the losses so generated are not usually a serious detriment, this characteristic does give rise to problems when series connection is employed. The leakage resistances of capacitors produced exhibit quite large tolerances, and if the series-connected capacitors are to share the dc voltage equitably, then ballasting resistance must be connected in parallel with each unit, or parallel-connected group of units, to swamp variations in leakage. This results in increased cost and losses.

Second, the capacitance tolerance of electrolytic capacitors is very large. Typically, $-25\% + 100\%$ tolerance applies to manufactured units; close tolerance versions may be specified as $-10\% +50\%$. Such tolerances do not create very severe problems insofar as interface design is concerned; they do, however, make for tremendous difficulties in tailoring converter-feedback loop responses because of the corresponding wide variations in transfer functions.

Like almost all electrical components, electrolytic capacitors are thermally limited. They have quite high internal series resistances, which dissipate power when ripple current flows. The ripple-current capability of a given electrolytic is limited by the dissipation and temperature rise so caused, so that to accommodate converter ripple current, most electrolytic-based dc voltage-source interfaces have capacitance in excess of need. The series resistance involved is a function of frequency, and hence so is the ripple-current limit for a given electrolytic. Manufacturers publish curves of r versus f or permitted I versus f to help designers apply their products.

The impedance of an electrolytic capacitor is not simply comprised of capacitive reactance plus series resistance with leakage resistance in parallel. All practical capacitors posess parasitic inductance, and electrolytics might be said to have more than their share. Because of this, individual units have self-resonant frequencies, ranging from a few kilohertz to several tens of kilohertz. Above its self-resonance, an electrolytic capacitor exhibits a primarily inductive impedance and rapidly loses its effectiveness as a ripple-current bypass. Manufacturers generally lump series inductance and resistance together under the somewhat misleading term, *equivalent series resistance* (ESR), publishing curves of ESR versus frequency in their product application notes.

Finally, there is the matter of de-formation of electrolytic capacitors. The dielectric therein is in fact an oxide layer of modest thickness. The capacitor needs a polarizing voltage fairly close to its rated working voltage to maintain its dielectric layer. If it stands idle for a long period, or is subjected to long-term voltage stress well below its rating, the capacitor "de-forms" by losing part (or all) of its dielectric. This causes the capacitance to increase, and results in very large, possibly destructive, surge-reforming currents when voltage is reapplied.

When the disadvantages of electrolytics make problems too severe, or when the dc voltage is greater than 1 to 2 kV at most, other types of capacitor must be sought. Of the wide variety manufactured, many are totally unsuitable for power electronics use because their construction does not allow for the flow of significant levels of ac current. Among these are the various metallized-foil types and the majority of ceramic dielectric capacitors. The most popular high-voltage dc interfacing capacitors are those with paper-oil dielectric systems and internal construction suitable for use with substantial ac current flow. For the most part, this implies that the conducting foils making up the capacitors' plates are extended beyond the edges of the dielectric foil so that connection can be made to all, or most, of the foil edge. Capacitors in which connection is made to a single point, or tab, on the foil are generally not suited for applications where significant levels of ac current will flow. Where the frequency is high and required capacitance values low, mica capacitors are often used. They are more expensive than paper-oil, but have much lower losses.

Losses, which generally increase with frequency, are the major problem with paper-oil capacitors. Self-resonance is present, of course, but for typical units, it is generally at frequencies one to two orders of magnitude or more above those for electrolytics. Special low-inductance units may have self-resonant frequencies of several megahertz and up, lower than those typical of mica but high enough for the vast majority of applications. There is not much to be done as regards losses. Paper-oil losses are usually considerably lower than those of electrolytics at any frequency. From a few hundred hertz up, however, they are significant enough to pose a thermal problem to the capacitors themselves, even though converter system efficiency may not be significantly affected. Water cooling is introduced to larger unit sizes, by embedding water-carrying tubing in the capacitor structure. Not common in dc capacitors, this technique is employed extensively for higher frequency ac-rated paper-oil capacitors.

For strictly ac interface applications, electrolytics cannot be used and paper-oil, with extended foil construction, are used in most cases. Again, mica may be chosen for higher frequencies, if the capacitance needed is relatively small. With paper-oil, it is not possible to use dielectric capabilities effectively in units rated much below 1000 V. As a result, size, weight and cost tend to become excessive in lower voltage applications. A variety of modern plastic film dielectrics, including polycarbonate, Mylar, and polystyrene, are better under these circumstances, as they are smaller, are lower in cost, and have lower losses. Care must still be taken, however, to ensure that internal construction is adequate for the ac current flow.

Invariably, ac capacitors are rated in accordance with their volt-ampere capacity (*kVA rating*, for kilovolt-ampere rating, is the common term). Thus, the rating of a given unit is directly proportional to frequency, but as the frequency increases so will the losses. Hence, care must be taken not to overuse the component and exceed thermal limits. These limits are determined by case size, which defines the heat rejection for a given temperature rise, and the maximum permissible operating temperature of the dielectric system. Thus for paper-oil (or any other dielectric), there is a maximum kVA rating for a given unit size.

For dc applications, the energy-storage rating (CV^2) is often used as a measure of capacitor size and cost. It is, of course, analagous to the LI^2 rating for dc inductors. Both can be converted to an equivalent VA rating at some frequency f simply by multiplying by $\omega = 2\pi f$.

Finally, it should be observed that the dielectric systems of paper-oil capacitors for dc (with ac ripple current) and ac applications are somewhat different.

Thus, while in name and external appearance they may be the same, paper-oil capacitors designed for dc interfacing usage are not suitable for ac usage and vice versa. In most cases, prolonged application of unipolar voltage to an ac-rated capacitor will cause dielectric degradation and eventual breakdown. By the same token, repeated reversals of potential cannot be tolerated by paper-oil capacitors designed for dc applications.

6.8 Electromagnetic Interference and Filtering

The primary concerns in interfacing switching power converters are the largest amplitude unwanted components in their dependent quantities. As shown in the converter analyses of Chapters 3 through 5, these are typically oscillatory components with frequencies not very far separated from those of the wanted components. The dependent quantities contain infinite series of unwanted oscillatory components, some of which are of very high frequency, albeit generally they are then of low amplitude. These components are of concern, not because they might significantly disturb converter operation, but because they can disturb other equipment.

The interfacing components and filters used to enable the converters to work in proper fashion normally have little influence on very high frequency, unwanted components. Capacitors possess self-inductance, inductors possess self-capacitance, and both therefore are self-resonant. The self-resonant frequencies of components, or component assemblies, large enough to fulfill primary interfacing requirements are generally low enough that the filters so created do not significantly attenuate, and may even enhance, unwanted components that have frequencies from 100 kHz up. In some instances, even lower frequencies may not be attenuated.

Thus, the higher frequency unwanted components tend to be passed directly to the converters' sources and loads. Despite the influence of converter imperfections, such as finite-device switching speed and commutation overlap (see Chapters 7 and 8), practical converters do generate high-frequency unwanted components. Although many sources and loads are very tolerant of these in the very low amplitudes at which they usually occur, they can nevertheless be the cause of serious problems. This is because of their potential to interfere with the operation of communications and data-processing equipment, which operates at similar frequencies and works with very low signal levels. Such interference can occur as a result of direct coupling resulting from the flow of unwanted components from the converters in load or source leads common to the other equipment, or by radiation from converter load or source leads. It rarely happens that direct radiation from the converter equipment produces disturbances, since levels are inherently low, and the normal steel enclosures provide good attenuation.

EMI resulting from converter currents flowing in source and load connections is, in principle, subject to two modes of attack for the purpose of its reduction. The converter may be redesigned to lower the inherent level produced. This is difficult to do at best; more often, it is simply not possible. The alternative is to provide special filtering specifically to attenuate those unwanted components (i.e., usually the higher frequencies) that can cause, the problems. The design of such filters is quite straightforward from a strictly electrical standpoint. Although the terminating impedances may not be quite so well defined as the designers might wish, they can usually arrive at a simple T or π or ladder network with Tschebychev or Elliptic response to serve the purpose.

Physical realization is not so easy. Inductors capable of carrying high currents are difficult to produce with very high self-resonant frequencies; so are capacitors capable of supporting high voltages. Beyond this, the geometries of the filter layout and connection-filter-source arrangement are critical to successful application. The art of component design and filter equipment layout for EMI reduction is an extremely specialized one. Difficulties escalate rapidly as converter power level goes up, and at the very highest power levels, there are no known completely successful techniques. Most texts on the subject (EMI and its reduction) are of little help, since they mostly address the problem of the communications equipment itself. Some do give valuable insight into appropriate layout practices, and into techniques available for reducing radiated EMI.

Radiated EMI can be substantially reduced without filtering. Making sure that load or source-connecting lines are balanced with respect to ground affords some benefit. Having them twisted tightly together produces about 24-dB reduction over well separated leads, and laying in steel conduit produces about 60-dB reduction. It is also often easy to shield the susceptible equipment.

The design, including mechanical design, of EMI filtering should definitely be left to those specialists in that art who have demonstrated competence. Such specialists should also be consulted on the mechanical design and physical layout of the converter itself when EMI reduction is a requirement or likely to be one.

One of the more distressing aspects of EMI is the proclivity of converters to interfere with each other. They do not generally do this by direct power-circuit interaction, although they can if common supply or load connections are involved. Every converter, however, has a control that determines its switches' existence functions. These controls (see Chapter 10) take the form of small dedicated hybrid, or latterly digital, computers and are thus susceptible to converter-generated EMI. In fact, it is not uncommon for a converter to interfere with its own control, a most displeasing phenomenon.

6.9 Converters Operating with Discontinuous Current

The analyses of current-sourced converters in Chapters 3 and 4 predicated perfectly smooth dc currents at converter terminals. The discussions of interfacing in Section 6.2 revolved around the maintenance of continuous unidirectional current, giving rise to the concept of critical inductance. If the inductance is below critical, either by design or because the converter enters an operating region beyond that for which the inductor was designed, discontinuous current will result.

With one exception, it is rare to find converters operating with discontinuous currents over most of their design operating range. Thus, in the vast majority of cases, discontinuous current only occurs when the converter's average current is well below normal; as a result, any changes in dependent-quantity spectra are not significant, and usually a designer doesn't even bother to consider them. These changes arise because the switch existence functions become current-dependent once that current is discontinuous. If the converter ultimately interfaces a good voltage source or sink, analysis is fairly easy; if not, it becomes extremely cumbersome.

Consider a midpoint current-sourced ac-to-dc/dc-do-ac converter feeding (or fed by) a smooth dc voltage through a subcritical inductance. Current will begin to flow only when the source voltage connected to a closed switch exceeds the defined voltage, and will terminate when the total volt-seconds applied to the inductor become zero. The switch will then open, and until the next switch in sequence closes and its source voltage exceeds the defined voltage, the converter terminal voltage will equal the defined voltage. The converter dependent voltage and current can be established, since switch-existence-function bounds are established by considering one source—defined voltage and inductor volt-seconds. This is not as simple as it might seem—an iterative solution is called for, so that the average current produced by the converter matches that consumed by the load. However, the procedure is quite straightforward, and may be applied to any of the current-sourced converters.

When the final termination is not a good approximation to a voltage source, the current flow, occurring as a result of switch closure and ac source voltage application at that phase angle to the total load network, must be analyzed. Since there is almost invariably an initial voltage at the load, which must be considered unknown at the outset, iterative solutions of the time-domain expressions for the network are necessary until convergence is obtained—i.e., until the initial voltage postulated matches the voltage obtained from the analysis. The time-domain expressions needed are most readily obtained by using Laplace transforms to solve the differential equations for the system. If the load network is so complicated that its transfer function contains more than two poles (or zeroes), then computer simulation may be a preferable approach to the converter's analysis.

Apart from the ac regulators discussed in Chapter 5, the only case of much practical importance where discontinuous current exists is the half-wave rectifier of Figure 4-17. There the source impedance forms the subcritical inductance. Analysis is possible by the method outlined in this section; however, Terman

(1943) contains a comprehensive set of curves defining the performance of this circuit, and the two-phase version, which is often called a *full-wave center-tapped* rectifier.[5] There are some further derivatives of this circuit, the voltage-multiplying rectifiers, such as the Cockroft-Walton circuit,[6] that also have found considerable usage. It should be clearly understood that the dependent-current spectra of all converters operating with discontinuous current are so bad that they would not be considered for power levels above a few hundred watts. Occasionally, they are pressed into service at slightly higher powers, but the results are invariably less than satisfactory. Unless the ac source is extremely tolerant of converter malfeasance, any cost saving in the converter will be more than offset by the operational problems that develop.

References

1. Ainsworth, J. D. "Filters: Damping Circuits and Reactive Volt-Amps in *HVDC* Converters." In *High Voltage Direct Current Converters and Systems*, edited by B. J. McDonald Cory. London (1965).
2. Terman, F. E. *Radio Engineers Handbook*, pp. 30-37, New York: McGraw-Hill, 1943.
3. Terman, op.cit., pp. 47-64.
4. Grover, F. W. *Inductance Calculations, Working Formulas and Tables*. New York: Dover Publications, 1962.
5. Terman, op.cit., pp. 602-612.
6. Cockroft, J. D. and E. J. Walton. Further developments in the method of obtaining high velocity positive ions. *Proceedings of the Royal Society* A, **136:** 619, 1932.

Problems

1. A simple buck dc-to-dc converter is used to supply a constant 5-V output from a dc source ranging from 8.5 to 13.5 V. If the minimum output current is 5 A and the operating frequency is 200 Hz, calculate the critical inductance for the converter.

2. A designer decides to address the conversion requirement of Problem 1 with a three-phase harmonic neutralized buck converter, also switching at 200 Hz. What critical inductance will now be needed?

3. The converter of Problem 2 requires an interphase reactor in addition to the current-sourcing interface reactor. Using 200 Hz as the base rating frequency, calculate

a. the kVA rating of this interphase reactor
b. the kVA rating of the critical inductance of Problem 2
c. the kVA rating of the critical inductance of Problem 1

assuming the maximum output current to be 30 A and ignoring the effects of ripple current. Comment on the total inductive component ratings of the simple and harmonic neutralized converters.

4. Another designer decides to use a current-source-transfer buck-boost converter for the application delineated in Problem 1. With the operating frequency 200 Hz, calculate the critical inductance and its kVA rating. Did the designer make a good choice?

5. A three-phase bridge voltage-sourced dc-to-ac converter has defined ac current-source magnitudes (line or pole currents) of 50 A and a dc source voltage of 200 V. Calculate the voltage sourcing interface capacitor needed to hold the peak ripple voltage caused by the lowest order ripple-current component to 5% of the dc voltage for converter operating α's of π, $5\pi/6$, $2\pi/3$, and $\pi/2$ rad. (Assume all the ripple flows in the capacitor.)

6. A designer addresses the conversion requirements of Problem 5 with two bridges used to form a 12-pulse converter, parallel-connected at the dc terminals. What dc interfacing capacitance is needed?

7. A current-sourced dc-to-ac converter, six-pulse bridge, is connected to an ac system with 0.06 per unit pure inductive source impedance. The ac line voltage is 480 V, and the converter maximum dc output current is 500 A. Calculate the L and C values of a fifth harmonic trap filter to limit that harmonic voltage at the ac terminals to 1% of the supply voltage if the filter Q is

a. 50

b. 15

8. Calculate the leading reactive power supplied by the traps of Problem 8. At what operating α's will they balance the converter's reactive power demand, assuming similar traps are applied to all three supply phases?

7
Semiconductor Switching Devices

7.1 Introduction

It is a fact that the basic characteristics, interfacing requirements, and, to a large extent, commutation requirements (see Chapter 8) of converters *do not depend on the types of switch they use*. For the past two decades virtually the only switches used have been those developed through the use of bipolar silicon semiconductor technology. (Only recently have field-effect devices begun to play any role, and it is as yet an unimportant one.) Thus, it behooves the engineer concerned with switching power converters to acquire a thorough understanding of the behavior of bipolar silicon switches.

The switches in question are the diode (higher current versions are often termed *rectifiers,* a somewhat confusing practice), the thyristor, the thyristor's relatives (triac, gate-controlled switch, gate-assisted turnoff thyristor, reverse conducting thyristor, and reverse blocking diode thyristor), and the transistor. Their properties are discussed in some detail in this chapter. Those properties that are not obvious and that give designers many problems in converter design are emphasized. The more trivial imperfections, voltage, current, and thermal limitations are afforded shorter shrift. They are ''obvious'' in that any reasonable individual would expect the device to exhibit them, and they are generally well covered in manufacturers' handbooks.[1,2]

7.2 Diode

The conventional circuit symbol, diagrammatic structural representation, and static V-I curves for a P-N junction diode are shown in Figure 7-1. The nomenclature for its terminals, *anode* for that into which current flows and *cathode* for that from which current exits, is directly derived from the vacuum-tube diode. The vast majority of modern high-current diodes are diffused devices. They can be made N-base or P-base. In the first case, the silicon slice is N-type, and the manufacturer diffuses P (acceptor) impurity into one side to form the junction and establish the diode anode. In the second, the slice is P-type, and N (donor) impurity is diffused into one side to produce the junction and establish the cathode connection. Both types are manufactured; P-base, termed *reverse polarity,* are less common than N-base for a variety of reasons.

The static V-I characteristics display the trivial, or obvious, limitations of the diode. When carrying current, it exhibits a conducting forward drop. Since it is a temperature-limited device, with maximum permissible junction temperature

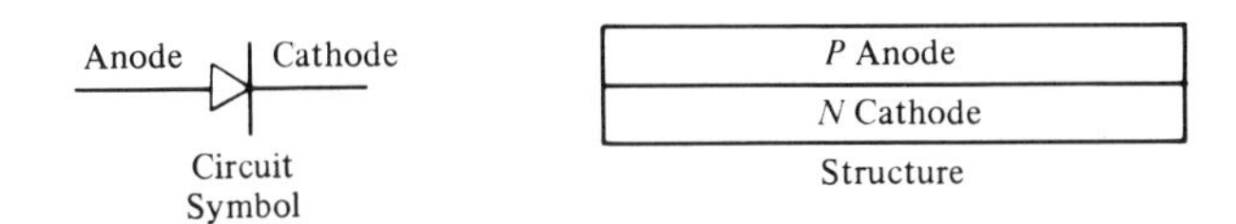

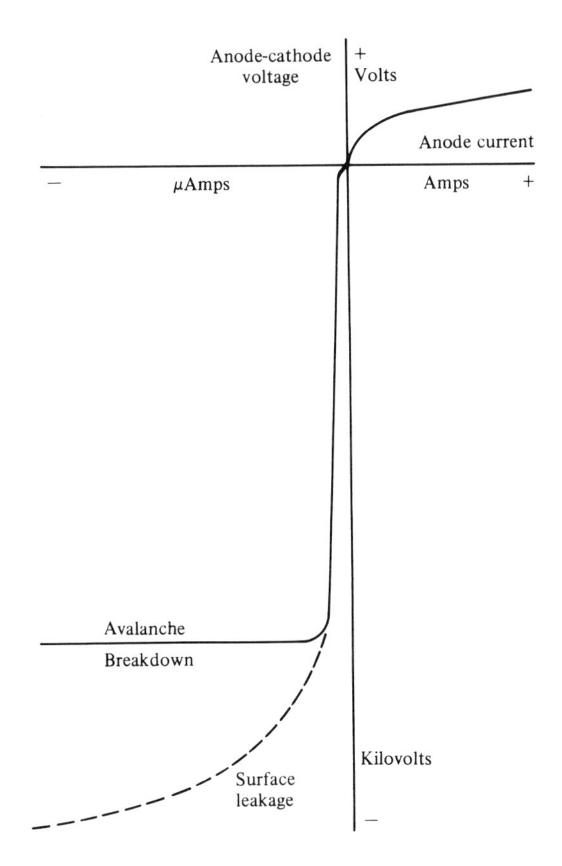

Figure 7-1. Diode.

typically from 250° to 300°C, the forward current limit is established conjointly by the dissipation of the device and the cooling technique applied. When blocking reverse voltage, the diode passes a small leakage current until its voltage limit is reached. The leakage current has two contributions: a small ''saturation leakage'' component that develops for very low reverse voltage and an essentially resistive addition thereto as the voltage is increased.

The reverse voltage limit for the diode may be set by either of two phenomena. The most common, in modern devices, is the avalanche voltage of the P-N junction. At this point, as illustrated by the solid line in the characteristic of Figure 7-1, the device current increases very rapidly for very little further increase in the device reverse voltage. The dissipation in this region is extremely high, and operation in avalanche is generally regarded as not permissible. For those devices where it is permitted, the so-called avalanche diodes, it is generally allowed only to a limited current level for very short time periods (1msec or less) as a means of circuit and device transient protection.

For diodes that have reverse voltage limit set by avalanche, the reverse current after breakdown is uniformily distributed across the silicon slice. High-voltage units may not be so limited, because a second phenomenon becomes critical before the junction avalanche voltage is reached. The electric field intensity at the edge of a diode, where the junction comes to the surface, is high. Also, the periphery, where the junction is exposed, provides an ingress for ionic impurities to modify or destroy junction characteristics. These problems are addressed in manufacture by tapering and beveling (and otherwise contouring in many instances) the slice edge all around the periphery and then applying an organic, nonconductive protective coating.

Unfortunately, no material is completely free from ionic impurities, nor is any organic material completely free from ionic effects under high electric fields. Hence, in high-voltage diodes, the surface coating may begin to contribute substantially to the leakage current as the reverse voltage is increased, as shown by the broken line of the V-I characteristic of Figure 7-1. When this happens, it is likely that this additional leakage current, and the dissipation it causes, will be confined to a small, isolated area of the diode periphery. The temperature rise there can be extremely rapid and large, since the lateral heat conductivity of the thin silicon slice is low. Thus, temperature in excess of that allowable at the junction can develop locally, and localized destruction of junction characteristics may take place. In such instances, the reverse voltage limit is set by the local surface dissipation allowed. The manufacturer uses testing, experience, and judgment to establish this.

Any dissipation caused by reverse leakage should be added to forward-conducting dissipation (and switching dissipation) to determine cooling requirements (see Section 7.7). However, the reverse-dissipation contributions of all but the very highest voltage, modern, silicon diodes are so small that they are generally ignored.

A very wide variety of types and sizes of diode are currently made. Lower current units, up to a few amperes rating, are generally encapsulated in epoxy resin and rely upon the leads to conduct heat away from the silicon slice. From about 10 A to 300 to 500 A ratings, up to about 1⅜-in. silicon-slice diameter, stud-mount packages are common. In these packages, heat is conducted from one side of the silicon slice by the hexagonal base of the package. Lower cost "press-fit" packages are used for certain high-volume applications with diodes having ratings up to a few tens of amperes. From 200 to 3000 A or more, from about 1 to 4 in. silicon-slice diameter, "hockey puck" or "flat pack" packages are employed so that heat may be conducted away from both sides of the slice by large area conductive "pole faces." These packages are illustrated in Figure 7-7. Single devices with reverse voltage-blocking capability up to some 6000 V are made. In the lower current ratings (a few amperes to a few tens of amperes), the upper voltage limit is generally 600 to 1000 V purely because of a lack of demand for higher voltage units; devices of higher voltage rating are made in this current range for some very high voltage applications needing series-connected strings. Medium-current diodes, from a few tens to a few hundreds of ampere ratings, are available with reverse voltage-blocking ratings typically up to about 2000 V; again, there has been little demand for higher voltages in the medium-power applications these devices serve. The high-current diodes are usually made in the highest possible voltage ratings.

The nontrivial, or nonobvious, limitations of the diode are caused by a property that is universal in bipolar silicon devices and is the root cause of all nonobvious limitations in all such devices. The phenomenon responsible is termed *stored charge;* a P-N junction, when conducting forward current, stores charge. For forward currents up to a certain level, while the charge-control equation for the device is valid, there are far more free carriers in the base region than are needed to sustain the current flowing.

The most important diode limitation that arises due to this effect is reverse recovery (see Figure 7-2a). A diode-carrying forward current stores charge, as has been stated. If the forward current is reduced to zero very slowly by external circuit action (the only way a diode current will change is through external circuit action), then the stored charge will dissipate, through normal recombination processes, and at zero current the device will be in equilibrium and assume the reverse-blocking, high-impedance state. If, however, as depicted in Figure 7-2a, the reduction of forward current is more rapid, then the stored charge does not have time to dissipate through normal carrier recombination, and there are still a great many free carriers in the base region when the current reaches zero.

The diode cannot assume the reverse-blocking state until these carriers are removed. In consequence, it remains low impedance, and the current builds up in the reverse direction. This "reverse recovery current," driven by external circuit potential, can reach magnitudes of one-third to one-half the original

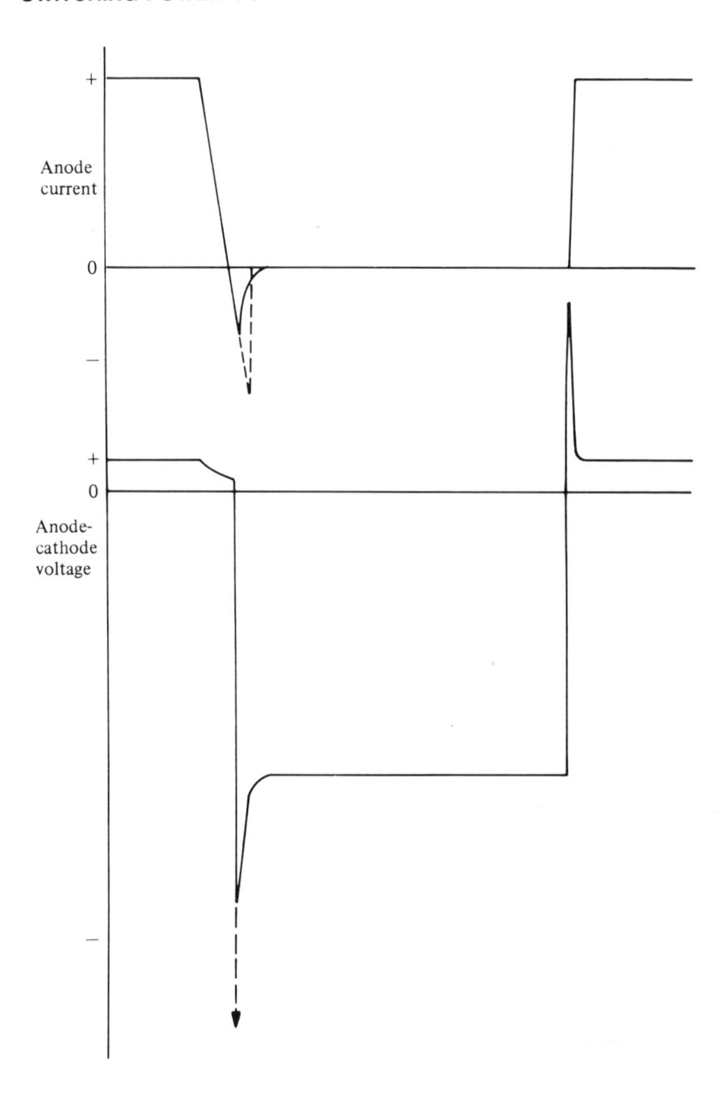

Figure 7-2. Diode-switching characteristics.

forward current if the rate of change of current, dI_R/dt, is high and the diode's stored charge considerable. It ends, or ceases to increase, when there are no longer sufficient carriers to sustain it. At this point, two radically different behaviors develop; which one occurs depends on internal diode parameters in a rather complex fashion that will not be discussed here. In one case, shown by the solid line in Figure 7-2a, the reverse-recovery current decays with an exponential characteristic having a time constant determined in part by the carrier lifetime in the diode's base region. Typically, the total decay time (four time constants) will be about equal to the total time taken for the reverse-recovery current to rise to its peak value; such a characteristic is termed a *soft* recovery. In the other case, depicted by the broken line in Figure 7-2a, the collapse of reverse-recovery current back to leakage levels is extremely rapid, typically occurring in less than 100 nsec. This characteristic, termed *snappy,* is most often found in P-base diodes; it is one of the reasons for their lack of popularity.

The rise and decay of the reverse-recovery current, of either type, is as such of no concern to the diode. The current is uniformly distributed across the junction and would not cause excessive dissipation or temperature rise were it not for external circuit action. In fact, during the period in which reverse-recovery current is increasing, the voltage on the diode remains in the forward direction—energy (from the stored charge) is being extracted from the device. The problems that arise because of reverse recovery arise because of the inevitable presence of inductance in the external circuit. As shown in Figure 7-2a, the external circuit may always be equivalenced by a potential, a small resistance, and an inductance. The potential is the reverse voltage that the external circuit seeks to apply to the diode. The resistance is invariably composed of assorted parasitic resistances, while the inductance may be purely parasitic. In many instances, however, it may contain a substantial bulk inductance placed in the circuit for such reasons as commutation, interfacing, dI/dt control, and so on.

The reverse-recovery current obviously must flow in this inductance. While it is increasing, it is primarily the inductance that supports the circuit potential and determines dI_R/dt. When the device "snaps" off, the reversal of current slope causes a reversal of LdI/dt; the resulting inductor voltage now adds to the circuit potential to cause a combined reverse voltage on the diode that may be (and usually is) very much larger than the circuit potential. If the diode characteristic is snappy, the rate of rise and final magnitude of the additional reverse voltage will be largely determined (1) by $LI^2_{rr}/2$, the energy stored in the inductance by the peak reverse-recovery current, and (2) by the self and stray capacitances associated with the inductance and the diode. When the diode has a soft recovery characteristic, the additional reverse voltage is generally truly LdI/dt, with stray capacitances having little influence.

Even if a designer could accurately predict the total reverse voltage that will develop, it is generally not possible to allow it. The primary reason is that device dissipation is very high, with a combination of high-current and high-reverse voltage; in addition, the cost of the extra device voltage capability is usually unacceptable. To control the reverse voltage, a "snubber" network is connected in shunt with the diode. It consists, in the vast majority of cases, of a simple series resistor–capacitor combination. Its purpose is twofold: (1) to moderate the dI/dt in the circuit inductance by providing an alternate path for the reverse-recovery current when the diode refuses to conduct that current any more, and (2) to provide a means (the capacitor) for absorbing the energy stored in the inductance without developing large voltages. The general principles and considerations surrounding the design of such networks are discussed in Section 7.8.

If the problems with reverse-recovery behavior seem pressing when considering an individual device, then note that they become even more so when diodes must be connected in series because total circuit voltage is beyond the capability of a single device. The stored charges of the individual devices in such an assembly invariably show a considerable tolerance. The device with least-stored charge will snap first, diverting the reverse-recovery current for the remainder of the series string into its own snubber. Two requirements then arise. The snubber must not restrict the subsequent recovery current to the extent that other devices in the string do not recover. Otherwise, those devices will not block reverse voltage, and others, including that snapping first, will probably be asked to support more than they are able to, with disastrous results. Also, the snubbers must not allow too much reverse-voltage accumulation on the earlier snapping diodes (those with lower stored charge) while the later snapping devices are completing recovery. This is because the initial reverse voltage on an early device after all diodes have recovered will be that accumulation plus its share of circuit potential, and the total must not exceed the individual diode's blocking capability.

The sizes of snubber networks needed to accomplish these ends generally considerably exceed those of snubbers in single device applications. Often, a series string with many devices must be split into sections, and each section must be supplied with an additional shunt R-C snubber to maintain adequate control of the transient reverse voltage on each device. The situation can be somewhat improved by using devices with inherently low stored charge and by selecting devices for tight tolerance on stored charge. Of course, low stored charge is also beneficial for single-device application.

Diode-stored charge can only be reduced by reducing carrier lifetime in the diode's base region. Unfortunately, this increases the forward conducting drop of the device at a given current density in the slice. There are three techniques available for lifetime reduction. In order of increasing effectiveness, they are: increasing the intrinsic resistivity of the base silicon by reducing its impurity doping level; introducing recombination centers by adding an additional dopant such as gold; and creating recombination centers after device fabrication by bombarding the slice with high-energy electrons. Increasing resistivity is not favored at all. Gold doping is widely used, but its effects on both forward drops and reverse leakage are such that it is generally limited to devices of less than 1000-V reverse-blocking rating. Electron bombardment is very effective, particularly for high-voltage devices, since it causes only a small increase in forward drop and has virtually no effect on leakage.

Devices treated to reduce stored charge are called *fast recovery* diodes. Whereas a normal 500-A device might have a stored charged under high dI_R/dt of perhaps 300 to 500 μC, a fast-recovery diode of the same voltage rating and with the same dI_R/dt will have only about 70 to 100 μC.

A rather special kind of diode, the Schottky barrier diode, has both much lower forward drop and much lower stored charge than does the silicon P-N junction diode. It is not a bipolar device, but uses a metal-semiconductor junction. Unfortunately, it has not yet been possible to make such devices with reverse-voltage capability higher than about 100 V. They do find extensive use in low-voltage, and especially in low-voltage high-frequency, applications, because of their excellent characteristics.

The other diode imperfection arising because of stored charge occurs when the diode acquires that charge. It is rarely seen because the dI/dt at diode turn-on is modest in the overwhelming majority of applications. When the dI/dt at turn-on is high, as illustrated in Figure 7-2b, it demonstrates its need to acquire the stored charge by acting as an inductance and transiently developing a forward drop of an order of magnitude or more higher than normal. This is unimportant as far as the diode itself is concerned. The excess drop is nondissipative and in any case is of such short duration (less than 100 nsec in most cases) that any dissipation caused would be negligible except in very high frequency applications. It is a matter of concern when the diode is used to effect overvoltage protection for other devices; if those devices are susceptible to damage from instantaneous overvoltage, then this characteristic of the diode may allow damaging overvoltage.

Despite its lack of controllability, the diode finds many uses in power converters. It is the obvious choice of switch for single-quadrant ac-to-dc converters, and it is used as the autocomplementary switch in all dc-to-dc converters. Because no single switching device possesses all the requisite capabilities, it is also used to provide one-half of the bidirectional current-carrying, unidirectional voltage-blocking switch needed in voltage-sourced ac-to-dc/dc-to-ac converters. In this application, it is connected directly or implicity in inverse parallel with a thyristor or transistor. Also, it finds quite extensive use in snubber networks, as discussed in Section 7.8.

7.3 Thyristor

The thyristor is currently the most important controllable switching device. As Figure 7-3 shows, it is a three-junction bipolar silicon device. The conventional symbol, also shown in Figure 7-3 along with the static V-I characteristics, represents the device as a diode with a control electrode (the *gate*) added. This device is fabricated by starting with N-type base silicon and diffusing P-type impurities into both surfaces of the slice. One surface is then masked, and N-type impurity is diffused into the other to form the gate-cathode junction. This

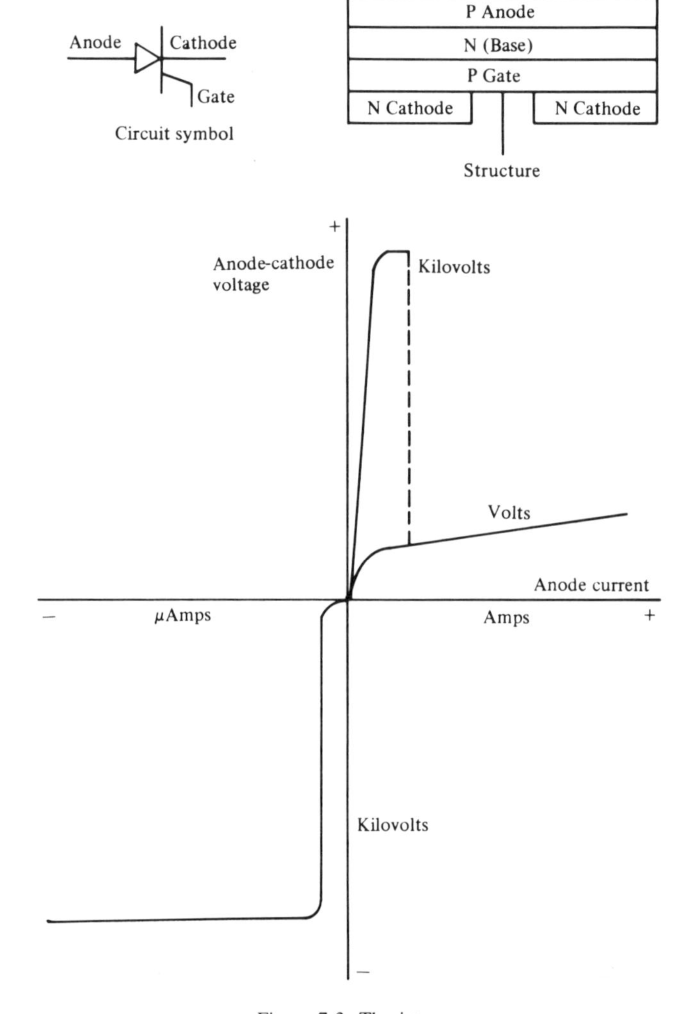

Figure 7-3. Thyristor.

junction is formed so that it has very litte, if any, voltage-blocking capability and plays no part in the blocking behavior of the device.

As the V-I characteristic shows, the thyristor will block voltage in either direction when no signal, or reverse bias, is applied between gate and cathode. Reverse voltage is blocked by the outer (i.e., anode) P-N junction, and the reverse-blocking characteristic is identical to that of a diode save that the saturation leakage is typically a little greater. Reverse-voltage limits are due to the same causes (i.e., bulk avalanche or surface leakage), and thermal considerations prohibit sustained operation in the bulk avalanche region. Forward blocking is provided by the inner forward-blocking N-P junction. The voltage limitation is again set by avalanche or surface leakage, but the device will not stay in forward blocking operation once its limit is exceeded. This is because once a certain forward-current level is exceeded in the gate cathode junction, regenerative two-transistor action begins between the lower P-N-P part of the structure and the upper N-P-N portion. As a result, the device latches into a conducting state in which it behaves much like a forward conducting diode, albeit with a somewhat higher forward drop. Once the thyristor enters this state, it can only be restored to the blocking state by external circuit action reducing the current to zero. Elementary textbooks are fond of saying to some small current called the *holding current*. This is true only as long as the dI_R/dt in the device is extremely low; it is of no practical importance whatsoever, although values for the holding current are given on almost every thyristor data sheet. More significant is that the device cannot initially sustain the conducting state unless a certain minimum anode to cathode ''latching'' current is available from the external circuit. This current is typically an order of magnitude or so greater than the holding current, and is quoted on very few data sheets.

Like diodes, thyristors are made in a wide variety of current and voltage ratings. The largest devices presently are made from 3-in. diameter slices and have current ratings of 1500 to 2500 A. Thyristors are packaged in the same manner as diodes of corresponding ratings, except that their packages make provision for the third connection, the gate lead. Examples are shown in Figure 7-7.

The unique controllable feature of the thyristor comes about because forward current injected into its gate-cathode junction while the device is blocking forward voltage causes it to switch to and maintain its diodelike conducting state. The amount of gate current needed ranges from a few milliamperes for thyristors rated for a few amperes to perhaps 1 to 1.5 A for a device rated at 2000 A and directly gated (many thyristors are indirectly gated, as will be explained shortly). The gate-cathode voltage needed typically ranges from about 0.5 V to about 3 V. Once the device is gated on, however, the gate electrode is unable to exert control over the forward current. Provided the external circuit drives latching current through a thyristor, it will remain in the conducting state until external circuit action forces it to return to the blocking state.

The thyristor has imperfections and limitations pertaining to both turn-on and turnoff. When gate current is first injected, there is a period called *delay time* during which no change of state occurs. This is due to the finite time taken for carriers to be generated and swept across the base region. Once a sufficient carrier concentration acts in the anode junction, the ''anode-voltage fall time'' begins, and the device changes state, becoming conductive after that time is over. However, it is not yet fully turned on. The gate drive and consequent regenerative turn-on action succeeds in bringing only a small area of the device into conduction. Typically, a thin line segment, or multiple segments, along the gate-cathode periphery will be on at this time (if very lucky, all the gate-cathode periphery may be on). This is reflected by a very much higher than normal forward drop, which is a dissipative drop, at any given current.

The subsequent spread of the conducting plasma (charge-storing carrier concentration) across the remainder of the device area is rather slow. Many efforts have been made to measure and characterize the plasma-spreading velocity, the most recent using infrared scanning techniques. Most results indicate a velocity less than 1 mm/usec, so that a 2-in.-diameter device gated at its center takes more than 25 μsec to achieve the fully on condition and exhibit normal forward drop. The spreading velocity is affected by many things, including current density in the area already on at the instant of measurement.

This turn-on spread phenomenon gives rise to the dI/dt limitation of the thyristor. Because the on area is initially small, high current immediately after gating will cause high forward drop, excessive local dissipation, and temperature rise, and local destruction of junction characteristics. Devices are observed to fail in two different, but related, modes when the rate of rise of forward current at turn-on, dI/dt, exceeds safe levels. If the dI/dt is very much greater than that which the thyristor can safely tolerate, immediate destruction of a small area on the gate-cathode periphery occurs. If the dI/dt is only somewhat greater than that tolerable, then a larger area fanning out from the gate-cathode periphery will eventually fail after many repeated operations. This "slow burn" mode is most vexatious, for a device may operate successfully for a considerable time before the failure occurs.

All three turn-on imperfections are found to be highly dependent on gate drive. The higher the amplitude and faster the rise time of the gate drive of a directly gated device, the shorter will be its delay time, the faster will its anode-voltage collapse, and the greater will be the amount of gate-cathode periphery initially turned on. With gate drive currents roughly 3 to 10 times those needed to initiate turn-on, and rise time less than 0.5 μsec, delay times typically 2 to 5 μsec for directly gated thyristors can be shortened to 0.1 to 0.5 μsec. Anode-voltage fall times reduce from 1 to 2 μsec to 0.05 to 0.25 μsec, and the extent of initial turn-on is increased considerably. However, the gate drive does not significantly influence turn on spread. Thus, the only way to reduce turn-on time is to increase the gate-cathode periphery and reduce the distance the plasma must travel by using complex gate-connection patterns of the types shown in Figure 7-4. These have the profound disadvantage of requiring much higher gating currents, and it is very difficult to transmit currents of more than 1 or 2 A with the rise time required across the insulation barrier necessary between the thyristor and the control circuit that provides its gate drive.

This disadvantage is overcome by building indirectly gated thyristors. Variously termed *dynamic gate, accelerated cathode emission,* and *amplifying gate,*

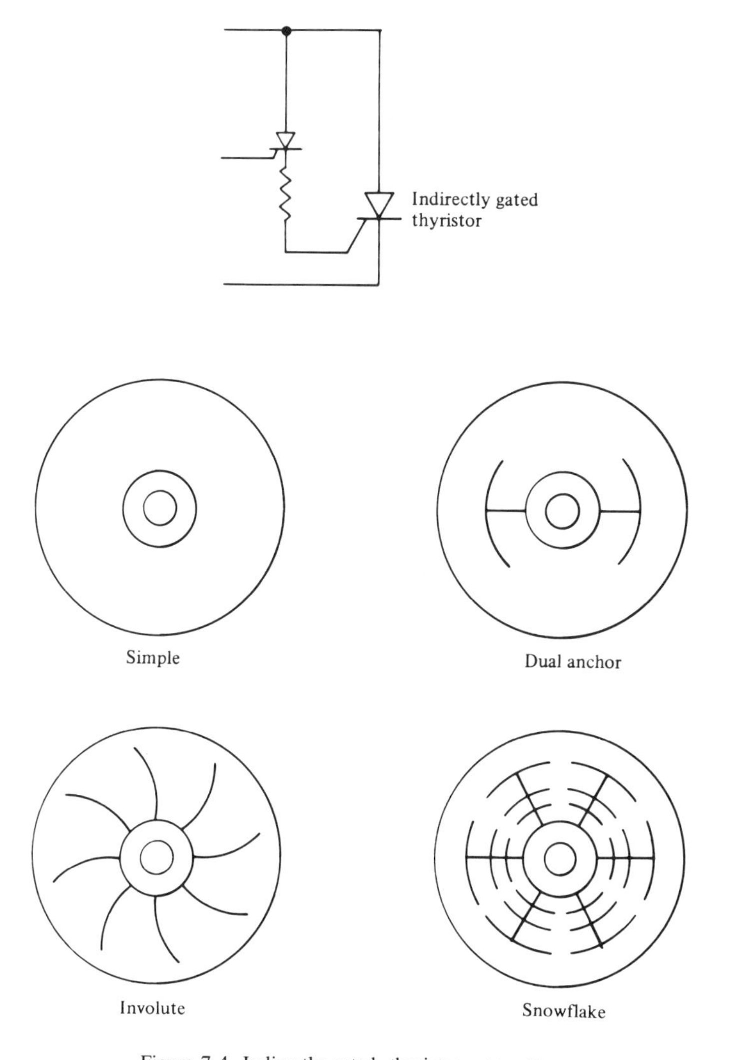

Figure 7-4. Indirectly gated, thyristor-gate patterns.

these devices essentially employ a small pilot thyristor built into the same silicon slice to provide gate current from the snubber circuit or the main power circuit (usually the former). This type of thyristor, depicted in Figure 7-4 (although the conventional symbol is unchanged from that of the directly gated version), needs only a small external gate signal for turn-on. Typically, 150 mA will gate a 2 or 3 in. slice with a complex gate structure. The rise time does not have to be particularly swift either, since the rise time of gate current to the main structure is determined by the anode-voltage fall behavior of the pilot thyristor and the dI/dt capability of the snubber or the external circuit.

Resistance is deliberately built into the anode-cathode path of the pilot device so that it will carry very little current once the main thyristor is on. The main-device gate structure may be a simple circle, for thyristors intended for low-frequency applications, or one of the more complex geometries when higher frequency applications are predicated. Operating frequency is significant because high frequency and high dI/dt invariably go hand in hand, although low-frequency operation is also often accompanied by high dI/dt. High-frequency operation can also lead to a situation where the switch conduction time is so short that even a thyristor with a complex gate structure is never fully on. In fact, infrared studies have shown that if the current, and hence current density, is low enough, a large (2-in.-diameter or more) thyristor may not achieve the fully-on state even when conducting steady direct current.

When a thyristor does not achieve the fully-on state, its forward drop at a given current is inevitably higher than normal. As a result, its dissipation is higher than normal, and its current rating must be reduced in order that the junction temperature limit, typically 125°C, sometimes 150°C, not be exceeded. This means, of course, that the current-carrying capabilities of a thyristor steadily decline as operating frequency is increased. Thyristor manufacturers generally publish curves of permissible maximum peak and average current versus frequency, particularly for devices intended for high-frequency and high dI/dt applications. Such a collection of data is usually for two current waveshapes: (1) half-sinusoidal, common in voltage-sourced ac-to-dc/dc-to-ac converters and (2) trapezoidal, the approximation to rectangular that arises because of restricted dI/dt. Trapezoidal current waves occur in practice in dc-to-dc, current-sourced ac-to-dc/dc-to-ac, and ac-to-ac converters.

Manufacturers also publish dI/dt ratings; for modern devices, these range from 100 to 800 A/μsec. The dI/dt rating of a thyristor is never an absolute limit. The dI/dt that the device will tolerate depends on two major circuit variables, repetition frequency, and the current to which the dI/dt obtains. The manufacturers rating is usually quoted as being to the device average current rating or some small (one and a half to two times) multiple thereof. If turning on to higher currents, the dI/dt must be reduced, but if the current reached at turn-on is lower than that stipulated, then higher dI/dt can be allowed. This is fortunate since, in most practical applications, the R-C snubber circuit connected in shunt with the thyristor will generally provide current at very high dI/dt when the device turns on. This current, or a portion of it, is used as gating current in indirectly gated devices. Empirical evidence shows that most thyristors will tolerate such ''dump'' currents of up to 10% to 15% of their peak current rating at virtually infinite dI/dt.

The use of indirectly gated thyristors greatly improves turn-on behavior, as has been discussed. However, two cautionary notes are in order. Such a device relies on its snubber, or the external circuit, for its gating current in the main power structure. If neither the snubber nor the main circuit can provide the requisite current with a sufficiently short rise time, turn-on will be rather poor and *dI/dt* capability will be much reduced. In practice, this usually translates into a requirement that considerable circuit voltage be present when the device is to be turned on; indirectly gated devices will generally exhibit much degraded turn-on when the forward blocking voltage applied at the instant of turn-on is low. Also, it behooves the designer not to supply excessive gate-triggering current to the pilot structure. Too large an injected gate current may cause spillover into the main structure's gate, with consequently poor turn-on. Although the devices are now constructed in a way that minimizes this possibility, it is still a danger to be considered, and for safety, no more than twice the gate current required to trigger (quoted in manufacturers' data) should be supplied.

One great disadvantage of indirectly gated thyristors is the extension of delay time that can occur. Obviously, such a device will exhibit the cascaded delay times of its pilot and main structures. Typical total delay times can range from 5 to 8 μsec. This is not good even in applications where devices are used individually. Where they are series-connected to achieve high-voltage capability, it can become a major problem, arising not so much because of the absolute value of the delay time but because of the greater spread of delay time from device to device that inevitably accompanies the larger absolute values.

If, in a series-connected string, some devices turn on early, then the existing circuit voltage will be applied to those still not on—the latecomers. It is necessary to prevent the voltage on any device from reaching the forward breakover value, for turn-on by forward breakover is almost invariably over a very restricted area of the device and leads to lowered *dI/dt* capability. Thus, the device snubber networks must, in such a case, restrain the rate of rise of forward voltage across the late devices so that they turn on from gate drive before their forward breakover voltage is reached. This becomes increasingly difficult as the number of devices in series increases and as the devices' delay-time spread increases. Thus, thyristors for series string applications must often be selected to rather tight delay-time tolerance, perhaps allowing a 1-to-2-μsec spread.

Delay-time spread is also a matter of concern when thyristors are to be connected in parallel for current capacity requirements beyond those of a single device. If some devices turn on early, they will collapse the voltage applied to the remainder, making them susceptible to poor turn-on with reduced *dI/dt* capability. Some isolating inductance is generally used (often just the stray bus inductance is enough) to transiently support the voltage for late devices. Again, the larger the number of devices and the greater their delay-time spread, the more acute the problem becomes.

The thyristor also possesses a body of turn-off imperfections. Since in the conducting state it behaves much like a diode, and its gate electrode has no influence on the forward current, only external circuit action can bring that current to zero and prepare the device for resumption of its blocking states. Like the diode, the thyristor stores charge while conducting, and it exhibits the same reverse recovery behavior. Although universally N-base, thyristors tend to be somewhat snappier than N-base diodes, and thus require somewhat larger snubbers to contain transient reverse voltage. Just like the diode, the thyristor recovers to the reverse-blocking state as it snaps in the reverse-recovery process. However, there is a further time delay before the thyristor is capable of blocking forward voltage again.

This arises because, in essence, reverse-recovery current for the anode junction is forward current in the inner forward-blocking junction of the thyristor. Thus, the reverse-recovery current creates stored change in the forward-blocking junction, which is left with free carriers after reverse recovery is completed. If forward voltage is then applied, these carriers create a ''displacement current,'' which acts as a triggering current in the gate-cathode junction, and the thyristor will turn back on.

Unfortunately, neither the external circuit nor the gate can do much to remove these unwanted carriers. Hence, it is necessary to wait until their concentration decays by natural recombination before applying forward voltage. The time taken is also called, to the confusion of many, *recovery time*. It is distinguished from the reverse-recovery time by using the appellations *anode recovery time* and *gate recovery time* and the symbols t_{rr} and t_q, respectively.

Gate recovery time can be shortened by employing the same techniques as are used to shorten anode recovery time—gold doping or electron bombardment—with the same trades as regards blocking voltage and conducting drop. Typical gate recovery times range from 10 μsec for small (few tens of amperes, few hundred volt rating) fast-switch devices through 60 to 80 μsec for the larger fast-switch thyristors to perhaps 300 μsec or more for large high-voltage devices not intended for fast switching applications.

The gate recovery time of a thyristor depends on a number of circuit and device variables. It is, of course, dependent on dI_R/dt, since that influences the peak reverse-recovery current and the resulting carrier concentration in the forward-blocking region. It is, therefore, dependent on device forward-conducting stored charge. It is also slightly dependent on the reverse voltage applied to the thyristor during the period, since reverse voltage exerts a weak influence on the carrier recombination rate. It is most strongly dependent on the rate of reapplication of forward voltage, or dV/dt. This is because the magnitude of the displacement current created by a given carrier concentration is essentially proportional to the voltage rate of change. For many years, this dependence gave rise to the belief that gate recovery time was strongly dependent on reverse voltage. This belief arose because, as will be seen in Chapter 9, circuits that apply much reverse voltage typically exert very moderate reapplied dV/dt

stresses, whereas those that limit the reverse voltage generally result in very severe reapplied dV/dt.

Even if a thyristor has not been conducting for a very long time and has only equilibrium carrier concentrations in the junction regions, there is a dV/dt, the "critical dV/dt," which will produce sufficient displacement current to turn the device on. This critical dV/dt is usually an order of magnitude or more greater then the reapplied dV/dt capability of the device after forward conduction and reverse recovery. Typical manufacturers ratings for the latter range from 20 V/μs for thyristors not intended for fast switching applications to 200 to 400 V/μs for fast-switch units. However, like dI/dt, reapplied (and critical) dV/dt ratings are not absolute limits. A dV/dt rating is only limiting under the conditions of test—forward current level before turnoff, dI_R/dt, and forward voltage to which the dV/dt obtains. In general, higher forward current, higher dI_R/dt, and higher reapplied forward voltage will all reduce the dV/dt withstand capability and perhaps increase the gate recovery time t_q. As might be expected, the converses will enhance dV/dt capability and may reduce t_q.

The enhancement of dV/dt capability as a result of reduction of reapplied voltage is very important, since in many applications (see Chapters 8 and 9) there exists an initial rise to perhaps 10% to 20% of device-blocking rating, which is at a very high dV/dt. This results from an inverse parallel-connected diode or thyristor's reverse-recovery current in circuit inductance acting on the R-C snubber network of the devices. The resulting dV/dt is limited only by stray capacitances, including device capacitances, and may reach values of 1000 V/μs or more.

The reapplied dV/dt capability of modern devices is so much higher than that of earlier thyristors because of the fabrication technique termed *emitter shunting*. This is done by creating a large number of small-area ohmic paths shunting the gate-cathode junction of the thyristor. They are geometrically disposed so as to provide paths for dV/dt-produced displacement current that do not allow it to act as gate-triggering or avalanche-inducing current. Their disposition is also such as to minimize their impact on required gate-triggering current. However, an emitter shunted device does exhibit higher gate-current demand than one not so treated, and this provides further motivation for the use of indirect gating. There is one further disadvantage attached to the use of emitter shunting; it creates regions in the N-base where stored charge is not too well removed by reverse-recovery current, and can lead to high transient reverse-blocking dissipation at device snap-off.

The gate recovery time required by the thyristor leads to a number of difficulties or restrictions in its application. In dc-to-dc converters, it creates a finite limit on maximum duty cycle; since the recovery time is essentially fixed, this limit grows increasingly irksome as operating frequency is increased. In current-sourced ac-to-dc/dc-to-ac converters and in ac-to-ac converters, it restricts approach to the inversion end stop (i.e., the condition $\alpha = -\pi$ rad) and thus limits the maximum output voltage magnitude to a value below that theoretically available. In voltage-sourced ac-to-dc/dc-to-ac converters, it represents "wasted" time. The switch in those converters should be either closed or blocking unidirectional voltage. Any time spent blocking voltage of the opposite polarity, or waiting to apply the unipolar voltage while not conducting current, detracts from the effectiveness of the converter and sets an upper operating frequency limit.

It is obvious that the duties of a snubber network used with a thyristor expand somewhat over those of the same type of network used with a diode. In addition to containing transient reverse voltage excursions, the snubber for a thyristor must restrain reapplied dV/dt, provide gating current for an indirectly gated device, restrain dV/dt and contain transient forward voltage on late devices in a series string, and not produce excessive dump currents at thyristor turn-on. Not surprisingly, all these sometimes conflicting requirements lead to difficulties in the design of snubbers for thyristors (see Section 7.8).

Despite its assorted limitations and imperfections, the thyristor has found extensive application in switching power converters. It is a natural device to apply in ac regulators (Section 5.5) and current-sourced ac-to-dc/dc-to-ac converters operating with phase-delay control. Because it has proven possible to make both high-current and high-voltage thyristors and because it is not yet possible to make other semiconductor switches with similar ratings, the thyristor is also pressed into service in the higher powered (above a few kW rating) versions of all types of switching power converter.

7.4 Thyristor Relatives

The switching limitations of the thyristor have led to many attempts to develop related devices not suffering those deficiencies. The efforts have been spurred both by the desirable aspects of the thyristor (it is much less susceptible to overcurrent and overvoltage damage than the bipolar or field-effect transistors) and by the consistent inability of semiconductor manufacturers to produce other types of switch with similar voltage and current ratings or with similar silicon utilization.

Unfortunately, the efforts to improve thyristorlike device behavior have met with limited success. Such efforts have been largely concentrated on improving turnoff behavior, but whatever success has been achieved has generally been offset by limitations arising from the structural changes made to obtain the benefits sought.

7.4.1 Triac

The *triac* does not represent an attempt to improve thyristor switching characteristics. Instead, it represents recognition of the fact that many applications require fully bilateral switches. The thyristor, of course, has bidirectional voltage-blocking, unidirectional current-carrying capability. To make a fully bilateral switch, two thyristors connected in inverse parallel are needed. Such an arrangement necessitates two heat sinks and two gate-drive circuits; thus, device appliers and developers theorized that if back-to-back thyristors, both controlled by the same gate, could be constructed on a single silicon slice, benefits would be obtained.

The triac is just such a device. It does indeed comprise two inverse parallel-connected thyristors diffused in the same silicon slice. As a result of the structure, the same gate serves to control both even though it is located at the anode junction of one. Moreover, either polarity of gate current will trigger the device on in either direction. However, switching and other behavior suffer in several ways.

First, so much cathode area is lost to both devices that the total rms current-carrying capacity is typically somewhat less than that of the simple thyristor built on the same slice. Second, the reduction in cathode area makes it difficult to construct complex gate structures without unacceptable further loss in current-carrying capacity. As a result, most triacs are directly gated and evidence rather low dI/dt capabilities. Third, reverse-recovery current for either thyristor in the structure is forward current for the other. This, together with the commonality of junction regions and carrier concentrations, tends to make t_q's long and reapplied dV/dt capabilities low. These factors have combined to make the popularity of triacs low except in low-power ac regulator applications. They have not been well suited to general ac-to-ac converter use, nor have high-power versions proven competitive with inverse parallel-connected thyristors for ac regulators.

7.4.2 Gate-Controlled Switch

The gate controlled switch (GCS), or gate turnoff thyristor, has received considerable attention since the middle 1960s. Obviously, the idea of a thyristor structure, in which the gate electrode would retain control over the forward current in order to reduce that current to zero on application of reverse gate current, is attractive. It was observed early in the thyristor era that directly gated devices with complex gate structures exhibited some turnoff gain—i.e., that reverse gate current could reduce and in some cases completely quench the anode-cathode current. This observation led to intensive efforts for over a decade to develop a usable device with this property.

To date, however, such efforts have not been totally successful. It has become clear that significant turnoff gain at high levels of forward current is only possible if the reverse voltage-blocking capability is largely destroyed. The turnoff gain (ratio of forward current to reverse gate current needed to quench it) has remained low (no more than 10 to 20), leading to difficulties in developing adequate gate-drive circuits. Moreover, there always exists a value of forward current, usually well below the peak current-carrying capacity of the slice, at which the gate loses the ability to quench conduction regardless of how much reverse gate drive is supplied. To further compound the gate-controlled switch problems, reapplied dV/dt capabilities have generally proven quite low, and it has been very difficult to fabricate large-area high-current devices, and high-voltage devices.

Thyristor-switching behavior, particularly in regard to t_q, has steadily improved, and power transistors have steadily increased their voltage and current capabilities. Understandably, gate-controlled switches have not made much headway as far as commercial applications are concerned. For the most part, laboratory exploitation that has occurred has been in relatively low-power systems.

7.4.3 Gate-Assisted Turnoff Thyristor

Those working with directly gated thyristors in the early days of the thyristor observed that the presence of reverse gate current during reapplication of forward voltage sometimes enhanced dV/dt withstand capability. This observation, coupled with the lack of complete success in developing gate-controlled switches, has led to the development of a group of devices that use this property: gate-assisted turnoff thyristors (GATT). Fairly good success has been achieved; the reverse-blocking property is generally abandoned, but devices of up to about 1000-V forward blocking and a few hundred ampere ratings have been marketed with t_q's on the order of one-quater to one-half those of conventional thyristors and with reapplied dV/dt capabilities varying from 3 to 10 times those of the corresponding thyristors. The t_q improvements are usually obtained whether or not reverse gate drive is applied, but the dV/dt enhancement is obtained only if reverse gate current flows during the period of forward voltage reapplication. Required reverse gate current levels are typically of the same order as forward gate current required to trigger; since these devices are directly gated, these gate currents are several times larger than those for conventional, indirectly gated thyristors in the larger current ratings. This leads to some difficulties with gate-drive circuits (see Chapter 10).

To date, GATTs have been fairly popular in high-frequency applications ($f > 10$ kHz) where the conventional thyristor's switching behavior either makes operation impossible or distinctly restricts circuit possibilities. However, the restricted ratings range of GATT's has confined them to low-to-medium-power applications, and their application costs have limited them largely to "cost no object" military, airborne, and spaceborne equipment.

7.4.4 Reverse-Conducting Thyristor

Two observations, one device-related and one circuit-related, led to the development of the reverse conducting thyristor (RCT). It has been observed and exploited in the GCS and GATT that killing a thyristor's ability to block reverse voltage enables the device designer to shorten t_q significantly and, to some extent, to improve dV/dt capability. It can also be observed that in a great many dc-to-dc converter and voltage-sourced ac-to-dc/dc-to-ac converter applications, the thyristor has an inverse parallel-connected diode associated with it, and thus reverse-blocking capability is not necessary in those circuits. Also, in the higher powered versions of such circuits, the stray inductance inevitably existing in the thyristor-diode loop can cause problems.

These factors can be exploited—and have been—by constructing the thyristor and inverse parallel-connected diode together on the same silicon slice. Such devices can be made with voltage capabilities as high (several kilovolts) as fast-switching thyristors and can be built on slices of any diameter available for thyristor construction. They have only had limited commercial success, however, because of two inherent problems. One is quite obvious—building the diode on the same slice increases cost and reduces thyristor current-handling ability; application cost for a given thyristor current requirement, generally the dominant current requirement, is higher. Second, reverse-recovery current for the diode has the same character in the slice as forward current in the thyristor. Unless great care is taken to "isolate" the two devices in the slice, carriers can migrate into the thyristor structure during diode reverse recovery and may lengthen t_q and reduce dV/dt capability.

7.4.5 Reverse-Blocking Diode Thyristor

The reverse blocking diode thyristor (RBDT) suffers from a most unfortunate nomenclature. Viewed as a thyristor, it blocks forward, not reverse, voltage; in the conventional thyristor "reverse" direction, it acts like a diode. The RBDT possesses no gate connection, being an adaptation of the earliest four-layer three-junction devices ever made, *Shockley diodes* and *Dynistors*. An RBDT is triggered into the conducting state by bulk avalanche engendered by momentary overvoltage applied while in the blocking state. It is constructed so that the avalanche is uniform, covering virtually the whole device area, and extremely fast. The result is a device with very high dI/dt capability and high-frequency current capability essentially not derated at all compared with its dc capability. However, it tends to have long t_q and rather poor reapplied dV/dt capability.

This device finds use in specialized applications requiring the conduction of very short duration, high-amplitude current pulses at modest repetition rates. (Radar pulse modulators are the classic example.) While some thyristors and GATTs have been used for these applications, the RBDT provides a better match of switch characteristics to application needs. It is not well suited to the more general range of switching power converter applications.

7.4.6 Light-Fired Thyristor or Light-Activated Silicon Switch (LASS)

The association of infrared radiation with the carrier phenomena in silicon semiconductors has been mentioned previously. Two completely different considerations have led to attempts to exploit this association by using radiation to trigger thyristors from the forward blocking to the conducting state. First is the difficulty of making conventional thyristors (directly or indirectly gated, with sufficiently high dI/dt capability, that are sufficiently free from turn-on spread limitations) to serve those applications for which the RBDT was developed. Second is the host of problems associated with providing electrical gate drive to thyristors operating at high voltage, particularly in large series strings. The difficulties of supplying sufficient current at adequate rise time and with proper coincidence in timing across the required insulation barriers are indeed formidable.

Attempts have been made to engage both types of application with thyristors designed to be gated on by infrared radiation rather than by low-level electrical signals. Such light-fired thyristors are often called *light-activated silicon switches* (LASS). The attempts have been plagued by two difficulties. First, for the thyristor to be radiation-triggered, its package must permit radiation ingress yet retain hermeticity to ensure device integrity. Second is the lack of sufficiently reliable and long-lived means for producing the radiation. Nonetheless, light-fired thyristors show promise of achieving some success, particularly in high-voltage series string applications where economic benefits are potentially substantial.

Presently, solid-state (gallium arsenide) laser diodes are used to provide the firing radiation for the LASS in such an application. Isolation is easily provided by the fiber-optic link carrying the radiation from sources at or near ground potential to the thyristors at elevated potentials. It has not yet proven possible to build high-voltage, high-current thyristors with sufficient sensitivity for triggering to be accomplished by the more reliable, longer-lived, and cheaper simple-junction photo-emitting diodes, nor does it seem likely that this will be possible in the near future.

For short-duration pulse-current applications, ruby rod lasers have been used as the firing medium. Here, the aim is not so much to "simulate" gate triggering, but to generate carriers by photon bombardment directly in the forward-blocking junction region and to create a uniformly distributed avalanche-induced turn-on akin to that of the RBDT. Not surprisingly, the cost and technical difficulties associated with this endeavor have distinctly limited its commercial success.

7.5 Transistor

The transistor is the oldest of the controllable silicon switches. In power switching converter applications, it has run a poor second to the thyristor for a number of reasons including its delicacy, limited available ratings, and relatively low power gain.

The attraction of the transistor is that it is a fully controllable switch. Apply sufficient base drive, and it switches from the blocking to the conducting state; remove the base drive and conduction ceases, and the transistor reverts to the forward-blocking state. The conventional circuit symbols for NPN and PNP transistors, diagrammatic structures, and typical static V-I curves are shown in Figure 7-5. When used in switching power converters, only the blocking (zero or reverse base drive) and saturation (or near saturation) regions of the V-I curves are of interest. Of most interest, in fact, is transistor behavior when switching between these states, depicted in Figure 7-6.

Consider first the behavior at turn-on. When base drive is initially applied, virtually nothing happens until t_d, the delay time, has elapsed, just like the thyristor's behavior. Once sufficient carriers have been swept into the collector junction region, the rise time, t_r, begins. Analogous to a thyristor's anode-voltage fall time, this is the finite period over which the transistor changes state. If the rate of rise of current in the circuit can be as high as the transistor's changing impedance allows, very high instantaneous dissipation will exist during this interval. In practice, this is rarely the case, since stray or intentional inductance always present in the circuit loop usually slows current rise so that transistor dissipation remains comparatively low.

If the high dissipation at turn-on does exist, a *forward-biased second breakdown* may ensue. This occurs when excessive temperature in the collector region causes local avalanche phenomena to develop, which cause further temperature increases and degradation or destruction of junction characteristics. Device manufacturers publish *safe operating area* (SOA) curves delineating the collector V-I–time boundaries safe for particular devices with forward base drive.

The transistor does suffer, mildly, from a turn-on spread effect. However, since base-emitter geometries for switching power transistors are invariably complex interdigitated structures and since the plasma-spreading velocity is much

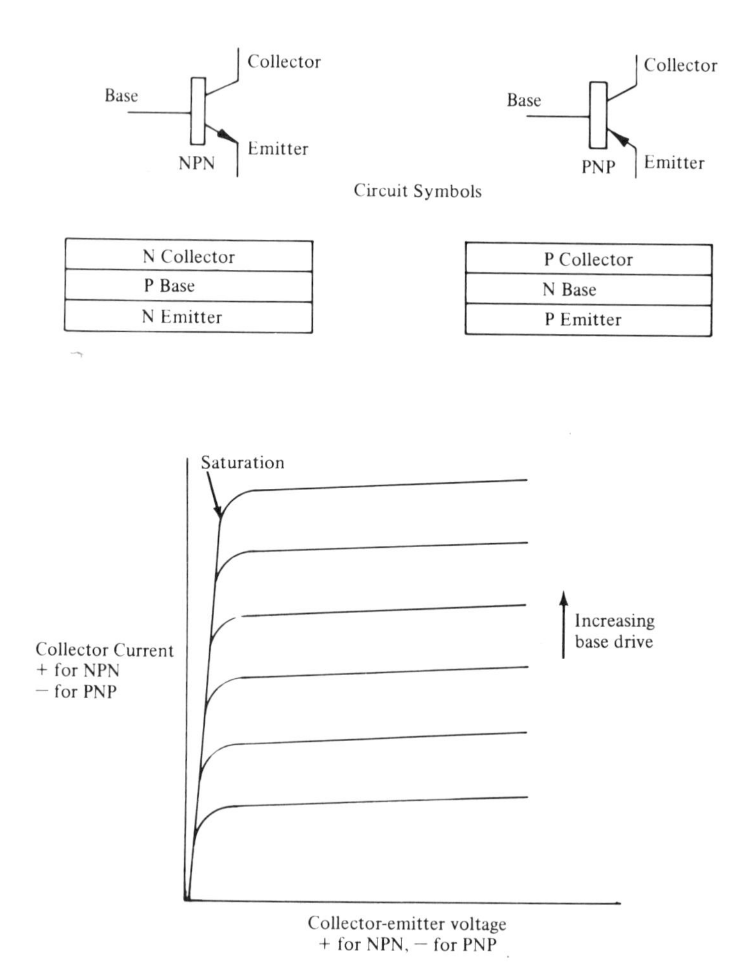

Figure 7-5. Transistor.

greater than for thyristors, the phenomenon is of little concern and is largely ignored.

The behavior at turnoff depends to a considerable extent on conditions existing during conduction. If the transistor is held saturated (i.e., with the base-collector junction forward biased) by excessive base drive, then a great deal of charge is stored and the storage time, t_s, will be long. Much less charge is stored, and much shorter t_s's are achieved, if the conducting transistor is held just out of saturation. This can be done by diverting excess base drive through a "clamping" diode to the collector. The price paid is, of course, higher conducting dissipation. In any event, t_s will be quite long if base drive is simply removed,

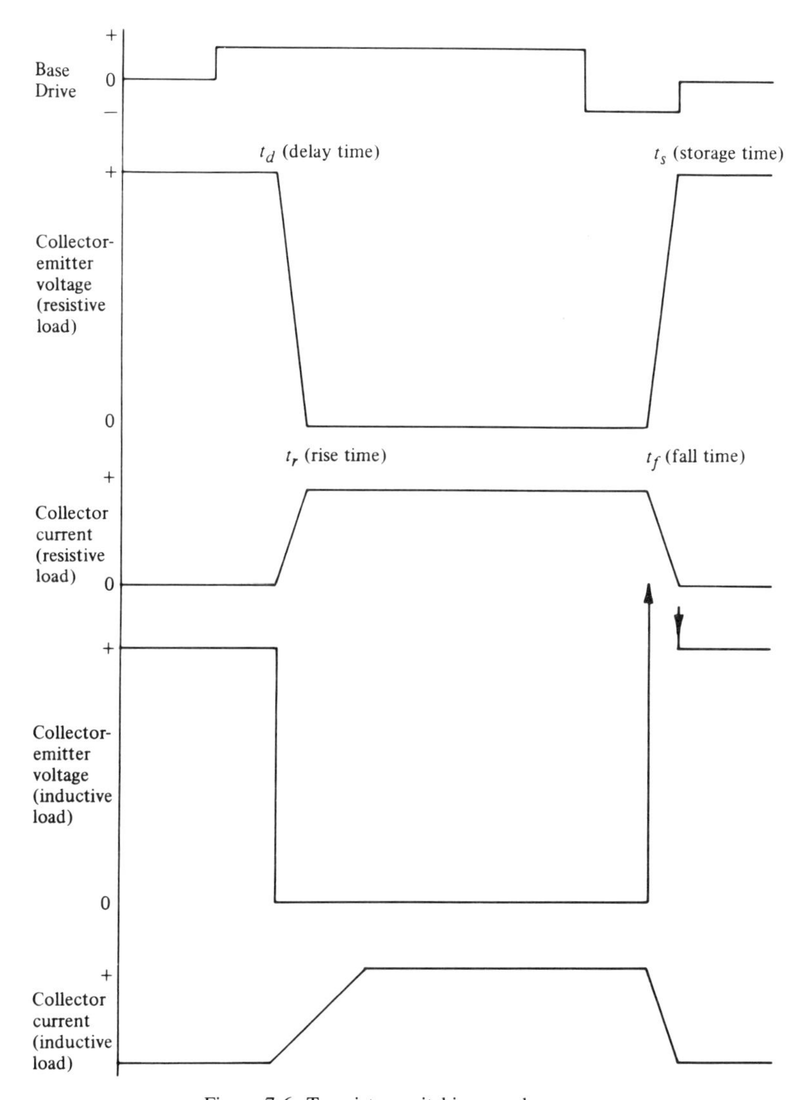

Figure 7-6. Transistor switching synchrograms.

since the stored charge in the base region is then dissipated only by natural carrier recombination. To achieve good switching performance from a transistor, reverse base drive ranging from 50% to 100% of the forward base drive current is required. Care must be taken, however, to terminate this current once the transistor enters its fall time region, for if the base emitter junction is avalanched, device destruction will result.

Once the storage time, t_s, has elapsed, the transistor changes state over a finite period known as the fall time, t_f. Note that the transistor controls the rate of change of current over this period regardless of the behavior of the external circuit. (At least, to a first order; what actually happens is that the transistor changes resistance very rapidly, rather like a snappy diode after reverse recovery.) Thus, the instantaneous dissipation over the fall time can be extremely high. The same circuit inductance that moderates dissipation during t_r will ensure its maximization during t_f, for it will add an LdI/dt contribution to the circuit voltage and develop high collector voltage over the whole of the fall time. *Reverse-biased second breakdown* may ensue; this destructive phenomenon apparently has two possible causes, either excessive temperature, as for the forward-biased case, or direct avalanche injection at the collector due to high-field and high carrier concentration coincidence. Either way, degradation or destruction of junction characteristics occurs.

In most practical applications, reverse-biased second breakdown is a much greater threat than is forward-biased second breakdown because of inevitable, if not deliberate, circuit inductance. Some device manufacturers have begun to recognize this fact, somewhat belatedly, and they now publish appropriate SOA curves. Circuit designers have recognized the fact for some time. In most cases, they have applied a snubber in order to limit the rate of rise and maximum excursion of collector voltage during the fall time, in exactly the same fashion as diode and thyristor reverse voltages are limited at snap-off. Often the transistor snubber is accompanied by a hard maximum voltage clamp—the supply, a zener diode, or a very large capacitor—which is switched in automatically by a diode. In such a case, the diode's turn-on transient may produce disastrous results.

During the fall time, current distribution in a transistor does not remain uniform. In a "reverse of turn-on spread" phenomenon, usually termed *pinch effect,* the current gradually constricts into a small area removed from the base-emitter periphery. This current crowding is a major cause of the high local temperature or carrier concentration that can lead to reverse-bias second breakdown.

These switching imperfections displayed by the transistor create application problems beyond those affecting the transistor itself, the breakdown phenomena due to switching dissipation. Most obvious is that the switching dissipation, even when within SOA, adds to total device dissipation. In high-frequency applications, the switching losses can become the major part of total device losses, acting to restrict transistor current-carrying capability just as dI/dt and turn-on spread effects limit that of the thyristor. Also, unit-to-unit discrepancies in the switching times lead to difficulties with series and parallel connection of transistors, just as delay-time and stored-charge differences do for thyristors. In high-frequency applications, one additional effect of switching-time discrepancies may be felt even when single transistors are employed. In ac-to-dc/dc-to-ac converters, an "unbalance" in the switching times of transistors, particularly delay and storage times, can result in a small unwanted dc component being present in ac dependent quantities. Where the quantity in question is voltage, and is applied to a transformer or reactor, saturation can result—hence, the need for air gaps in the cores of converter magnetic components even when excitation is nominally ac only.

Despite their imperfections, switching power transistors are now quite widely used in converters. They are restricted to low-power applications (up to a few kilowatts) in the main because it has not yet proven possible to make very high voltage or very high current transistors and, to some extent, because the gross $V \times I$ rating of a thyristor fabricated on a given silicon slice can be, typically, 10 or more times that of the transistor made from the same slice. One factor that restrains transistor usage is low power gain—the β (collector-to-required-base-current ratio) of high-current devices typically ranges from 5 to 15. In addition, under circuit fault conditions, when collector current rises, the base current may become inadequate to hold the device in or near saturation, and excessive dissipation can occur, with the resultant device degradation or destruction. The thyristor, being a latching device, is limited only by the thermal capacity of its silicon and surroundings as to the overcurrent it can safely absorb (see Section 7.7).

Transistors are used in dc-to-dc converters and voltage-sourced ac-to-dc/dc-to-ac converters (with inverse parallel-connected diodes to provide bilateral current capability, of course). Being incapable of reverse voltage blocking, they are rarely used in current-sourced ac-to-dc/dc-to-ac converter or in ac-to-ac converters. They are most favored when power levels are low and operating frequencies high. As power levels increase or operating frequencies decrease, thyristors become increasingly competitive and eventually clearly superior, both in cost and efficiency. This situation is not likely ever to change; what happens, on a continuing basis as both types of device undergo evolutionary development, is that the power-level boundary for transistors increases and so does the frequency boundary for thyristors.

7.6 Field-Effect Transistors

Field-effect (MOSFET) transistors, operating by the effect of an electric field on the conductivity of doped (N or P-type) silicon, have been the staple of the communications and data-processing semiconductor industries for quite some time. Until recently, they found no application in power switching converters because of rating limitations and relatively high conducting drops. Of late, this situation has changed, and power MOSFET devices are beginning to compete with bipolar transistors in the very lowest power (a few watts to a few hundred watts at most) converter applications.

Two technology changes have led to the emergence of power MOSFETs. First, the realization that whereas single large-area MOSFETs are extremely difficult to construct, numerous smaller ones may be fabricated on the same silicon slice, or die, and will have characteristics well enough matched for direct parallel connection to achieve the current ratings needed for low-power switching converters. Second, the development of neutron-bombarded silicon, which has vastly superior resistivity uniformity, has allowed the fabrication of higher voltage MOSFETs with reasonable conducting drops. Although still generally worse in this regard than bipolar transistors, MOSFETs can now deliver acceptable performance.

The attraction of the MOSFET lies in its switching behavior—in fact, its superior speed (or bandwidth) is the major reason for its having virtually completely supplanted bipolar devices in signal-processing applications. The MOSFET achieves superior switching performance because it does not store charge in the semiconductor material in the same way as all bipolar devices do. In consequence, charge storage effects are largely absent from its switching behavior. A word of caution is in order, however; the metal-oxide-semiconductor structure from which the MOSFET takes its name is capacitive. Charge must be delivered to this capacitance to render the channel (semiconductor) conductive and must be removed to restore the high resistance state (*blocking* is not really a valid term to use with MOSFETs). In large effective area devices, the charge movement required can become substantial, and the gate drive required may become difficult to produce if very fast switching is sought.

A gate drive current is, of course, merely a transient capacitor charging or discharging current when a MOSFET is used. Its average power gain far exceeds that of bipolar transistors, which is another attractive feature.

Whether further evolutionary development of MOSFETs will spread their range of application in switching power converters remains to be seen. Based on present evidence, it seems likely that they will encroach on bipolar transistor use in the lower power, higher-frequency converter applications. However, they are unlikely to threaten the thyristor in its most entrenched application areas, high-power lower frequency converters.

7.7 Thermal Considerations in Use of Semiconductor Switches

All the semiconductor switches in use are temperature-limited devices. All dissipate power while operating; conducting drop is usually the major contribution to total dissipation, but switching losses can become significant in some cases, as discussed earlier. To maintain safe operation, the heat generated by device dissipation must be carried away from the silicon at a rate sufficient for the maximum permissible junction temperature (in MOSFETs channel temperature) not to be exceeded. The situation is complicated by the fact that the silicon slice on which the device is fabricated is of such small thickness, and hence of such low mass and thermal capacity, that significant temperature fluctuation can take place therein during an operating cycle. Small area devices may evidence significant temperature excursions over times as short as a few microseconds (as can small areas of larger devices; witness dI/dt burnout in thyristors). Even the very largest area devices will exhibit significant temperature fluctuations over periods of a few hundred microseconds to a few milliseconds.

Thermal problems in connection with semiconductor power switching devices are treated as analogous to electrical problems. The concepts of thermal resistance (reciprocal of thermal conductance) and thermal capacitance (mass × specific heat) are introduced and used with power dissipation taken as being analogous to current and temperature rise analogous to voltage in the equivalent thermal (electrical) circuit. This admittedly simplistic view completely ignores the diffusive nature of heat flow. However, for most situations, it is very simple to apply; moreover it yields results that are accurate enough for all engineering purposes. Even heat-transfer specialists are willing to admit that formulating and manipulating the appropriate sets of partial differential equations, which can be solved only by using finite-element analysis techniques on a digital computer, is wasted effort when the requisite results can be so simply achieved by the electrical circuit behavior analogy.

A device's primary electrical and thermal interface with the outside world is its housing or encapsulation. Most silicon devices are first mounted, via hard-solder bonding, on a thin molybdenum disc. This "moly" provides needed mechanical support as well as good electrical and thermal conductivity. Since molybdenum's thermal coefficient of expansion is nearly equal to that of silicon, the moly also relieves cyclic lateral thermal stresses that might fracture the silicon slice if it were directly bonded to copper or some other material with an expansion coefficient very different to that of silicon.

In all packages other than the plastic encapsulations of the very smallest devices, packages that are pictured in Figure 7-7, the next interface is with copper alloyed with beryllium. The beryllium is added for improvement of mechanical properties without too severe a reduction in thermal and electrical conductivity. In the smaller devices and packages, the moly is hard-soldered to the beryllium copper mounting pedestal within the package. In larger units (housing devices greater than about 15 mm in diameter), pressure maintains the contact. In stud mount packages, the pressure is generated by a Belleville washer enclosed in the package; hockey-puck packages need the external pressure created by the clamps used to hold their heat sinks. It cannot be emphasized too strongly that a hockey-puck packaged device cannot be used without adequate external pressure; in fact, many device tests may fail to give proper data or may damage the device if the pressure is not applied.

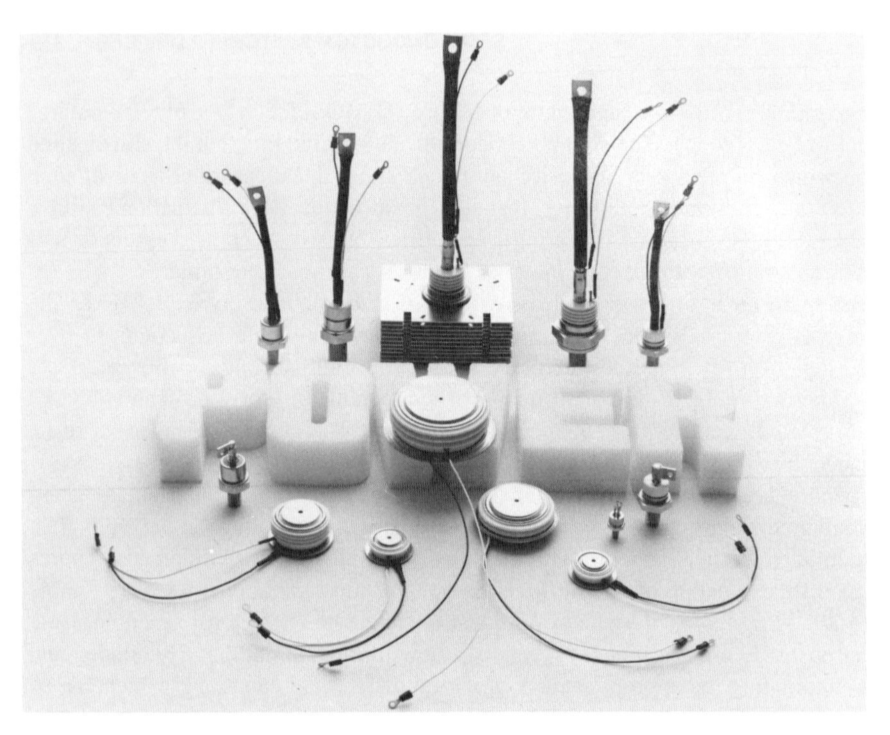

Figure 7-7. Switching-device packages. *Courtesy Westinghouse Electric Corporation Semiconductor Division, Youngwood, Pa.*

The beryllium copper studs or pole faces of these packages provide both means for making electrical contact with the device and means for allowing the heat generated by device losses to flow out toward the ambient surroundings or coolant, which will ultimately absorb that heat. Device manufacturers thermally characterize a given switching device in a given package in two ways. They universally give the thermal resistance, junction to case, of the device. With dimensions of °C/watt, thermal resistance indicates the junction rise to be expected over case-surface temperature for each watt of steady dissipation in the device, and therefore provides the information necessary for calculating "average junction temperature rise" if the average power dissipation is known or can be calculated. Often, the manufacturer will also give a "case-to-sink" thermal resistance figure, to account for the inevitable temperature difference that will arise when heat flows from the package to the device's heat sink through a purely mechanical contact maintained by pressure alone. These two thermal parameters are commonly symbolized by θ_{JC} and θ_{CS}, respectively.

Manufacturers also publish curves of "transient thermal impedance" for their devices. Such a curve is a plot of the junction temperature rise above (usually) heat-sink surface temperature when 1 W is dissipated in the device for varying time periods. Generally, these periods range from about 1 msec to several seconds, at which it becomes asymptotic to the steady-state thermal resistance, junction-to-case plus case-to-sink. The transient thermal impedance, $\theta(t)$, allows a designer to calculate junction temperature fluctuations within an operating cycle and hence to ensure that peak junction temperature remains within specified limits. Methods of calculation, for both average and peak temperatures, are explained by Rice (1970) and Grafham and Hey (1972) and will not be detailed here.[1,2] Note that a digital computer can greatly facilitate the determination of peak temperature excursions when time-varying currents (and hence power dissipations) are involved, since it permits rapid and easy summation of a great many terms in an expression of the form

$$\Delta T = W_1(\theta_{n1} - \theta_1) + W_2(\theta_{n2} - \theta_2) + W_3(\theta_{n3} - \theta_3) \ldots$$

where the W's represent constant power approximations over very short time intervals and the θ's are appropriate transient thermal impedances.

Cooling techniques for the packaged devices convey the heat from the device's case surface to some coolant. Until recently, ambient air was by far the most common coolant. Higher current devices and higher powered converters have now begun to make extensive use of a circulating liquid coolant.

When air cooling is used, it may be either by natural or forced convection. In natural convection, the device is mounted on a heat sink; it may be a flat metal (copper or aluminum) plate, but it is usually an aluminum extrusion with multiple fins. Heat escapes to the surrounding air both by convection and by radiation, and despite the low surface temperatures typical of such sinks (40° to 100°C), the radiative contribution to the heat exchange is usually significant and quite often dominant. This cooling technique is usually limited to devices dissipating no more than 100 to 200 W. Beyond this, the required sink area becomes so large that it becomes uneconomic. In addition, the "spreading resistance" of the heat sink steadily increases as the fin area becomes larger, thus making the sink less efficient. Spreading resistance is the effective average thermal resistance of the heat sink material between the area of device mounting and the fin surfaces exposed to the ambient cooling air. It can only be reduced, for a given fin configuration, by increasing material thickness and thus weight and cost. Examples of typical heat sinks intended for natural (or "free") convection cooling of devices are shown in Figure 7-8.

For devices dissipating from upwards of 100 W to several hundred watts, forced convection is most often used. Fans are used to blow (or suck) air at high velocity over the fins of heat sinks to which the devices are attached (or between which hockey puck devices are mounted). Since convective heat transfer to the highly turbulent airflow always dominates, fin spacing is usually much closer on these sinks than on those for natural convection cooling, as can be seen in Figure 7-8. The web of such a sink, the device-mounting surface, is usually quite thick in order to reduce spreading resistance to the numerous close-spaced fins.

Figure 7-8. Switch packages and heat-sink assemblies. *Courtesy Westinghouse Electric Corporation Semiconductor Division, Youngwood, Pa.*

Grafham and Hey (1972) give numerous nomographs and the basic (approximate) heat-transfer equations governing the behavior of both natural- and forced-convection-cooled heat sinks.[2] Heat-sink manufacturers (including some device manufacturers) publish data on the thermal resistances exhibited by their products. The heat sink is usually ignored in transient temperature calculations; its thermal capacity is typically so large that its surface temperature is essentially invariant for short time periods despite fluctuating device dissipation. If need be, the transient thermal impedance of a sink can be approximated by considering its steady-state thermal resistance and its thermal capacitance (its mass times the specific heat of its material), as a parallel circuit into which the power dissipated by the device flows. However, this approximation will be increasingly optimistic as the time of interest grows shorter, because it does not allow for the heat diffusion in and across the sink. In such cases, consult a specialist in the heat-transfer field or make measurements.

When device losses exceed a few hundred watts, forced-air convection cooling becomes uneconomic because of increases in heat-sink size, cost, and weight, and decreases in heat-sink effectiveness. It is replaced by another form of forced convection cooling, forced circulating-liquid cooling. Although occasionally applied to stud-mount devices, this technique is normally reserved for hockey-puck devices, clamping them between chill blocks as depicted in Figure 7-8. Within these blocks are bored or cast passages to carry the liquid coolant. To achieve low thermal resistance (high-film coefficient), the flow must be highly turbulent (Reynolds numbers in excess of 10,000 are commonly used). The fluid most commonly used is water because of its excellent thermal properties. To avoid losses and other problems, including corrosion, its resistivity must be kept high, generally above 100 kΩ-cm, by using continuous deionizers. On occasion, the water used is process water available at the site of converter application. More commonly, a dedicated closed-loop water system is provided to cool the devices. One further stage of heat exchange is then needed, via a water-to-air heat exchanger or a water-to-water heat exchanger using low-quality process water as the final heat-absorbing medium.

The flow rates needed depend on device size and dissipation and on chill-block internal configuration. Typically, from 2 to 7 gal/min may be used, and such high rates and the corresponding high Reynolds numbers can give rise to erosion problems of the chill blocks, fittings and interconnecting insulating tubes or hoses. Where freezing might be a problem, glycol or methanol is used as antifreeze. Both degrade heat-transfer properties somewhat. In addition, methanol is flammable and glycols corrode copper—the most common chill-block material—if the dissolved oxygen content of the fluid lies in certain (common) ranges.

Recently, there has been a movement to use water-to-Freon heat exchangers as the next stage in circulating-liquid cooling systems. The Freon is the working fluid of a heat pump, permitting both chilling and heating of the water in the primary loop. Heat-pump technology is well-established, low-cost, and highly reliable, and its ability to chill cooling water below the sometimes rather high temperatures of ambient air or process water is a decided benefit to power switching devices.

A variety of other cooling techniques have received some attention, particularly in relation to high-voltage, high-power applications. Oil spray has been used as the cooling medium in airborne applications, where it is the medium for generator cooling. Total oil immersion has been used for HVDC valves, and total immersion in boiling and recondensing Freon has also been employed on rectifier equipment. (The latter technique has found more widespread use with both airborne and spaceborne electronic equipment.) It has been suggested, but not yet attempted, that a two-phase Freon system that uses circulating Freon in the chill blocks as the evaporator of a refrigeration system, might be the most economical and efficient technique of all for high-power converters.

No discussion of semiconductor cooling problems and approaches could be concluded without mentioning overload considerations. In the vast majority of switching power converters, the switching devices must be able to carry short-duration overload currents in the event of load, source, or converter faults. The device cooling system is not usually of much help in such instances. The short duration of overload current, typically at most a few milliseconds, allows no time for heat to diffuse much beyond the device housing. Thus, the increased power dissipation due to the overload current must be absorbed by device and case thermal capacitance without damaging temperatures developing.

In this regard, diodes, thyristors, and thyristor relatives are quite capable. Since their forward conducting drop does not increase rapidly, even for currents much greater than normal, the excess dissipation is limited, and sufficient thermal capacitance is available to absorb it. Moreover, the operation of fault-interrupting devices (fuses, breakers, and sometimes static interrupters) often leaves a situation in which the switching devices are not required to exhibit normal characteristics, particularly voltage-blocking characteristics, for some time after the fault current ends. In such cases, considerably higher than normal operating temperatures are permitted under the fault. Such nonrepetitive fault currents, for which manufacturers publish magnitude-versus-time ratings, sometimes allow junction temperatures in excess of 300°C in thyristors and up to 400°C in diodes. Even repetitive fault-current ratings for thyristors, which predicate reverse but not forward voltage post fault blocking, generally allow junction temperatures somewhat over 175°C to be achieved.

Transistors are not so favorably regarded. Higher than normal collector currents can easily cause much higher than normal conducting drops, since the transistor can pull well out of saturation. Thus, the device's SOA may be exceeded, with destructive results. Since β (the current gain) invariably decreases as collector current increases beyond rated level, it is difficult to prevent this. Transistors are therefore generally regarded as delicate, as compared to thyristors and diodes, because of their susceptibility to overload-induced damage.

7.8 Considerations in Snubber Circuit Design

It has been seen that snubber circuits are required whenever semiconductor switches are applied. The design of a snubber for a specific device in a particular circuit involves the detail parameters peculiar to that situation and is invariably a compromise between conflicting requirements. It is difficult to generalize snubber design procedure. Nonetheless, there are considerations common to all situations, particularly to component selection for snubbers.

The need for and the functions of snubbers are created and defined, not only by switching device imperfections, but also by the inevitable presence of inductance in practical switching converter circuits. As stated previously, the inductance may be totally parasitic. It rarely is, for usually inductance must be introduced to the circuit, as mentioned in Sections 7.2 and 7.3. Since the inductance often plays a part in creating the need for a snubber and is always in circuit, it should not be surprising to find that it exerts a strong influence on snubber design.

There are usually two basic sets of circuit behavior to consider in snubber design. First, how does the circuit, snubber included, behave when the device being "snubbed" changes state from low impedance to high impedance? Second, the converse—what is the behavior when the device changes state from high impedance to low impedance?

In the first case, the snubber is usually attempting to protect the device from the consequences of device imperfections. In the second case, it is more usual to find that the device needs to be protected from the snubber, or, perhaps more accurately, that the snubber must be designed so as not to damage the device that it must protect.

Most snubbers are simple series resistor-capacitor networks connected in shunt with the device(s) they serve. Their function during and after a low-to-high-impedance-state transition of the device is universally to limit both the rate of change and magnitude of the voltage excursion then occurring. Since the circuit loop involved invariably contains potential sources, inductance with initial current and the snubber, the first requirement is that the L-R-C series network formed by circuit inductance and the snubber be heavily damped. The damping provided by the snubber R should approach or exceed critical; otherwise, the network will generate parasitic oscillations that both disturb converter operation and tend to produce larger than normal voltage excursions (overshoots).

When the device undergoes a high-to-low-impedance transition (i.e., turn-on), the snubber capacitor is charged to the circuit voltage blocked by the device prior to turn-on. It will then "dump" that charge by producing discharge current through the resistor and the device. Obviously, the magnitude and rate of rise of the dump current must be within the capabilities of both the resistor and the device. This requirement generally calls for a snubber resistor at least as high in value, but often higher, as that demanded by the damping requirement.

Where it is not possible to effect an appropriate compromise on values for a simple *R-C* snubber, more complex networks may be used. The requirements for a large *R* often give rise to problems with the initial voltage "kick" produced from device reverse-recovery current "trapped" in circuit inductance. These can be avoided by using an appropriately poled diode in shunt with the snubber R, giving rise to the fairly common diode–*R-C* network. The problem then becomes one of underdamping of the snubber-circuit inductance combination. It may be solved by introducing parallel-connected resistance across the snubber capacitor to provide the damping, but it rarely is because of the high losses usually associated with that technique (the resistor so connected will dissipate power all the time the switching device is blocking voltage of the polarity involved). More often the problem is ignored (i.e., the voltage overshoot is tolerated), or a clamping voltage is introduced through another diode to positively limit voltage excursion.

The capacitors used in snubber circuits are subjected to high-amplitude short-duration current pulses. Generally of quite modest capacitance (0.05 to 1.0 μF), extended foil construction is necessary to permit this sort of duty. While dc-type dielectrics are used in many cases, since snubber capacitor potential is often unipolar, high rms and high peak current capability are necessary. Also, to permit the high dI/dt transients, snubber capacitors must possess very low self-inductance and very high self-resonant frequencies. Although some of the capacitors made primarily for filter applications are capable of meeting snubber requirements, manufacturers make special types, with paper oil, plastic film, or polycarbonate dielectric systems, specifically for such usage.

The prime requirement for resistors used in snubber circuits is that they too have very low self-inductance. If not, the snubber will not serve to control voltage rate of rise when the circuit wishes to transfer device reverse-recovery current or forward current to the snubber. A secondary requirement is for high transient-current capability—i.e., high peak-dissipation rating. This virtually eliminates carbon composition resistors from consideration in all but the lowest power converter circuits. Thus, wire-wound resistors are almost universally used. To achieve the low inductance, nonmagnetic forms and conductors are necessary, and either of two special winding configurations are used. A bifilar winding, consisting of two conductors wound together and connected in series (start-finish-finish-start), is sometimes used. However, it is not popular for "noninductive" resistors because the full voltage stress applied to the resistor appears between the two start turns. More common is the Ayrtron-Perry winding, consisting of two overlaid, counterwound conductors connected in parallel. In addition to very low self-inductance, resistors thus produced have much lower self-capacitance than bifilar-wound resistors.

References

1. Rice, L. R., ed. *SCR Designers Handbook.* Second edition. Youngwood, Pennsylvania: Westinghouse Electric Corporation, 1970.
2. Grafham, D. R. and J. C. Hey. (Editors). *SCR Manual* (Fifth Edition). Syracuse, New York: General Electric Company, Semiconductor Products Department, 1972.

Problems

1. A diode with a stored charge of 150 μC is used in a circuit that produces a dI_R/dt of 50 A/μsec. Calculate the peak reverse-recovery current and recovery time if

a. The device is snappy, with instantaneous reduction of reverse-recovery current from peak to zero.
b. The device is soft, and reverse-recovery current decay occurs at a constant rate of 50 A/μsec after the peak is reached.

2. Calculate the peak turn-on and turnoff dissipations of a transistor used to switch 10 A on and off with a 200 V collector supply if $t_r = t_f = 2$ μsec and the collector load is

a. purely resistive
b. highly inductive with collector voltage clamped at 300 V

(For both *a* and *b*, assume linear or instantaneous voltage and current transitions during switching.)

3. If the transistor of Problem 2 is operated at a 1:2 duty cycle (on time = one-half of the cycle time) at

a. 100 Hz
b. 1000 Hz
c. 10,000 Hz

and has a conducting drop of 1.5 V and $t_d = t_s = 5\ \mu$sec, what are the contributions of conducting and switching losses to total device dissipation?

4. Discuss those limitations of the thyristor that most hinder its use in high-frequency converters.

5. Explain why the snubbers for series-connected thyristors are usually larger than those for individually applied devices.

6. State at least two types or combinations of actual semiconductor switches that can be used to realize each of the following functions, and briefly discuss the relative merits of the alternatives.

a. Unidirectional current-carrying, bidirectional voltage-blocking
b. Bidirectional current-carrying, unidirectional voltage-blocking
c. Fully bilateral.

8

Commutation in Switching Power Converters

8.1 Introduction

The phenomenon of commutation is fundamental in the operation of switching power converters. It is not associated with any particular switching device or class of devices, nor does it differ according to switching-converter operating modes. The term *commutation* is borrowed from the field of electrical machines—the behavior during, and requirements for, commutation in a switching power converter exactly parallel those in a rotating machine. A single commutation in either comprises:

1. Transfer of current from one conducting path to the next in sequence
2. Transfer of voltage connection from one terminal (or terminal pair) to the next in sequence
3. Quenching, or turning off, one switch (commutator segment in a machine) and initiating conduction in the next in sequence

In ideal converters with ideal switches, commutations occur freely, instantaneously and without problems. In practical converters with real switches, commutations can give rise to severe problems. Inductance present in real circuits, combined with device imperfections, causes commutations to take place over finite, nonzero time periods, during which converter-dependent quantities may depart significantly from the ideal. In addition, the unwillingness of real switching devices to participate in a commutation must be overcome in many instances. This chapter is concerned with basic commutation behavior; Chapter 9 deals with commutating circuits used to circumvent switching device shortcomings where they present problems.

8.2 Commutation in DC-to-DC Converters

Much of the discussion of commutations in this section applies to all types of converters. The dc-to-dc converters are used as a vehicle for introducing the subject because they are very simple and they require both possible commutations for successful operation. The buck (voltage-sourced) converter of Figure 8-1a is used as the prime example.

Consider first the situation when the diode is carrying the output current I_S and the switch S is open. The next event in the sequence of converter operations is for S to be closed and the diode to cease to conduct. If S is closed, a commutation will take place instantaneously as the source voltage, V_S, reverse-biases the diode and forces the current therein to zero at infinite dI_R/dt. Assuming the

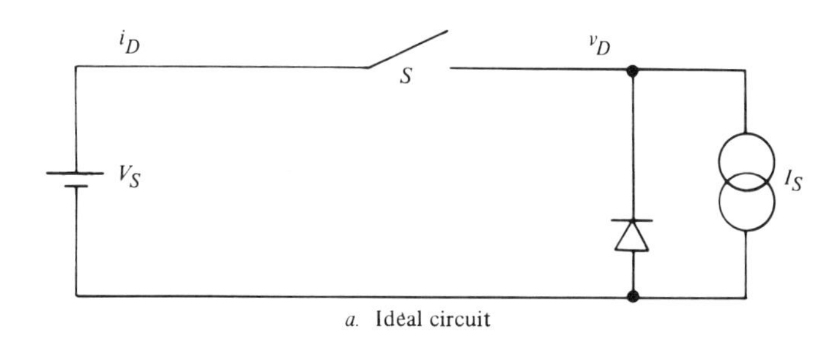

a. Ideal circuit

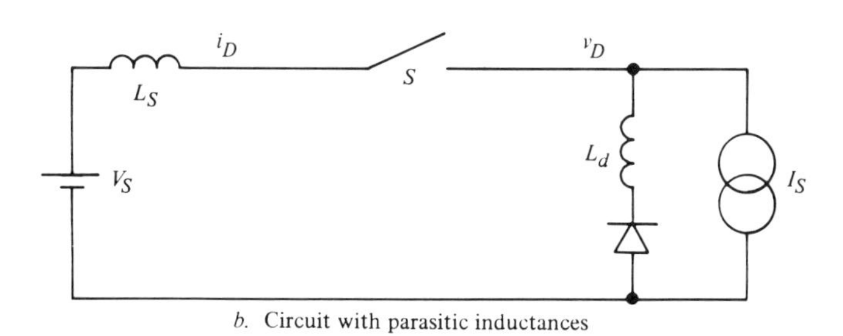

b. Circuit with parasitic inductances

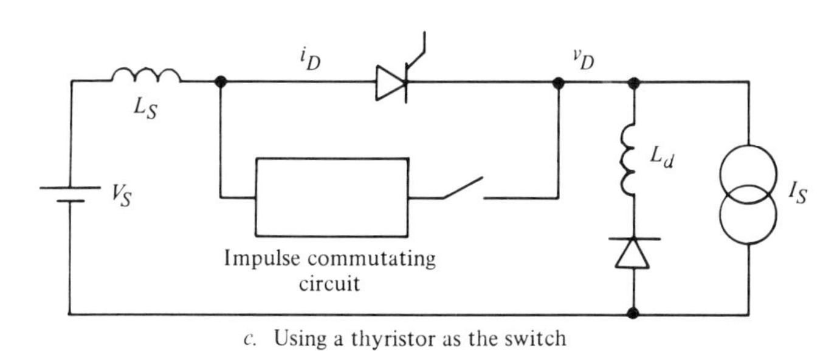

c. Using a thyristor as the switch

Figure 8-1. Buck dc-to-dc converter.

diode to be ideal (i.e., to have no stored charge), the switch S merely picks up the output current I_S at infinite dI/dt.

Now consider the more practical circuit of Figure 8-1b, with the same commutation occurring. L_d and L_s may be purely parasitic inductances, but it is likely that one or both are bulk inductance, introduced intentionally to limit dI/dt. Now when S is closed, V_S is applied to L_s and L_d in series, creating a dI/dt for S and dI_R/dt for the diode of $V_S/(L_s+L_d)$. Note that the output voltage, v_D, will not rise instantaneously to V_S but to $V_SL_d/(L_s+L_d)$, a value it will hold (assuming the diode drop is negligible) until diode snap-off. Observe that at diode snap-off, the peak reverse-recovery current, I_{rr}, of the diode is "trapped" in both L_d and L_s. In being forced to decay by the diode snap, this trapped current will create voltage transients across both L_d and L_s. Since S is closed and can support no voltage, and the voltage source V_S will allow no transients at the input terminals, the transient voltages will appear in the dependent voltage v_D and the diode reverse-blocking voltage. Dependent voltage v_D will jump to $V_S + L_s dI_{rr}/dt$, while the diode reverse voltage will be $V_S + (L_s+L_d)dI_{rr}/dt$ if no snubbers are present. Both potentials decay to V_S as I_{rr} and dI_{rr}/dt decay to zero.

It can thus be seen that the presence of inductance causes the commutation to take a finite, nonzero time and disturbs v_D over that time. The length of time and degree of disturbance increase with increasing inductance, and i_D is clearly also disturbed from the ideal.

This time is called the *commutation overlap time*. Both the old and new current paths, and both the incoming and outgoing switches, conduct simultaneously until current transfer is completed; this phenomenom is called *commutation overlap*. The appellation is used most frequently in connection with current-sourced ac-to-dc/dc-to-ac converters, where the disturbance in dependent quantities can be very marked (see Section 8.3). However, it should be recognized that commutation overlap occurs for all source commutations in all practical converters.

This commutation, in both the ideal and practical cases, occurs automatically when S is closed—no other action is necessary. It does so because the source voltage V_S is of the correct polarity to drive the current in the diode to zero and then "open" the diode by reverse-recovering and reverse-biasing it. In this book, all such commutations are called source commutations. They can occur in all types of converter. In the literature at large, they are called natural, line, or load commutations. All are essentially the same, occurring whenever a voltage source connected to a converter switch automatically forces the desired current transfer, voltage reconnection (to the voltage connected to the closing switch), and switch opening. Consideration of all other dc-to-dc converter circuits—the boost, current- and voltage-source-transfer buck-boost converters—shows that source commutations occur therein whenever the controllable switch is closed while the autocomplementary (diode) switch is conducting. In the boost converter, the defined *load* voltage forces the source commutation; in the current-source-transfer buck-boost, both load and source voltages do so, while in the voltage-source-transfer buck-boost, the transfer-source voltage acts to force the commutation.

Now consider Figure 8-1a while S is closed and conducting I_S. The next event in the converter's cycle of operations is for S to open so that I_S no longer flows in V_S but is instantaneously transferred to the diode. At the same time, of course, v_D will change instantaneously from V_S to zero (neglecting diode-conducting drop). No source potential exists to force this commutation. It occurs because the current source, I_S, forces it to do so when S opens. As soon as S opens, assuming instantaneous action thereof, I_S must transfer to the diode to satisfy KLC, and the source voltage, V_S, will be impressed on the open S with infinite dV/dt. This type of commutation is called an *impulse commutation* herein because, in the majority of practical circuits, it is accomplished with the aid of an artificially produced voltage impulse. In most other literature, such commutations are usually called *forced commutations;* this is an unfortunate choice of terminology—the commutation is no more forced, in this or any other type of converter, than the source commutation previously discussed.

Consider the same event in the more practical converter of Figure 8-1b. It can be seen immediately that instantaneous opening of S will create infinite dI/dt, and hence infinite voltage transients, across both L_d and L_s since KLC still demands that I_s instantaneously transfer to the diode. Having S take a finite time to effect the transition from zero to infinite impedance will obviously make the voltage transient finite, as will placing a snubber network in shunt with S. Note, however, that S must support, transiently, a voltage greater than V_S when it opens. This is the impulse, created by LdI/dt in L_s and L_d, that was referred to in describing the terminology. Obviously, v_D and i_D will be disturbed, and the commutation will take a finite time to complete, being finally considered complete only when the current in L_s is zero, the voltage on S is V_S, the current I_S flows in L_d, and the diode with v_D equal to zero (neglecting diode-conducting drop).

Clearly, impulse commutations occur in all dc-to-dc converters whenever the controllable switch opens and the diode is forced to conduct. In each case, the controllable switch must transiently support a voltage *greater* than the circuit's source potential, which is the ideal forward-blocking voltage impressed upon it.

Now suppose the switching device used to implement the controllable switch function in a dc-to-dc converter is a bipolar transistor, a MOSFET, or a GCS (operating at a current level at which gate turnoff is possible). All requirements for the impulse commutation will then be met when the switching device is turned off by control action provided the device can withstand the transient impulse voltage and the dV/dt at which the snubber in shunt with the device allows the transient to rise. But suppose now that the converter is to operate at such a high current or with such high source voltages that transistors, MOSFETs, and GCSs are not available to meet the demands. Of course, devices could be series-connected, parallel-connected, or both. However, a thyristor would seem a better candidate, except that there is no means for effecting the impulse commutation once a thyristor is turned on.

If the thyristor is to be used, some means must be found to force its anode-cathode current to zero in order for the impulse commutation to occur. Since no such means exist in the basic circuit, something more must be introduced. That something is called an *impulse-commutating circuit* herein. (In the literature, it is sometimes called a *force-commutating circuit*.)

In all converters, all impulse-commutating circuits work in essentially the same fashion. The thyristor current must first be diverted into an alternate path provided by the impulse-commutating circuit, so that the thyristor may regain reverse-blocking and subsequently forward-blocking characteristics. Turnoff of the thyristor *does not and cannot complete the commutation*. That must await the subsequent action of the impulse-commutating circuit.

Whatever the impulse-commutating circuit comprises, it will act only when yet another switch is closed (Figure 8-1c). In any dc-to-dc converter, this switch must be an additional auxiliary device, usually another thyristor, and the result is an auxiliary impulse-commutating circuit. In ac-to-dc/dc-to-ac and ac-to-ac converters, impulse-commutating circuit action is sometimes initiated by the closure of the next converter switch in sequence; it is then termed *auto-impulse commutation*. Thus, impulse commutations, while always involving an impulse-commutating network, take two forms, *auxiliary* and *auto* (i.e., automatic). Source commutations are *always* auto-commutations.

Once the thyristor current is diverted to the path provided by the impulse-commutating circuit, the network therein must act to develop the voltage impulse necessary to effect current transfer to the proper converter path, the diode in a dc-to-dc converter, and complete the commutation. While doing this, it must also arrange for the turnoff of any auxiliary switch or switches used to initiate and further its action. Specific circuit details, operational descriptions, and analyses will be presented in Chapter 9. For now, it will merely be stated that in the majority of cases, the circuits used are simple L-C series-resonant circuits. Pulse-forming networks would generally create a much closer approximation to the V-I characteristics needed, except that the terminating impedances are usually neither well defined nor constant.

The disturbances of dependent quantities occurring during an impulse commutation involving a thyristor are usually far more significant than those when transistors, MOSFETs, or GCSs are used. Also, the finite time taken to complete commutation is generally much longer. With the fully controllable devices, that time is largely a function only of I and dI/dt, i.e., of circuit voltages and circuit inductance. With a thyristor, it must obviously be at the very least equal to t_q. Even if dI_R/dt is allowed to be infinite, t_q must elapse before the impulse-commutating circuit may allow forward-blocking voltage to be reapplied to the thyristor, and the full-circuit voltage must be present thereon for commutation to be complete.

Where impulse commutation is required, a thyristor alone cannot fulfill the switch requirements. An impulse-commutating circuit must be added to enable the thyristor to serve, and thus the thyristor and the circuit together form the switch. This is not a common view in the literature; all too often, the thyristor is called the *switch*, and the commutating circuit is regarded as an appendage to the converter circuit, rather than the part of the switch that it really is. Viewing commutating circuits as integral parts of switches using thyristors and thyristor relatives has two distinct benefits. It permits the relationships between converters and converter circuits to be perceived clearly, and avoids the attempts, all too common in the literature, to classify converter circuits according to the means of commutation and the commutating circuits. Variations in commutation and commutating circuits are purely and simply variations in commutation and commutating circuits. They have little influence on converter performance and no influence on converter function. The classification of commutating circuits, discussed in Chapter 9, has absolutely nothing to do with the general classification of converters discussed in Chapter 1.

The terms *commutation overlap* and *commutation overlap time* (or *angle*) are rarely used in connection with impulse commutations. It should be recognized that the same phenomenon does occur: old and new converter current paths and incoming and outgoing switches conduct simultaneously until current transfer is accomplished. Moreover, the disturbances of dependent quantities during impulse commutations are typically much greater than those during source commutations in any given converter, particularly when thyristors and impulse commutating circuits are used as the switches.

Finally, for both source and impulse commutations, it is the circuit inductance in combination with switch imperfections that gives rise to dependent-quantity and converter operational disturbances. In both cases, the inductances in the old and new converter current paths demand that sufficient volt-seconds be impressed to effect current transfer. In source commutations, the source voltage(s) provide these volt-seconds over the commutation overlap time. In impulse commutations, the volt-seconds are produced by the transient voltage impulse, which may be created either by the action of a fully controllable switch (and its snubber) or by the impulse-commutating circuit used together with a thyristor (or thyristor relative, other than the GCS). In these cases, circuit designers often have a choice to make. They can make the transient voltage impulse large in amplitude, so that the requisite volt-seconds are provided in a short time, or make it small in amplitude and suffer extended commutation overlap. The first choice obviously increases switching-device peak voltage stresses and transient magnitude disturbances in converter dependent voltage(s). The second minimizes switch stress but allows converter dependent-quantity disturbance to persist longer, tending to make high-frequency operation difficult and to alter dependent-quantity spectra in undesirable ways even in low-frequency converters. The proper compromise is often very difficult to achieve; some considerations surrounding the choice are discussed in Chapter 9.

8.3 Commutation in Current-Sourced AC-to-DC/DC-to-AC Converters

Consider now the ideal midpoint current-sourced converter of Figure 8-2a, at the time when S_n is closed and carrying I_S while applying v_n to that current. The next event in the sequence of converter operations will be for S_{n+1} to close and S_n to open. This commutation involves the transfer of the current, I_S, from the path $v_n - S_n$ to the path $v_{n+1} - S_{n+1}$ and the transfer of the dc terminal voltage connection from v_n to v_{n+1}. If the switches are fully controllable, and S_n is

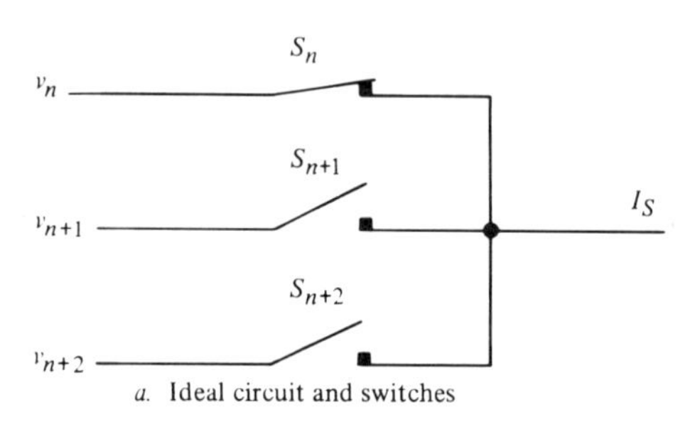

a. Ideal circuit and switches

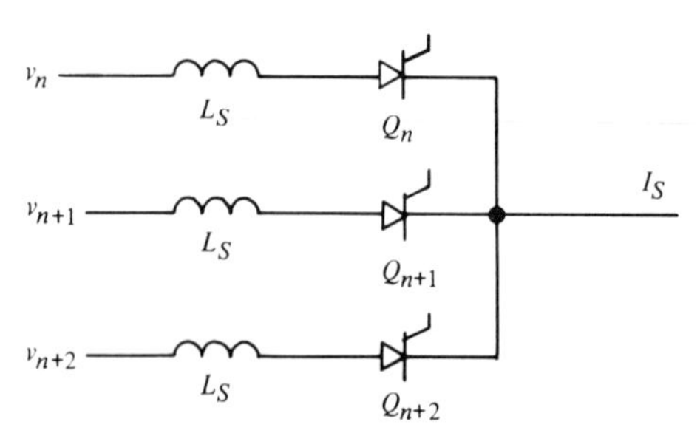

b. Circuit with source inductances and thyristors as switches

Figure 8-2. Current-sourced converter.

opened instantaneously as S_{n+1} is similarly closed, then the commutation will take place instantaneously at that time regardless of whether v_{n+1} is more positive or more negative than v_n. Thus, this commutation, in this ideal converter with fully controllable switches (ideal GCSs, for example), might be thought to have the same nature whether α is negative positive.

This is not the case. If the switches open automatically when the current they carry goes to zero, as most semiconductor switches do, then when α is negative and v_{n+1} is more positive than v_n, the difference between the two source voltages will act to effect the commutation. Thus, when α is negative and v_{n+1} is more positive than v_n, a source commutation occurs. However, if α is positive and v_{n+1} is more negative than v_n, the difference between the incoming (v_{n+1}) and outgoing (v_n) source potentials *opposes* the commutation; an impulse voltage, equal to that difference, must be created by opening S_n for the commutation to take place. Thus, when α is positive, commutations in the converter are impulse commutations.

The same will hold true for a negative group, i.e., when I_S is reversed. Since all current-sourced converters are made up of combinations of midpoint groups, the commutation relationship is valid for all—operation with α negative (phase-delay control) makes all commutations source commutations; operation with α positive (phase advance) requires that all commutations be impulse commutations. If CAM, PWM, or programmed waveform control is used, then both types of commutation will occur. This is so because some commutations will have the positive-type-wave/positive-current or negative-type-wave/negative-current character associated with lagging converter operation, while others will have the positive-type-wave/negative-current and negative-type-wave/positive-current character indicative of leading converter operation.

These situations do not change when source inductance and practical switches are introduced, as depicted in Figure 8-2b. For negative α, source commutations still occur since the difference between incoming and outgoing voltages is of appropriate polarity to provide the volt-seconds necessary to effect commutation. Thyristors can obviously be used to fulfill the switch requirements, as shown; where operation at $\alpha = 0$ is all that is desired, diodes may be substituted, and the simple rectifier, the basis from which most literature describes source (naturally) commutated circuits, emerges. However, if α is positive, then impulse commutation will be necessary. Moreover, the impulse voltage must then exceed the difference in incoming and outgoing source voltages by a sufficient amount and for a sufficient time to provide the requisite volt-seconds. In the worst case, $\alpha = \pi/2$ rad, the impulse voltage must of necessity exceed the peak line voltage of the midpoint group's ac supply.

Clearly, when α is positive and impulse commutation is needed, thyristors will no longer meet the switch requirements. Fully controllable switches, with snubbers, may be used provided that they otherwise meet the needs—unidirectional current-carrying, bidirectional voltage-blocking capabilities. Since transistors definitely do not possess reverse voltage-blocking capability, and GCSs often do not, diodes must be connected in series with them to implement the required switches. Such implementations are rarely found; impulse-commutating circuits are usually used with thyristors to create the switches needed.

The source inductance present in practical current-sourced converters is often substantial. Such converters are frequently fed from isolating transformers having leakage reactances ranging from 0.04 to 0.2 per unit (4% to 20%). In such cases, the commutation-overlap angle can become large, and the disturbances to dependent components very significant. Calculation of the commutation overlap angle for a source commutation is quite simple, for the expressions

$$v_{n+1} = V\cos(\omega_s t - \pi/M)$$

and

$$v_n = V\cos(\omega_s t + \pi/M)$$

can be used to define the incoming and outgoing source voltages, respectively, where M is the number of phases. The difference is

$$v_{n+1} - v_n = 2V\sin(\pi/M)\sin(\omega_s t)$$

and for a perfectly smooth defined current, I_S, flowing at the dc terminal and source inductance L_s in the ac lines, the equation

$$2L_s I_s = (1/\omega_s)\int_{\alpha}^{\alpha+u} (v_{n+1} - v_n)\,d\omega_s t$$

must hold, where α is the delay angle and u the commutation-overlap angle. This is so because sufficient volt-seconds are needed to force the current from I_S to zero in one source inductance while forcing it from zero to I_S in the other. (Reverse recovery is neglected; this is not unreasonable, at least not in low-frequency converters, since dI/dt's are usually very low.) The commutation-overlap angle is thus defined by

$$\cos\alpha - \cos(\alpha + u) = \omega_s L_s I_s / V\sin(\pi/M) \tag{8.1}$$

It can be seen from Equation 8.1 that u has its maximum value when $\alpha = 0$ and grows large again as α approaches π; the minimum commutation overlap angle will be observed for α in the vicinity of $\pi/2$. Also, it can be seen that the existence of commutation overlap prevents the phase-delay-controlled converter, with thyristors as the switches, from reaching the inversion end-stop operating point.

The commutation-overlap angle, defined by Equation 8.1, can be usefully expressed in terms of the per unit, rather than absolute, reactance of the ac voltage source-feeding the converter. The per-unit reactance, x, is defined by

$$x = \omega_s L_s/(V_e/I_e)$$

where V_e is the rms phase voltage, equal to $V\sqrt{2}$, and I_e the rated rms line current. I_e is related to I_S; for a midpoint converter, the relationship is simply $I_e = I_S/\sqrt{M}$, while for a bridge it is $I_e = I_S\sqrt{2/M}$. Substituting in Equation 8.1 yields

$$\cos\alpha - \cos(\alpha+u) = x\sqrt{M/2}/\sin(\pi/M)$$

for a midpoint converter and

$$\cos\alpha - \cos(\alpha+u) = x\sqrt{M}/2\sin(\pi/M)$$

for a bridge. In the important three-phase case, $M = 3$, this latter relationship reduces to

$$\cos\alpha_3 - \cos(\alpha+u)_3 = x$$

Of course, the commutation-overlap affects the converter dependent quantities. Of particular interest is its effect on the wanted component of the dc terminal voltage. This occurs because at the instant of switching, the dependent voltage does not become v_{n+1}, as it would in the ideal case, but assumes the average value of the two voltages, $(v_{n+1} + v_n)/2$, for the duration of the commutation overlap. There is thus an instantaneous loss of dc terminal voltage given by

$$\Delta v_D = (v_{n+1} - v_n)/2$$
$$= V\sin(\pi/M)\sin\omega_s t$$

The volt-seconds lost to v_D are given by the integral of Δv_D over the limits α to $\alpha+u$, and are thus

$$\Delta VS_D = L_S I_S$$

from the equation used in deriving Equation 8.1. In the midpoint converter, there are M commutations per cycle, giving a total volt-second loss per cycle of

$$\Delta VS_{D1} = ML_S I_s$$

and the average voltage loss is thus

$$\Delta V_{DD} = f_s ML_S I_s$$

so that the wanted component of the dc terminal voltage with a commutation overlap angle u becomes

$$\begin{aligned} V_{DD,u} &= V_{DD,0} - f_s ML_S I_s \\ &= (M/\pi)\sin(\pi/M)\cos\alpha - f_s ML_S I_s \end{aligned}$$

The form of this equation shows that insofar as the average dc terminal voltage is concerned, the commutation overlap creates an effective source resistance of $f_s ML_S$; some care is needed in using this concept, however, for it is a "phantom" resistance—*no* power dissipation is involved.

Substituting for $L_S I_s$ from Equation 8.1 and putting $f_s = \omega_s/2\pi$ yields

$$V_{DD,u} = (M/\pi)\sin(\pi/M)[\cos\alpha + \cos(\alpha + u)]/2 \tag{8.2}$$

Equation 8.2 carries the apparent implication that the displacement factor of a converter with commutation overlap angle u will be $[\cos\alpha + \cos(\alpha+u)]/2$ rather than $\cos\alpha$. Detailed analysis of the dependent current shows this to be almost so. Substitution in Equation 8.2 of the expression for a three-phase bridge u in terms of x, the source per unit reactance, yields

$$V_{DD,u,3} = (3\sqrt{3}V/\pi)(\cos\alpha - x/2)$$

Note that the change in V_{DD} is negative whether the converter is operating in the rectification or inversion quadrant. Thus, in the rectification quadrant, the effect of commutation overlap is to reduce the magnitude of the wanted component in the dc terminal voltage, but in the inversion quadrant the effect is to increase that magnitude.

As discussed earlier, whenever impulse commutation is necessary in a current-sourced ac-to-dc/dc-to-ac converter, the commutating-impulse voltage must exceed the difference between the source voltages over the commutating period.

It follows that the critical design point for impulse commutation will be when α (phase advance) is in the vicinity of $\pi/2$ rad, when the voltage difference is greatest. Thus, for impulse commutation the maximum commutation overlap with a given design will occur for α close to $\pi/2$, and commutation overlap will decrease as α tends to zero or π. This is the reciprocal of the behavior observed when α is a phase delay and the converter is source-commutated. Moreover, it is easily deduced that the commutation-impulse contribution to the dependent voltage of a positive group is always positive. Hence, when impulse commutation is used, the magnitude of the wanted component in the dc terminal voltage is increased in the rectification quadrant and decreased in the inversion quadrant—reciprocal behavior once more. Therefore, converters intended to be complementary cannot be truly so once commutation overlap exists. Phase-delayed and phase-advanced converters, the former source-commutated and the latter impulse-commutated, will have different average dc terminal voltages if operated at the same absolute α's. Hence, to be parallel-connected (with reactor isolation to support ripple voltage difference) at their dc terminals, they must be operated at slightly different α's, and their quadrature current demands will not be equal.

In the rectification quadrant, the α of the impulse-commutated converter will need to be greater than the absolute value of α for the source-commutated converter, resulting in a net leading quadrature-current. In the inversion quadrant, the converse applies, and a net lagging quadrature-current demand will arise.

Analysis of the impact of commutation on the unwanted components in converter dependent quantities is very tedious. Computer aid is needed to effect numerical solutions of the transcendental equations that arise, there being no closed-form solutions available. Some general observations on the results of such analyses, and on the qualitative deductions that can be made by observing the effects of commutation on dependent-quantity waves, are all that will be given here.

Observing the dependent-voltage waves, it is deduced that for source-commutated converters, the volt-second content of the ripple voltage is increased for α in the vicinity of zero or π but decreased for α in the vicinity of $\pi/2$. Analysis of the spectra reveals that the amplitude of components whose frequencies are odd harmonics of Pf_s (including Pf_s, the ripple fundamental), with P being the converter pulse number, are generally increased, whereas those of components with frequencies that are even harmonics of Pf_s are generally decreased. For impulse-commutated converters, ripple volt-second and unwanted component magnitude effects are highly dependent on the specific impulse commutating-circuit design in use. There is a general tendency toward behavior reciprocal to that evidenced by source-commutated converters, as might be expected.

It is evident from the dependent-current waves of source-commutated converters that commutation overlap reduces their displacement factors, i.e., increases their lagging quadrature-current demand. Analysis reveals that the total harmonic distortion also tends to be increased, but selectively. Unwanted components with frequencies $(2k-1)Pf_s \pm 1$ generally have their amplitudes increased; those with frequencies $2kPf_s \pm 1$ generally have decreased amplitudes. In impulse-commutated converters, the effects are again dependent on impulse-commutating circuit implementations. However, examination of dependent-current waves suggests, and analyses confirm, that the displacement factors are increased so that leading quadrature-current supply is reduced, and the effects on unwanted component magnitudes again tend to be reciprocal to those in source-commutated converters.

Before leaving the subject of commutation in current-sourced ac-to-dc/dc-to-ac converters, it is well to mention the possible existence of "overlapping overlap." In a source-commutated converter, the overlap angle u defined by Equation 8.1 clearly increases as L_s increases. If L_s, or x, is large enough, then u may equal or exceed π/M. Once this happens, the analytic treatment given here is no longer valid. Analysis becomes much more laborious; Kimbark reports that one effect is to drastically increase the "phantom" output resistance of the converter.[1] Obviously, a similar phenomenon can arise in impulse-commutated converters. In bridge configurations, as Kimbark discusses, overlapping-overlap effects first develop when u exceeds $\pi/2M$, for the commutations of each of the complementary groups forming the bridge then influence those of the other.

8.4 Commutation in Voltage-Sourced AC-to-DC-/DC-to-AC Converters

It was noted in Chapter 4 that with the exception of the special two-phase converter depending on coupled ac current sources, all voltage-sourced ac-to-dc/dc-to-ac converters consist of a number of identical "poles" together forming a bridge configuration. Since all the poles in a converter operate in identical fashion save for phase displacement matching that of their ac current sources, it suffices to study commutation in one pole alone, or a "half-bridge" of the single-phase case.

An ideal version of such a pole is represented in Figure 8-3a. The bidirectional current-carrying, unidirectional voltage-blocking switches required are implemented by connecting ideal diodes in inverse parallel with fully controllable ideal switches. Consider now the case when α, which is not a control variable but the converter's operating phase angle with respect to its load (or source) is lagging (see Section 4). Then the defined ac current leads the converter's dependent ac voltage. If the upper switch is conducting, then on approaching the

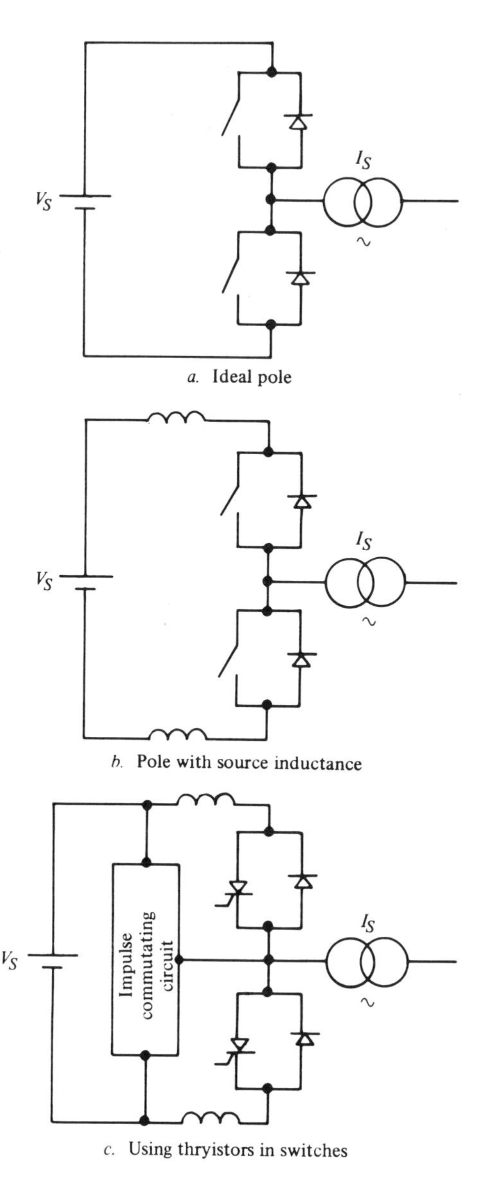

Figure 8-3. Voltage-sourced converter pole.

end of a positive half-cycle of the dependent voltage, the current will be flowing via the diode of the upper switch, and the fully controllable element thereof will be open.

The next event in the sequence of pole (converter) operations is for the upper switch to open and the lower to close. The commutation should transfer the defined ac current from the upper to the lower switch, and transfer the pole dependent-voltage connection from the positive to the negative supply terminal. This commutation will occur as a source commutation when the controllable element of the lower switch is closed, for the dc source voltage will then act to instantaneously reduce the upper-switch-diode current to zero and open the upper switch by reverse-biasing the diode.

The introduction of inductance (Figure 8-3b) does not change this stituation. Of course, it does make for a finite dI_R/dt in the diode and dI/dt in the controllable element, and hence creates commutation overlap. During this overlap, the pole's dependent voltage will be at the average of the positive and negative dc source terminal potentials (presuming the upper and lower source inductances to be equal) and hence is at zero potential with reference to the ac neutral of the converter's dependent ac voltage set (see Section 4-4).

It is quite clear that, given a leading load on the pole, the same type of commutation occurs at the end of a negative half-cycle of the dependent voltage. The current, having once more reversed, will then flow in the lower switch diode, and a source commutation will occur when the controllable element of the upper switch is closed. Hence, as long as a voltage-sourced ac-to-dc/dc-to-ac converter is operated with a lagging displacement factor, or presented with a leading load, all commutations will be source commutations effected by the action of the defined dc source voltage. In such cases, thyristors can as well fulfill the requirements of the controllable switch elements as fully controllable switching devices. In fact, the thyristor possesses a redundant coapability—reverse voltage blocking.

The inductances, both parasitic and intentionally used for dI/dt limiting, in the dc source loops pf voltage-sourced converters are usually much smaller than those attributable to the ac voltages feeding current-sourced converters. As a result, commutation overlaps are generally much shorter and are usually ignored by designers. Many do not even know they exist, for the actual use of source commutation in voltage-sourced converters is rare. This is because it is very difficult to guarantee that the load will be leading, or the converter's operation will be lagging, over the whole range of operating conditions faced in an application.

When the load is lagging, or the converter operates with a leading displacement factor, impulse commutation is necessary. This is evident if the commutation first considered is reconsidered for this condition. Now the current still flows via the controllable element of the upper switch at the time when the commutation must take place, and so that element must open, to create a voltage impulse, when the lower switch is required to conduct. Of course, the diode of the lower switch will conduct in this case. This commutation is identical to that occurring in the buck dc-to-dc converter when its controllable switch is opened.

The controllable element of the lower switch will not have to be closed until the defined ac current reverses. Most practical controls, however, close it at the instant of commutation, since it is not generally easy to predict when current reversal will occur.

As in all cases discussed so far, the voltage impulse required is greater than the source potential when inductance is present, and for the same reason. The impulse must provide the volt-seconds required to effect current transfer in the circuit inductances, and again it may be either of large magnitude and short duration or lower magnitude and longer duration, at the designer's choice.

Being fully controllable, transistors, GCSs, or MOSFETs can be used to implement the controllable elements of voltage-sourced converter switches provided proper snubbers are used to control impulse-voltage magnitudes and rates of rise. Thyristors can only be used in converters with lagging loads if impulse-commutating circuits are added, as depicted in Figure 8-3c. As seen in Chapter 9, the inverse parallel-connected diode then often plays a role in the commutating circuit as well as in the main converter circuit. As a result, in many circuits, it no longer appears directly in parallel with the thyristor, but rather is isolated by inductance, which is part of the impulse-commutating circuit. This has caused many to erroneously believe that the diode is part of the commutating circuit and not part of the converter's switch. It is both, but its primary function is always to provide the reverse current path needed by the converter switch. Involvement in the impulse commutation circuit is incidental.

Before closing this discussion, two important points will be reasserted. First, inductance, parasitic or otherwise, in the ac load or source does not directly influence commutation in voltage-sourced ac-to-dc/dc-to-ac converters. This is because such inductance is not present in the circuit loops between which commutations take place. Second, it is the *dc voltage source* that effects source commutations when they are possible and acts to prevent them when they are not. Even when a voltage-sourced converter is connected, via an appropriate current-sourcing interphase, to an ac voltage set, those voltages play no part in and have no direct influence on the converter's commutations.

8.5 Commutation in AC-to-AC Converters

The strong relationship between ac-to-ac converters (other than the ac regulators) and current-sourced ac-to-dc/dc-to-ac converters suggests that commutation processes should also be very similar. This is the case; source commutations are effected by the defined ac source voltages whenever conditions permit, and impulse commutations must be used whenever they do not. Clearly, whenever the wave type/current product at the converter's defined current terminal associated with the commutation is positive, source commutations will occur. Whenever that product is negative, impulse commutations are needed.

In other words, these converters also obey the general rule linking commutation and converter displacement factor, which has been seen to apply to both current- and voltage-sourced ac-to-dc/dc-to-ac converters. However, they obey it on the basis of their instantaneous displacement factors. Whenever the instantaneous displacement factor that an ac-to-ac converter presents to its defined ac voltage source is lagging, source commutation takes place. Whenever that instantaneous displacement factor is leading, impulse commutation is necessary. Since an ac-to-ac converter generally exhibits both leading and lagging instantaneous displacement factors in its cycle of operations, both types of commutation are usually found in the converter's sequence.

It is worthwhile to explore this point a little further, for the commutation requirements of the various ac-to-ac converters with different loadings can readily be deduced. Consider first the UFC. In Chapter 5, it was seen to develop a positive-type dependent-voltage wave whenever the slope of that voltage's wanted component was positive, and a negative-type wave whenever that slope was negative. Thus, with unity power factor loading, equal time periods existed, in an output cycle, of positive wave type/current product and negative wave type/current product. Hence, one-half of the commutations would be source commutations and one-half impulse commutations. If the load is made lagging, then a preponderance of negative wave type/current product develops, and thus impulse commutations will outnumber source commutations. For a purely reactive lagging load, the wave type/current product is always negative, and then all commutations must be of the impulse variety. Obviously, leading loads produce complementary results—source commutations will outnumber impulse commutations, and with a pure reactive leading load, all commutations can be source commutations.

Since the SSFC is the complement of the UFC, it exhibits complementary behavior in regard to commutation too. With resistive (or regenerative) loading, equal numbers of source and impulse commuations develop. With lagging loads, a preponderance of source commutations develops, and a purely lagging load gives rise to a wholly source commutated converter. Leading loads produce a preponderance of negative wave type/current product, and thus a requirement for a majority of impulse commutations. A purely reactive leading load results in a totally impulse-commutated SSFC.

The two complementary converters operated with the nonlinear modulating function—the positive and negative converters—always produce positive-type and negative-type dependent-voltage waves. Hence, they always have equal numbers of source and impulse commutations regardless of the displacement factor presented at their defined current terminals. Therefore, so does the UDFFC. However, the CDFFC, created by forming a consecutive composite dependent-voltage wave, will have a preponderance of source or impulse commutations according to whether its input displacement factor is made lagging or leading.

It was seen that when the switching angle of the consecutive composite wave was made equal to the load phase angle, the input displacement factor was always lagging, regardless of lead displacement factor. This is because the wave type/current product is then always positive; as a result, all commutations are source commutations. Hence, the name *naturally commutated cycloconverter* (NCC) applied to a converter operated in this fashion.

The defined currents for ac-to-ac converters are, of course, time-varying. Since they vary sinusoidally at the converters' output terminals, the commutation overlaps vary from commutation to commutation. Pelly gives an excellent qualitative discussion of the effects on dependent quantities in the NCC.[2] Gyugyi develops the analysis, using existence functions, for all types of ac-to-ac converters; a complex and rather tedious operation, it will not be pursued here.[3]

8.6 Converters Without Commutations

There exists a class of converters that do not have true commutations at all. (The ac regulators of Section 5.5 are the prime examples.) These circuits all use thyristors, or close relatives of thyristors, as their switching devices. The current driven through the thyristor by the external circuit has a natural zero, or attempted reversal, after some period. Of course, this causes the thyristor to reverse-recover and reassume its reverse-blocking state. The circuit must also provide reverse voltage for a time at least equal to t_q; the thyristor will then regain its forward-blocking capability and be in the "off" state until the next controlled turn-on.

There is no current transfer associated with this behavior, only a cessation of current flow. There is no transfer of voltage connection either, only a circuit interruption. No other switch closes, just one switch opens. Thus, the events in these converters are not commutations. Historically, they have been confused with source commutations because in the case of the ac regulators, the same agencies—ac voltage sources—are responsible for their occurrence. However, there is a distinct difference. In a true source commutation, two ac voltages, upon the closing of a switch, immediately act so as to force the commutation to occur by applying volt-seconds to the sources' own inductances. In the case of *natural quenching,* as it will be termed here, only one voltage is involved, and it produces switch turn-off because the current it causes to flow in any load network inevitably has a zero at some time following switch closure and current establishment.

As stated, the ac regulators are the prime example of this type of converter. Even though they can use phase-delay control and their thyristors are quenched by the (indirect) action of the ac supply voltages, *they are not related to phase-delay-controlled ac-to-dc/dc-to-ac converters* in any way, except that both use thyristors.

Currents with natural zeros can also be produced by the action of dc voltage sources on underdamped series-resonant circuits. In fact, this is the basis for most impulse-commutating circuit designs. Converters using this technique have achieved some popularity because it is possible to use thyristors at much higher frequencies than is the case with impulse commutation. This is because the thyristor's t_q, which is not useful time in an impulse-commutated converter and ultimately so detracts from performance that the converter is virtually useless, becomes merely a part of the normal operating cycle in circuits using natural quenching.

Because of an apparent relationship to voltage-sourced impulse-commutated converters, these converters have been dubbed "resonantly commutated" by many. They have no direct relationship to the true voltage-sourced converter other than that they use thyristors as switches and are fed by dc voltage sources. The function of the resonant circuit(s) used in such a converter is most definitely not to commutate—it is to shape the current wave subsequent to thyristor turn-on so that a natural current zero will occur.

The first such converter ever to be used is depicted in Figure 8-4. It is called the *series inverter* in most of the literature, and appears in all of the four circuits shown. These circuits all operate in much the same fashion. Those of Figures 8-4a and b differ only in that the latter impresses a dc potential (one-half of the dc supply potential) on its pair of tuning capacitors in addition to the ac voltage developed. The circuits of Figures 8-4c and d also impress dc potentials (one-half the dc supply voltage) on their capacitors; they are simply ac Thévenin equivalents to 8-4b. All operate by simply gating the thyristors in sequence at or below the resonant frequency of their series-tuned circuits; these circuits are formed by the internal tuned inductors and capacitors and the *resistive* loads. Using Laplace transform methods to formulate and solve the differential equations established after switch closure (thyristor turn-on), analysis is quite simple. Using the circuit of Figure 8-4a, let C, the tuning capacitor, be charged to some potential E_o and thyristor $Q2$ be off when thyristor $Q1$ is turned on. Then the equation

$$E_S/2s = E_o/s + \bar{i}\,(sL+1/sC+R)$$

will define the Laplace transform, $\bar{i}$, of the current i which flows in $Q1$, $L1$, C, and R following $Q1$'s turn-on. This transposes to

$$\bar{i} = [(E_s/2-E_o)/L]\{1/[(s+\alpha)^2+\omega^2]\}$$

where $\alpha = R/2L$ ($L = L1 = L2$) and $\alpha^2+\omega^2 = 1/LC$. Now the Laplace transform of the capacitor voltage is

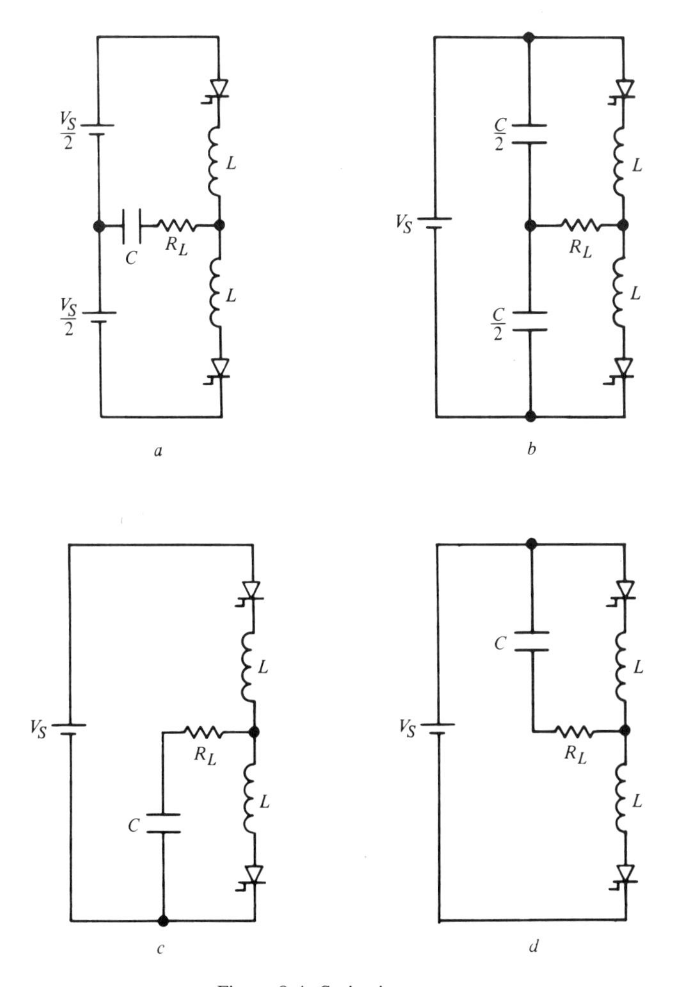

Figure 8-4. Series inverter.

$$\bar{e}_c = E_o/s+\bar{i}/sC$$

$$= E_o/s+(E_s/2-E_o)\{1/s-(s+2\alpha)/[(s+\alpha)^2+\omega^2]\}$$

which inverts to give

$$e_c = E_o+(E_s/2-E_o)(1-\epsilon^{-\alpha t}\cos\omega t-(\alpha/\omega)\epsilon^{-\alpha t}\sin\omega t)$$

The transform for the current inverts to give

$$i = [(E_s/2 - E_o)/\omega L]\epsilon^{-\alpha t}\sin\omega t$$

This current clearly goes to zero when $\omega t = \pi$, at which time $Q1$ will turn off. At that time, the capacitor voltage is given by

$$e_{c\pi} = E_o + (E_s/2 - E_o)(1 + \epsilon^{-\alpha\pi/\omega})$$

If the converter is in steady-state operation, the relationship

$$e_{c\pi} = -E_o$$

must hold, yielding

$$E_o = -E_s(1 + \epsilon^{-\alpha\pi/\omega})/[2(1 - \epsilon^{-\alpha\pi/\omega})]$$

as the peak negative voltage on the capacitor. The peak positive voltage is, of course, equal in magnitude. The half-sinusoidal current pulse is of magnitude

$$I = E_s\epsilon^{-\alpha t}/[(1 - \epsilon^{-\alpha\pi/\omega})\omega L]$$

Now observe that $\alpha t = Rt/2L = R\omega t/2\omega L = (1/2Q)\omega t$ where Q is the quality factor, $\omega L/R$, for the tuned circuit (neglecting losses in C, $L1$, $L2$, and the thyristors). The peak current, I_p, can be shown to occur at $t = \arctan(\omega/\alpha)$ or $\arctan\,(2Q)$

$$I_p = E_s\epsilon^{-\arctan(2Q)/2Q}/[(1 - \epsilon^{-\pi/2Q})\omega L]$$

Assuming $Q2$ is turned on immediately after $Q1$ turns off, the power delivered to R is given by

$$W_R = \{E_s^2/[Q^2R(1 - \epsilon^{-\pi/2Q})^2]\}(1/\pi)\int_0^{\pi} \epsilon^{-\omega t/Q}\sin^2\omega t\; d\omega t$$

$$= \{E_s^2/[2\pi QR(1 - e^{-\pi/2Q})^2]\}[4Q^2/(1 + 4Q^2)](1 - e^{-\pi/Q})$$

To see that turning $Q2$ on immediately after $Q1$ turns off (i.e., operating at the resonant frequency $f = \omega/2\pi$) is possible, consider that the potential on C as $Q1$ turns off is given by

$$e_{c\pi} = E_s(1 + \epsilon^{-\pi/2Q})/[2(1 - \epsilon^{-\pi/2Q})]$$

and therefore the reverse bias applied to $Q1$ at this instant is $E_{c\pi} - E_s/2 = E_s\epsilon^{-\pi/2Q}/(1 - \epsilon^{-\pi/2Q})$. This reverse-bias voltage will decline to zero as the capacitor voltage follows the equation given with signs reversed:

$$e_c(Q2 \text{ on}) = (E_o - E_s/2)[1 - \epsilon^{-\alpha t}\cos\omega t - (\alpha/\omega)\epsilon^{-\alpha t}\sin\omega t]$$

The reverse bias is $e_c - E_s/2$; for $Q = \infty$ (no power transfer), the reverse-bias time extends to $\omega t = \pi/2$ (one quarter of a cycle). For the finite Q's for which the circuits are practical, in the range 1.5 to 4, the alloted ωt_q is found to range from 0.36π to 0.44π. For thyristors with t_q's of 15 μsec, operation at 12 kHz is possible—which is far in excess of the maximum frequency of operation using standard impulse-commutation techniques.

If operation is at the resonant frequency, then the current waveform in (and hence voltage waveform across) the resistive load is a low-distortion content near sinusoid. Power control can be achieved by controlling E_s, keeping the current and voltage waves of the load near sinusoidal, or by reducing operating frequency. In the latter case, the load's current and voltage distortion rapidly increase. Under such control, the capacitor will retain its voltage from the time one thyristor turns off until the other is turned on. The power at some operating frequency f_o, where $f_o < f$, is

$$W_{f_o} = (f/f_o)W_f$$

while, where dc voltage controls the power at some supply voltage E_o, $E_o < E_s$, the power is

$$W_{E_o} = (E_o/E_s)^2 W_{E_s}$$

Operation much above the resonant frequency, f, is not possible. This is because a rapidly rising shoot-through current $(E_s/2\omega L)\omega t$ develops when $Q1$ and $Q2$ are both on. For a very small increase of f_o above f, this cuts recovery-time allotment substantially by delaying current zero in the thyristor attempting to quench. The end result is "commutation failure" with $Q1$ and $Q2$ both conducting and a large (fuse or breaker cleared) fault current occurs.

Loads outside the design Q range present a problem for these series inverters. The highest permissible Q is a function of the thyristor voltage rating relative to E_s, for the denominator in the voltage expressions, $1 - \epsilon^{-\pi/2Q}$, clearly decreases quite rapidly as Q is increased. The lowest Q that can be tolerated (highest ohmic value of R) is determined by the reduction in t_q and ultimately by approach to critical damping ($Q = 0.5$) and consequent failure of the current to exhibit natural zeros. The power delivered for a given peak switch voltage has a maximum at $Q = 1.33$, but no more than 2% drop in power is observed with designs ranging from $Q = 1.07$ to $Q = 1.67$.

A nonresistive load also presents a problem to these inverters. Any reactive component in the load, intentional or parasitic, modifies the tuned circuit parameters and changes both the resonant frequency and the Q. If the load is fixed (for example, an antenna for a VLF transmitter), its reactive contributions can be "designed in" to the converter's tuned circuit. Variable reactive loading is difficult to contend with.

These series inverters of Figure 8-4 (and the full bridge versions thereof, which are sometimes employed) are clearly single-quadrant dc to single-phase-ac converters. Polyphase output versions are rarely used, but can be made quite simply by using multiple poles, or half-bridges, as they are with conventional voltage-sourced converters. Since perfect balance between three tuned circuit-load combinations is almost impossible to maintain, retention of the neutral connection (to the dc source center tap) is highly desirable.

A rather curious use of such a polyphase scheme, which has found some application, is shown in Figure 8-5. It was mentioned previously that reducing operating frequency below resonance extends t_q for the thyristors. If the t_q allotted at $f_o = f$ is kT, where $T = 1/f$, then the t_q at $f_0 = f/(2n-1)$, where n is integer, is $(n+k)T$. Thus, by allowing $2n-1$ inverters to feed a common load, as depicted in Figure 8-5, operating each at $f_o = f/(2n-1)$ and appropriately phase-staggered, the load in question can be fed a continuous sinusoid at the resonant frequency. This permits operation at much higher frequencies—five stages with 15 μsec devices could, using $k = 0.18$, deliver power at 145 kHz. However, the economic penalties are severe; a multiplicity of thyristors and resonant-circuit components are involved, and the thyristors suffer severe derating. This is because they conduct short-duration current pulses (half-sinusoids

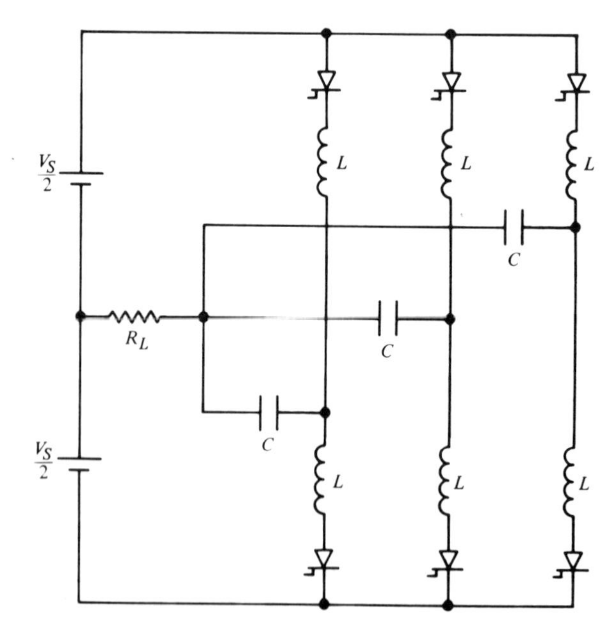

Figure 8-5. Time-sharing ("Gatling gun") series inverter for higher output frequency.

of duration corresponding to half-cycle of the load frequency) at low-duty cycle. Therefore, simple transistor converters have tended to displace this "Gatling gun" version of the series inverter.

Before leaving the discussion of simple series inverters, it should be observed that the half-bridge versions of Figures 8-4a and b do in fact form the basis for most modern impulse commutating circuits. These are discussed in Chapter 9, but it can be mentioned here that many of the ills of such commutating circuits have their roots in the same basic behavioral patterns that so restrict the series inverter.

The extremely restricted capabilities of the series inverter, as regards load tolerance and single-quadrant operation only, led to considerable effort to develop improved versions in the early days of thyristor converter research. It is perhaps ironic that the most successful of these (Figure 8-6) was discovered by accident after it had almost been given up for a lost cause.* Simple series inverters are occasionally parallel-loaded, i.e., their loads are connected in parallel with their tuning capacitors (bridge or true half-bridge versions only) rather than in series. The idea of adding diodes in inverse parallel with the thyristors, and thus covering the converter to a two-quadrant voltage-sourced converter (of sorts), arose because of the success that had come in 1961 to Wagner's parallel inverter through the same modification.[4] However, no matter what manipulations were initially performed, the series inverter thus modified refused to run—it did nothing but indulge in repeated commutation faults. Then, an error resulted in operating frequency being set to about 2/3 resonant frequency when it had

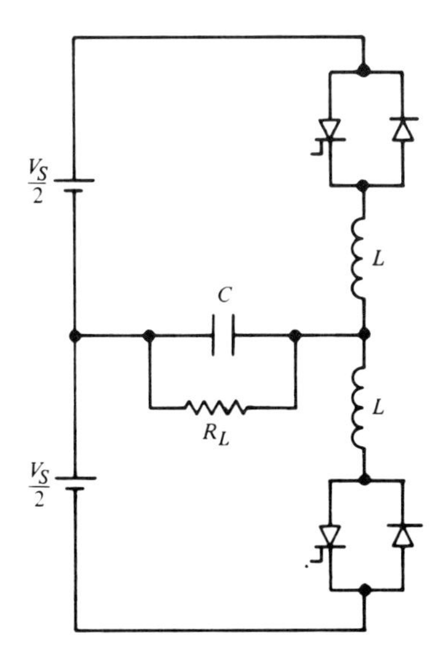

Figure 8-6. "Resonant feedback" inverter.

* The author was personally associated with the work.

previously been set at or close to resonance. The results were both startling and impressive. From an intractable inoperative beast, the converter was suddenly transformed into a useful and extremely versatile performer.

This deceptively simple circuit is remarkably difficult to analyze. Unlike the series inverter from which it is derived, there are no closed-form solutions available for operating parameters. Computer analysis or simulation are the only ways to achieve comprehensive performance data, other than by experimental investigation: Mapham must be credited with the first public revelation of a comprehensive analysis.[5] Its performance may be summarized as follows.

The fully loaded $Q(R_L\sqrt{C/L})$ should be in the vicinity of 3, and the optimum f_o/f ratio is about 0.65. The converter will then act as a dc-to-ac converter generating a near sinusoidal voltage with less than 5% total harmonic distortion, across C and R_L. The rms value of this voltage is close to 0.7 times the dc source voltage at full load and rises by no more than 10% (with a constant dc source voltage) as R_L is increased to ∞, i.e., for open-circuit operation. Decreasing R_L much below that corresponding to a Q of 2 will cause the circuit to fail in the "shoot through" condition. Reactive loading, for X_{CL} or X_{LL} values from ∞ down to the full-load-design value of R_L, has little effect. The reactive load component merely shifts the f_o/f ratio slightly, causing some slight variations in output-voltage magnitude and total harmonic distortion. The addition of series capacitance feeding the load will make the converter operable from open- to short-circuit load conditions at the expense of some increase in load-voltage regulation and distortion, provided that the f_o/f ratio with short-circuit loading does not exceed about 0.92. A word of caution is in order when this technique is used—reactive loads that resonate with the series capacitor at the operating frequency can create low enough operating Q's for failure to occur. Of special interest is the fact that the t_q afforded the thyristors is about $1/3f_o$ at full load, enabling quite high-frequency operation—22 kHz with 15 μsec devices.

Inevitably, severe penalties are attached to the achievement of such sterling performance. The volt-ampere rating of the tuning capacitor is about three times that of the load, and each tuning reactor has a volt-ampere rating about 1.8 times that of the load. The thyristors see currents that are close to half-sinusoids at the *resonant* frequency with maximum amplitudes about three times the peak load current, while the diodes, at no load, see similar currents with peak amplitude 2 to 2.5 times the load current. Device voltages are also considerably magnified (two times or more) above dc supply voltage. Also, because of the direct inverse parallel connection of the diodes and the relatively large series inductance (compared to many impulse-commutating circuits), reapplied dV/dt for the thyristors is severe and voltage overshoot (due to diode reverse-recovery current) is difficult to contain.

Despite these disadvantages, the circuit has enjoyed considerable popularity in medium-to-high-power applications in the moderately high frequency range (3 to 30 kHz). Its popularity has declined as transistors of adequate size to serve the applications have become available.

Finally, the naturally quenched dc-to-dc converter of Figure 8-7 must be mentioned. Because it permits relatively high-frequency operation with a thyristor as its switching element, it has achieved fair popularity where power levels are beyond the range of available transistors and where size and weight are key considerations. Of course, these latter push for high-frequency operation because of the reduction achieved in interfacing component size. However, polyphase transistor converters using interphase reactors (see Chapter 3) can be very competitive, because the resonant circuit components of the converters of Figure 8-7 are by no means negligible additions.

The analysis of this circuit, using Laplace transform techniques as for the series inverter, is quite straightforward. Assuming that a smooth defined current I_S flows from the capacitor to the load (via an interface consisting of at least a current-sourcing inductor), if the potential on the capacitor when the thyristor is turned on is postulated to be E_{co}, then the defining equation for the current pulse is

$$E_s/s = (sL+1/sC)\bar{i}+E_{co}/s-I_S/s^2C$$

or

$$\bar{i} = [(E_s-E_{co})/L]/(s^2+\omega^2) + I_S[1/s-s/(s^2+\omega^2)]$$

where $\omega^2 = 1/LC$ and the resonant circuit, thyristor included, is presumed to have negligible loss resistance.

Inverting produces

$$i = [(E_s-E_{co})/\omega L]\sin\omega t+I_S(1-\cos\omega t)$$

which clearly does not again reach zero until ωt is somewhat greater than π. The equation

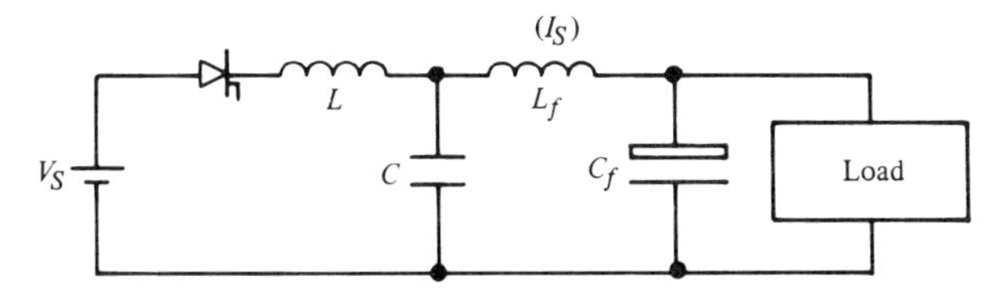

Figure 8-7. "Resonantly commutated" dc-to-dc converter.

$$[(E_s - E_{co})/\omega L I_S]\sin\omega t - \cos\omega t = -1$$

is of the form

$$(1/y)\sin\theta - \cos\theta = k$$

and such equations are easily given closed-form solutions by putting $\theta = a+b$ where $a = \arcsin(y/\sqrt{1+y^2}) = \arccos(1/\sqrt{1+y^2})$. This substitution yields

$$b = \arcsin(ky/\sqrt{1+y^2})$$

Now the capacitor voltage during thyristor conduction is defined by

$$\begin{aligned}\bar{e}_c &= E_{co}/s + \bar{i}/sC - I_S/s^2C \\ &= E_{co}/s + (E_s - E_{co})[1/s - s/(s^2+\omega^2)] - I_S/s^2C\end{aligned}$$

which inverts to

$$e_c = E_{co} + (E_s - E_{co})(1-\cos\omega t) - I_S\,\omega t/\omega C$$

Substituting the appropriate value of ωt, which is $a+b$ from the foregoing, gives the value of e_c when the thyristor turns off. During the "off" period, the capacitor voltage declines by the contribution $-I_s\omega t/\omega C$ and must be equal to E_{co} when the thyristor once again turns on. A simple iterative computer program will converge on the appropriate value of E_{co} for any given "off" period, and the mean (wanted) value of e_c and thyristor recovery time allotment are readily enough determined from the expressions once the value of E_{co} is established.

Most literature refers to the circuit in Figure 8-7 a *resonantly commutated chopper*. *Chopper* is the inelegant terminology widely used for dc-to-dc converters, particularly the buck converter; while the resonant circuit is used to cause thyristor current to quench naturally, no true commutation occurs. Like the ac regulators and the series inverter and its relatives, this dc-to-dc converter has no commutations.

8.7 Commutation Failures in Converters

It has been assumed here, in every instance, that by whatever means commutation was attempted, it was always successful. In practical converters, this is not always the case, and commutation failures take place. As the term implies, a commutation failure occurs when the desired current transfer, voltage reconnection, and switch opening and closing are not accomplished. The consequences vary widely, ranging from minor and temporary perturbations of converter operation to catastrophic faults needing fuse or breaker clearing to protect converter devices, sources, and loads from damage.

There are three possible causes of converter commutation failure: converter control malfeasance; converter switching device failure; and the failure of the converter to afford its thyristors, when used, adequate recovery time or reapplied dV/dt control. Since the results are generally the same no matter what the cause, the treatment here will focus mainly on the effects of a converter switch failing to open in the normal sequence of operations.

Consider first the buck dc-to-dc converter. If the controllable switch thereof, transistor or thyristor, does not turn off when it should, then the dependent voltage remains at V_S when it should become zero. If this is merely a transitory switch malfunction, and proper operation is restored at the next attempted turn-off, only a very minor perturbation of v_D results. The longer the condition persists, the greater the disturbance of v_D; if the condition is permanent, then v_D and V_{DD} both become V_S. This is not usually tolerable, and it must be cleared by some means. Depending on the load's sensitivity to voltage, a "crowbar" switch may be closed at the converter output terminals to force v_D to zero, thus creating a truly catastrophic fuse-clearing fault to disconnect V_S. Alternatively, a contactor may simply be opened to disconnect V_S before v_D adversely affects the converter's load. It should be noted that commutation failure in a buck converter rarely threatens damage to the converter's switching devices. This is not true of all the dc-to-dc converters; their detail behavior differs and depends considerably on interfacing.

The boost converter, when its controllable switch fails to open, remains in the state with v_D equal to zero when it should be V_S. Now the realization of the input current source, I_S, is almost universally from a real dc voltage source through an interfacing inductor. Hence, I_S will rise, and if the commutation failure persists, it will threaten the faulted switch unless a fuse or contactor quite rapidly opens to clear the fault. The buck-boost converters both exhibit behavior similar to the boost—a rapidly rising source current—since their controllable switches also carry defined currents that in practice, are created by inductor approximations. Fault currents then rise with slope V_S/L amperes per second when the actual source voltage is continuously impressed on the interfacing inductor L.

Current-sourced ac-to-dc/dc-to-ac converters exhibit behavior that depends on their mode of operation. Consider first a source-commutated converter operating in the rectification quadrant. If a supposedly outgoing thyristor fails to turn off, then the supposedly incoming thyristor cannot succeed in turning on (except in the event of complete thyristor failure). The result is that the dc terminal remains connected to an ac supply voltage that is going toward the inversion region of converter operation. Should the condition be temporary, and cleared at the next similar commutation attempt, there is merely a minor temporary reduction in both dc terminal voltage and current. If the condition is permanent, the dc terminal voltage becomes ac, and thus the dc current output will reduce to zero. As in the case of a buck dc-to-dc converter, although operation must be terminated by some protective means, converter devices are not generally threatened.

If the converter is operating in the inversion region, however, the commutation fault is more serious. The ac voltage to which the dc terminal remains connected when commutation fails now takes the converter into the rectification region. This causes the current in the sourcing inductor to increase rapidly because converter terminal voltage and actual dc source voltage are then additive in their effects on the inductor current. The failure can be ''ridden out'' (i.e., cleared at the next commutation attempt) at the expense of a usually considerable transient increase in the dc current. Attenatively, it must be interrupted by some protective device. This is not an easy task, for the interrupting agency must absorb all the energy stored in the current-sourcing inductor. For a source-commutated converter, it should be observed that a commutation fault in the inversion region *initially* results in a *reduction* in the dc current. This is because the voltage wave, which should have been outgoing, actually carries on all the way through inversion end-stop before finally reversing slope and taking operation back toward, and finally into, the rectification quadrant. Thus, the occurrence of a commutation failure can be sensed immediately because the slope of the dc ripple current does not change sign when commutation is attempted; if the commutation is successful, the slope will change sign.

In impulse-commutated current-sourced converters, commutation faults may be due to either ac source-voltage disturbances or commutating circuit inadequacy. If the fault occurs in the rectification region, there is a short-lived transient increase in the dc current, as the voltage wave progresses to rectification end-stop, followed by a decrease as that wave carries the converter into the lagging inversion quadrant. If sustained, the fault results in permanent connection of a single ac phase to the dc terminals, just as in the source-commutated converter. If the fault occurs in the inversion region, then the voltage wave that remains connected carries the converter rapidly into the rectification region, causing an even more rapid rise in dc terminal current than occurs in the source-commutated converter. With the impulse-commutated converter, there is no ''early warning signal'' for the fault in the inversion region—the nonreversal of ripple slope occurs now for a fault occurring in the rectification quadrant.

Commutations faults in ac-to-ac converters are identical to those in the current-sourced ac-to-dc/dc-to-ac converters from which the ac-to-ac converters are derived.

Commutation faults in voltage-sourced ac-to-dc/dc-to-ac converters can be truly catastrophic. Any commutation fault results in the switches of a converter pole producing a near short circuit across the defined dc source voltage. Since the inductance in the path is usually very low, the rate of rise of fault current is usually very high. There is no way of ''riding through'' such a condition—external interruption must be invoked. In the vast majority of cases, high-speed current-limiting fuses are used to avoid damage to the converter's active devices. Recently, some use has been made of impulse-commutated dc interrupters—in essence, buck dc converters used for one commutation only—so that repetitive, automatically resettable fault-clearance can be obtained. However, the design of such interrupters is difficult and they are not overly reliable. Also, they are usually unable to prevent the energy from the converter's dc sourcing capacitor from dumping into the switches involved in the failure, and that is often sufficient to destroy them.

References

1. Kimbark, E. W. *Direct Current Transmission*. Volume I, pp. 80–105. New York: John Wiley & Sons, 1971.
2. Pelly, B. R. *Phase Controlled Converters and Cycloconverters*. Chapter 13. New York: John Wiley & Sons, 1971.
3. Gyugyi, L. *Generalized Theory of Static Frequency Changers*. Ph.D. thesis, University of Salford, Salford, England, 1970.
4. McMurray, W. and D. P. Shattuck. A silicon controlled rectifier inverter with improved commutation. *AIEE Transactions* **80**(1), 1961.
5. Mapham, N. An SCR inverter with good regulation and sine wave output. *IEEE/IGA Proceedings* (April–May 1967).

Problems

1. Explain the differences between source and impulse commutations.

2. The relationship of wave type and current to displacement factor and of wave type and current to commutation lead to displacement factor–commutation relationships for ac-to-dc/dc-to-ac and ac-to-ac converters. Derive and state the relationships of displacement factor and commutation.

3. Calculate the commutating overlap angle, u, and the dc terminal voltage witholding. ΔV_{DD}, for a six-pulse bridge current-sourced converter operated from 60-Hz 480-V line, with a source impedance of 0.06 per unit and a smooth dc output current of 100 A if the delay angles are

a. 0
b. $\pi/3$
c. $\pi/2$
d. $2\pi/3$

What is the maximum permissible delay angle if no recovery time is needed for the switches?

4. A voltage-sourced converter pole is operated from a dc voltage of 400-V with a defined ac current of magnitude 100-A. If the total dc source reactance is 100 μH and the converter is operated at lagging displacement angles of 0, $\pi/6$, and $\pi/3$ rad, calculate

a. The commutation overlap angles
b. The resulting modifications (from ideal) to the amplitude of the wanted component of pole dependent voltage. (Assume that the dependent current has a constant value during the overlap.

5. Explain why a source-commutated dc-to-dc converter cannot be constructed.

6. For the converters of Problems 3 and 4, calculate the actual operating displacement angles when the effects of the commutation overlaps are accounted for. (Use the approximation deriving from the dependent voltage expression in the current-sourced case.)

9
Impulse-Commutating Circuits

9.1 Introduction

Impulse-commutating circuits are needed to make thyristors viable switches, or parts of switches, in a great many converters. DC-to-dc converters all require that a thyristor have an impulse-commutating circuit if it is to be the converter's controllable switch. Current-sourced ac-to-dc/dc-to-ac converters require impulse commutation if operated with phase-advance control or if CAM, PWM, or programmed wave-form control techniques are used; so do voltage-sourced ac-to-dc/dc-to-ac converters. The ac-to-ac converters, excluding the NCC, exert a general demand that thyristors be equipped with impulse-commutating circuits if they are to be used as the switching devices in those converters.

All impulse-commutating circuits operate on the same basic principle. *Two* commutations are engendered in the circuit in order to effect the one that is necessary to converter operation. The first commutation effects a transfer of current from the thyristor that is to be turned off as a part of the converter commutation to the impulse-commutating circuit. The second commutation then transfers the current from the impulse-commutating circuit to the incoming converter-circuit branch. Universally, impulse-commutating circuits use a capacitor or, occasionally, several capacitors, as a surrogate voltage source to effect both commutations. This is called the *commutating capacitor*. The capacitor voltage at the instant commutation is engendered is always such as to effect a source commutation of the current from the outgoing converter thyristor into the capacitor and the remainder of the impulse-commutating circuit. The converter current is accompanied in most instances by a *commutating impulse current* that flows in the circuit as a result of the initial capacitor voltage. These currents then act to change the capacitor voltage until it is able to effect a source commutation transferring the converter current into the incoming converter-circuit path.

Impulse-commutating circuit action is invariably initiated by closing a switch. The switch used is almost invariably a thyristor; there would be little point in using a fully controllable switching device to initiate impulse commutation when thyristors are used in the main load-bearing paths of the converter. In such a case, the fully controllable switches might just as well be used as the converter switches. In dc-to-dc converters, the thyristor used to engender commutating-circuit action must obviously be an extra, auxiliary, thyristor. In ac-to-dc/dc-to-ac converters and ac-to-ac converters, this is also often the case; such circuits are called *auxiliary impulse-commutating circuits* because they use auxiliary thyristor switches not involved in load-bearing duty in the converter. However, it is also possible to devise ac-to-dc/dc-to-ac and ac-to-ac converter circuits in which turning on the appropriate load-bearing thyristor initiates the action of the impulse-commutating circuit that will effect commutation of the converter current into that thyristor. As mentioned in Chapter 8, such arrangements are called *auto* (for automatic) *impulse-commutating circuits,* or more usually *auto-commutating circuits*.

Although the same basic double commutation principle applies to all impulse-commutating circuits, detail operation varies quite widely. However, two fundamental types of behavior are observable. They have come to be known as *hard* and *soft commutation,* but this terminology is not particularly apt. Hard-commutating circuits subject the outgoing thyristor to severe dI_R/dt stress followed by reverse voltage application. However, they impose only moderate reapplied dV/dt stress to the thyristor, since its voltage during the commutation period is the commutating capacitor voltage or some fraction thereof. Soft-commutating circuits exhibit exactly complementary behavior. They produce quite modest dI_R/dt stress and no reverse bias voltage, since a diode is connected directly in inverse parallel with the converter thyristor. (As a result, soft-commutating circuits are not usually employed in current-sourced ac-to-dc/dc-to-ac converters or in ac-to-ac converters. When they are, an additional thyristor is needed in the ac-to-dc/dc-to-ac converter switches to provide the bilateral current path during the commutating period.) The reapplied dV/dt stress created by soft-commutating circuits is severe.

The remainder of this chapter will describe and discuss a selection of impulse-commutating circuits commonly used in a variety of converter circuits. Analytic techniques for designing and optimizing the impulse-commutating circuits are introduced, and the final section deals with the passive component requirements of the circuits.

9.2 Analysis of Resonant Circuit Behavior

Since most impulse-commutating circuits are based on simple resonant circuits, it is well to review the behavior of such circuits before proceeding farther. The circuits in question generally consist of an *L-C-R* series combination. Usually, whenever a switch is closed to activate the circuit or a change of switch states occurs to produce a transient in the circuit, there are an initial capacitor voltage, here designated E_{co}, and an initial inductor current, here designated I_o, in addition to a dc forcing voltage that will be called V_s. Occasionally, a damping resistor is found connected in parallel with the capacitor or the inductor; this case will also be treated.

The time-dependent parameters of interest are almost invariably the circuit current and the capacitor voltage. Simplified commutating-circuit analyses often treat the circuits as lossless (i.e., infinite Q), but since the equations for that case are readily derived from those for the more general cases, that procedure is adopted here. The Laplace transform equation defining the current in the basic L-C-R_S series circuit can be written

$$V_s/s = E_{co}/s + \bar{i}/sC + \bar{i}R_S + sL(\bar{i} - I_o/s)$$

which translates to

$$\bar{i} = [(V_s - E_{co})/L]\{1/[(s+\alpha)^2+\omega^2]\} + sI_o\{1/[(s+\alpha)^2+\omega^2]\}$$

where $\alpha = R_s/2L$ and $\alpha^2+\omega^2 = 1/LC$. In commutating circuits, only the underdamped case need be considered. This is the case where $1/LC > \alpha^2$ and ω is in consequence real. Inverting the current transform equation yields

$$i = [(V_s - E_{co})/\omega L]\epsilon^{-\alpha t}\sin\omega t + I_o[\epsilon^{-\alpha t}\cos\omega t - (\alpha/\omega)\epsilon^{-\alpha t}\sin\omega t]$$

as the time-domain equation for i. Setting $R_s = 0$ puts $\alpha = 0$ and simplifies the equation to

$$i_o = [(V_s - E_{co})/\omega L]\sin\omega t + I_o\cos\omega t$$

In this case, the first current zero after $t = 0$ occurs when

$$\omega t_o = \arctan[-\omega L I_o/(V_s - E_{co})]$$

If both I_o and $V_s - E_{co}$ are positive or negative, this angle lies in the second quadrant ($\pi/2 \leq \omega t_o \leq \pi$); if one is negative and the other positive, it lies in the first quadrant ($0 \leq \omega t_o \leq \pi/2$).

The transform equation defining the capacitor voltage is

$$\bar{e}_c = E_{co}/s + \bar{i}/sC$$

which yields

$$\bar{e}_c = E_{co}/s + (V_s - E_{co})\{1/s - (s+2\alpha)/[(s+\alpha)^2+\omega^2]\} + (I_o/C)\{1/[(s+\alpha)^2+\omega^2]\}$$

Inverting gives the time domain equation as

$$e_c = E_{co}\epsilon^{-\alpha t}[\cos\omega t+(\alpha/\omega)\sin\omega t]$$

$$+ V_s[1-\epsilon^{-\alpha t}\cos\omega t-(\alpha/\omega)\epsilon^{-\alpha t}\sin\omega t]$$

$$+ (I_o/\omega C)\epsilon^{-\alpha t}\sin\omega t$$

Putting Q at infinity, $\alpha = 0$, and this simplifies to

$$e_{co} = E_{co}\cos\omega t+V_s(1-\cos\omega t)+(I_o/\omega C)\sin\omega t$$

Substituting the value of ωt, which is ωt_o, for the current zero gives the capacitor voltage at the current zero. To solve for the time of capacitor voltage zero, which is often of interest, an iterative computer program is necessary in the general (damped) case since there is no closed-form solution. The ideal case ($Q = \infty$) has a closed-form solution, as was shown in Chapter 8.

When a resistor R_p appears in parallel with either the capacitor or the inductor, it is usual to consider that R_s is zero to simplify matters. With this assumption, the defining equations for currents with the resistor in parallel with the capacitor become

$$V_s/s = E_{co}/s+\bar{i}_c/sC+sL(\bar{i}-I_o/s)$$

$$E_{co}/s+\bar{i}_c/sC = \bar{i}_R R_p$$

and

$$\bar{i}_R+\bar{i}_c = \bar{i}$$

Transposing gives

$$\bar{i}_R = E_{co}/sR_p+\bar{i}_c/sCR_p$$

and hence

$$\bar{i}_c = \bar{i}/(1+1/sCR_p)-E_{co}/[sR_p(1+1/sCR_p)]$$

Substituting in the first equation yields

$$(V_s-E_{co})/s+E_{co}/[s^2CR_p(1+1/sCR_p)]+LI_o = \bar{i}\{sL+1/[sC(1+1/sCR_p)]\}$$

Transposing gives

$$\bar{i} = (V_s/R_p)[1/s-(s+2\alpha)/D]+[(V_s-E_{co})/L](1/D)$$
$$+ sI_o/D+(I_0/CR_p)(1/D)$$

where $D = (s+\alpha)^2+\omega^2$, $\alpha=1/2CR_p$, and $\alpha^2+\omega^2 = 1/LC$. Inverting yields

$$i = (V_s/R_p)[1-\epsilon^{-\alpha t}\cos\omega t-(\alpha/\omega)\epsilon^{-\alpha t}\sin\omega t]$$
$$+ [(V_s-E_{co})/\omega L]\epsilon^{-\alpha t}\sin\omega t$$
$$+ I_o\epsilon^{-\alpha t}\cos\omega t+(I_o\alpha/\omega)\epsilon^{-\alpha t}\sin\omega t$$

Putting $Q = \infty$ ($R_p = \infty$, $\alpha = 0$) reduces this to

$$i_o = [(V_s-E_{co})/\omega L]\sin\omega t+I_o\cos\omega t$$

as before. The capacitor voltage is defined by

$$e_c = V_s-L\,di/dt$$
$$= V[1-\epsilon^{-\alpha t}\cos\omega t-(\alpha/\omega)\epsilon^{-\alpha t}\sin\omega t]$$
$$+ E_{co}\epsilon^{-\alpha t}\cos\omega t+(I_o/\omega C)\epsilon^{-\alpha t}\sin\omega t$$

Again, putting $\alpha = 0$ reduces this to

$$e_{co} = V_s(1-\cos\omega t)+E_{co}\cos\omega t+(I_o/\omega C)\sin\omega t$$

as before.

The analysis for R_p in parallel with L, assuming $R_s = 0$, is similar and yields

$$i = [(V_s-E_{co})/\omega L]\epsilon^{-\alpha t}\sin\omega t + [(V_s-E_{co})/R_p]\epsilon^{-\alpha t}[\cos\omega t-(\alpha/\omega)\sin\omega t]$$
$$+ I_o\epsilon^{-\alpha t}[\cos\omega t-(\alpha/\omega)\sin\omega t]$$

where $\alpha = 1/2CR_p$ and $\alpha^2+\omega^2 = 1/LC$ as before.

Solving also for e_c gives

$$e_c = (V_s-E_{co})[1-\epsilon^{-\alpha t}\cos\omega t+(\alpha/\omega)\epsilon^{-\alpha t}\sin\omega t] + (I_o/\omega C)\epsilon^{-\alpha t}\sin\omega t$$

Again, putting $R_p = \infty$, $\alpha = 0$ reduces to the ideal-case equations, and no closed-form solutions are available for the times of current or voltage zeros in either of the parallel damped cases.

9.3 Impulse-Commutating Circuits for DC-to-DC Converters

One of the most common hard-commutating circuits applied to dc-to-dc converters is shown, in somewhat idealized form, in Figure 9-1a. Although shown applied in a buck converter, it may be used in any of the dc-to-dc converters since its function is to make a thyristor a fully controllable switch, not to serve any particular converter.

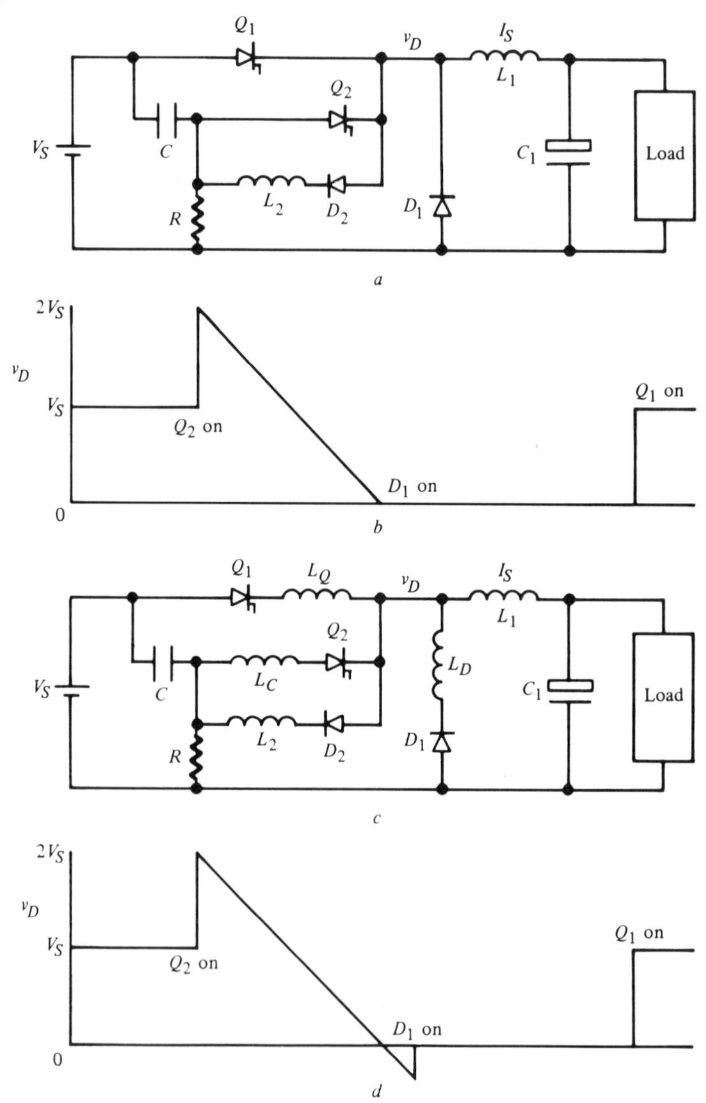

Figure 9-1. A dc-to-dc converter, hard-commutating circuit.

Operation is quite straightforward. Assume that while the main thyristor, $Q1$, is conducting, the commutating capacitor, C, is charged to the source potential V_S with the plate connected to the auxiliary thyristor, $Q2$, positive. Then, when $Q2$ is turned on, an immediate source commutation will take place. $Q1$ will be reverse-recovered with infinite dI_R/dt, and V_S will be applied to it as reverse bias. The current, I_S, now flows in C and $Q2$ after this first commutation. The constant current flow depletes the charge on C and eventually reduces its voltage to zero. The rate of change of voltage is linear, and the equation

$$CV_S = I_S t_{qa}$$

defines the recovery time, t_{qa}, allotted $Q1$ or, alternatively, the value of commutating capacitance, C, needed to produce a given t_q for $Q1$. The flow of I_S in C and $Q2$ continues, with C now charging to acquire a potential the reverse of that it initially possessed. When the voltage on C reaches V_S, the joint cathodes of $Q1$, $Q2$, and the freewheel diode $D1$ will arrive at zero voltage. The result is a second commutation, the current transferring into $D1$ and ceasing to flow in C and $Q2$. $Q2$ will recover, and provided that it is allowed, adequate t_q will reestablish forward-blocking capability.

The dependent voltage, v_D, is depicted in Figure 9-1b. Like the vast majority of hard-commutating circuits, its initial excursion when commutation is engendered is of the opposite polarity to the excursion required for the converter commutation.

With the commutation complete, C is left charged to V_S but with the wrong polarity for subsequent commutation. Hence the presence of D_2 and L_2, which are reversing components. When $Q1$ is again turned on in the sequence of converter operations, a source commutation occurs and I_S transfers from $D1$ to $Q1$ with v_D changing from zero to V_S. In addition, a resonant discharge of C takes place via $Q1$, $D2$, and L_2. Only a single half-cycle of current at the natural frequency of L_2 and C ($f = 1/2\pi\sqrt{L_2C}$) can flow, since $D2$ will reverse-recover on the attempted current reversal at the end of the first half-cycle. As a result, the potential on C is reversed, following a cosinusoidal excursion as the half-sinusoid current pulse flows, and the capacitor voltage is V_S of the correct polarity for the subsequent commutation once this reversal is complete (assuming no losses in L_2, C, $Q1$, and $D2$).

The resistor R is present to allow C to become charged to V_S when the source voltage is initially connected to energize the converter. Sometimes R is then removed from circuit by a switch, which may be electromechanical or solid state, to eliminate it as a loss-producing component during converter operation. Once C has charged through R, the first time $Q1$ is turned on to begin converter operation, a reversal takes place and commutation can procced subsequently in the manner described.

This circuit has a number of disadvantages even in the idealized form. To begin, the total time taken to complete a commutation is a function of I_S, the converter's load current. If that current varies over a wide range, as it does in many applications, then so will the total commutation time t_c. Now, given a main thyristor with a reverse-recovery-time requirement t_q and a converter with source-voltage range V_{S-} to V_{S+}, minimum to maximum, and load current range I_{S-} to I_{S+}, the defining equation quoted previously yields

$$CV_{S-} = I_{S+}t_q$$

or

$$t_q = CV_{S-}/I_{S+}$$

The minimum commutating time t_c, is given by

$$t_{c-} = 2t_q$$

since a further change of potential equal to V_S volts must occur on C before commutation is complete.

If V_S increases, t_c will also increase; if I_S decreases, t_c will further increase, yielding the maximum commutation time, t_{c+}, as

$$t_{c+} = (V_{S+}/V_{S-})(I_{S+}/I_{S-})t_{c-}$$

Taking as an example a source-voltage range of 1.5 to 1, a load current range of 5 to 1, and a thyristor t_q of 15 μsec, t_{c-} would be 30 μsec but t_{c+} would be 225 μsec. Observe now that if $Q1$ were again turned on just as the commutation ended, the effective A for the converter would be unity (i.e., $V_{DD} = V_S$). This is because the commutating impulse adds in the period $t = 0$ to $t_{qa}(V_C = 0)$ a volt-second integral equal to that withheld over the period t_{qa} to t_c. Hence, for the converter to exert control over V_{DD}, the "off" time for $Q1$ must be longer than t_{c+}, and the upper operating frequency of the converter is severely restricted.

A further undesirable consequence of the extension of t_c is the increased perturbation of v_D that obviously occurs. The volt-seconds generated over the period $t = 0$ to t_{qa} of the commutating period will clearly increase the positive ripple current excursion in the current sourcing inductor L_1. Since an equal increase in the negative excursion must occur as a result of the corresponding volt-second withholding, the critical inductance requirement is increased.

The extension of t_c that occurs at low load current is sometimes reduced by connecting a diode and inductor in inverse parallel with Q_1. When this is done,

C discharges resonantly through the path that is provided whenever Q_2 is turned on to initiate commutation. Thus, with zero load current, t_c is simply a half-period of the natural frequency of the oscillatory circuit consisting of C and the inductance introduced. As I_S is increased, t_c and t_{qa} do decrease, of course, but the ratio of maximum to minimum times can be greatly reduced. However, in addition to the extra costs incurred by introducing the diode and inductor, the capacitance of C must be increased to provide a given recovery time at maximum I_S. This is because the total current in the capacitor is now

$$i_c = I_S \cos\omega t + (V_S/\omega L_3)\sin\omega t$$

where L_3 is the inductor introduced and $\omega^2 = 1/L_3C$.
Thus, the capacitor voltage is

$$e_c = V_S \cos\omega t - (I_S/\omega C)\sin\omega t$$

which becomes zero earlier, for a given C and I_S, than e_c if L_3 is not present.

The second major disadvantage of this commutating circuit is associated with the reversal of potential on C to "reset" the circuit ready for each commutation. The half-sinusoidal current impulse which accomplishes the reversal flows in Q_1, and hence the duration of the impulse is the minimum on time for Q_1. To maintain the fullest possible control range for the converter, this time must be short compared to the time of one cycle of converter operation. However, the shorter the reversing time, the greater will be the peak impulse current, since a fixed amount of charge ($2CV_S$) must be moved. The impulse current occurs as Q_1 turns on, and represents dI/dt and turn-on peak current stress for the thyristor well beyond that associated with the converter load current.

To prevent this problem, D_2 is sometimes replaced by a thyristor so that the reversing current impulse can be delayed until Q_1 is fully on. Obviously, this increases the minimum on time for Q_1 and, not incidentally, the circuit cost.

A practical version of the circuit, shown in Figure 9-1c, possesses inductances in all current paths. Inductance is needed, in any event, to moderate dI_R/dt and dI/dt for all the active devices. While parasitic inductances are sometimes sufficient to accomplish the dI/dt limitations needed, bulk inductance usually has to be added. The best position for such an addition is that designated L_Q in Figure 9-1c. Inductance so placed limits dI_R/dt and dI/dt for all the devices, whereas inductance in the position of L_D only limits dI/dt for Q_1 and dI_R/dt for D_1. Inductance placed at L_C limits only dI_R/dt for Q_1 and dI/dt for Q_2. Apart from limiting the rates of change of the various currents, and thereby creating commutation overlaps, circuit inductance has other undesirable effects. Some have been discussed in Chapter 8; one new effect found in this circuit is that the presence of L_C and L_D means that the capacitor, C, must provide sufficient volt-seconds to effect the second commutation, from C, L_C, and Q_2 to D_1 and L_D. To do so, the capacitor voltage must become greater than V_S, and v_D will be less than zero for some period, as depicted in Figure 9-1d. The commutation time is lengthened, and output current ripple is further increased.

The capacitor overcharge is usually not serious. In fact, it is often regarded as beneficial since the energy acquired by C during the second commutation overlap may wholly or partially offset the reversal losses. This makes the commutating voltage (the voltage on C immediately prior to a commutation) closer to V_S than would be the case without the overcharge due to the presence of L_C and L_D, particularly in low-voltage converter applications. The acquisition of surplus energy by C is, however, representative of another common aspect of impulse commutating circuits—the "trapped energy" phenomenon. The effect is perceived to arise from the energy "trapped" in L_C by the load current at the instant commutation should be complete. While perhaps beneficial in this circuit, it can be the bane of many others.

A common soft-commutating circuit used in dc-to-dc converters is shown in Figure 9-2a. Again, its function is to make the thyristor, Q_1, a fully controllable switch, and, although it is shown applied to the buck converter, it may be used in any of the dc-to-dc converters. (It can also be used in a variety of other converters, but it seldom is since other circuits prove more efficient.) The diode D_3, in direct inverse parallel connection with Q_1, plays no part in converter operation of course. In this instance, it is not a part of the switch required by the converter, but truly a part of the commutating circuit. Operation of the circuit, which is quite straightforward, will now be described.

Assume that while Q_1 is conducting, C, the commutating capacitor, is charged to some potential greater than V_S with the plate connected to the anode of Q_1 positive. Before commutation is to be initiated, Q_2 is turned on so that the charge on C will reverse by a resonant discharge through Q_2 and L_C. This action is confined to the commutating circuit and has no influence on the converter whatsoever. However, when this reversal is completed (i.e., when the half-sinusoidal current pulse in Q_2 terminates and Q_2 recovers), another reversal will immediately begin through L_C, D_2, and Q_1. It might seem strange to view Q_1 as conducting this impulse current, and in truth it does not. The resonant circuit (C and L_C) is, however, unable to detect this—as long as Q_1 is on, the circuit is closed and the impulse will flow. In the process, the load current is transferred from Q_1 to C, L_C and D_2, effecting the first commutation. The reason for the presence of D_3 should now be clear. Once Q_1 is extinguished, D_3 provides a path for the "surplus" commutation impulse current over and above the load current. The converter circuit undergoes no disturbance whatsoever until the commutating impulse current once again falls to the level of the load current. As depicted in the waveforms of Figure 9-2b, the impulse can then no longer follow its half-sinusoidal form, since Q_1 and D_3 are both off ("open") and the

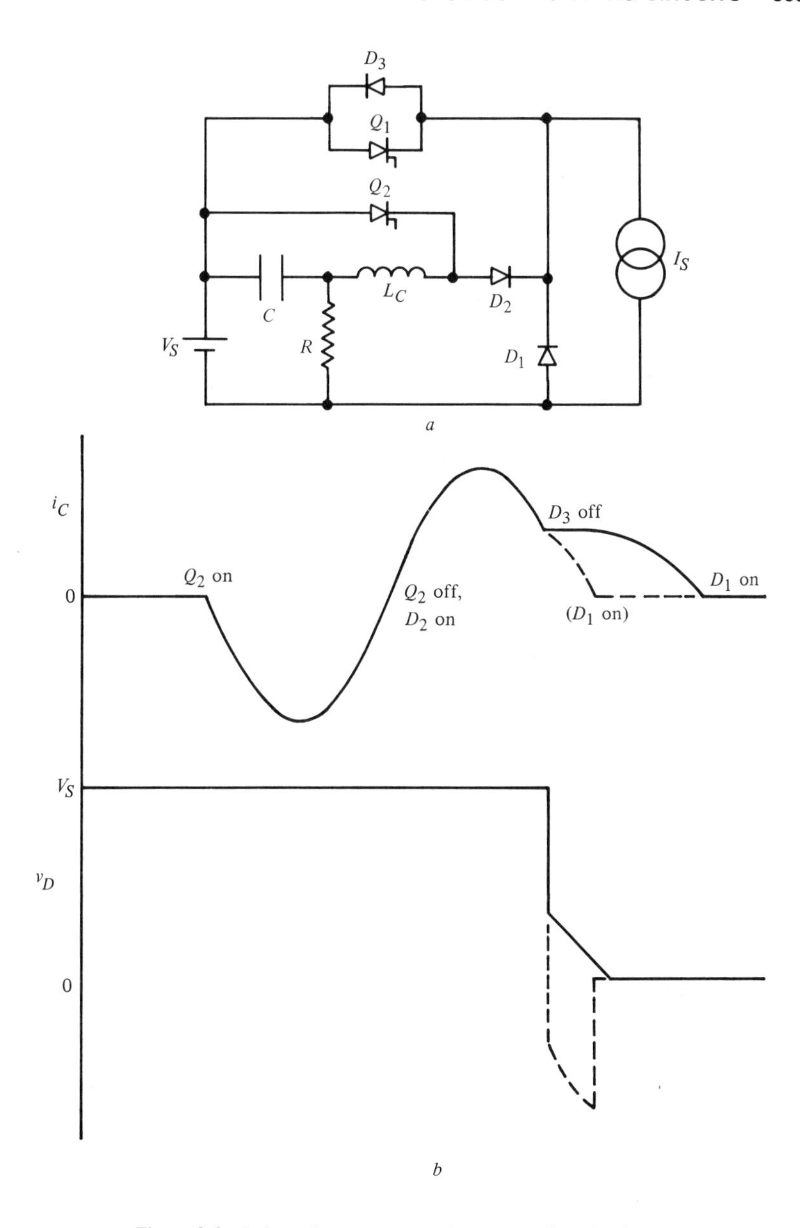

Figure 9-2. A dc-to-dc converter, soft-commutating circuit.

load current (presuming effective current sourcing) refuses to change. As a result, the current in C, L_C, and D_2 becomes "frozen" at I_S. The voltage on L_C promptly collapses, since there is no longer a dI_C/dt to sustain it; the recovery time afforded Q_1 is, of course, equal to the time for which D_3 conducts (i.e., the time for which the commutating impulse current exceeds the load current).

As shown, the voltage on C at this point has reversed but has by no means completed its reversal (if I_S is not zero), and hence the value of v_D drops suddenly from V_S to V_S+v_C, where v_C is negative. Two situations can arise. If v_C, not withstanding incomplete reversal, has absolute value greater then V_S, then the second commutation begins immediately. If not, then v_C and v_D both decrease at a linear rate until v_D reaches zero (minus the threshold voltage of D_1 in the practical case), and then the second commutation begins. In either event, C will acquire considerable overcharge in the process of providing sufficient volt-seconds to commutate the load current out of L_C and into D_1.

If the resonant circuit is sufficiently low Q, or the converter current well below rated, or R of a value allowing discharge of C back to or close to V_S during Q_1's on period, then the process simply repeats with the latter case prevailing. However, the overvoltage reached on C at completion of the second commutation can reach values ranging from about 1.1 to 1.7 times V_S, depending on per-unit load current and the particular circuit design. If none of the stipulated conditions exist, then the overcharge is incremented with each cycle of operation. If the circuit Q were infinite and R not present, then the voltage on C would, in theory, increment with each successive commutation and ultimately become infinite too. In practice, the finite Q of the resonant circuit limits the voltage buildup. This is because, eventually, the losses suffered in each reversal and commutation impulse exactly balance the energy gained by C in each second commutation.

If R is not present, the final voltage on C can still reach uncomfortably high values. For practical circuit Q's, ranging from 20 to perhaps 100 (rarely quite so high, never higher), the peak voltage on C at maximum converter current will range from 1.5 to about 3.5 times V_S, depending also on circuit design parameters. R is placed in circuit not only to allow C to charge to V_S initially, prior to converter operation, but also to ensure that C will discharge via R and V_S, losing at least some of the overcharge, between commutation. During this process, one-half the energy removed from C is returned to V_S and one-half is dissipated in R.

It is not usual to design so that C discharges back to V_S; at maximum converter current, peak capacitor voltage of some 1.2 to 1.5 times V_S is generally allowed, because such a design minimizes the total losses, including the trapped energy losses, in the circuit.

Design optimization is tedious and can only be approached effectively using the current and voltage expressions developed in Section 9.2 with a digital computer to execute iterative solutions. There is no single choice of L_C and C values for the circuit. Unlike the hard commutating circuit described previously, where a simple expression gives the unique value of C required in a particular case, the soft commutating circuit's requirements can be satisfied by an infinitude of $C-L_C$ combinations. This can be seen if a dimensionless parameter is defined as

$$x = V_{CC}/\omega L_C I_S$$

where V_{CC} is the voltage on C when the first commutation begins, ω the natural angular frequency of the circuit, and I_S the maximum current to be commutated. Assuming I_S to be constant over the commutating interval and the circuit to have infinite Q, the recovery time allotment is defined by

$$\omega t_{qa} = \pi - 2\arcsin(1/x)$$

Clearly x must be greater than unity. Equally clearly, for any value of x selected, a corresponding value of ω (ω_x, for example) can be derived if t_{qa} is set equal to the t_q required by Q_1. Also, under the above assumptions,

$$x = (V_{CC}/I_S)\sqrt{C_x/L_{Cx}}$$

and

$$\omega_x = 1/\sqrt{C_x L_{Cx}}$$

where C_x and L_{Cx} are the values of L_C and C that will satisfy circuit requirements. Simple manipulation yields

$$\begin{aligned} L_{Cx} &= V_{CC}/x\omega_x I_S \\ &= V_{CC}t_q/\{xI_s[\pi - 2\arcsin(1/x)]\} \end{aligned}$$

and

$$\begin{aligned} C_x &= xI_S/\omega_x V_{CC} \\ &= xI_S t_q/\{V_{CC}[\pi - 2\arcsin(1/x)]\} \end{aligned}$$

The parameter x may thus be regarded as a unique single variable permitting circuit design to any desired criteria. A designer may wish to find the lowest cost, lowest loss, or lowest weight combination satisfying a given converter's needs. Using x as the prime variable, the L_{Cx} and C_x values can be established and the circuit's ranges of parameters determined. The procedure is not quite so simple as outlined above, of course; computer assistance is required to determine $V_{CC}-x$ combinations, and the exact expression for $\omega_x t_q$ has no simple form. Proper application of numerical analytic techniques is essential to a practical design.

It should be observed that the dimensionless parameter x is simply the ratio of peak commutating-impulse current to the maximum current to be commutated. Its application is not restricted to this circuit—in fact, it can be applied to nearly all impulse-commutating circuits, both hard and soft. The only exceptions are those hard-commutating circuits of the type depicted in Figure 9-1, which do not generate a commutating-impulse current.

A variety of other impulse-commutation circuits exists that have been used, or proposed for use, in dc-to-dc converters. Each is a variant of one of the two described here.

9.4 Impulse-Commutating Circuits for Voltage-Sourced AC-to-DC/DC-to-AC Converters

The voltage-sourced converters are treated first for two reasons. First, like the dc-to-dc converters, they commutate ''against the wishes'' of a dc voltage source, and hence the circuits used bear a strong relationship to those described in Section 9.3. In fact, these two circuits, and their variants, can be used in ac-to-dc/dc-to-ac converters. That they are not is attributable to the more elegant and efficient circuits that can be realized if the total commutation needs of a converter pole are addressed by a single impulse-commutating circuit.

Second, the historical development of converters and their applications has led to great emphasis upon (and much money spent on) the development of impulse-commutating circuits for voltage-sourced ac-to-dc/dc-to-ac converters. Indeed, since the dawn of the thyristor era over 20 years ago, designers have zealously sought the ''ideal commutating circuit.'' In truth, the ideal commutating circuit does not exist. A rational consideration of the true requirements for commutation and the means available to meet them furnishes adequate proof of that fact.

It is in ac-to-dc/dc to-ac converters that auto-impulse commutation is first encountered. The so-called McMurray-Bedford circuit,[1] depicted in Figure 9-3a, was the very first impulse-commutating circuit to find widespread application in thyristor converters. Its operation is now described. An obvious variant, using a center-tapped dc supply, operates in an essentially similar manner. The circuit is in one respect typical of the vast majority of impulse-commutating circuits for voltage-sourced ac-to-dc/dc-to-ac converters: it provides the means for making both thyristors in the pole controllable switches. Thus, it enables both converter commutations to be accomplished—commutating the ac current from the positive dc bus to the negative and vice versa. Since these commutations are electrical ''mirror images'' of each other, only the first will be described.

If Q_1 is conducting, the point A is at potential V_S with respect to the reference bus, the supply negative. Thus, C_2 (which is equal in capacitance to C_1) is charged to V_S while C_1 has no charge, and the current I_S flows via L_1 which is

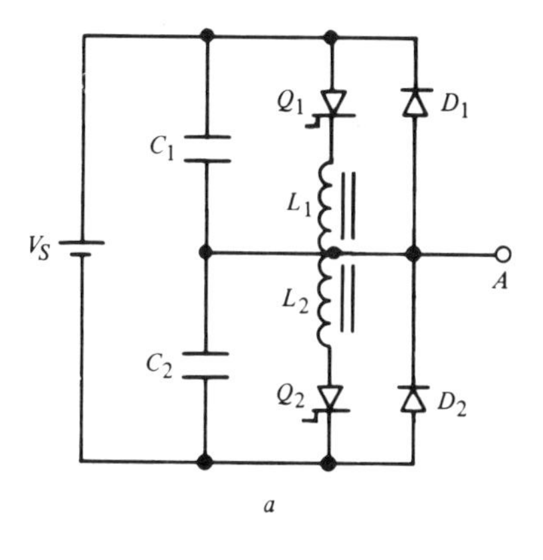

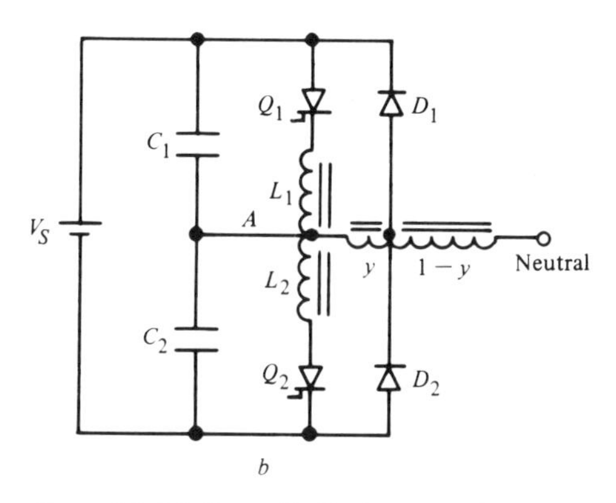

Figure 9-3. McMurray-Bedford commutating circuit.

equal in inductance to L_2 and perfectly coupled thereto. Now if Q_2 is turned on, V_S is impressed across L_2. As a result, V_S is induced in L_1 so that the cathode of Q_1 assumes potential $2V_S$ and Q_1, having suffered reverse recovery at infinite dI_R/dt, is reverse-biased by V_S. (In practice, the leakage inductance of L_1 and L_2 is usually sufficient to moderate the dI_R/dt adequately to satisfy device needs; if not, additional bulk inductance is introduced.) The converter load current I_S, at whatever level exists at the time of commutation, transfers to C_1 so that the first commutation is effected. I_S ceases to flow in L_1 but appears in L_2 so that the ampere-turns of the combined inductors remain the same and no Ldi/dt contribution appears in the voltages. This I_S flows in C_2, of course.

The Thévenin equivalent for the impulse circuit now becomes a capacitance equal to the parallel combination of C_1 and C_2 with an initial potential V_S discharging through L_2, which has an initial current I_S and with a constant current drain of I_S in shunt with L_2. Neglecting resonant circuit losses, the currents and voltages are

$$i_c = (V_S/\omega L)\sin\omega t + 2I_S\cos\omega t$$

$$i_{L_2} = (V_S/\omega L)\sin\omega t + I_S(2\cos\omega t - 1)$$

$$e_c = e_{L_2} = V_S\cos\omega t - 2\omega L I_S \sin\omega t$$

$$= V_S\cos\omega t - (2I_S/\omega C)\sin\omega t$$

where $L = L_1 = L_2$, $C = C_1 + C_2$, and $\omega^2 = 1/LC$.

Reverse bias for Q_1 ends with $e_c = V_S/2$, for at that time the voltage at the cathode of Q_1 is V_S, equal to the voltage at its anode. Hence t_{qa} is defined by the solution of

$$V_S\cos\omega_s t_{qa} - 2\omega L I_S \sin\omega t_{qa} = V_S/2$$

Putting $V_S/2\omega L I_S = x$ yields

$$\omega t_{qa} = \arcsin(x/\sqrt{1+x^2}) - \arcsin(x/2\sqrt{1+x^2})$$

$$= G(x).$$

The commutating interval ends when $e_c = 0$, for at that point the second commutation will occur with I_S transferring to D_2. This occurs when

$$\omega t_c = \arctan(x) = \arcsin(x/\sqrt{1+x^2})$$

At t_c, the current in L_2 is given by

$$i_{L_2} = (V_S/\omega L)\sin\omega t_c + I_S(2\cos\omega t_c - 1)$$

$$= 2x^2 I_S/\sqrt{1+x^2} + I_S(2/\sqrt{1+x^2} - 1)$$

$$= I_S(2\sqrt{1+x^2} - 1)$$

The nondimensional parameter x introduced here has the same form as that previously seen in the analysis of the soft-commutating circuit for the dc-to-dc converters. However, in this case, x is one-half the ratio of the peak commutating-impulse current ($V_S/\omega L$) to the load current (I_S).

Now observe that if $x = V_S/2\omega L I_S$, $\omega^2 = 1/LC$, and $\omega t_{qa} = G(x)$, then the values of L and C required to provide a given t_q at given values of V_S and I_S can be derived from

$$xG(x) = (V_S/2\omega LI_S)\omega t_q$$

or

$$L = V_S t_q/2xG(x)I_S$$

$$x/G(x) = (\omega CV_S/2I_S)/\omega t_q$$

or

$$C = 2xI_S t_q/G(x)V_S$$

and the nondimensional parameter, x, is a unique design-defining variable.

The major disadvantage of this circuit should be quite obvious. The current in L_2 when commutation ends, $I_S(2\sqrt{1+x^2}-1)$, is trapped and will decay only as a result of losses incurred while in circulation via L_2, Q_2, and D_2. Thus, the trapped energy

$$W_t = (L/2)I_S{}^2(2\sqrt{1+x^2}-1)^2$$

$$= (V_S I_S t_q/4)(2\sqrt{1+x^2}-1)^2/xG(x)$$

is dissipated at each commutation. The function of x involved can be shown to have a minimum (rather flat) at $x = 1.14$, when $W_t = 1.95V_S I_S t_q$. The total component rating (combined inductor and capacitor ratings) needed can also be expressed in terms of x. Using $LI^2/2+CV^2$, the expression

$$R = 2W_t + V_S I_S t_q x/G(x)$$

is obtained. This has a rather flat minimum at $x = 1$; a reasonable design compromise is $x = 1.07$, since both functions dependent on x are quite flat in the vicinity of their minima. However, there is one further consideration. The ratio of maximum to minimum commutation time is also a function of x; at the maximum current to be commutated, $\omega t_{c-} = \arcsin[x(\sqrt{1+x^2})]$, *as was seen. At zero current,* $\omega t_{c+} = \pi$, for a full half-cycle discharge of C via $L2$ will then occur. The ratio, t_{c+}/t_{c-}, progressively reduces as x is increased. At $x = 1$, it is 4; for $x = 2$, it is 2.84, and at $x = 5$, it is 2.29. If very short turnaround time is to be maintained at light loads, then higher than optimum values of x may be needed; the consequences, of course, will be increased trapped energy losses and a more costly design.

The trapped energy problem of the McMurray-Bedford pole can be alleviated to some extent by using the arrangement shown in Figure 9-3b. The transformer winding may belong to a completely auxiliary energy recovery transformer, but is more usually simply the primary of the transformer coupling to the pole's load. In this circuit, it is necessary for the pole point (A) to be at $-yV_S/2$, where y is the transformer turns ratio indicated in Figure 9-3b, before D_2 will conduct. As a result, this voltage is impressed on $L2$ when the second commutation (from C into D_2) takes place, and the current in L_2 decays much more rapidly. Moreover, the decay is no longer wholly dissipative. The trapped energy is now in part returned to the dc voltage source, in part returned to C and subsequently to the load. A full analysis of the circuit is given in Bedford and Hoft (1964).[1]

Although the efficiency of the McMurray-Bedford pole can be improved by this trapped-energy recovery technique, the proportion of the energy recovered is not usually as high as might be desired. Rarely is much more than half the trapped energy recovered in practical circuits, mainly because parasitic inductances (particularly the leakage inductance of the transformer winding) make the recovery process distinctly nonideal and cause operational disturbances that significantly increase losses. Since y must be small, if serious disturbance of the pole's dependent voltage is to be avoided, these problems are quite intractable. As a result, this autocommutating pole has fallen into disuse, and auxiliary impulse-commutating circuits now enjoy wide popularity, since they do not generally suffer from trapped-energy problems.

Figure 9-4 shows one of the most common circuits used when hard commutation is desired. The commutating inductors, L_1 and L_2 of Figure 9-4, are equal in value and may be coupled or not as the designer wishes. Q_1 and Q_2 are the main load-bearing thyristors; Q_{1a}, Q_{1b}, Q_{2a}, and Q_{2b} are the auxiliary (commutating) thyristors; C is the commutating capacitor; D_1 and D_2 are, of course,

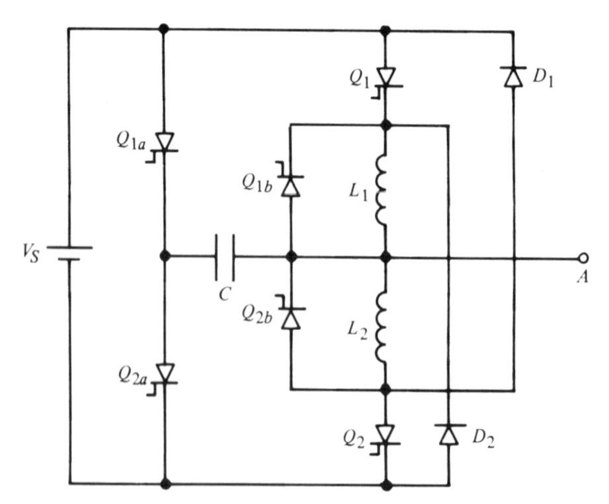

Figure 9-4. Hard-commutating circuit with no trapped energy.

the diodes used to give the pole's switches bilateral current capability. As in the McMurray-Bedford pole, they are not in direct inverse parallel connection with the thyristors because hard commutation is desired and they are involved in commutating circuit action.

Again, the two converter commutations needed are both accomplished by this one circuit, and they are electrical mirror images so only one will be described. For variety, that involving the turnoff of Q_2 is chosen. If Q_2 is conducting, then point A is at $-V_S/2$ with respect to the pole's neutral point and I_S is flowing via L_2 and Q_2. Assume C to be charged to V_S and the plate connected to the anode-cathode junction of Q_{1b} and Q_{2b} to be negative. To initiate commutation, Q_{2a} and Q_{2b} are gated on. The voltage on C, V_S, is immediately applied to Q_2 as reverse bias, and the first commutation, I_S now flowing via L_2, Q_{2b}, C, and Q_{2a}, is accomplished. At the same time, a resonant commutating impulse current path is established via D_2, L_1 and L_2, and C. The equations for the various currents and the capacitor voltage are readily found to be, ignoring resonant-circuit losses,

$$i_c = i_{L2} = (V_S/2\omega L)\sin\omega t + I_S\cos\omega t$$

$$i_{L1} = i_{L2} - I_S$$

$$e_c = -V_S\cos\omega t + (I_S/\omega C)\sin\omega t$$

where $L = L_1 = L_2$ and $\omega^2 = 1/2LC$.

The reverse-bias period for Q_2 clearly ends when $e_c = 0$, giving

$$\omega t_{qa} = \arctan(x) = \arcsin(x/\sqrt{1+x^2})$$

where $x = V_S/2\omega L I_S$. Thus,

$$L = V_S t_q / 2x\arctan(x) I_S$$

and

$$C = x I_S t_q / \arctan(x) V_S$$

are the required values of L and C.

The second commutation, transferring I_S into D_1, cannot occur until the voltage on the capacitor reaches $+V_S$. This occurs when

$$\omega t_c = 2\arctan(x)$$

at which time

$$i_c = i_{L2} = I_S$$

and

$$i_{L1} = 0$$

Thus, there is no energy trapped in either L_1 or L_2 when the second commutation occurs, and C is charged to V_S, precisely the potential thereon when commutation was initiated. Parasitic inductances, and any bulk inductances introduced to limit dI/dt and dV/dt, modify behavior slightly since they will trap small amounts of energy, which will be dissipated.

The first disadvantage of this pole is the use of four commutating thyristors. There exists an alternative configuration using only two commutating thyristors and a four-winding coupled reactor; however, the peak voltage stress on each of the two thyristors then used is $2V_S$, as opposed to V_S on each of the four used in the configuration depicted, and the reactor is difficult and costly to produce. The second disadvantage is that the diodes, D_1 and D_2, are exposed to a peak reverse voltage stress of $2V_S$. However, the main thyristors, Q_1 and Q_2, are only exposed to V_S as a peak voltage stress. In view of the fact that all soft-commutating poles expose main load-bearing thyristors to peak voltage stress in excess of the defined dc source voltage, this may be considered an advantage. Third, the commutating inductors, L_1 and L_2, are in the main load-current path. This makes them more expensive than reactors in many soft-commutating circuits, and causes higher losses. Finally, the recovery of the auxiliary thyristors is rather tenuous, and they are exposed to severe dV/dt. This is not a serious problem when the operating frequency is low, but becomes one when it is high.

The total passive component rating required by this pole is

$$R = CV_S{}^2/2 + 2(LI_{LMAX}{}^2/2)$$

where I_{LMAX} is the peak current in L_1 and L_2, and is equal to $I_S\sqrt{1+x^2}$. Thus,

$$R = (V_S I_S t_q/2)(2x + 1/x)[1/\arctan(x)]$$

which has a rather flat minimum at $x = 1.28$ when $R \simeq 1.84\ V_S I_S t_q$. As for the McMurray-Bedford, the total commutation time at $I_S = 0$ is defined by $\omega t_{c+} = \pi$, so that

$$t_{c+}/t_{c-} = \pi/2\arctan(x)$$

which is again a decreasing function of x, having the value 2 for $x = 1$. Thus, in this respect, this pole is significantly better than the McMurray-Bedford, and in fact compares not unfavorably with many soft-commutating poles.

Historically, this pole was developed later than many of the soft-commutating poles, largely because it required four commutating thyristors. Device costs eventually fell and so the lack of overvoltage stress on main thyristors, coupled with the voltage limitations of fast-switching thyristors, made the pole a fairly attractive candidate in applications with dc source voltages in excess of a few hundred volts. However, the use of passive components is poorer than in most soft-commutating poles, and the losses and costs resulting from commutating reactors in the main load current path are not welcome. Hence, soft-commutating poles, with implementation much eased by the development of high *dV/dt* capability thyristors, are favored by many.

Although the McMurray-Bedford pole does not suffer from a prevalent disadvantage of hard-commutating poles, the second pole described does. When commutation is engendered, the pole voltage undergoes an instantaneous transition in the opposite direction to that which the commutation ultimately produces. When the pole voltage is at $-V_S/2$ with respect to neutral and at zero with respect to the negative dc bus, it becomes $-V_S$ with respect to neutral immediately Q_{2a} and Q_{2b} are turned on. Similarly, when the pole voltage is at $+V_S/2$ with respect to neutral it becomes $+V_S$ when Q_{1a} and Q_{1b} are turned on. Thus, the disturbance of the dependent voltage for the pole is quite marked and grows more serious as operating frequency is increased.

The first soft-commutating pole to be developed was the McMurray (after its inventor) depicted in idealized form in Figure 9-5. Unfortunately, it suffers from a severe trapped-energy problem, which can be seen quickly when analysis is essayed and should be evident from comparison with the soft-commutating cir-

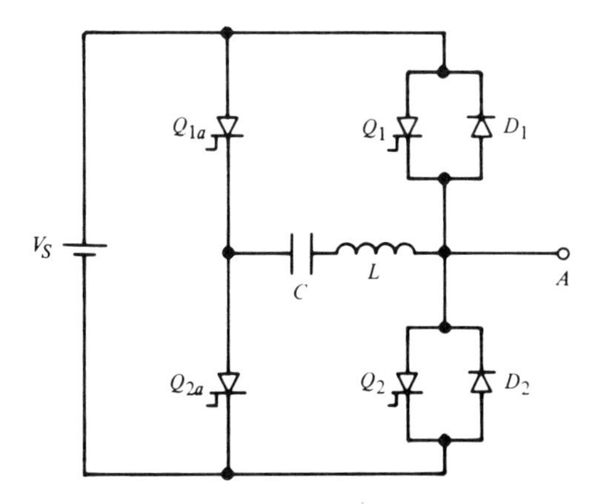

Figure 9-5. McMurray commutating circuit.

cuit for dc-to-dc converters described in Section 9.3. In fact, that circuit is derived directly from the McMurray pole, or rather from the "clamped McMurray" pole, which is discussed in this section.

Assume Q_1 of Figure 9-5 to be conducting I_S and C, the commutating capacitor, to be charged to V_S with the plate connected to L positive. Commutation is engendered by turning on Q_{1a}; C then discharges resonantly through L, and Q_1 is reverse-recovered when the commutating impulse current reaches the value of I_S. D_1 conducts the "surplus" commutation impulse current, and the converter is not aware that a commutation is in progress until the half-sinusoidal impulse current tries to fall below I_S. If, as is usually the case, the pole's ac current source will not allow it to do so, the current in Q_{1a}, C, and L remains at I_S. The potential at A then falls immediately to V_S+e_c, with e_c negative, but, C having started at V_S, smaller in magnitude than V_S. The voltage on C decreases linearly until point A reaches zero, at which time the second commutation occurs with I_S transferring into D_2. It cannot do so, however, until sufficient volt-seconds accumulate on L to commutate I_S out of L. In the process, C is charged to a voltage greater in magnitude than V_S—in fact, to $-(V_S+I_S\sqrt{L/C})$—acquiring the energy trapped in L by I_S when the voltage at A reaches zero.

Subsequent commutations depend greatly on circuit parameters. If these are such that the voltage V_S+e_c is not less than zero or $|e_c|$ is not greater than V_S when the reverse-recovery period for Q_1 or Q_2 ends, then operation is as described on each and every commutation.

If, however, $|e_c| > V_S$, then the second commutation begins immediately. C still acquires charge due to the stored energy in L, and if the resonant circuit Q were infinite, the voltage on C would ultimately become infinite. For the finite Q's ranging from 20 to perhaps 100 in practical circuits, the peak voltage on C reaches values ranging from $\sim 1.5\, V_S$ to $\sim 3.5\, V_S$ when the pole is commutating maximum current, depending on Q and the design parameter $x = V_S/\omega L I_{S\text{MAX}}$. The peak voltage on C increases as Q increases, as might be expected, but decreases with increasing x (basically because $I_S\sqrt{L/C}$ is then a smaller fraction of V_S). At lower current levels, less voltage magnification is obtained, and at zero current there is none.

It is usually not possible for the designer to take advantage of the magnification in designing the circuit. Pole current at commutation can change very rapidly, and the voltage on C does not stabilize at the magnified value until many cycles ($\sim Q$) have elapsed. Hence, the voltage magnification created by the pole simply results in conservative operation under many circumstances and significantly increased losses in and ratings of the commutating circuit components. The devices, main and commutating, are also subjected to the increased voltage stress, and the diodes and commutating thyristors see higher impulse currents as a result of the pumping action.

The recovery-time allotment in the ideal case, first commutation, is defined by

$$\omega t_{qa} = \pi - 2\arcsin(1/x)$$

where $\omega^2 = 1/LC$. Thus,

$$L = V_S t_q / x[\pi - 2\arcsin(1/x)]I_S$$

and

$$C = xI_S t_q / [\pi - 2\arcsin(1/x)]V_S$$

With a voltage magnification factor M, so that the peak capacitor voltage is MV_S and the peak impulse current MxI_S, the component rating is

$$R = M^2 V_S I_S t_q[x/(\pi - 2\arcsin(1/x)]$$

If M is independent of x, R exhibits a very flat minimum at $x \sim 1.53$; since M decreases with increasing x, the actual design optimum is for x somewhat higher, and increasingly so with increasing circuit Q. Designs are best executed by using a digital computer to perform iterative solutions of the transcendental equations that arise, including tuned circuit damping, since it is generally necessary to design for commutation of a rising fault current. The algorithm is quite simple: assign x, guess ω, calculate the two intersections of the current to be commutated with the impulse current when $M = 1$ (first commutation), compare t_q with desired value, adjust ω, and iterate until desired t_q is obtained. Then iterate calculations of capacitor voltage when commutating maximum steady-state current until stable, to determine M and R. Iterate whole program for various x values, seeking optimum for application purpose (which may not be minimum R—many other criteria are used in practice).

The trapped-energy problems of the McMurray pole led to the development of the *clamped McMurray* circuit depicted in Figure 9-6. It might better be termed the *damped McMurray*, for the purpose of resistor R_C and diodes D_{1a} and D_{2a} is to produce a heavily damped resonant discharge of C subsequent to commutation, so that the trapped energy that C acquires is partly dissipated in R_C and partly returned to the defined dc voltage. If R_C is selected so that resonant circuit Q with R_C inserted is in the range $1/\sqrt{2}$ to $\sqrt{2}$, M will be restricted to somewhere in the range 1.2 to 1.4. Moreover, not all the energy trapped by L and initially transferred to C is dissipated—from 30% to 70% is returned to V_S. The disadvantage is that the pole's turnaround time is considerably increased.

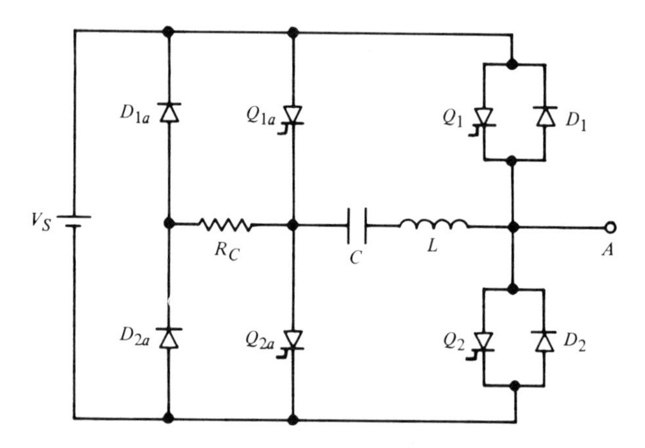

Figure 9-6. Clamped McMurray circuit.

In the simple McMurray, the maximum time from the initiation of a commutation to the condition of readiness for a subsequent commutation is one-half cycle time of the resonant frequency of the resonant circuit. Thus, the ratio of t_{c+} to t_q is

$$t_{c+}/t_q = \pi/[\pi - 2\arcsin(1/x)]$$

in the idealized case, or ~ 1.75 for $x = 1.6$. Ratios of 2 to 2.5 are common in practical McMurray circuits.

In most designs, the introduction of the damped discharge extends t_{c+} by more than one full cycle of the resonant frequency. Usually, at least one cycle of damped oscillation occurs, the first half-cycle via D_2, L, C, R_C, D_{1a}, for example, and then the second via D_{2a}, R_C, C, L, and Q_2 (which will be turned on to pick up I_S if that current reverses in the meantime). Since the natural frequency with heavy damping is considerably lower than that at high Q, turnaround time for the damped McMurray extends to 3.5 to 5 or more times t_q in practical designs, severely limiting operating frequency.

This problem led to attempts to develop a soft-commutating circuit that had neither trapped energy nor turnaround-time problems. The first attempts centered on the circuit depicted in Figure 9-7, which is clearly a relatively simple adaptation of the McMurray pole involving a Thévenin transposition of the impulse circuit.

The circuit of Figure 9-7 has an exact Thévenin equivalent in which the two capacitors are combined (parallel-connected) and returned to the midpoint of the defined dc source voltage. The behavior of both circuits differs from that of the McMurray though, as can be seen when analysis of the commutating behavior is performed. Consider the circuit with a center-tapped supply, and assume Q_1

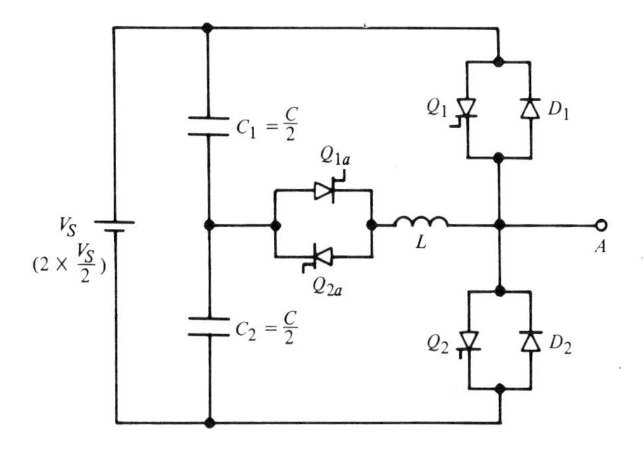

Figure 9-7. The "ac switched" modified McMurray circuit.

to be conducting I_S with C charged to V_S, with the plate connected to L being positive. A commutation is engendered by turning on Q_{1a}, when a resonant discharge of C via L, Q_{1a}, $Q_1 - D_1$, and $V_S/2$ begins, with the idealized defining equations

$$i = (V_S/2\omega L)\sin\omega t$$

and

$$e_c = (V_S/2)(1+\cos\omega t)$$

where $\omega^2 = 1/LC$.

This discharge continues until the second intersection of the impulse current and I_S, which occurs for

$$\omega t_2 = \pi - \arcsin(2\omega L I_S/V_S)$$

in the idealized case. Note that at this point, e_c is still *positive*, unless I_S is zero, in which case e_{c2} is zero. Thus, when the current in C and L freezes at I_S, the pole point is positive, and the second commutation cannot begin. The voltage on C now obeys the equation

$$e_c = e_{c2} - I_S\omega t/\omega C$$

and the pole voltage is, of course,

$$v_A = V_S/2 + e_c$$

Now if Q_2 is turned on before v_A reaches $-V_S/2$, a second commutating impulse will be generated, flowing via $V_S/2-C-L-Q_{1a}$ and Q_2. Since v_A is made $-V_S/2$ by turning on Q_2, D_2 can pick up the load current, I_S, once this commutating-impulse current falls below I_S. The impulse will end when the current in L and C reaches zero so that no energy will be trapped in L. The idealized defining equations are, with e_{c3}, the initial voltage on C,

$$i = [(V_S/2+e_{c3})/\omega L]\sin\omega t+I_S\cos\omega t$$

and

$$e_c = -(V_S/2)(1-\cos\omega t)+e_{c3}\cos\omega t-(I_S/\omega C)\sin\omega t$$

It is desirable, of course, that e_c should equal $-V_S$ when $i = 0$—i.e., when this second commutating impulse ends. It does so if $I_S = 0$, for $e_{c2} = e_{c3} = 0$, and ωt_4, where t_4 is the time at which the second commutating impulse current goes to zero, is π. For nonzero I_S, it can be made so by choosing the appropriate time for turning on Q_2. From the equation for the second impulse current

$$\begin{aligned}\omega t_4 &= \arctan[-\omega LI_S(V_S/2+e_{c3})]\\ &= \arctan(-y_1), \text{ for example.}\end{aligned}$$

Then,

$$e_{c4} = -(V_S/2)(1+1/\sqrt{1+y_1^2})-e_{c3}/\sqrt{1+y_1^2}-\omega LI_S y_1/\sqrt{1+y_1^2}$$

which can equal $-V_S$ if y_1 and e_{c3} have appropriate values. Although a closed-form solution for either can be extracted, it is so cumbersome that an iterative solution is preferable if computer facilities are accessible. If nonideal defining equations are used, as they should be for accurate analysis, then iterative solution is mandated.

The required value of e_{c3} gives the required time for turning on Q_2 from the relationship

$$e_{c3} = (V_S/2)(1+\cos\omega t_2)-I_S t_3/C$$

which obviously makes t_3 a function of I_S, since t_2 is a function of I_S and the second term of e_{c3} is also a function of I_S.

When the behavior of this circuit is analyzed in detail, for both ideal (infinite Q resonant circuit) and practical cases, a curious fact emerges. If *$\omega t_2+\omega t_3$ is maintained at π rad (i.e., if Q_2* is turned on one-half cycle of the commutating circuit's resonant frequency later than Q_{1a}, then the peak voltage on C stabilizes within 1% or 2% of V_S *regardless of* I_S, *the load current being commutated.* This rather pleasant behavior—no trapped energy and no commutating capacitor overcharge—should make the circuit a favorite. It has not done so because two commutating impulses flow, not one, and the second must be carried by a main load-bearing thyristor.

If Q_2 (or Q_1, for the other commutation of the pole) is turned on a little early, the peak voltage at which C stabilizes is a few percent higher than V_S. If the turn-on is delayed a little, then the peak voltage on C is a few percent lower than V_S. Thus, it is not necessary to have extreme accuracy in the control—a reasonable approximation to π rad delay from the turn-on of Q_{1a} (or Q_{2a}) will suffice.

All three soft-commutating circuits for voltage-sourced converters discussed so far suffer from a common disadvantage. It is necessary to introduce additional inductance between the dc voltage source and the pole's devices, main and commutating, for the McMurray and clamped McMurray, to provide means for dI/dt and dV/dt control. In most practical realizations, this inductance is split into two portions, one between the positive terminal of the dc voltage source and the anode of Q_1 (all three figures) and one between the negative terminal and the cathode of Q_2 (in clamped McMurray, the diodes D_{1a} and D_{2a} only are connected directly to the dc source). Often, the two portions of the dV/dt limiting inductance are coupled, and in many practical realizations of the poles, they are made saturable, since the inductance is only needed when the current in the dc loop is low. A configuration that avoids the need for these "spanning reactors," as they are sometimes called, is depicted in Figure 9-8. Its behavior is identical to that of the circuit of Figure 9-7, and it too has a Thévenin equivalent with the capacitance returned to the midpoint of the dc voltage source.

It is interesting to compare the passive component ratings required by these circuits, and to compare them with those of the McMurray, clamped McMurray, and McMurray-Bedford and the hard-commutating circuit without trapped energy. In the circuit of Figure 9-7, x is defined by

$$x = V_S/2\omega L I_S$$

and ωt_{qa} by

$$\omega t_{qa} = \pi - 2\arcsin(1/x)$$

The peak current in L, which determines L's energy storage rating, is only very slightly higher than xI_S, and decreasingly so as x increases, while the peak voltage on C may be taken as V_S. Hence

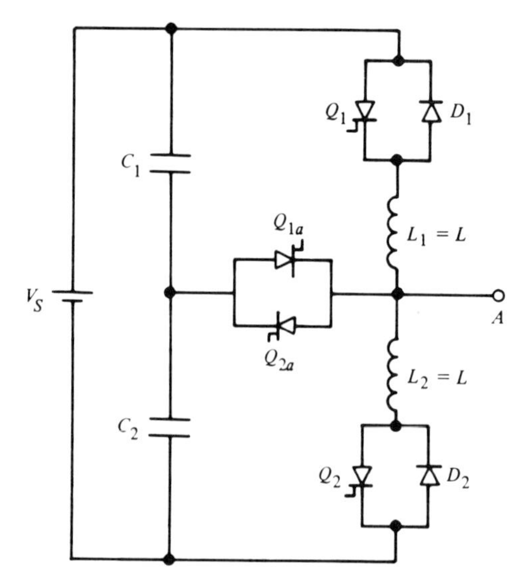

Figure 9-8. Further modified McMurray circuit.

$$L = V_S t_q / 2x[\pi - 2\arcsin(1/x)] I_S$$

and

$$C = 2x I_S t_q / [\pi - 2\arcsin(1/x)] V_s$$

giving $R = 1.25\ V_S I_S t_q x / [\pi - 2\arcsin(1/x)]$

which minimizes for $x = 1.53$. The circuit of Figure 9-8 produces the same L and C values, but since it requires two inductors, the multiplier 1.25 in R becomes 1.5 for that circuit.

Hence, the passive-component rating comparison can be drawn up as follows, with $V_S I_S t_q$ as a common factor in all cases:

- McMurray-Bedford: $(2\sqrt{1+x^2}-1)^2/2xG(x) - x/G(x)$

  with $G(x) = \arcsin(x/\sqrt{1+x^2}) - \arcsin(x/2\sqrt{1+x^2})$

  minimum, at $x = 1$, 6.3

- Hard-commutating circuit without trapped energy (Figure 9-4): $(x + 1/2x)[1/\arctan(x)]$

  minimum, at $x = 1.28$, 1.84

- McMurray: $M^2x/[\pi - 2\arcsin(1/x)]$

  where M is the circuit Q dependent multiplier V_C/V_S, V_C the peak capacitor voltage. Minimum, at $x = 1.53$, $0.891M^2$—for a typical M of perhaps 2.3, minimum is 4.71.

- Clamped McMurray: Same as McMurray; typical M, however, 1.3 giving minimum of 1.50.

- Circuit of Figure 9-7: $1.25\,x/[\pi - 2\arcsin(1/x)]$ minimum at $x = 1.53$, 1.114.

- Circuit of Figure 9-8: $1.5\,x/[\pi - 2\arcsin(1/x)]$ minimum at $x = 1.53$, 1.363.

It is also interesting to compare the ratio of turnaround time to t_q for the various circuits, for this factor provides an upper operating frequency limit for any pole. The expressions, with values for minimum rating designs, are:

McMurray-Bedford:	$\pi/G(x)$, 7.41
Circuit of Figure 9-4:	$\pi/\arctan(x)$, 3.46
McMurray:	$\pi/[\pi - 2\arcsin(1/x)]$, 1.83
Clamped McMurray:	$\simeq 3\pi/[\pi - 2\arcsin(1/x)]$, $\simeq 5.5$
Circuits of Figures 9-7 and 9-8:	$2\pi/[\pi - 2\arcsin(1/x)]$, 3.66

The advantages of circuits without trapped energy over those with it are clearly evident, as is the general component rating advantage of soft-commutating circuits over hard-commutating circuits. These comparisons are, however, very simplistic. They do not consider such factors as device stresses or the need for dI/dt and dV/dt control and should not be taken as indicative of any one best circuit.

One of the key problems in the design of impulse-commutating circuits for voltage-sourced converters arises because, in many applications, the dc voltage is not fixed but varies over quite a wide range. The design equations show, for any circuit, that the commutating capacitor and inductor must have the values appropriate for the lowest dc voltage that will be encountered (and, of course, the highest current to be commutated). Thus, impulse-current magnitude will vary, increasing over the design value as the dc voltage increases from its minimum value. Since commutating circuit losses are largely I^2R losses, the efficiency will drop sharply as the dc voltage increases. The wider the range of dc voltage over which the pole must operate, the more acute the problem becomes. As a result, many designers choose to use a controlled ac-to-dc converter or a dc-to-dc converter to regulate the dc voltage supplying a voltage-sourced ac-to-dc/dc-to-ac converter. The extra costs and losses incurred by such a step may be offset by the savings in costs and losses in the voltage-sourced converter's commutating circuits.

One further point must be considered. The hard-commutating circuit of Figure 9-4 and the McMurray circuit of Figure 9-5 confine their commutating impulse currents entirely to commutating circuit loops—no parts of the impulse currents flow in the dc source voltage. Split capacitor circuits, like those of Figures 9-7 and 9-8 and the McMurray-Bedford pole of Figure 9-3, circulate one-half of their commutating impulse currents via the dc source voltage, and center-tapped dc supply versions impose all their commutating impulse currents on that voltage. Since the design of the voltage source interface may be profoundly influenced by the presence or absence of commutating impulse current burdens, the behavior of the circuits in this respect must be considered in any comparative evaluation.

9.5 Impulse-Commutating Circuits for Current-Sourced AC-to-DC/DC-to-AC Converters and AC-to-AC Converters

Much less time and effort have been expended on commutating circuits for current-sourced converters than on those for their voltage-sourced brethren. This is partly because it has rarely proven necessary to apply impulse commutation to current-sourced converters, but also it is partly a consequence of the formidable problems encountered when such application is attempted.

AC-to-ac converters, being closely related to the current-sourced converters, have similar problems and are implemented with similar (if not identical) commutating circuits. In part, the problems of current-sourced-converter impulse-commutating circuits arise because the switches therein must block reverse voltage. This obviates the use of soft-commutating circuits unless thyristors, not diodes, are used as the inverse parallel-connected switch elements carrying the surplus impulse current, and the cost and complexity of so doing are generally considered unacceptable. The ac-to-ac converters using thyristors have them inverse parallel-connected, directly or implicitly, and hence can use soft-commutating circuits. However, most implementations have been with hard-commutating circuits, and all current-sourced-converter commutating circuits used or proposed are of the hard variety.

More serious than this is the unavoidable trapped energy problems that arise in both types of converter. The energy is not trapped in commutating circuit components, but in the inevitable and usually substantial reactances that form the bulk of the source impedance of practical defined ac voltage sets to which the converters connect. This trapped energy is the cause of commutation overlap when the converters are source-commutated, of course, and creates severe problems for impulse-commutating circuits.

Consider the simple circuit of Figure 9-9, showing an impulse-commutating circuit that might be applied to one "pole" of a current-sourced bridge converter. Q_1 and Q_2 are the main load-bearing thyristors, Q_{1a} and Q_{2a} are auxiliary (commutating) thyristors, C is the commutating capacitor, and Q_3 and Q_4 are the thyristors of the bridge pole next in sequence. The defined ac voltages, v_1 and v_2, each have purely reactive source impedances L_S as shown—an idealized representation that is not too far from reality since the reactance-to-resistance ratio (X/R) for most practical ac voltage sources is large.

Suppose now that Q_1 is conducting I_S into v_1 and that an impulse commutation is to be engendered so that I_S will be transferred to v_2 with Q_1 being turned off and Q_3 turned on. If an impulse commutation is necessary, v_2 is more positive than v_1 at this time. Assume C is charged to some voltage V_0 with the plate connected to the junction of Q_1's cathode and Q_2's anode positive. Then, when Q_{1a} is turned on, there will be an immediate hard commutation of Q_1, and I_S will transfer to Q_{1a} and C. However, it will continue to flow into v_1 via L_{S1}, and cannot be induced to participate in the second commutation, transferring to Q_3, L_{S2}, and v_2 for some time yet because the potential difference between point A and v_2 is even more inhibitory for such a commutation than it was prior to turning on Q_{1a}.

The recovery-time allotment for Q_1 is clearly defined by

$$I_S t_{qa} = CV_o$$

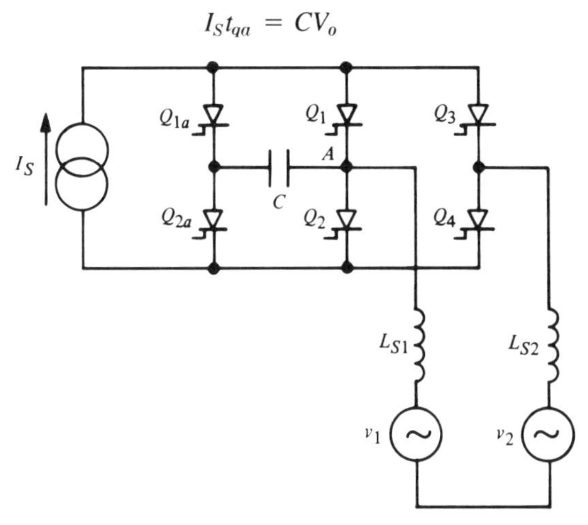

Figure 9-9. Hard-commutating circuit for current-sourced converter pole.

The second commutation cannot begin until I_S has flowed in C for a sufficient length of time to make the voltage at point A equal to that of v_2—i.e., until $v_C = v_2 - v_1$ with v_C of opposite polarity to V_0, the initial voltage on C. At that time, Q_3 can be turned on, and the second commutation can begin. It will not conclude, however, until sufficient volt-seconds have accumulated on L_{S1} and L_{S2} to extract the energy trapped in L_{S1} by I_S and build up the current to I_S in L_{S2}, thereby trapping the energy in L_{S2}. In effect, a major duty of the commutating circuit is to transfer current, and hence energy, from one source reactance to the next in sequence. In so doing, the commutating capacitor C will acquire a significant charge, as can be determined easily by analysis. Once v_C becomes equal to $v_2 - v_1$ and Q_3 is turned on, resonant circuit action takes place around the loop $v_1 - L_{S1} - C - Q_{1a} - Q_3 = L_{S2} - v_2$ until Q_{1a} turns off (the current goes to zero in L_{S1} and v_1) at which time I_S will be flowing via Q_3 into v_2. The idealized defining equations are:

$$i_1 = I_S \cos\omega t$$

$$i_2 = I_S - i_1$$

$$e_c = v_2 - v_1 + (I_S/\omega C)\sin\omega t$$

where i_1 is the current in Q_{1a}, C, L_{S1}, and v_1; i_2 is the current in Q_3, L_{S2}, and v_2; $\omega^2 = 1/2L_SC$; and e_c is the capacitor voltage. The simplifying assumption that v_1 and v_2 remain constant during the second commutation has been made. This is not the case, of course, but the analysis for the actual behavior of v_1 and v_2 is quite cumbersome and beyond present scope.

Clearly, i_1 goes to zero for $\omega t_2 = \pi/2$, at which time the capacitor voltage is given by

$$\begin{aligned} e_{c2} &= v_2 - v_1 + I_S/\omega C \\ &= v_2 - v_1 + 2\omega L_S I_S \\ &= v_2 - v_1 + \sqrt{2L_S/C}\, I_S \end{aligned}$$

For steady-state operation, this should equal V_0, since the complementary commutation, from Q_2 to Q_4, will begin, with C charged to this voltage, when Q_{2a} is turned on. It is easily seen that the minimum V_o occurs when $v_2 - v_1 = 0$ (which occurs at the converter operating boundaries $\alpha = 0$ and $\alpha = \pi$), and thus

$$C\ 2\omega L_S I_S = I_S t_{qa}$$

gives the minimum t_{qa}. To provide a given recovery time, t_q, the relationship $\omega t_q = 1$ must hold, or

$$C = t_q^2/2L_S$$

giving the voltage augmentation on the capacitor as

$$\Delta V_C = 2\omega L_S I_S = 2L_S I_S/t_q$$

If the supply frequency is f_s, angular frequency $\omega_s = 2\pi f_s$, and the per-unit reactance x is $\omega_s L_S I/V$, where

$$v_1 = V\cos\omega t$$

(i.e., V is the crest value of the defined ac source voltage) and

$$I = (4/\pi)\sin(\pi/M)I_S$$

is the crest value of the fundamental component of the current flowing in v_1, with M the number of phases in the bridge, then

$$\Delta V_c = \pi x V/[2\omega_s t_q \sin(\pi/M)]$$

For the important three-phase case, this reduces to

$$\Delta V_c = \pi x V/\sqrt{3}\omega_s t_q$$

at $x = 0.05$ (on the low side for most practical sources) and with $t_q = 60$ μsec, ΔV_c for a 60-Hz supply is $4V$. Thus, the peak voltage stress on the converter's thyristors, and on C, would be ~ 3.3 times the peak line voltage of the supply.

It follows that long t_q's are the only way to keep peak voltages reasonable—at a t_q of 240 μsec, ΔV_c is V and the worst-case stress is only $2.73V$, which is much more tolerable. The higher the per-unit source reactance, the longer the design allotment of recovery time must be to limit peak voltage to reasonable bounds. Fast recovery devices are of no benefit to impulse-commutated current-sourced converters, or, by extension of the same arguments, to ac-to-ac converters, unless ω_s is high or x (per-unit reactance) is very low.

The energy-storage rating for C, which is $CV^2_{CMAX}/2$, is

$$R = C(V+\Delta V_C)^2/2$$

$$= VI_S t_q\ (2/\pi)\sin(\pi/M)\omega_s t_q[2\sin(\pi/M)+x\pi/\sqrt{3}\omega_s t_q]^2$$

which minimizes when $\omega_s t_q = \pi x / 2\sqrt{3}\sin(\pi/M)$ and $\Delta V_C = 2\sin(\pi/M)V$, the peak line voltage of the supply.

A number of commutating circuits based on the arrangement of Figure 9-9 have been proposed. They differ mainly in the implementation of the commutating thyristors. Figures 9-10 through 9-12 show arrangements suitable for use with three-phase bridges. That of Figure 9-10 uses only a single pair of commutating thyristors, but three capacitors. Figure 9-11 shows an arrangement using only one capacitor, but needing a total of eight auxiliary thyristors. The circuit of Figure 9-12 is interesting because six of the total of ten auxiliary thyristors involved are the main load-bearing thyristors of a source-commutated converter, the configuration being one of "complementary" current-sourced converters. All these circuits can, with appropriate extensions, be used with ac-to-ac converters.

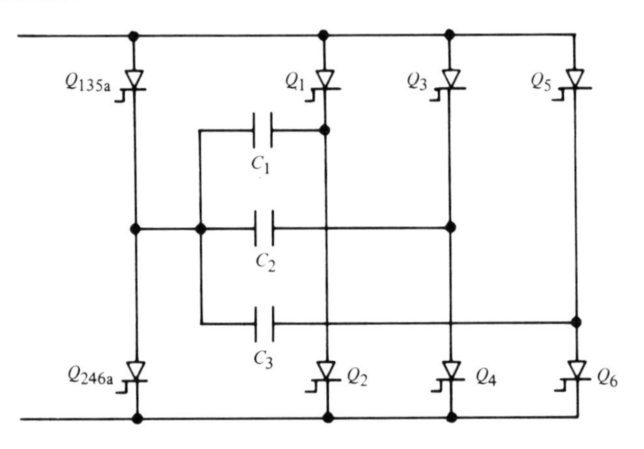

Figure 9-10. Three-capacitor circuit for three-phase bridge.

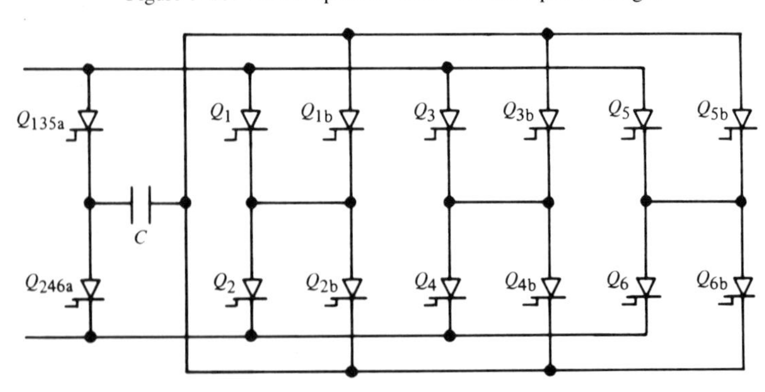

Figure 9-11. Single-capacitor circuit.

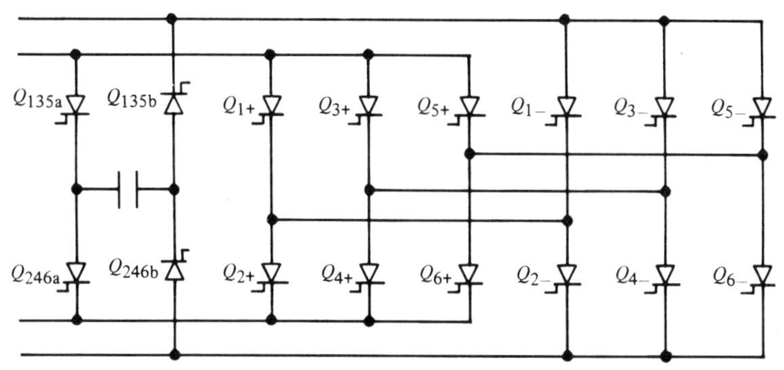

Figure 9-12. Complementary converter circuit.

None of these circuits has acquired any popularity in the commercial world because of their use of auxiliary thyristors. The circuit depicted in Figure 9-13, which is essentially auto-commutating, is found most often in actual use. In this circuit, the two midpoint groups of the bridge commutate independently, and the diodes serve to "lock" the voltage on the capacitors after a commutation and to ensure that the peak commutating voltage is never less than the peak line

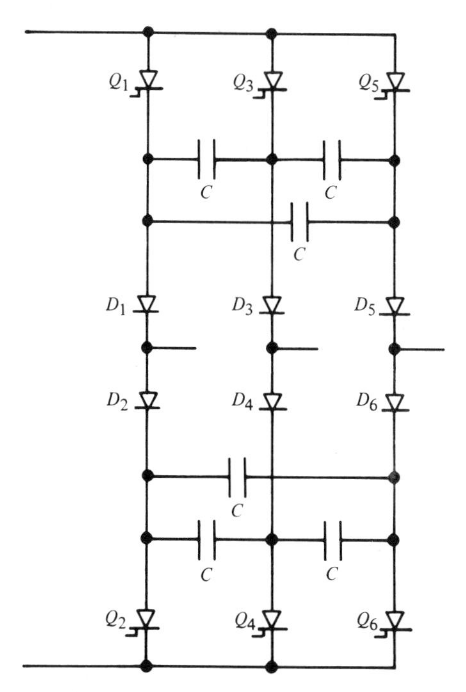

Figure 9-13. Auto-commutating circuit.

voltage of the supply. If $v2-v1+\Delta V_C$ is less than the peak line voltage, then the diodes ensure that the composite C, $3C/2$, charges to the peak line voltage between commutations; if $v2-v1+\Delta V_C$ is greater than the peak line voltage, the diodes ensure the composite capacitor voltage remains at that level until the next commutation. Despite the autocommutating ability of the circuit, the extra losses incurred by having the diodes in the main current paths are far from negligible, and it is highly probable that the circuit of Figure 9-10 is both more economical and more efficient.

These circuits all share the gross disadvantage of the dc-to-dc converter of Figure 9-1—turnaround is a function of the dc terminal current, since it is the only current acting to change capacitor voltage. The modification to improve matters is essentially the same—a resonant reversing circuit is added. For these commutating circuits, inverse parallel thyristors must be used as the switches in the reversing circuit; thus, the option of timing the inception of reversal to get greater consistency of peak capacitor voltage as I_S varies is available, but is rarely used. A good discussion of the design effects of such reversing circuits can be found in Gyugyi and Pelly (1976).[2]

As stated, ac-to-ac converters bring back the option of soft commutation provided the converter's switching devices are appropriately controlled. This is a point of some significance, as the discussion in Section 9-4 indicates that soft-commutating circuits generally need lower passive component ratings than do their hard-commutating relatives. Unfortunately, this is not the case when the ac source reactance dominates commutation requirements; the soft-commutating circuit is soft only for the first commutation, as it turns off a main converter thyristor and assumes conduction of the converter current. Once the transition stage is entered, while the capacitor is accumulating the necessary voltage for the second commutation to begin, and subsequently while the second commutation is proceeding, the circuit to all intents and purposes is equivalent to a hard-commutating circuit. There is thus little point in a detailed discussion of soft-commutating circuits for ac-to-ac converters. An excellent dissertation can be found in Gyugyi and Pelly (1976).[2]

9.6 Components in Impulse-Commutating Circuits

The basic passive components of all commutating circuits are the commutating capacitor and commutating inductor. Depending on converter application and commutating circuit configuration, two (or more) of either may be used for a pair of converter thyristors. However, they then are merely fragments of a composite component. In all cases, the duties are quite similar. Short-duration current pulses must be passed through the components, and rapid voltage changes take place across them. In consequence, the actual components used differ quite markedly in detail design from those used for converter interfacing duties. Components designed for general-purpose use are simply not suitable for commutation circuit applications and will almost invariably fail if subjected to commutation-circuit-imposed stresses.

Commutating capacitors are almost always subjected to full voltage reversals, so that ac dielectrics are mandated. When split C arrangements such as those of Figures 9-7 and 9-8 are used, the capacitors are also subjected to continuous dc stress, compounding dielectric problems. Even though the rms current in a commutating capacitor is often moderate, if the commutation rate (operating frequency) is low, the instantaneous current is always high and the rate of change of field within the capacitor is very high compared to that experienced by capacitors in ac interfacing. As a result, extended foil construction is mandated, and the dielectric systems of commutating capacitors must be suitable for high-frequency ac application even though commutation rate will be low to moderate. Manufacturers make special capacitors for this type of duty. At the higher voltages, dielectric systems are almost universally paper-oil, but for voltages lower than about 1000 V, many polycarbonate and other plastic-film dielectrics are used. It should be noted that standard ac capacitors will almost always fail if used in commutating circuits, and the types made for the application should be selected for use.

Similar considerations arise with commutating inductors, which are usually "home brewed." Some see only commutating impulse currents; some also have to carry main load current. In the latter case, the load current is generally a thermal consideration only—it does not define magnitizing force, where magnetic cores are used, nor is it overly significant in conductor selection. Because of the high amplitude and short duration of commutating impulse currents, skin and proximity effects are prime considerations; Litz conductors are used in many, if not most, commutating reactors. Where coupled reactors are needed, magnetic cores must be used, for it is impossible to get good coupling between air-cored reactors. When air-cored reactors can be used, they usually are since the peak magnetizing force developed by the commutating impulse generally negates the effect of a magnetic core as far as volume and weight reductions are concerned.

A key point in commutating inductor application and design, which is often overlooked, is that insulation stresses are much more severe than in a standard ac inductor. This because whenever a commutation is engendered, the reactor is subjected to a very fast rising voltage wavefront. This also can cause problems with transient voltage distribution within the reactor. Care must be taken that stray capacitance from the reactor to the reference plane for the commutating voltage is very low; otherwise, one or two reactor turns will have to support the full initial-voltage transient. However, the self-capacitance (turn-to-turn capacitance included, of course) of a commutating reactor must be low if the component is to serve its purpose properly, for its self-resonant frequency must be much higher than the commutating-circuit resonant frequency. Since transient voltage distribution is a function of the ratio of turn-to-ground (reference plane) to turn-to-turn capacitance, and worsens rapidly as that ratio becomes other than miniscule, the constraints are severe. This is another reason for not using magnetic cores in commutating reactors, for a core inevitably increases the stray capacitances associated with the reactor.

References

1. Bedford, B. D., and R. G. Hoft. *Principles of Inverter Circuits*. Chapter 7, pp. 190–206. New York: John Wiley & Sons, 1964.
2. Gyugyi, L. and B. R. Pelly. *Static Power Frequency Changers*. Chapter 8. New York: John Wiley & Sons, 1976.

Problems

1. A simple commutating circuit of the type depicted in Figure 9-1 is to be used in a buck dc-to-dc converter operating from a source voltage ranging from 210 to 280 V with a dc output current range of 10 to 50 A. The main thyristor is specified to have a recovery time (t_q) of 25 μsec. Calculate

a. The value of the commutating capacitor needed.
b. The value of the reversing inductor needed to keep the peak reversing current to no more than the maximum dc current, and the reversing time then extant.
c. The value of reversing inductor needed to make the reversing time no longer than $2t_q$, and the peak reversing current then extant.

2. For the converter and commutating circuits of Problem 1, calculate the absolute maximum operating frequencies if the output dc voltage is to be

a. fixed at 180 V
b. variable from 30 to 150 V

3. The designer of the converter of Problems 1 and 2 decides to incorporate an ancillary diode-inductor discharge path in shunt with the main thyristor. If it is designed so that t_{c+}, the maximum commutation time, is 100 μsec, what values of commutating capacitor and auxiliary inductor will be needed?

4. A voltage-sourced ac-to-dc converter pole is to be operated from a dc source voltage ranging from 200 to 350 V with a defined ac current of 100 A magnitude and main thyristors having $t_q = 30$ μsec. If the worst-case commutation occurs with $\alpha = \pi/2$, calculate the values of L and C needed in the commutating circuit if the configuration is

a. McMurray-Bedford (Figure 9-3a) with $x = 1.07$
b. Figure 9-4 with $x = 1.28$
c. Figure 9-8 with $x = 1.53$

5. For the pole designs of Problem 4, calculate the turnaround times and total passive component ratings.

6. Explain, using appropriate numerical examples, why thyristors with very short t_q are of little benefit to an impulse commutated ac-to-ac converter operating with 60-Hz input and output. What sort of ac-to-ac converter could this be?

10
Control for Converters

10.1 Introduction

Since semiconductor switches were first used, converter controls have undergone an evolution parallelling that of the communications and data-processing electronics from which they spring. Controls cause the switches in a converter to assume the existence functions pertinent to their application. In the early days of power semiconductors, converters could only hope to compete in modest power-level applications, and costs had to be low. Out of necessity, controls had to be fabricated using discrete components and devices, and in order to be low-cost had to be very simple.

As time passed, the situation changed. Power switching devices grew in capability, so that converters could penetrate application areas where control costs were not quite such an overriding consideration. At the same time, the emergence of small-and medium-scale integrated circuits allowed the realization of more control sophistication per unit cost, hastening the development of better controls and control techniques. Further expansion of switching device capability has now propelled converters into very high power application areas where control costs, no matter how high they may seem when considered in isolation, are virtually negligible. Coupled with continuing low-power device development, this has led to the development of very sophisticated control schemes, each essentially a small dedicated hybrid computer, serving those applications' needs. The recent emergence of large-scale integrated circuits, particularly the microprocessor, has caused the development of fully digitized converter controls with capabilities beyond the wildest dreams of those who struggled to implement "one transistor" controls in earlier years.

A converter's control in fact consists of four basic inter-communicating and interdependent functional subdivisions. As depicted in the block diagram of Figure 10-1, these are:

1. The *gate or base drive "channels,"* which provide "amplification" of low-level logic signals and isolation so that the switching devices have appropriate level control signals impressed.
2. The *logic section,* which is the heart (or perhaps the brain) of the control. This function responds to stimuli from (3) and (4) and feeds pulse patterns, usually low-level replicas of existence functions, to the gate or base drive channels so that the converter performs the desired function.

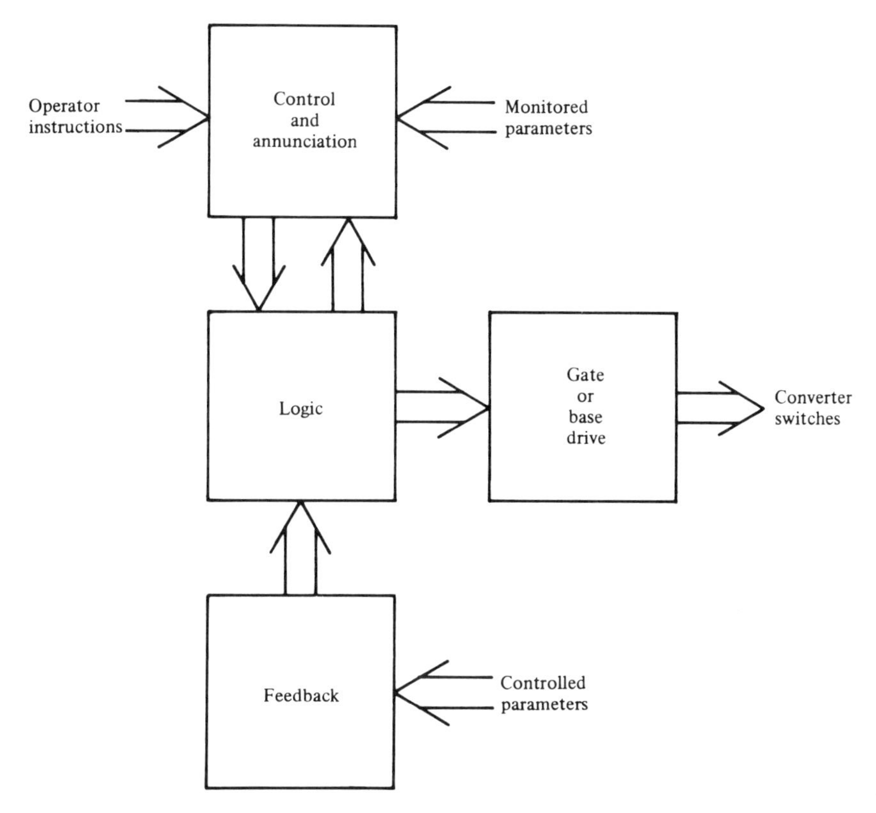

Figure 10-1. Block diagram of converter control.

3. The *feedback loop(s)*. Until recently, the feedback loops were always implemented through analog circuitry. They provide the signal processing whereby closed-loop control of converter parameters is achieved. In addition to the normal duties of feedback circuitry in linear systems, the feedback loops must "translate" the results so that the logic will understand and respond correctly.
4. The *control and annunciatory section* provides the means for the converter to communicate with and receive instructions from its operators, be they human or machine.

This chapter offers a glance at converter controls. Problems and design principles for gate and base drives, the most independent and, as far as semiconductor technology is concerned, the most unyielding of the control functions, are discussed briefly in Section 10.2. The logic control techniques typically used with a variety of converters are given similar treatment in Section 10.3, and feedback control is briefly covered in Section 10.4. Some aspects of digitization and microprocessor-based control schemes are introduced in Section 10.5.

10.2 Gate and Base Drives

Except in the very lowest power converters, gate and base drive channels are still implemented with discrete component and device technology. This is because

1. the power levels required are generally not conducive to the effective use of integrated circuit technology.
2. the high level of electrical isolation usually required between control logic and power circuitry demands a certain "robustness" of the drive channels.
3. the large amplitude and high speed of voltage and current transitions at the switch can feed back, with disastrous effects, to low-level energy-sensitive circuits.

Throughout this section, *thyristor* is used to include most of the thyristor's close relatives as well as the thyristor itself. Since the thyristor is a latching device, a logical conclusion is that its gate drive need be only a single short pulse whenever the device is to be turned on. That is not always the case; indeed, unless the thyristor's conduction period is short and the turn-on dI/dt is high, it is rarely the case. Two factors work to create a demand for extended or repeated gate drive. The obvious device-related factor is latching current—gate drive must be present until the anode current rises above the latching level for the thyristor. More subtle is the tendency of many circuits to produce short-duration oscillatory currents, caused by parasitic inductances and capacitances, whenever switches close or open. The result can be the unwanted turnoff of thyristors in those circuits unless gate drive is maintained.

To supply single short-pulse gate drive with isolation up to a few hundred volts, a single transistor and pulse transformer can provide the channel required. The transformer core must be allowed to reset properly (flux return to zero between pulses) and, in the process, must not produce excess reverse-gate cathode voltage. Inverse parallel-connected diodes are often used at thyristor gate-cathode junctions to prevent this. In the early days of thyristor applications, the blocking oscillator configuration was quite popular since it made definition of the pulse length a function of gate-drive channel design without the need for external defining logic. In addition, it could provide trains of pulses whenever the controlling input bias was above the design threshold. Because of the stresses a blocking oscillator imposes on its switching transistor, it lost popularity as integrated circuit costs fell and the implementation of pulse-duration control by external logic, usually a monostable multivibrator, became attractive. As the required pulse duration of a single pulse drive becomes longer, problems with the pulse transformer increase. It becomes larger, and thus more costly and more difficult to reset. One way of alleviating the latter problem somewhat is to invert the mode of operation. In "energy storage" drives, the transistor is on when no gate drive is required, drawing a resistor-limited magnetizing current through the pulse transformer primary. When gate drive is needed, the transistor is turned off, and the gate drive (magnetizing current transfers to the secondary and becomes gate current) allows core reset. However, the problems of making a transformer with leakage reactance low enough to get the necessary gate-current rise time and sufficiently low self- and interwinding capacitance to avoid undesirable resonance and feedback effects become acute for pulse durations greater than 100 to 200 μsec and unmanageable for pulse durations much greater than 0.5 msec.

It was popular in earlier times, and still is in some quarters, to approximate a continuous or extended gate drive by means of a "picket fence" drive. This provides a train of pulses, each of duration that the designer hopes is sufficient to ensure latching and with mark-space ratios ranging from 1:3 to 1:10. Such a pulse train is naturally generated by a free-running blocking oscillator not inhibited by external bias—hence the early popularity of the approach. Two dangers exist when such a gate drive is employed. First, it is common that the pulse train and the thyristor switching rate are asynchronous. If great care is not taken, the first gate-drive pulse delivered may be incomplete, even to the extent of risking dI/dt damage to the thyristor. Synchronization at each thyristor switching is mandatory and often difficult to achieve. The second problem is that the resolution of the pulse train may not be adequate. Should the thyristor be turned off by some power circuit phenomenon at an inopportune time, it will be turned on again by the next pulse in the train. If the spacing between pulses is too great, then power circuit malfunction can result; in many cases, little or no delay in refiring is tolerable, and picket-fence drives are then totally unacceptable. Even where some delay can be tolerated, the resolution of the picket fence must be tailored to the power circuit's needs.

In certain cases, continuous gate drive is mandatory, and in almost all instances where thyristor conduction exceeds a few hundred microseconds, it is highly desirable. It is both prohibitively expensive and unconscionably difficult to provide long-term continuous drive by the simple transistor-transformer route. If dc drive is required, it is physically impossible to provide it via a pulse amplifier. The technique in common use to provide extended drive, including dc drive, is simple carrier drive. A power oscillator is used to generate a moderately high frequency (generally from 15 to 50 kHz) carrier, usually square wave, which is applied to transformers when controlling transistor switches are on, but not when they are off. The carrier is simply rectified at the transformer secondary to provide gate drive.

The transformers are thus small, relatively cheap, and easy to make with low leakage inductance and low parasitic capacitances. Such a drive is very expensive for a single thyristor, but becomes more attractive as the number of thyristors increases since a single power oscillator can serve them all, or a substantial number of them. However, it is sufficiently more costly than single-pulse or picket-fence drives to discourage its use except where necessary.

Transistors require continuous drive. They are, however, generally operated at frequencies high enough that simple pulse amplifier techniques can be used. Because of their relatively low power gain, and variable current gain, emitter current feedback is often used to relieve the base drive circuit of much of the burden of holding the transistor on and to ensure that the transistor will neither pull out of nor be driven too deep into saturation while conducting. Transistors also require reverse base drive at turnoff; this is generally provided by making the base drive circuit symmetrical, producing as much reverse base current as forward. Care must be taken so that the reverse base current does not avalanche the transistor's base emitter junction after the storage time, either by limiting the available voltage (usually not a reliable technique) or by providing a defined voltage alternate path in shunt with the base emitter once that junction recovers. Diode-zener diode combinations are sometimes used; simple resistors are more common.

Thyristors can also benefit from reverse-gate bias, and GCSs and GATTs require it. Even if a thyristor's switching behavior is not affected in any way by reverse-gate bias, the power circuit can benefit substantially because of the reduced risk of turn-on due to spurious noise-related signals. With simple pulse-amplifier drive channels, or blocking oscillators, the only reverse bias available will be that incidentally provided by core reset. The high-frequency carrier drive can be adapted to supply reverse bias continuously whenever the device is off, at the cost of an additional transformer and secondary rectifier per thyristor. For GATTs, the drive circuit must be made symmetrical so that the reverse gate current needed for dV/dt enchancement can be provided. GCSs make for considerably greater difficulties—their reverse gate-drive requirements at turnoff are typically an order of magnitude or more greater than forward gate-drive requirements at turn-on. Thus, they need quite high power (tens to hundreds of watts) gate-drive circuits in which the reverse-drive short pulse is dominant, and the forward drive is almost incidental.

Both thyristors and transistors benefit from high-speed transitions in their drive currents. They usually require that the initial current level be somewhat higher than that maintained through the conduction period or the greater part of the pulse duration. Typically, the high initial drive can be allowed to decay on a 2-to-5-μsec time constant. The simplest way to implement this is to design the circuit with greater voltage capacity than necessary for the steady-state drive and insert a resistor shunted by a capacitor in series with the base or gate. The connection may be direct, at the transformer secondary, or coupled, in the primary circuit. On energizing, the capacitor affords no initial impedance and allows the fast rising overdrive. However, it charges, directing drive current into the resistor, which then limits that current to the final desired value. One benefit of this approach is that when the gate- or base-drive source shuts off, the capacitor will provide short-duration reverse bias. Often, this can be arranged to be sufficient for a GATT, and occasionally it can be made adequate for a transistor.

More complex schemes, involving local switches in the immediate gate or base network, have been used, particularly with high-frequency carrier gate-drive circuits. They are not generally favored.

Once the required isolation level between gate drive and thyristor exceeds a few hundred volts, it becomes increasingly difficult to make pulse or carrier transformers with the necessary insulation and the low-leakage reactances needed for fast rise-time pulse transfer. The same problem would arise with transistor-base drives, of course, but as yet transistors are not capable of high-voltage operation and are not used in such circumstances. As a result, the "conventional" gate drive circuits are abandoned for high-voltage applications, and more complex arrangements are used. These are inevitably more expensive than the simple circuits used at low voltage levels, and thus have provided a primary motivating force for the development of radiation-triggered thyristors.

The most common arrangements for providing gate drive to thyristors at high voltage are two-channel schemes. One channel carries the energy required for gate drive, and stores it in a local repository, either capacitor or inductor, at the gate locale. If speed is of no concern, it is relatively easy to design transformers, particularly current transformers, that have the necessary insulation. However, the transformer for an individual gate drive is small, and insulation requirements dominate its cost if the thyristor operating voltage is more than a few kilovolts. Hence, cascaded transformers are often used, with one transformer of quite moderate insulation supplying the energy for two to eight contiguous gates, its primary being fed by a larger transformer serving many more thyristors and bearing most of the isolating and insulation burden. Multilevel cascades, involving three or more transformations, may be found in very high voltage systems.

Where single pulse drive can be used, the thyristor's snubber networks are often used as the source of gate-drive energy, eliminating one channel of the gate-drive scheme. This technique needs care in the design of the pick-off circuitry whereby the energy is transferred to the gate, because, when the thyristor is supporting maximum voltage, its snubber capacitor will be charged to that voltage and stores more energy than the gate, or the local gate circuitry, can safely absorb. Also, a thyristor so gated cannot be turned on from low anode-cathode voltages, because there is insufficient snubber capacitor voltage to produce the gate drive.

The second channel of a high-voltage gate-drive scheme involves a switch, usually transistor, at the gate locale to control the delivery of the gate drive and the means for providing the low-level signal controlling that switch. Transformer-coupled transfer of those signals has been used because, when the energy level is low, it is relatively easy to provide sufficient voltage-forcing to overcome high leakage reactance and produce fast rise times despite transformer problems. Modern systems, however, use optical coupling. The control (logic) output is delivered via light-emitting diodes, gallium arsenide, or gallium phosphide types, which radiate in the infrared region and couple to fiber-optic links. These provide the electrical isolation needed, and deliver the radiation to photoreceptors, silicon PIN photodiodes, or phototransistors, which are part of the local switch at the gate. Such schemes are easy to implement, modest in cost, and highly reliable. Nonetheless, this use of radiation to indirectly gate thyristors has led to the exploration of systems using thyristors directly gated by infrared radiation, since the junction regions of a thyristor will generate carriers under photon bombardment.

These efforts have met with limited success. It has not proven possible to make large-area thyristors sufficiently radiation-sensitive to be triggered on by the radiation levels produced by normal light-emitting diodes. It has proven difficult, in fact, to make the thyristors sufficiently sensitive to operate from the radiation levels produced by laser diodes. These pose two problems when predicated for use in converter gate-drive schemes. First, their own reliability and life, which have been notoriously poor and short to date. Technology improvements have ameliorated this problem, and ultimately it should disappear. Second and more serious, laser diodes themselves need very short-duration high-amplitude current pulses to excite them—75 A for 150 nsec is not atypical. The circuitry needed to produce this excitation is as complex, costly, and inefficient as a standard two-channel electrical drive scheme using optical coupling of control signals. Thus, there is little incentive for the radiation-fired thyristor's use in high-voltage applications, nor will there be until more reliable and more easily excitable radiating sources can be found.

10.3 Logic Circuitry in Converter Controls

As stated in the introduction to this chapter, it is through the control logic that the converter switches' existence functions are implemented. When fully controllable switches are in use, the logic controls the base- or gate-drive channels to turn on a switch at the beginning of the unit-value period of its existence function and to turn it off at the end of that period. When thyristors are used, this cannot be the case. In all autocommutated converters, including source-commutated converters, a thyristor's existence function becomes unity when the thyristor is gated on and becomes zero once more when another thyristor is gated on. Thus, it is only necessary that the control logic maintain the proper sequence of gate drive pulses to the converter thyristors.

In auxiliary impulse-commutated converters, still other factors come into play. A thyristor, or thyristor switch, existence function still assumes unit value whenever the thyristor is gated on, unless it did so previously as a result of an inverse parallel-connected diode assuming conduction. However, the thyristor-switch existence function is now returned to zero when or sometime after an auxiliary commutating thyristor is gated on. Depending on commutating-circuit configuration, there will be a prescribed minisequence of gate drives that the logic must produce, repeated within the main sequence of converter-switch operations.

When logic was implemented by discrete component technology, because of economic and technical considerations, many elegant and simple control schemes were devised for converters. Although designers have long been freed of the constraints that originally forced adoption of simple techniques, many techniques warrant continued usage. Only the advent of the microprocessor, which is incapable of reproducing certain control functions because of its discretized and serial nature, has caused changes in the best control concepts.

A major interface for the control logic is that with the feedback loop, for through that interface the converter is made to perform its application function. One needs only a nodding acquaintance with feedback control theory to realize that there exists a great advantage in having the converter, as seen by the feedback system, appear to be a linear amplifier. The creation of that appearance has long been a subsidiary goal of control logic design, and it has been achieved in many instances.

Single-quadrant dc-to-dc converters having only one controllable switch are a good example of how simple control logic can be. They illustrate several basic principles common to a great many converter controls. In such a converter, control of the dependent quantities is achieved by controlling the relative duration of the unit- and zero-value periods of the controllable switch existence function. Constant frequency operation is not universal, but is the most common operating mode. A very simple way of generating the existence function is depicted in Figure 10-2a. A sawtooth wave is generated and compared with an adjustable dc level, with the intersections of the two defining the bounds of the unit-value and zero-value periods. As shown, the existence function has unit value whenever the sawtooth is more positive than the "reference"; the complementary definition could be implemented equally well, with the unit-value period of the existence function corresponding to those periods when the sawtooth is more negative than the reference. Also, the sawtooth of Figure 10-2a is positive-going, with negative-going reset. One intersection of sawtooth and reference always occurs at sawtooth reset. If the implementation shown is used, the trailing edges of the existence function's unit-value periods are fixed in time; if the complementary implementation is used, the leading edges of those periods are fixed in time.

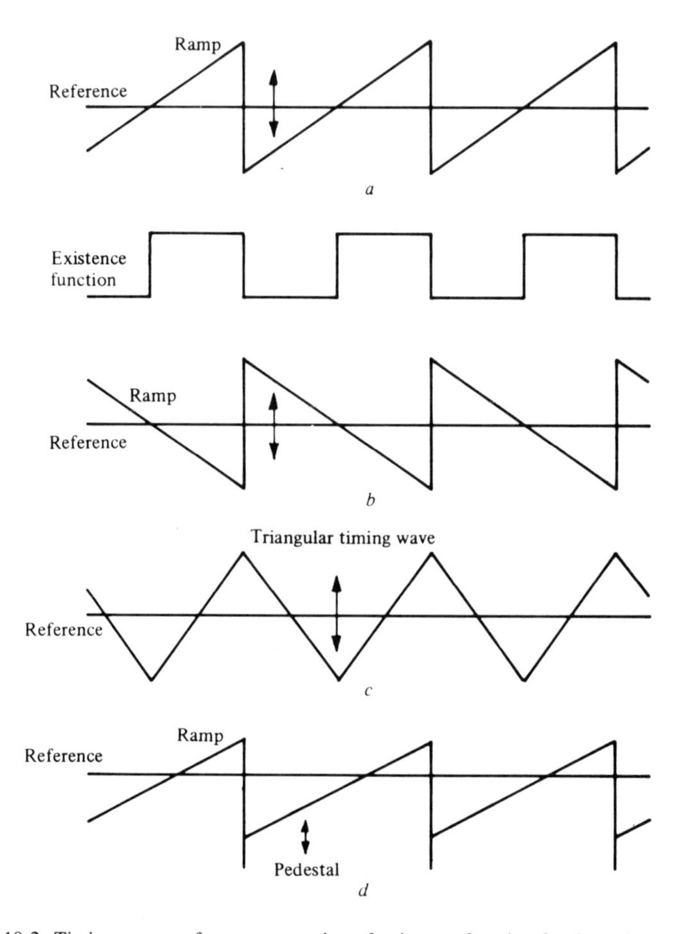

Figure 10-2. Timing wave–reference generation of existence function for dc-to-dc converters.

The general technique illustrated is "timing wave–reference wave" generation of existence functions. The sawtooth could be made negative going, with positive reset, as depicted in Figure 10-2b, in which case the existence functions for given criteria would be the complements of those generated by a positive-going sawtooth with the same criteria and reference. Alternatively, a triangular wave can be used for timing, as illustrated in Figure 10-2c. In this case, neither edge of the existence function's unit-value period is time-fixed—the center is. Figure 10-2d shows yet another possibility, which in the past was widely used for a number of converter types. Termed *ramp and pedestal,* it comprises a sawtooth intersecting a fixed dc reference. Variation of the period between intersections is achieved by varying the dc level, or pedestal, upon which the sawtooth is generated. It should be obvious that any periodic wave with periods in which the amplitude is not time-invariant can be used as the timing wave. Almost all dc-to-dc converters use one of the techniques shown in Figure 10-2; other converters use more complex timing and reference waves.

How would such a scheme behave if the reference wave (dc in these cases) exceed the bounds of the timing wave? Since there would be no intersections, no existence-function pulse train would be generated. Hence, the existence function would be either permanently unity or permanently zero, depending on the manner in which it was normally generated and the direction of excess of the reference wave. In a buck converter, this might be regarded as no great disaster—the result is simply loss of control, until the reference returns within timing-wave bounds, of converter-dependent quantities. In a boost converter or either of the buck-boost converters, such a condition could prove disastrous. Recall that current sources are fabricated by using inductors in conjunction with real voltage sources. In such a situation, a permanently-on switch will lead to very high current in the inductor, with destructive results.

This scenario makes clear the need for end-stops in converter controls. End-stops are the means whereby the basic existence-function generating circuitry is prevented from ever losing the capability to generate an existence-function pulse train. They ensure that a pulse train will always be generated, no matter what the attempted excursions of the reference with respect to the timing wave.

In the examples shown in Figure 10-2, it is conceivable that end-stops could be implemented simply by limiting, through the use of stiff voltage clamps, the permitted excursions of the reference. This is not recommended. It requires very accurate control of timing-wave amplitude and clamp levels, which is often difficult and sometimes impossible. Moreover, it is not possible to approach the "theoretical end stops" (the ideal limiting case of infinitesimally short unit- or zero-value periods) very closely if reference clamping is adopted, for the variations that will occur in timing-wave amplitude and clamp level must be accounted in the design. What is worse, perhaps, is that the closeness of approach to theoretical end-stops will vary as timing-wave and reference clamps vary over their tolerances, and will vary from unit to unit of a number of such controls made.

A much better end-stop implementation is shown in Figure 10-3. Reference clamps are still used, but beyond the maximum bounds of the timing wave and with little concern for their accuracy. End-stop pulses are added to the timing wave at its resets to ensure two intersections with the reference under any conditions. Obviously, the approach to theoretical end-stop is now defined by the time duration of these pulses. They can be generated by triggered astable multivibrators, or be selected pulses from a synchronous clock. Moreover, end stops are guaranteed provided that the pulse amplitude can be guaranteed to exceed clamp levels. This is not difficult to do, and so it proves to be a very secure means for end-stop generation.

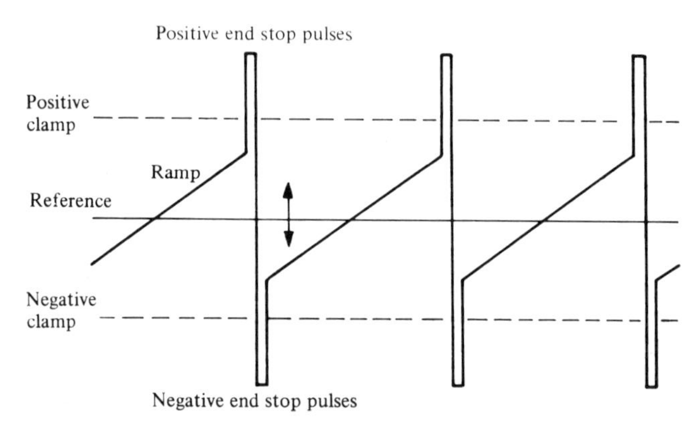

Figure 10-3. End-stops produced by adding pulses to timing wave.

All the control techniques illustrated in Figure 10-2 result in a linear variation of existence-unit value period with the reference level (or pedestal level, in the case of Figure 10-2d). For the buck converter, this results directly in the "linear amplifier" relationship between the reference voltage and the wanted component of the converter's output (dependent) voltage. For the boost and buck-boost converters, it does not since their transfer functions (ratios of average output voltage to average input voltage) are $1/(1-1/A)$ and $-(1/A)/(1-1/A)$, respectively. These translate to $1/(1-kR_{r})$ and $-kR_{r}/(1-kR_{r})$, respectively, where k is a constant incorporating the slope(s) of the timing wave and R_{r} is the reference or pedestal voltage. Thus, if linear amplifier performance is required of these converters, specific nonlinear timing waves must be used. This is not often done, since generation of the functions required is not trivial. Modern electronics do make it feasible and rather easily so in digitized control systems.

Current-sourced ac-to-dc/dc-to-ac converters also were generally controlled by ramp and pedestal techniques in the early days of discrete component control logic. For a single-phase converter, the ramp has to be synchronized to the ac supply voltage with a reset each half-cycle. Phase delay (or, rarely, advance) is determined by the time delay from ramp reset to the subsequent intersection with the reference as shown in Figure 10-4. It was found that maintaining balanced delays in the two half-cycles was quite difficult, and that those difficulties grew if three independent ramps were used for three-phase operation. Also, the overall converter transfer function was nonlinear because the phase delay, α, is a linear function of R_{r} with such a technique, and the converters' average dc terminal voltage is a function of $\cos\alpha$.

The difficulties of phase-to-phase and half-cycle-to-half-cycle balance of the delay angle, α, can only be addressed by very careful circuit design and component selection in any timing wave-reference scheme for producing phase delay

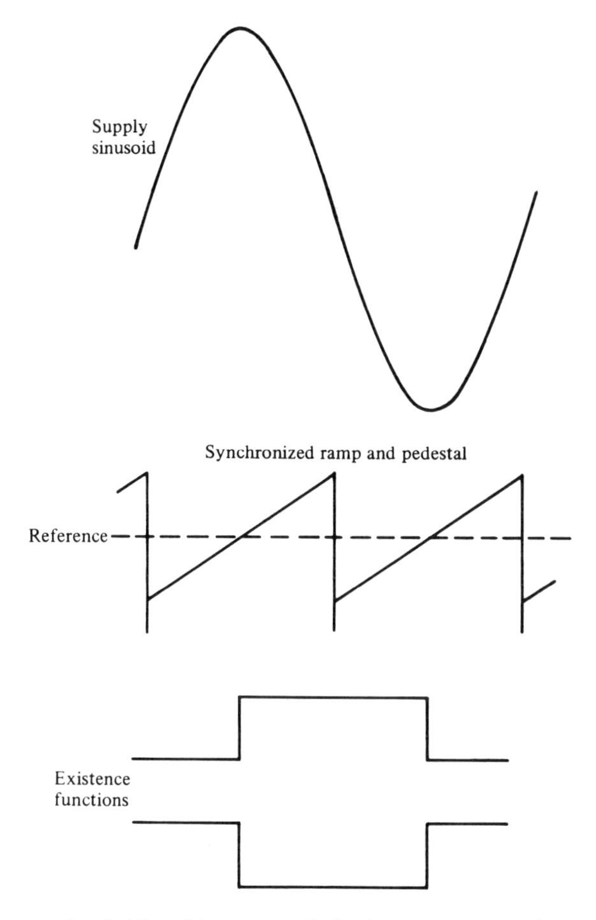

Figure 10-4. Ramp and pedestal used to generate single-phase current-sourced converter existence functions.

(or advance) control. The linearity of the transfer function can be assured by using the cosine crossing technique illustrated in Figure 10-5. There, the timing wave is derived from the ac voltages feeding the converter, being a composite (consecutive composite) of appropriate fragments of cosinusoidal waves derived from the sinusoidal voltage waves. When the intersections of this wave are used to establish α, they create the relationship.

$$\alpha = \arccos(R_{1'}/V_t)$$

where V_t is the peak value of the timing wave. If the dc reference, $R_{1'}$, is exchanged for a cosinusoidal reference wave given by

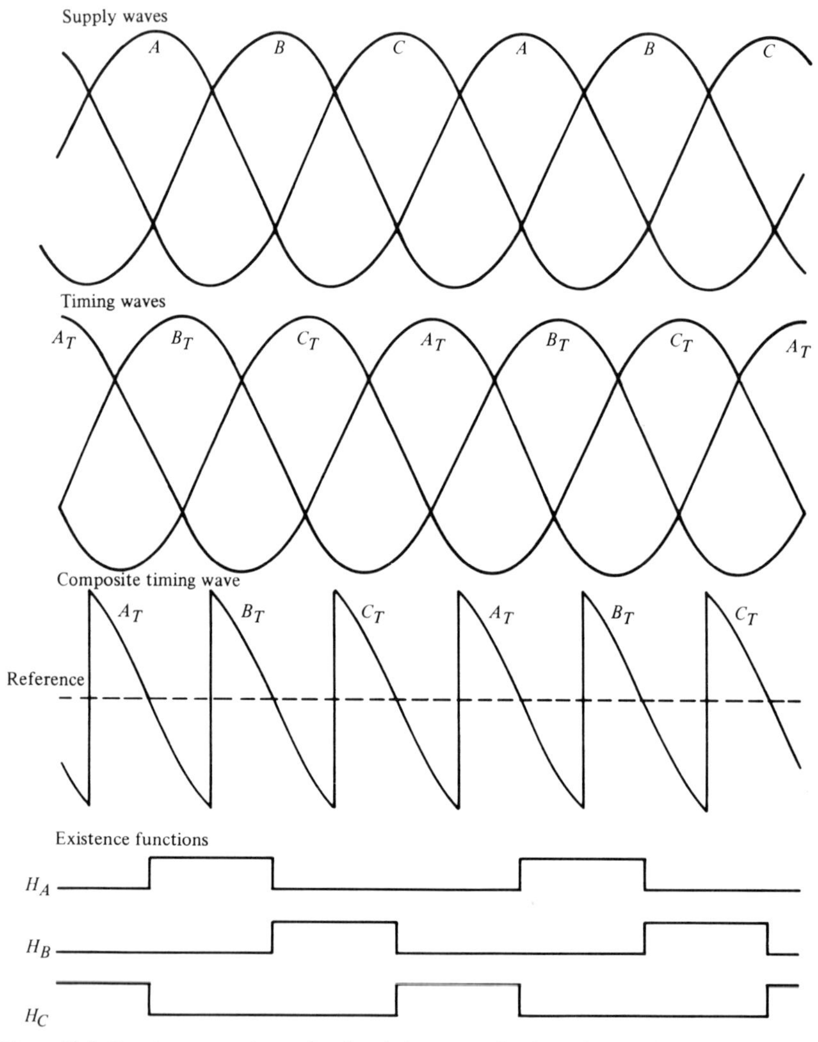

Figure 10-5. Development and use of cosine timing waves for three phase midpoint group existence functions.

$$R = xV_t\cos\omega_0 t$$

then the nonlinear modulating function

$$M(x,\omega_0 t) = \arccos(x\cos\omega_0 t)$$

results. Thus, this control technique is eminently suitable for ac-to-ac converters using that modulating function.

It is plagued, however, by electrical noise problems. Unlike the linear timing waves previously discussed, which are generated entirely within the control logic and hence can be made essentially noise-free, the cosine timing waves must be derived from the actual source voltages. They are, in cosequence, potentially subject to a great deal of noise. Appearing as random, small, amplitude disturbances on the waves, the noise creates random errors in α, which can be most disturbing to the converter that the scheme is intended to control. As a result—and because of the phase-to-phase and half-cycle-to-half-cycle balance problem—the phase-locked-loop control scheme illustrated in Figure 10-6 is often used for current-sourced converter and ac-to-ac converter control. This is of considerable interest, for it contains a feature common to many other control schemes—namely, distributive generation of the switch existence functions. In steady-state operation, the oscillator runs at N times the ac supply frequency, and its output is counted down by the ring counter to produce, after appropriate combinational logic operations, the desired control signals for the switch gate- or base-drive channels. Thus, perfect balance of α's, phase-to-phase and half-cycle to half cycle, is assured provided that the voltage set itself is balanced. The oscillator is of the voltage-controlled variety—i.e., its frequency is controlled by a control terminal voltage level and is usually linearly proportional to that voltage. It is held phase-locked to the ac supply system by feedback control, and therein lies one of the principal disadvantages of this technique—it is impossible, with the basic phase-locked-loop control of Figure 10-6a, to operate the converter-system open loop; feedback must be present for the phase lock to exist. This problem can be overcome by using the "pseudo-converter" scheme illustrated in Figure 10-6b. There, a small-scale model of the converter, or of some fraction of the converter, is used to provide the feedback necessary for phase lock. The main converter system can then be run open loop, and its transfer functions explored, before implementing the full closed-loop controls for the system.

When control is exerted on a phase-locked-loop controlled converter (i.e., when it is desired to change α), the oscillator control voltage changes transiently so as to produce a transient frequency shift. If the oscillator frequency is increased, a progressive phase advance obviously results, while if the frequency is reduced a progressive phase delay results. This indicates another disadvantage of the phase-locked-loop scheme, for the change in phase delay is proportional to the integral of the oscillator's control voltage change. Thus, a $1/s$ term is introduced into the overall transfer function, not generally a desirable circumstance. In the event an ac-to-ac converter is to be controlled, the difficulties depend on the modulating function involved. For a UFC or an SSFC, no problems arise since the derivative of the modulating function is constant. For all converters using the nonlinear modulating function at other than $x = 1$, the difficulties are considerable, since the derivative of $\arccos(x\cos\omega_s t)$ is not an easy function to implement.

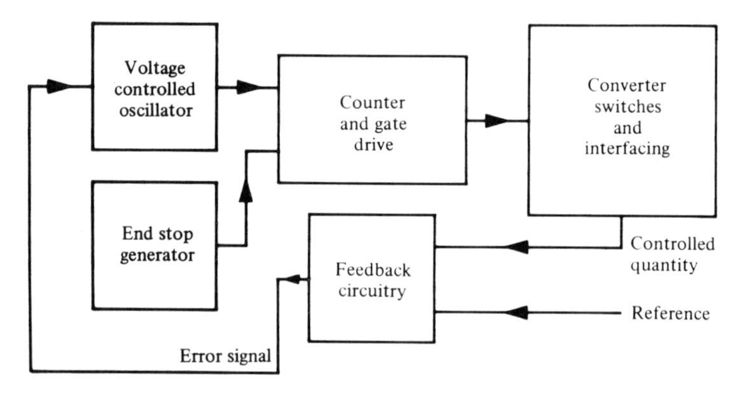

a. Simple phase lock

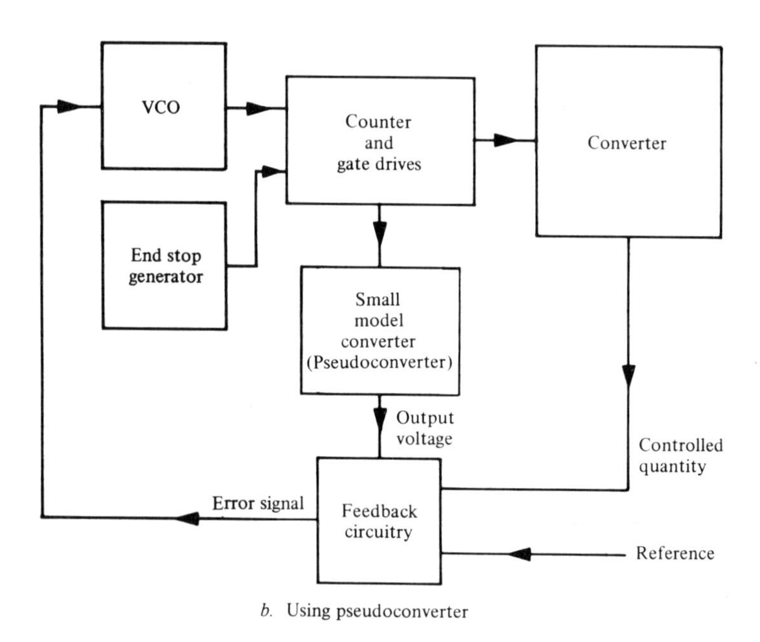

b. Using pseudoconverter

Figure 10-6. Phase-locked loop controls.

Both timing wave–reference wave and phase-locked-loop controls need end-stops, of course. For phase-delay-controlled converters, the inversion end-stop (α approaching $-\pi$) is crucial since overrunning it will result in a commutation failure with potentially severe fault consequences. Overrunning the rectification end-stop is usually more tolerable; if extended gate drive is used, an overrun almost equal to the switch conduction time must occur before the converter ''knows'' the end-stop has been violated. With single-pulse gate drives, overrunning the rectification end-stop results in ''single phasing,'' and, if maintained, in eventual natural shutdown of the converter. For timing wave–reference controls, end-stops can be accomplished simply by reference clamping, as in the dc-to-dc converter case, but with the same difficulties. More often, the added pulse technique of Figure 10-3 is employed. In the early days, a variety of ingenious circuit techniques, many based on saturating transformer or saturable reactor behavior, were devised for end-stop generation. The capabilities of modern electronics have, perhaps unfortunately, removed the need for such ingenuity, and straightforward pulse-generation techniques are almost universal. Phase-locked-loop controls present a more difficult problem, since the loop provides no inherent means of producing or introducing end-stops. As a result, most practical controls using phase lock introduce end-stops by converting to cosine crossing control as the end-stop is reached. They do this by having in place an auxiliary cosine crossing control, functional only at the end-stops, which takes over from the oscillator as the pulse input to the ring counter when the end-stops are reached. This complication is, of course, yet another drawback for phase-lock controls; it is perhaps surprising that they have achieved such widespread popularity in view of the many problems they introduce.

There is one other fact worth considering regarding timing wave–reference controls for current-sourced ac-to-dc/dc-to-ac converters. Although linear timing waves do not give linear converter transfer functions, the departure from linear is only large in the vicinity of the theoretical end-stops. If practical end-stops are set, of $\pi/12$ in advance of (inversion) and delayed from (rectification) the theoretical end-stops, then a linear timing wave is in fact a reasonably close approximation to the ideal cosine timing wave over the practical control range. The converter's transfer function will not be highly nonlinear, and no major difficulties will occur in closed-loop stabilization.

In approaching the design of control logic for voltage-sourced ac-to-dc/dc-to-ac converters, the view has traditionally been one of pole requirements rather than individual switch existence functions. Distributive logic is almost universally used to ensure accurate half-cycle to half-cycle balance within a pole and accurate phase displacement between pole-dependent voltage waves in polyphase converters. Thus, the basis for most voltage-sourced converter controls becomes that depicted in Figure 10-7a. It comprises an oscillator running at $2N$ times pole operating frequency, where N is the number of poles in the converter, and a ring counter producing N identical phase-displaced square waves representing the desired (basal) pole output voltages. For an autocommutating pole design, all that is necessary is that these square waves control the gate (or base) drives of the switches. When auxiliary impulse commutation is used, a subsidiary ''pole logic'' circuit is needed as shown at Figure 10-7b. Using monostable multivibrators or other means of generating appropriately time-displaced signals,

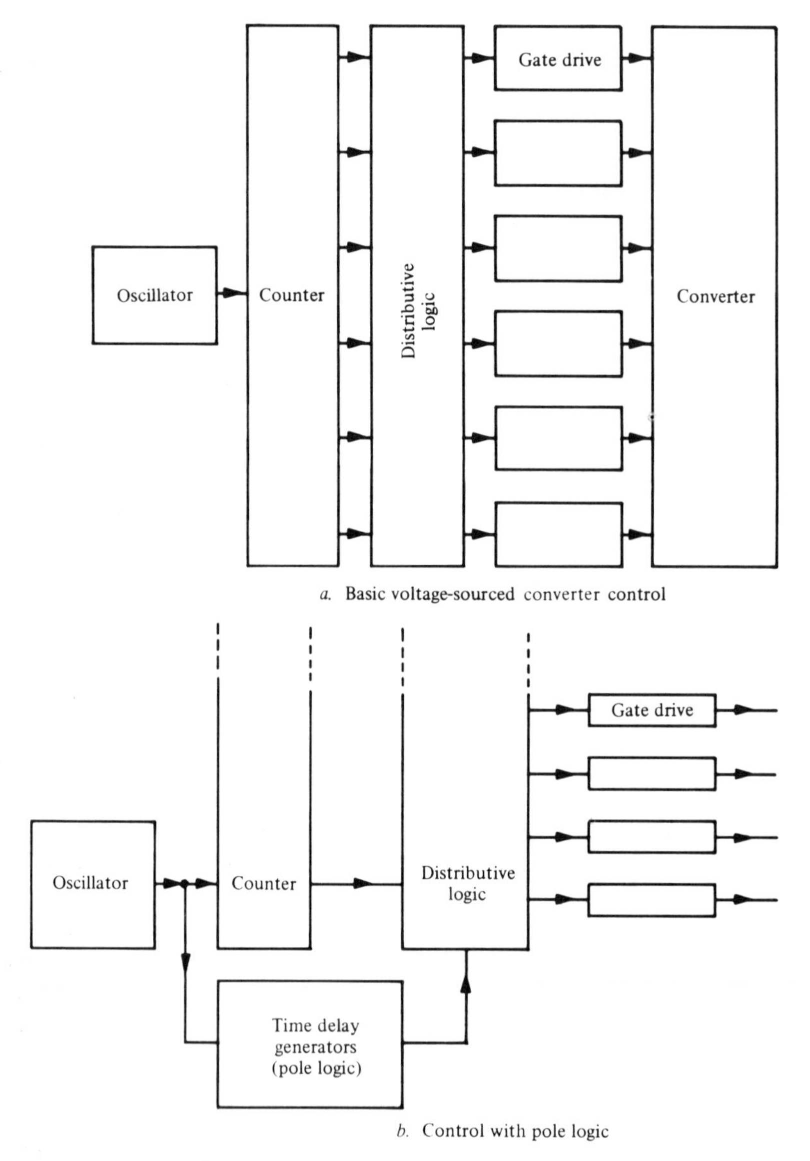

a. Basic voltage-sourced converter control

b. Control with pole logic

Figure 10-7. Voltage-sourced converter controls.

this circuit sets up the sequence of gate drives needed at each commutation and is invoked, via distributive logic, at each square-wave transition.

When phase-displacement control of wanted component magnitude is used, the same basic control technique is employed. To generate the phase displacement, a scheme like that depicted in Figure 10-8 is common. A double-frequency

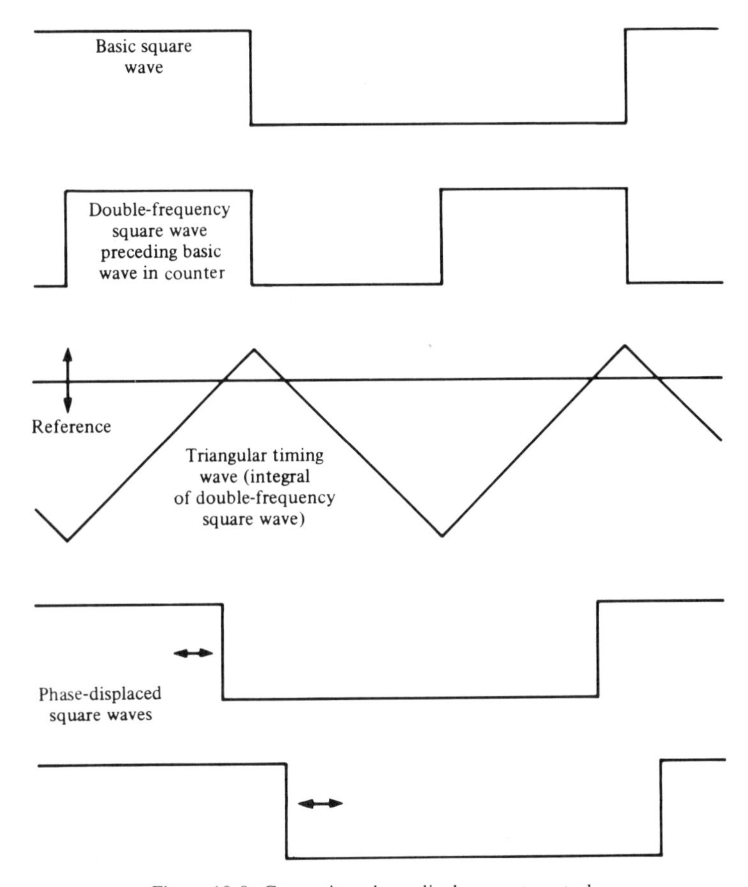

Figure 10-8. Generating phase-displacement control.

square wave developed by the basic scheme is used to synchronize a triangular timing-wave generator. The intersection of this with a reference is used to define the square waves providing control functions for the two converters; pole logic is now common to all poles in both converters.

Simple PWM control is most often produced using the triangular-timing-wave/sinusoidal-reference-wave technique depicted in Figure 4-33. Pole logic is added whenever auxiliary impulse commutation is used, and is invoked at each switching transition via distributive logic. CAM and programmed waveform generally involve rather different control philosophies, as now discussed.

Simple CAM, using only one or two notches in each pole wave and not involving waveform programming, is also usually produced using analog-based control circuitry. Typical schemes for developing single- and two-notch CAM are illustrated in Figure 10-9. They involve the generation of triangular timing

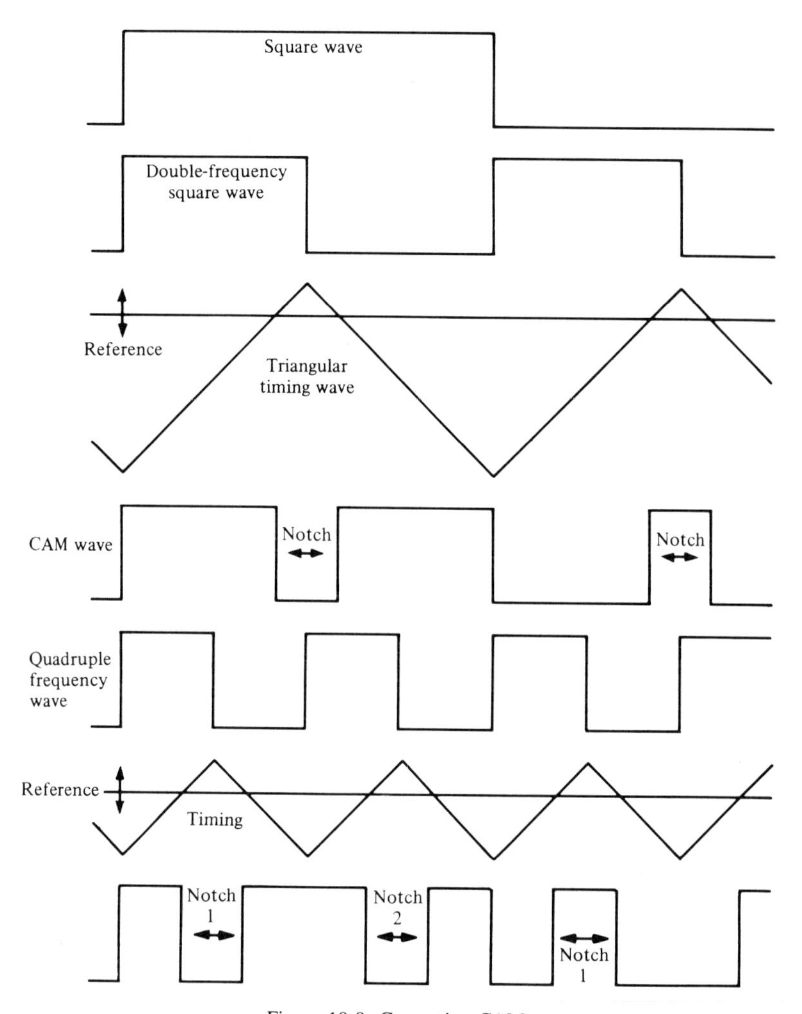

Figure 10-9. Generating CAM.

waves bounded by the basic pole controlling square waves. The intersections of these timing waves and a dc reference potential define the notch switching times; again, pole logic is used and its function appropriately distributed when, as is most often the case, auxiliary impulse commutation is in use.

The development of programmed waveform techniques for combined wanted-component magnitude and spectral characteristic control necessitates a completely different approach to converter control. No longer are there any simple analog means for generating the pulse patterns required. These must be stored in a memory of some sort, and retrieved as required to produce the desired converter performance. Implementation of these techniques became possible only with the development of large-scale integrated circuits. Solid-state memories with costs and access times low enough to permit their use became available, and it is on those memories that the controls depend. The basic control functions remain, of course, the set of square waves defining pole half-cycle bounds and the pole logic. The memory stores a large number of pulse patterns coded as binary number sequences. The appropriate sequence—i.e., memory address(es)—is determined by feedback loop or external control variables. It is read out cyclically and translated into transition times for the pole, which are then implemented through normal logic and the gate-drive channels. There are many concepts usable for the coding-decoding of patterns necessary in this type of control. One of the most elegant is to store the pattern as a series of numbers representing the time delays between successive pole transitions in a half-cycle and to have these numbers loaded into a binary counter, clocked at an appropriate rate many times the converter operating frequency, which calls for a pole transition each time it counts to zero.

Programmed waveform is not the only control technique to benefit from modern electronic technology. The much simpler converter function of an ac regulator with pure inductive load, which has considerable application importance (see Chapter 11), has also enjoyed great benefit. The wanted-voltage component of this regulator is given by

$$v_{DD} = (1-2\alpha/\pi-\sin 2\alpha/\pi)V\cos\omega_s t$$

(See Chapter 5.)

To achieve a linear overall transfer function, it is necessary to develop the appropriate and rather complex functional relationship between α and R_V, where R_V is a controlling reference voltage. To do this using discrete component and device technology would present a formidable technical problem and be prohibitively costly. Using analog integrated circuitry, it becomes manageable (but not trivial) technically and quite reasonable in cost.

Finally, we will briefly consider *hysteresis* (''bang-bang'') and integral control schemes. The hysteresis technique, often called *optimal control*, by optimistic practioners, is a very simple concept deriving from timing wave–reference PWM implementation. In a hysteresis controller, a filtered converter quantity is continuously compared with a reference, its wanted component. Converter switching is initiated whenever the difference reaches some arbitrarily established hysteresis value. Optimal it most certainly is not; unless the switching to wanted-frequency ratio is high, the spectral results are absolutely dreadful. Since simple PWM is much easier to implement, it is difficult to understand why such a technique would ever gain credence. Proponents claim faster response to transient disturbance of converter load conditions, but this is at best a dubious argument.

Integral control is somewhat similar in concept. Such a control looks at the integral of the difference between a raw-converter dependent quantity and a reference that is its wanted component, and invokes converter switching when that integral reaches some present hysteresis value. Although possessing some of the disadvantages of simple hysteresis control, integral control has one major saving grace. Modified versions thereof can be used to great advantage with ac-to-ac converters normally using the nonlinear modulating function. This subject is discussed in detail by Gyugyi and Pelly (1976)[1] and will not be pursued further here.

10.4 Feedback Control of Converters

The logic and gate-drive control functions, together with some oversight from the external control inputs for start/stop sequencing, permit a converter to execute its basic function. It is through the use of feedback control that most converters are made to satisfy the needs of an application. However, only the quantities sensed for feedback implementation are highly application-dependent. In general, the behavior of a converter under such control is only slightly dependent, and usually only during transitory operating periods, on the specific application.

Basically, feedback is applied to converters in the same way it is applied to linear systems. System output quantities, which may or not be electrical depending on the application, are sensed and compared with desired values. The resulting errors are amplified and used to adjust converter (amplifier) operation by acting as the main continuous input to the control logic. The scheme is depicted in Figure 10-10 and produces the familiar equation

$$O = ABR/(1+AB)$$

where O is the output quantity, R the reference, the desired value of O, and A and B are the transfer functions of the converter-output network and the feedback loop, respectively. This gives $O \simeq R$ if $AB >> 1$.

Even in completely linear systems complications arise because A and B are generally not wholly real, nor can $AB >> 1$ be maintained over the frequency domain (or over the totality of the complex plane). For switching power converters, two additional complications arise. First, a converter dependent quantity will respond instantaneously to changes in the defined quantity from which it derives. In practical converters, this translates into dependent voltage responding immediately to defined voltage change. In many instances, such changes can be very rapid, and considerable transient perturbation of output conditions may occur before the feedback loop catches up. This problem is relatively easily addressed by using a "feed forward" addendum to the control, which causes

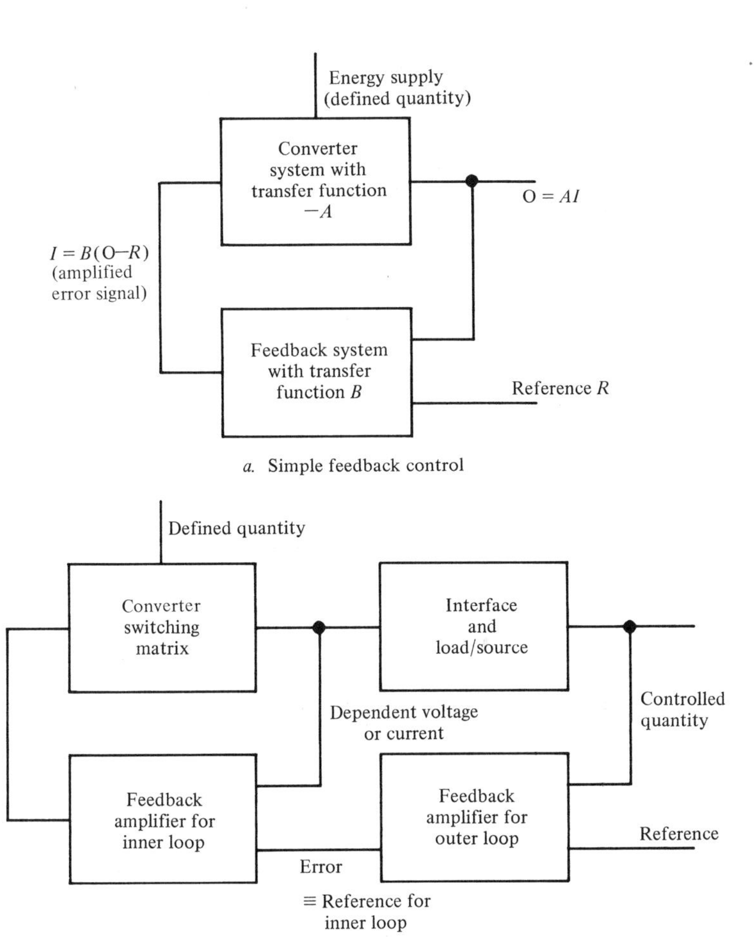

Figure 10-10. Converter feedback loops.

the converter logic to react immediately to changes in the defined (source) voltages, but allows the feedback loop to take over later. No stability problems are encountered with the feed-forward loop, since it involves only a direct transfer function. However, its time-domain behavior must be carefully tailored to mesh with that of the feed-back loop; otherwise, it may induce stability problems in that loop.

The second problem arises when the frequency of interest in the feedback loop begins to approach the switching frequency of the converter. So long as loop disturbance frequencies are much lower than the switching frequency, the converter and its logic control may be treated as a linear amplifier, if the control is so designed, or as a nonlinear device with a continuous transfer function if the control does not linearize the system. This condition is often met in practice, since the open-loop response of the converter system is often dominated by interfacing filters and load components having transfer functions with time constants very much longer than converter switching times or natural frequencies very much lower than the converter switching frequency.

When this is not the case, the approximations obtained by treating the converter and its logic as continuous (in the time domain) devices are no longer valid. They may produce serious errors in loop analyses performed by Bode or root-locus techniques, which are quite adequate when the frequency relationships are suitable. Analysis of loop behavior is then only possible using the techniques applicable to sampled data systems. These are well developed by Jury (1964)[2] and will not be pursued in any detail here.

It is very common to find multiple loop control in converter systems. Converter designers generally do not feel comfortable if the primary variable under loop control is not a converter electrical quantity, or if it is not the "right" quantity. This results in configurations such as that of Figure 10-10b, where an inner current-feedback loop provides the primary loop control, and will continue to control, within its limits, regardless of the behavior of the outer loop that controls voltage, speed, torque, temperature, physical displacement, or whatever other system variable it is ultimately desired to control. Such a scheme also permits much better implementation of overriding control safeguards, such as current limiting in the instance shown, since the inner-loop response can usually be made much faster than that of the outer loop.

A rather different kind of two-loop control problem often arises when a converter having a dependent ac voltage set is to be interfaced with a set of ac voltages. The interface is usually almost purely inductive impedance. The result is that active (real) power flow between converter and the ac voltage set is dependent on the quadrature component of converter voltage (wanted component) with respect to the set, and reactive power flow depends on the real (in-phase) component of converter voltage.

The converter control will, typically, be able to adjust the magnitude of its wanted component of voltage and the phase angle thereof with respect to the defined set. Both of these variables affect both the real and quadrature components of the wanted component, and hence both active and reactive power flow. Hence, there exists a situation in which two orthogonal system variables, active and reactive power, are controlled by two orthogonal converter variables (magnitude and phase) but orthogonality of the controls is not maintained. A simple transposition, however, restores control orthogonality, which is highly desirable if both real and reactive power are to be controlled.

10.5 Digitized Control Schemes for Converters

The advent of the microprocessor has had a great impact on converter control philosophy, particularly for the more sophisticated converter functions in the lower operating-frequency range. Two inherent, related problems arise when a microprocessor is made the heart of a converter real-time control scheme, quite apart from any difficulties that may be experienced as a result of microprocessor speed limitations.

A microprocessor is, like any digital computer, a serial processing device. That is to say, it performs operations in sequence, one after the other, and cannot perform multiple tasks simultaneously as can analog circuitry and multiple-path digital logic. Also, the clock which controls a microprocessor, and its attendant peripherals, memories, input-output ports, and other, must run at a rate that will enable the microcomputer so formed to perform multiple operations, in series, between converter switchings. In practice, clock rates of several megahertz are needed even for quite low frequency converter controls. It is almost impossible to synchronize a clock running at such a frequency with external events, such as a set of defined ac source voltages or the basic operating frequency oscillator for a dc-to-dc or voltage-sourced ac-to-dc/dc-to-ac converter. Hence, microcomputer and converter operations must generally be asynchronous. Now the control must produce certain events in the converter at specific times with respect to the aforesaid external world events. If it is running asynchronously, it can only do so as a result of interrupts generated in synchronism with those events, and those interrupts usually have to be nonmaskable.

An additional problem arising when microprocessor control is used is that the feedback loop(s) must now be treated as sampled-data-control systems under all circumstances. Moreover, they will usually be asynchronous sampled-data systems, a class presenting considerably more analytic and implementation difficulties than their synchronous relatives.

The software package for a microprocessor-based converter control usually contains three program blocks corresponding to the logic, feedback, and annunciatory/control block functions (Figure 10-1) of a conventional hybrid control. Control logic sections of the program are activated immediately by externally generated interrupts and suspend the operation of other subprograms—they are the top-priority tasks for the microprocessor and its peripherals. Next in line come the feedback subprograms. They are usually also activated by interrupts, when external quantities are sampled, but are subject to masking or interruption by the control logic routines. Lowest in priority are the routines that communicate with and receive information from the operator's world, for the overall system can stand considerable delay regarding such informaiton transmission or implementation of commands.

These monitoring and sequencing routines are often programmed as a contiguous block of instructions through which the microcomputer loops continuously. There usually are subroutine branches and subloops within, but the basic format is to have the microprocessor endlessly perform the communication and command chores subject to interruption at any time. The feedback-loop routines then become interrupting subroutines to which the microprocessor is diverted whenever a data sample is taken. Samples are usually taken at uniform time intervals, since a sampled-data feedback system that is not only asynchronous but also indulges in nonuniform sampling is indeed a monster. The feedback routines themselves are subject to the absolute priority interrupts that divert the microprocessor to the control logic routines. These may themselves be ordered in a hierarchy, for certain protective functions may need immediate attention above and beyond the normal converter switching sequence.

Those interested in pursuing this subject further should first acquire a thorough understanding of the basic character and architecture of digital computers, and then arm themselves with a thorough grasp of numerical techniques.

References

1. Gyugyi, L. and B. R. Pelly. *Static Power Frequency Changers*. Chapter 7, pp. 294–306. New York: John Wiley & Sons, 1976.
2. Jury, E. I. *Sampled-Data Control Systems*. Huntington, New York: Robert E. Krieger, 1964.

11
Converter Applications

11.1 Introduction

This chapter will provide a brief discussion of some of the most important application areas for switching power converters. It also furnishes some appreciation of the limitations—and the strengths—of converters in various applications. The compound or cascaded converters so often found in practice are also discussed.

The author had some difficulty deciding whether to puruse this subject through converter classification or application classification. The former, of course, is the essence of all the foregoing text. However, in some ways, the latter seems more appropriate for this particular discussion. Nevertheless, converter classification for the basic framework was retained. Any other approach merely leads back to the confusion previously existing in the field—a hodgepodge of application-related circuits having little structure and no order.

11.2 Applications of DC-to-DC Converters

The dc-to-dc converters are used whenever a controlled dc source is required, but the primary electrical energy source provides uncontrolled dc. The applications split into two major areas by load definition: dc machine drives and electronic loads. The latter also splits into two—communications and data-processing equipment requiring low voltage (typically 5 to 15 V) supplies of modest power level and converter loads, dc-to-ac converters, at higher voltages and over a wider range of power levels. Prime sources also split into two major areas. There are the true dc sources, batteries, fuel cells, solar photovoltaic arrays, and dc generators (rotating machines). There are also many instances in which a dc-to-dc converter is fed from an ac-to-dc converter—a rectifier set.

For supplying communications and data-processing equipment, the dc-to-dc converter must produce a closely regulated, wanted component of output voltage with very low ripple content. Also, it should not pass to the load any source-borne transient overvoltages, since even momentary excessive voltage can be fatal to the loads' delicate circuitry. Output dc current levels generally range from a few amperes to a few hundred amperes, with power levels never exceeding a few kilowatts.

The dc-to-dc converters used are invariably single-quadrant. Buck, boost, and buck-boost topologies are all employed; which one is used depends on the relationship of source-voltage range to the regulated output voltage required. Where the source-voltage range is under the designer's control, (the designer can then pick the battery, fuel cell, PV array, generator, rectifier output voltage range, subject only to the tolerances imposed by source characteristics), the buck configuration is usually used. Regulation is achieved, of course, by varying the control variable A of the active switch existence function, no matter which converter topology is used, under closed-loop feedback. Ripple content of dependent quantities is limited by appropriate interface design. Because of the modest power levels of this converter application, and the consequent reduction in interface-component size, cost, and losses with increasing frequency, most of the converters so used are run at frequencies above 10 kHz. It is noteworthy that as voltage and current levels are increased, the frequency of minimum interface size decreases (see Chapter 6). Hence, higher power converters should not be operated at such high frequencies, or nonoptimum designs will arise.

The use of such high frequencies mandates the choice of transistors as the active switches in these converters. Until recently, bipolar devices were the universal choice. Of late, power MOSFETs have become viable in the lower power versions, and they promise to expand their applicability with the passage of time.

When a dc-to-ac converter is the dc-to-dc converter's load, output voltage from the dc-to-dc converter (which is, of course, input voltage for the dc-to-ac converter) typically lies in the range from a few tens to a few hundreds of volts. Occasionally, applications requiring several kilovolts are encountered. This application arises because of certain dc-to-ac converter characteristics. If that converter is current-sourced, then it is almost certain to be source-commutated at its ac interface (see Chapter 8). As a result, it exerts lagging quadrature current demand on its ac voltage source, created by the load and the ac voltage interface used. If the dc source has a wide voltage range, the converter's α will depart markedly from $-\pi$, the inversion end-stop, for the lowest source voltage, since the design can only allow α to approach $-\pi$ at maximum dc source voltage. Thus the maximum lagging quadrature-current demand may be severe if a source-commutated current-sourced dc-to-ac converter is fed directly from an uncontrolled dc source. If restrictions on the quadrature-current demand apply, corrective measures are necessary. One way in which such correction can be achieved is to regulate the dc voltage input to the converter so that it always operates with α as close to $-\pi$ as is feasible, and thereby minimize its inherent quadrature-current demand.

When a voltage-sourced dc-to-ac converter is the end processor of the electrical energy, rather different considerations enter. Such a converter will almost invariably be impulse commutated, and at power levels higher than a few kilowatts is likely to be implemented with thyristors and impulse-commutating cirucits. As discussed in Chapter 9, impulse-commutating circuits for voltage-sourced ac-to-dc/dc-to-ac converters must be designed to meet commutation requirements at the lowest anticipated input dc voltage. Commutating impulse currents are then linearly proportional to the dc voltage supplied, and commutating circuit losses increase almost as the square of the ratio of extant to minimum (design limit) dc voltage. As a result, commutating circuit losses can become embarrassingly large at the high voltage end of the range of a wide-tolerance dc source. Regulating the dc voltage feeding the converter by using a dc-to-dc converter eliminates this problem. However, a word of caution is in order. The problem is eliminated from the dc-to-ac converter, but not from the system. It simply transfers to the dc-to-dc converter, and the designer must be careful not to jump from the frying pan into the fire.

There is yet another aspect of the use of dc-to-dc converters as the supplies for dc-to-ac converters, one that also merits attention when a dc-to-dc converter is fed from an ac-to-dc converter. If the switching frequencies of the converters are asynchronous (i.e., not integer-multiple-related), sideband unwanted components will appear in dependent quantities. Some sidebands will have frequencies lower than those of wanted components, and if of significant amplitude can seriously disturb converter operation unless interface designs are revised to account their presence. Also, sidebands may enter the feedback loop(s) of the converters and create even greater disturbances of system operation, particularly when their frequencies are in the vicinity of the loop crossover frequency. In consequence, it is eminently preferable that converter operating frequencies be integer-multiple-related, so that the sidebands become harmonics of the lower of the two or three frequencies involved. When this cannot be the case, one frequency should be much higher than the other so that low-frequency sidebands will be of very small amplitude.

Because voltage-sourced dc-to-ac converters tend to have better efficiency and lower cost when operated from dc voltages higher than those commonly available or readily obtainable from the primary dc or ac energy supplies used, it is common to find boost converters feeding them. When current-sourced dc-to-ac converters are fed by dc-to-dc converters, any of the dc-to-dc configurations may be found; however, the buck converter is most common. There are other good reasons for these usages, arising from the fault behavior of the dc-to-dc converters. If the active switch of a boost converter fails open (it will eventually through the action of protective devices, such as fuses, even though initial failure is almost always permanently closed), then the converter's output is simply connected directly and permanently to its input. If a buck converter's active switch fails open, the converter's output voltage becomes zero. A voltage-sourced impulse-commutated dc-to-ac converter will suffer commutation failure if its dc source voltage is reduced to zero, but it may continue functioning, at least for a sufficient time to allow orderly shutdown, if that voltage maintains a value not too far below its design operating point. Conversely, a current-sourced source-commutated dc-to-ac converter is least likely to suffer commutation failure when the dc terminal voltage is reduced to zero, and will if that voltage is of too great a magnitude. Hence, the choices of boost converter for voltage-sourced and buck for current-sourced give the best chances of noncatastrophic dc-to-ac converter shutdown when dc-to-dc converter malfunctions occur.

The dc-to-dc converters used as dc voltage or current sources for dc-to-ac converters are almost invariably single-quadrant converters. Powers range from a few to several hundred kilowatts, with occasional examples of several megawatts being found. Transistors may be used as the active switches in these converters at the lower power levels, but thyristors are more common, being essentially mandatory above 5 to 10 kW or so. Thus, the operating frequencies tend to be low, typically no more than a few hundred hertz, and the higher power applications quite often are addressed with harmonic neutralized converters to reduce interface costs and losses. When impulse-commutated thyristors are used as the active switches in a harmonic neutralized dc-to-dc converter, a single impulse-commutating circuit may be made to serve them all.

The use of dc-to-dc converters to supply dc machines arises in two application areas, one quite well exploited and the other as yet not a truly commercial venture. Both are traction applications for the machines. The first—that already commercialized—is in rapid-transit rail systems where the cars are driven by dc motors, with the power picked up from a third-rail dc distribution on the track. The second is in the drive for electric automobiles, where the dc source is a battery, or battery–dc generator combination in the case of so-called hybrid vehicles.

Before discussing the converters used, a description of the basic principles is in order for those not familiar with the behavior of separately excited dc motors. At constant field, the induced armature voltage of a motor is proportional to its speed of rotation. The torque it produces is always proportional to the product of armature and field currents, and hence is proportional to armature current when field current is fixed. The armature current is simply the difference between applied and induced armature voltages divided by armature resistance. It follows that the speed of a motor with a given load is almost proportional to the applied armature voltage, since the *IR* drop of the armature resistance is usually small.

When a motor is running, the armature circuit possesses two kinds of stored energy. Some energy is stored in the armature's magnetic field, being a consequence of armature reactance. Much more is stored as kinetic energy in the mass of the rotating armature and the load the motor is driving. To brake the motor, and the load, to rest, this energy must be extracted or dissipated. That stored in the magnetic field is rapidly eliminated if the applied voltage is reduced to zero or less, for then the armature current will rapidly decay. If this is all that is done, however, the motor will subsequently keep turning, and its load will keep moving until the kinetic energy that was stored is dissipated in assorted frictional losses. In both the applications presently under consideration, this is highly undesirable—the railcar or automobile may travel a great distance, or impose great burdens on its friction brakes, after armature current is reduced to zero.

To extract the kinetic energy, the motor must be induced to develop braking torque. It will do so only if (1) armature current is made to flow in the reverse direction to that in which it flowed while the motor was running, or (2) the field current is reversed while the armature current continues to flow in the same direction. Field time constants are usually quite long, so that reversing field current is a lengthy procedure, not the preferred technique. Buildup of reverse armature current is the technique generally used. In many instances, particularly the transit railcar drives, this is done simply by connecting resistors to the armature as a load to absorb the kinetic energy—so-called dynamic braking. Regenerative braking can be accomplished if the converter supplying the armature can operate so as to reverse the current and if the dc source to which the converter is connected can absorb the energy. The battery of an electric automobile is well able to do so, but the third-rail distribution of a transit system often cannot; hence, the frequent use of dynamic braking in that case.

Transit-car drives are usually several hundred horsepower, and third-rail distributions range from 600 to 1000 V. Since the commutators of dc motors are limited in voltage capability, and since very low applied voltages are needed to limit armature current to a safe value when accelerating from rest, buck converters are universally used in this application. Because of the dc voltage and power levels, thyristors with impulse-commutating circuits are used as their active switches. It is rare to find harmonic neutralization used, although many specific implementations of such drives could benefit thereby. "Two-quadrant" converters of the type depicted in Figure 3-12 are often used so that dynamic braking may be accomplished under converter control. They are necessary, of course, when regenerative braking is used.

Buck converters are also used in experimental electric vehicle drives, despite the much lower dc voltages and powers that are used. Again, the major reason is the need for very low armature voltages when accelerating from rest. In this application, dc source (battery) voltage usually lies between 50 and 100 V, with power at most a few tens of horsepower. Until recently, the power levels forced the use of thyristors as active switches, but the emergence of large power transistors has caused designers to shift to them. Regeneration is virtually mandatory in an electric vehicle drive if acceptable overall energy efficiency is to be achieved, and hence "two-quadrant" arrangements of the type depicted in Figure 3-12 are quite common. However, in both the automobile and transit-car systems, the converter's reverse-current capability is often created by reconfiguring, using electro-mechanical switchgear. What was a buck converter between dc source and motor armature is simply rearranged to become a boost converter between armature and source (or resistors, where dynamic braking is used). This is done for economic reasons, and the practice is likely to die out as switching devices and their accessories become still cheaper.

In both applications, and in many others served by dc machines, the constant horsepower operating region is often exploited. Maximum torque, given by maximum armature and field currents, is generally needed only for initial acceleration. For high-speed operation, lower torque can be tolerated. The higher speed and lower torque can be achieved by weakening the field (reducing field current) once the armature reaches rated voltage. If the field is then weakened, the rotational speed must increase to keep the armature current under control by maintaining the induced armature voltage. The torque will be lower because the field current is lower, and if constant armature current is maintained with a constant applied voltage, the horsepower developed by the motor remains constant. Field weakening is used to provide a speed increase of about 50% over normal motor rating in most instances; commutator limitations and mechanical considerations generally inhibit attempts to increase speed much beyond that level.

11.3 Applications of Current-Sourced AC-to-DC/DC-to-AC Converters

By far the most common artifact of the switching power converter industry, current-sourced ac-to-dc/dc-to-ac converters are used in a wide variety of applications. Their popularity is due to a number of factors, not the least of which is that source-commutated versions are easily realized with thyristors as their switches, and were earlier realized with mercury-arc tubes.

There are two major applications of considerable interest, dc machine drives and high-voltage dc transmission (HVDC). In addition, this section will afford discussion to the use of such converters as dc sources for a variety of applications, and to their latter-day use in ac machine drives as Adjustable Current Inverters (ACIs).

Current-sourced converters are used as the armature voltage supplies in variable-speed dc machine drives in much the same way as dc-to-dc converters are. Of course, the primary energy source in these cases is the public-utility ac voltage distribution. If true regenerative braking is needed, then a four-quadrant converter must be used unless field reversal is employed. Power levels range from a few tens to a few hundred horsepower, with occasional examples of 1000 hp or more. The very lowest power realizations sometimes use single-phase bridge or two-phase midpoint (often called *single-phase center-tapped* or *single-phase full-wave* in the literature) configurations, but the simple three-phase six-pulse bridge dominates. Only in the very highest power equipment is harmonic neutralization used; both the VAR demand and the ac harmonic injection of the converters have customarily been ignored, allowed to burden the ac voltage supply. This cavalier attitude must perforce change as the concentration of such equipment increases. Steel-rolling mills are the largest volume customers for four-quadrant drives, with elevators providing a relatively minor market. In both cases, the full reversing capability of the drives is used in addition to regenerative braking; the switching power converters have replaced Ward-Leonard sets in these applications.

HVDC was first implemented in the 1950s, with controlled mercury-arc rectifiers. Thyristors were first used in the mid-1960s, and now completely dominate, but plasma-tube switches may yet reconquer the field. The application is characterized by extremely high powers and voltages. Although the first HVDC links were a few tens of megawatts capacity with dc voltages of a few tens of kilowatts, capacities now generally exceed a gigawatt, and the most common dc voltage is 400 kV (the highest to date is 533 kV). Links are usually bipolar with respect to ground, so that unipolar operation with ground return can be maintained in the event of partial equipment failure. To minimize the cost, size, and VAR supply of harmonic filters, 12-pulse operation with series-connected bridges is maintained for both dc voltage polarities in most links; some of the earlier and smaller installations make do with single six-pulse bridges for each polarity, necessitating the installation of filters for the six-pulse harmonics if unipolar operation under partial outage is desired.

Current-sourced converters also find wide usage as dc power supplies. The uncontrolled single-quadrant variety are often used to supply dc-to-dc converters feeding electronic and converter loads. Controlled two-quadrant converters are sometimes used in the same situation, eliminating the additional power-processing stage, but are not overly popular because it is usually possible to achieve faster closed-loop system response with the dc-to-dc converter present. This is because that converter can be made to switch at a rate considerably in excess of the supply frequency in many instances. Uncontrolled single-quadrant two-phase versions and single-phase bridges are widely used to supply dc to moderate power loads that happen to be uncritical of the dc voltage regulation; high-power amplifiers are typical examples. Consumer electronic products make extensive use of the "voltage-sourced" single-phase single-quadrant half-wave rectifier, and the voltage-multiplying cascade connections thereof.

One minor application of the controlled two-quadrant current-sourced ac-to-dc converter has been as the dc supply for single-phase current-sourced dc-to-ac converters feeding parallel-tuned induction-heating loads, giving a compound converter—ac-to-dc followed by dc-to-ac—arrangement fulfilling an ac-to-ac conversion function. Such compounds are common because of the inability to meet the technical and economic constraints of applications with true ac-to-ac converters using presently available switching devices. The induction heating application generally has output frequencies ranging from 180 Hz to 10 kHz, with power levels from several megawatts at the lower frequencies to 50 to 150 kW at the higher frequencies. The arrangement is usually a three-phase controlled two-quadrant or single-quadrant converter, fed by the industrial 440- or 660-V plant distribution, producing dc voltage ranging from 0 to 600 or 900 V and providing the dc current source for a single-phase bridge dc-to-ac converter commutated by the voltage developed across the parallel resonated load. Water cooling is prevalent in this equipment because water cooling is almost always used for the induction-heating work coil and the shunt-connected compensating (tuning) capacitors.

Observation of the operating features of such equipment, coupled with the economic difficulties encountered in applying voltage-sourced converters, has led to a larger market than induction heating for the double current-sourced converter compound. This is the variable-speed ac machine drive market. The equipment used is generally a two-quadrant, current-sourced, three-phase bridge, ac-to-dc converter feeding a two-quadrant, current-sourced, three-phase bridge, dc-to-ac converter that supplies the stator of a synchronous ac motor with variable-frequency, variable-amplitude (via control of the dc-to-ac converter) currents. Commutation of the dc-to-ac converter is achieved by the induced voltage set of the machine except from rest to about 20% of rated speed, when that voltage is insufficient. Starting and initial acceleration is usually accomplished by controlling the ac-to-dc converter in order to produce a "PWM" current in the dc interface, eliminating commutations in the dc-to-ac converter. This is feasible because low motor speed means low-frequency operation of the dc-to-ac converter, with one cycle extending over several to many cycles of the ac voltage supply to the ac-to-dc converter.

Since this application involves extensive asynchronous operation of the two converters, the dc current-sourcing interface serving both converters must be designed with due attention to the inevitable sideband unwanted components that develop. System application is generally limited to drives of several hundred horsepower because wound rotor (field) synchronous motors are not cost-competive with dc machines at lower power ratings, particularly when run over-excited to provide for the commutation needs of a current-sourced source-commutated converter feeding their stators.

A traditional market for high-power current-sourced ac-to-dc converters, controlled and uncontrolled, is the electrochemical industry. Since the days of mercury-arc tubes, zinc and aluminum pot lines and large electrolytic-plating bath systems have been fed by converters. The dc terminal voltages are low to moderate, from a few tens to at most a few hundred volts, but dc current requirements range from tens to hundreds of thousands of amperes. Because of this and because of cost pressures, the three-phase three-pulse midpoint converter is the near universal building block in the extensively harmonic neutralized arrangements used. Bridge configurations are rare in these applications. Despite the decreasing benefits and increasing complexities of very high pulse numbers in practical converters, it is not uncommon to find 24, 48, and even 96 pulse converters employed. In fact, harmonic neutralization is used to avoid the problems attendant on direct parallel connection of switching devices (thyristors in the modern versions). The double three-phase midpoint converter, with interphase reactor between neutrals, is extensively used as the basic six-pulse connection because of its economic advantage.

Impulse-commutated current-sourced converters have found but little usage to date. Attempts have been made to commercialize compound converters using a source-commutated ac-to-dc converter feeding an impulse-commuted dc-to-ac converter supplying the stator of a squirrel-cage induction motor. Economic obstacles have limited market penetration. The formidable technical difficulties attending the impulse commutation of current-sourced converters (see Chapter 9) make their widespread use unlikely in the future.

A general application area, which is still in the embryonic state, may yet create a greater market for current-sourced conversion equipment than all others combined. Many of the power generation and energy-storage technologies being touted for future use in utility systems are low-to-medium dc voltage; therefore, they need conversion equipment in order to interface with the ac transmission and distribution network. It seems likely that current-sourced dc-to-ac converters will provide the cheapest, most efficient, and most reliable means for achieving such interfaces, particularly for the higher power realizations of such systems. Thus, the future may see many such converters used, at power levels ranging from several to several hundred megawatts or even a few gigawatts, to connect batteries, fuel cells, magnetohydrodynamic generators, superconducting magnetic energy stores, solar photovoltaic arrays, and other exotic technologies, to utility systems.

11.4 Applications of Voltage-Sourced AC-to-DC/DC-to-AC Converters

Almost all of the voltage-sourced converters in use are impulse-commutated. Those that are not are converters in which no commutations occur, converters in which series-resonant circuits are used to wave-shape device and dependent currents. Many, if not most, voltage-sourced converters are part of compound converter systems. This use arises largely because ac-to-ac converters, with limited exceptions, have not been implementable due to the lack of appropriate switching devices for and the difficulties of applying impulse commutation to them.

Two major application areas have emerged, one quite well exploited and the other having difficulty in penetrating the marketplace because of economic factors. Well established is the use of voltage-sourced converters as the final stage of ''Uninterruptible Power Supplies'' (UPS). UPS arise because supply continuity to many loads in commercial and industrial establishments is critical. Computers are the prime example, dumping core memory and otherwise malfunctioning when short-duration outages occur and being very susceptible to malfunction or even damage from ac line-borne transients. The UPS provides a means for ensuring continuous and, in some cases, transient-free operation of critical loads. The equipment consists of a controlled ac-to-dc converter, a battery, and the dc-to-ac converter.

Two types of installation are found. For ''supercritical'' loads, the UPS operates continuously and continuously feeds the loads, isolating them completely from the utility supply. Redundant converters are used to guarantee infinitesimally small probabilities of converter supply outage. In some instances, bypass contactors are used so that the utility may be used to supply the loads directly in the event of an ultimate converter catastrophe. The second type of installation, used with loads more tolerant of momentary outages and utility-system transients, operates with a bypass switch so that the utility normally supplies the loads. The converter is maintained idling on line—spinning reserve, if you will—and supplies the loads whenever the utility suffers an outage and the bypass switch opens. Bypass switches may be electromechanical but are more usually solid-state; when used with continuously supplying UPS, they are always solid-state.

The battery of a UPS system is designed to provide for continued operation of the UPS while it is the sole source for the loads. Sometimes, this means providing sufficient energy storage for the maximum anticipated duration of the utility outage. More often, the battery merely provides sufficient capacity (15 minutes or so) for the start-up of local alternative generation (typically, an onsite diesel generator) to substitute the utility.

Equipment ratings range from a few kilowatts single-phase to several hundred kilowatts three-phase, with occasional examples in the megawatt region. Supplied from 120-V single-phase (low-power) or 208/480/660-V three-phase distributions, dc link voltages are mainly 125 or 250 V because of battery limitations. The ac output voltages are always required to be high-quality (low-distortion) sinusoids, and harmonic neutralization is commonly employed in the higher power units. PWM has lately gained acceptance in the lower power units; earlier, ferroresonant transformers were a popular means of achieving both output-voltage regulation and waveform purity. Only recently have programmed waveform techniques been recognized as perhaps the best way of meeting the requirements for low- and medium-power converters.

More limited success has come in the application area of ac machine drives. Despite intense, if sporadic, activity over nearly two decades, the voltage-sourced converter induction motor drive has not yet proven competitive with dc drives because of excessive converter costs and losses. This led directly to the development of current-sourced converter synchronous machine drives, and their success makes it doubtful that high-power variable-speed induction motor drives ever will achieve major market penetration.

Both types of ac machine drive rely upon the same aspects of machine behavior. For both induction and synchronous machines, the speed may be controlled by controlling the frequency of stator excitation. In both cases, the magnitude of the excitation must also be controlled to maintain flux levels—too great an excitation will cause saturation of the machine's stator and rotor magnetic structure, and intolerably high magnetizing current demand. Voltage-sourced converters are used to supply such variable-voltage variable-frequency, excitation to induction motors. Being impulse-commutated, they do not need the induced voltage set of a synchronous machine for commutation, and they can tolerate the inevitable lagging power factor of the induction motor. Although harmonic neutralized converters with superimposed CAM have been used, most equipments have used PWM and, more recently, programmed waveforms. Powers have ranged from a few tens to a few hundred horsepower; ac traction, for rapid transit, has been a perennially proposed, but so far not widely implemented, application in the higher power range. In most instances, compound conversion is used in which the dc-to-ac converter is preceded by an ac-to-dc converter, controlled and current-sourced. Typically, dc link voltages range from 200 to 600 V, although lately there has been a strong tendency to use 1000 V or more in the higher powered versions.

The waveform purity requirements for the output of such a converter are not very stringent over most of the machine's operating range. From full speed down to 20% to 30% speed, induction motors will, with some derating to allow for increased joule heating, quite readily tolerate six-pulse waves, and worse, with minimal effects on motor dynamics. At very low speeds and excitation frequencies, this is not the case. The harmonic torques then developed from the harmonic excitation are close to synchronous (full-speed) torques, and *cogging* occurs. This is the development of motor and shaft oscillations or rotary pulsations, which are very disturbing to the load. There is also a grave danger, particularly with the larger machines, of a harmonic frequency coinciding with a mechanical torsional resonance of the system. A sustained coincidence can cause serious mechanical damage. As a result, much effort is usually spent in ensuring good waveform quality, or at least freedom from the most troublesome low-order harmonics, when in low-frequency operation. Programmed waveform techniques greatly facilitate this endeavor.

Such ac drives are capable of four-quadrant operation if the ac-to-dc converter is capable of two-quadrant operation, since the phase sequence of the dc-to-ac converter output is easily reversed through control action. Thus, they can be made to match the shaft performance of dc drives, and they have long been attractive because of the simplicity and ruggedness of the squirrel-cage induction motor, which requires far less maintenance than a dc machine. Unfortunately, the machine benefits have been more than offset by converter penalties in most cases, and these drives have found use mainly in circumstances where dc motors cannot be tolerated. Examples are applications involving explosive atmospheres, such as certain mining operations, and applications where machine maintenance must be very infrequent, such as nuclear-reactor control rod and pump drives.

A third application area for voltage-sourced dc-to-ac converters comes from certain dc-to-dc conversion needs. When the ratio of output-to-input voltage of a dc-to-dc conversion application exceeds about 3 to 1, the boost and buck-boost dc-to-dc converters become expensive, inefficient, or both because of poor switch utilization and interface design problems. The method traditionally used to circumvent these delinquencies is to use a dc-to-ac converter, getting the maximum transfer ratio required in the output transformer thereof, and then reconverting by means of a simple uncontrolled single-quadrant ac-to-dc converter (a rectifier). Control is usually accomplished by CAM in the dc-to-ac converter; occasionally, noncommutating dc-to-ac converters are used, with operating frequency variation employed to exert control. These compound converters are generally restricted to low-power applications (a few to a few hundred watts), and most often use transistors as the switching elements in their dc-to-ac sections. They are often called *dc-to-dc converters* in the literature. This is unfortunate for, while their overall function is dc-to-dc conversion, they bear no relationship to true dc-to-dc switching converters.

The combination of low-power levels and a natural desire for lowest possible cost has led to the development of "single-ended" versions of these compounds in which the dc-to-ac conversion function becomes obscured by the topological resemblance to the boost dc-to-dc converter. Examples are the "flyback" and "forward" converters.[1,2] It should be noted, however, that they are in truth compound converters, and, topology notwithstanding, they have no direct relationship to the boost dc-to-dc converter.

The use of compounded converters to achieve the dc-to-dc conversion has the advantage of providing transformer isolation, if desired, between input and output. The "dc transformer" is not, after all, an impossibility—at least not in the world of switching power converters.

11.5 Applications of AC-to-AC Converters

Many applications requiring the ac-to-ac conversion function are addressed by the use of compound converters—ac-to-dc followed by dc-to-ac, or ac-to-dc followed by dc-to-dc (for voltage or current control), followed by dc-to-ac. These are discussed in previous sections; this section will be devoted to those areas in which true ac-to-ac converters have been used.

To date, the only ac-to-ac converter to become more than a laboratory curiosity is the naturally commutated cycloconverter (NCC). As previously stated, this is largely because suitable switching devices for implementing the other converters, all of which require impulse commutation, do not yet exist, and the difficulties attending the application of impulse-commutating circuits to ac-to-ac converters are formidable. Ironically, the NCC was used long before the semiconductor switches were developed. The first efforts at commercialization came in Europe in the 1920s and 1930s, using grid-controlled mercury-arc rectifiers as the switching devices in converters to provide single-phase 16⅔- and 25-Hz outputs from three-phase 50-Hz distributions. The application was railroad electrification; European rail systems have long used single-phase overhead-catenary ac electric distribution for rail traction drives.

These early NCCs were intended to replace the rotary converters (motor generator sets) then in use for the ac-to-ac conversion needed. Although technically successful, they were not commercially successful for a variety of reasons, including cost and reliability problems. The advent of the thyristor brought the NCC into consideration once more, but for rather different applications. Variable-speed four-quadrant induction motor drives are one area in which some success has been achieved. The other, which has yet to reach its full potential but seems likely to soon become a major market, is the variable-speed constant-frequency (VSCF) power-supply system for aircraft.

Obviously, an NCC can provide the variable-frequency variable-voltage stator excitation needed for variable-speed induction motor drives. Its success in the application has been limited by two factors. First, the output frequency must always be lower than supply frequency, and the maximum output frequency obtainable with 50 to 60 Hz input is 20 to 30 Hz, using a six-pulse (three-phase bridge-based) NCC. Operation is possible at higher output frequencies, but the dependent-quantity spectra become unacceptable. Thus for a given hp, the motors are larger and more expensive than those made for 50- or 60-Hz full-speed operation, and have lower shaft speeds for a given number of poles. Also, a six-pulse NCC requires a minimum of 36 thyristors and a rather complex control, whereas a dc drive uses a minimum of six thyristors and a much simpler control, and compound-converter synchronous machine drives use a minimum of 12 thyristors and a not overly complex control. The NCC only becomes competitive in high-power applications where multiple (parallel) devices are needed in the compound converters, and then only when very low shaft speeds are needed.

Such an application is the ball-mill drive, of which a few installations have been made. A ball mill is a large steel drum containing a mixture of ball bearings (or in some instances gravel) and a substance such as paint pigment or certain chemicals, which must be ground exceedingly fine. The grinding is accomplished by rotating the drum slowly for a long time, and typical equipments require several thousand drive horsepower at shaft speeds of a few tens of revolutions per minute (rpm). Conventional machines need gearboxes to meet the application needs, and gearboxes for continuous transmission of such powers are fearfully expensive; this makes the NCC and low-frequency machine combination competitive.

A variant of the ac machine drive that has also found some success is the "field" control of wound-rotor induction motors. It is well known that the speed of an induction motor can be controlled by controlling its rotor (field) excitation. In a wound-rotor machine, a converter can be used to control rotor excitation frequency and hence motor speed. Moreover, the converter will then only handle S times the machine power, where S is the per unit slip (ratio of rotor excitation frequency to stator excitation frequency), and the slip can be made negative (i.e., the motor can be made to run above synchronous speed) by reversing the phase sequence of the rotor excitation. Thus, an NCC connected so as to excite the rotor with its output can be used to vary the speed of an induction motor. If the speed variation needed is small, the maximum output frequency required will be low, and of course, the NCC power rating will be much lower than the machine rating.

An application has been found for this type of drive, and a number of installations exist. Dragline excavators are large digging machines that are often found in remote locations served by long transmission lines from rather weak ac systems. The dragline draws very large power "pulses" of a few seconds' duration whenever its bucket bites into the earth (or other material being dug); this causes serious voltage droop at its connection point and considerable unpleasantness on the ac system at large. The dragline can be buffered by a flywheel energy-storage system using a wound-rotor induction machine drive. The installation consists of one or more large (several thousand horsepower) machines each driving a flywheel and each having its rotor excitation NCC controlled. While the dragline is drawing little power, swinging its boom to the discharge point, emptying its bucket and swinging back to dig position, the NCCs are used to accelerate the machines and flywheels to 4% or 5% above synchronous speed, drawing power from the ac supply and storing energy in the rotating masses of the flywheels. As the excavator begins to bite, the NCCs act to decelerate the machines to 4% or 5% below synchronous speed, causing them to become generators extracting energy from the flywheels and supplying the extra power the dragline needs. The power demand on the ac system can be smoothed out by this means, improving both the productivity of the excavator (because it is no longer ac-system-limited for digging power) and obviating the ill effects of power pulsations on the ac system.

A related application arises in nuclear fusion research, where the various plasma experiments in progress require very large peak pulse powers with short durations and long zero-power intervals. However, because the average power is generally much lower in these cases, much wider machine-speed variations are used—50% to 70% is not uncommon—and NCCs are not used. Instead, the induced rotor voltage of wound-rotor induction machines is converted to dc and then re-inverted into the ac system by a compound pair of current-sourced converters. Energy is stored by allowing acceleration, for example, from 30% speed to full speed during the idle period, then extracted by forcing deceleration during the pulse-power period. Again, the utility is spared most of the ill effects of the fluctuating power demand. Such equipments, called *slip recovery* drives, are occasionally found in more conventional variable-speed induction-motor applications.

The VSCF application has been under development for over 15 years, but has so far had very little commercial impact. It now seems that the equipment has matured sufficiently to effect market penetration, and a thriving business is anticipated. Conventional aircraft power supplies use hydraulic constant-speed drives, driven through a gearbox from the turbines (jet engines), to drive 400-Hz generators providing the bulk of the electrical power requirements. Turbine speed varies over about a 2-to-1 range, typically 12,000 to 24,000 rpm, and much aircraft electrical equipment is designed for use with fixed-frequency, closely regulated, 400-Hz ac distribution. The VSCF system allows the generator to be gearbox coupled to the turbine, so that generated frequency varies over a 2-to-1 range (typically 1000 to 2000 Hz). An NCC is used to convert the generator output to the constant-frequency 400-Hz supply needed. Power ratings range from an uncommon low of 20 kVA per unit (small fighter aircraft) to 60 to 100 kVA per unit (for commercial jetliners). The application is extremely weight-conscious, for obvious reasons, and is also envelope-conscious since the equipment must integrate well mechanically with aircraft structure.

There are some interesting possible future applications for the NCC arising from implementation of the so-called HF base version. It was shown in Section 11.3 that a current-sourced dc-to-ac converter's "source" voltage could be furnished by a tuned load in the form of a parallel resonant circuit. Obviously, the same is true for an NCC, using multiple-tuned circuits, and their frequency can (and in fact should) be substantially higher than the 50- or 60-Hz utility supply to which the defined current terminals of the converter are then connected via a current-sourcing interface. Now the tuned circuits do not have to be loaded; they can exist solely for the purpose of providing source voltages to effect commutation. In this case, no real power transfer will take place except to makeup for the losses in the tuned circuit components. However, reactive power can be supplied to (leading kVA) or extracted from (lagging kVA) the ac distribution—a static VAR generator (SVG) has been created. Currently, no such equipment is in use. As discussed in Section 11.6, somewhat cruder converters are employed as SVGs. The attractions of the high-frequency base, so named because of the relatively high operating frequency of the tank circuits (generally a few hundred hertz), are (1) the passive components should be smaller and cheaper than components furnishing the same reactive power directly at 60 Hz and (2) system response can be faster because of the higher converter operating frequency.

A very useful extension of this scheme is possible, although again no actual usage has as yet occurred. As discussed in Section 11.3, one use for HVDC links is to provide asynchronous ties between ac transmission systems. A disadvantage of HVDC, using current-sourced source-commutated converters, is that even though controllable and reversible power interchange is achieved, both ac networks suffer the lagging quadrature-current demand of a converter. Suppose that two NCCs are used as the intertie, with a common set of high-frequency base tuned circuits providing their source voltages and effecting their commutations. Then, not only is controllable and reversible real power exchange possible—so is reactive power "exchange." In fact, while real power flow between the ac systems is occurring, either or both systems can be burdened with lagging kVAs or have leading kVAs supplied, provided that total current in the individual converters is held within design limits. The technical attractions of such a scheme to a utility system are obvious. So far, however, it has not proven possible to predicate that it is economically competitive with HVDC, and the limited number of asynchronous ties installed to date has not encouraged development expenditures.

11.6 Applications for AC Regulators

The ac regulators, those simplest and crudest of all switching power converters, are the second most successful in application, in regard to both sales volume and installed power. Their low cost, coupled with a high demand for the services they can perform, accounts for their being surpassed only by current-sourced ac-to-dc/dc-to-ac converters. Like those for such converters, the applications for ac regulators are legion. A few of the most successful are highlighted.

In terms of numbers of equipments sold, the simple incandescent lamp dimmer outstrips all converters except the half-wave rectifier. Most examples use a triac as the switching device and phase delay to control the voltage applied to the almost purely resistive filaments of their incandescent lamp loads. Built to fit in standard wall-switch boxes or built into table lamps, power ratings range from 300 to 600 W at 115 V, 60 Hz in North America. Two design problems dominate these simple regulators. The first, and easiest to overcome, is the inrush into cold filaments, typically 7 to 10 times the hot (steady-state) current. The designer must insure that the triac has sufficient transient-current capability to withstand repeated short-duration currents of this magnitude. More difficult is the problem of radio interference. There are sufficient higher order harmonics generated in the currents flowing in house wiring when a dimmer is phased back to cause interference with AM radio reception unless adequate filtering is provided.

The same type of regulator is extensively used as a speed control for the universal motors used in small electrical appliances. Blenders, mixers, food processors, hand drills, sanders, etc; all have felt the impact of this rudimentary switching power converter. Cost, inherently low because of simplicity, has been driven far down by the high-volume production methods used.

Of industrial applications, by far the most important is the use of such regulators in controlling the power delivered to resistive heating elements. Many industrial heat processes use ovens and furnaces with such heating elements, and a good many liquid heating baths use submersible versions. Temperatures can be regulated closely by the power control exerted by the regulators, which replace electromechanical contactors in many cases. Single- and three-phase versions are found; the latter is almost always wye-switched; the three-thyristor–three-diode version is very popular. Power ratings range from a few hundred watts to the megawatt region.

If phase-delay control is used, two problems arise. Conducted interference, caused by the higher order harmonics generated, can be very troublesome. Also, poor power factor can be a problem since many customers are subject to utility demand charges that escalate with worsening power factor. The long thermal time constants of the heating elements, and the things they heat, allow the use of integral cycle control. This largely eliminates the threat of interference problems, since high-frequency sidebands are of extremely small amplitude. As shown in Chapter 5, it does not influence the power factor. However, the metering system used by utilities does not register the larger amplitude sidebands as reactive power demand, and hence the customers are saved the charges they should in fact incur. The only worry left is the possible impact on transformers, a hidden trap for the unwary. There is a deep-rooted belief in the industry that the integral cycle controller is unity power factor. However, nothing could be further from the truth, and sooner or later the electricity supply industry is going to find out that it is being duped by this equipment.

An important and relatively new high-power application for the ac regulators is the static VAR generator (SVG). This equipment, as applied to one phase of an ac supply, consists of a shunt capacitive branch, which continuously supplies reactive power, and a parallel connected inductive branch, which can absorb reactive power. The latter consists of an inductor with inverse parallel-connected thyristor switches as the controlling element. The inductor is generally sized so that at full current (full conduction in the regulator switch), it draws current equal to or exceeding that of the capacitive branch. As phase-delay control is applied, the inductor current reduces, and VARs are supplied. The configuration is always that of a delta-switched three-phase regulator with inductive load. The capacitive branch is usually split into a number of tuned harmonic filters to reduce harmonic injection to the ac supply when phase control is applied.

Two major applications have been established for the SVG. The simplest was the second to be exploited, transmission-line voltage support. Transmission lines in ac systems, particularly weak systems, often develop undesirably large voltage droops along their length. This can be prevented by applying capacitive compensation in series with the line, tuning out its inductive reactance, but this technique sometimes gives rise to subsynchronous resonance problems when lines are switched. It is also inflexible—the compensation is fixed. An alternative is to apply shunt-capacitive compensation distributed along the line. Using SVGs allows the compensation to be adjusted for line loading, greatly enhancing system stability in some instances. The SVG is usually transformer-coupled to the line, common practice being to make use of a delta tertiary on an existing transformer if possible. SVG voltages range from 13.8 to 34.5 kV, and ratings from a few tens to several hundred megavolt-amperes. The larger installations use switched capacitor banks, with thyristor switches, and reactor branches rated only to vernier control the VAR supply of one bank of the capacitive branch.

In the first application, which proved their technical and economic feasibility, SVGs are used to curb flicker produced by arc furnaces and compensate their reactive power demand. The arc furnace, which is becoming popular in the steel industry because it is efficient and essentially nonpolluting, is an appallingly ill-behaved load. It operates at poor lagging power factor, exhibits large random unbalanced power and reactive power demand variations, and is a prodigious harmonic generator. A properly designed and applied SVG can correct all these delinquencies. Power factor can be continuously corrected, with very rapid response, and the harmonics can be absorbed by the filters constituting the SVG's capacitive branches. If the current rating of the reactive branches is sufficiently greater than that of the capacitive branches, then the unbalance can also be dynamically compensated by using the SVG as a rapid-response Steinmetz balancer. Operations in this vein balances both active and reactive power demand, as is well known. While the most elegant analysis of the phenomenon is through the use of symmetrical components, a simple qualitative reasoning approach is also possible. First, any unbalanced, purely reactive loading can obviously be balanced and compensated for by connecting an appropriate set of unbalanced capacitive and inductive branches to the lines, and the SVG can clearly do that. Any unbalanced real loading on a three-wire system can be resolved into a balanced load plus additional single-phase loads across two pairs of lines; one on the AB line pair will be used as an example. The current phasor in the A line due to the single-phase load then leads by $\pi/6$ rad, that in the B phase lags by $\pi/6$ rad. If purely reactive loads are now connected on the BC and CA line pairs, capacitive on BC and inductive on CA, they create current phasors having the following phase angles:

- In the B line, leading by $2\pi/3$ rad.
- In the C line, one lagging by $\pi/3$ rad, one leading by $\pi/3$ rad.
- In the A line, one lagging by $2\pi/3$ rad.

Obviously, the resultant in the C line is an in-phase current. If the magnitudes are chosen properly, the lagging quadrature current created in the A line can be made to cancel the leading quadrature current developed by the single-phase real load. The leading quadrature current in the B line can be made to exactly cancel the lagging quadrature current in that line resulting from the single-phase real load. The result must be perfect balance and unity power factor. The required line-to-line quadrature-current loads are equal, each $1/\sqrt{3}$ times the real line-to-line current.

Clearly, an SVG with both capacitive and inductive line-to-line current-demand capability can perform this balancing act regardless of the whereabouts of the single-phase constituents of an unbalanced real load, and can respond rapidly to changes in the degree and nature of the unbalance.

Applying an SVG thus to an arc furnace has beneficial results. Power capacity is actually increased, because the voltage droop due to reactive power demand is eliminated. Harmonic injection is curbed, and, most significant to many other utility customers, the flicker (random voltage fluctuations) is greatly reduced. Finally, much to the utility's relief, negative-sequence currents are also greatly reduced.

References

1. Owen, H. A., Jr. T. G. Wilson, S. Y. M. Feng, and F. C. Y. Lee. A computer-aided design procedure for flyback step-up dc-to-dc converters. *IEEE Transactions on Magnetics* **8** (3): 289–91 (September 1972).
2. Jansson, L. E. A survey of converter circuits for switched mode power supplies. *Mullard Technical Communications* **12** (119): 271–278 (Mullard Ltd, England, July 1973).

Index